Structure, Function, and Regulation of Molecules Involved in Leukocyte Adhesion

Proceedings of the Second International Conference on:
"Structure and Function of Molecules
Involved in Leukocyte Adhesion II"

Held in Titisee, Germany
October 2–6, 1991

Co-Chairpersons:
Peter E. Lipsky, M.D.
C. Wayne Smith, M.D.

P.E. Lipsky R. Rothlein T.K. Kishimoto
R.B. Faanes C.W. Smith
Editors

Structure, Function, and Regulation of Molecules Involved in Leukocyte Adhesion

With 131 Illustrations

Springer-Verlag

New York Berlin Heidelberg London Paris
Tokyo Hong Kong Barcelona Budapest

Peter E. Lipsky, M.D.
Professor, Internal Medicine and Microbiology
University of Texas
Southwestern Medical Center at Dallas
5323 Harry Hines Blvd.
Dallas, TX 75235-8884

C. Wayne Smith, M.D.
Professor, Department of Pediatrics
Leukocyte Biology Section
Texas Children's Hospital
Clinical Care Center, Suite 1130
6621 Fannin. MC 3-3272
Houston, TX 77030-2399

Robert Rothlein, Ph.D.
Distinguished Scientist
Cellular Adhesion/Immunology;
Takashi Kei Kishimoto, Ph.D.
Senior Scientist
Cellular Adhesion/Immunology;
Ronald B. Faanes, Ph.D.
Director, R&D Scientific and Technical Operations
Boehringer Ingelheim Pharmaceuticals, Inc.
900 Ridgebury Road, Box 368
Ridgefield, CT 06877-0368

Cover Art: see Figure 5.6 for description

Library of Congress Cataloging-in-Publication Data
International Conference on "Structure and Function of Molecules
    Involved in Leukocyte Adhesion II" (1991 : Titisee, Germany)
    Structure, function, and regulation of molecules involved in
leukocyte adhesion : proceedings of the Second International
Conferences on "Structure and Function of Molecules Involved in
Leukocyte Adhesion II," held in Titisee, Germany, October 2–6, 1991
/ eds., Peter E. Lipsky . . . [et al.].
        p. cm.
    Includes bibliographical references.
    ISBN-13: 978-1-4613-9268-2   e-ISBN-13: 978-1-4613-9266-8
    DOI: 10.1007/978-1-4613-9266-8

    1. Leucocytes—Congresses. 2. Cell adhesion—Congresses.
I. Title.
    [DNLM: 1. Cell Adhesion Molecules—immunology—congresses.
2. Cell Adhesion Molecules—physiology—congresses. 3. Integrins—
physiology—congresses. 4. Leukocytes—physiology—congresses.
5. Lymphocyte Transformation—physiology—congresses.   QW 573 I602s
1991]
QR185.8.L48I65   1991
616.07'9—dc20
DNLM/DLC
for Library of Congress   92-2337

Printed on acid-free paper.

Production managed by Karen Phillips; manufacturing supervised by Vincent Scelta.
Typeset by Asco Trade Typesetting Ltd., Hong Kong.

9 8 7 6 5 4 3 2 1

Acknowledgments

The editors are indebted to Ms. Leah Neumaier for her organizational and administrative contributions to the preparation of the symposium and this publication. Professor Alan S. Rosenthal, Sr. Vice President of Scientific and Medical Affairs, BIPI USA, and Professor Franz Waldeck, Member of the Corporate Board and Head of Research and Development BI GmbH, are acknowledged for their foresight in supporting research activities which set the stage for the Second Titisee Symposium on Leukocyte Adhesion Biology. The contributions of Doctors of Philosophy, Hermann Fröhlich and Ms. Ingeborg Wernicke of the Boehringer Fonds in facilitating the arrangement of accommodations and special events at the Schwarzwald, which made the meeting a memorable occasion, cannot go unnoticed.

Participants

Donald Anderson, Andrea Apperl, Randall Barton, Anthony Berendt, Franz Birke, Barbara-Jean Bormann, Eric Brown, Robert Colvin, A. Benedict Cosimi, Nava Dana, Laurie Davis, Ronald Faanes, Carl Figdor, Alain Fischer, Herman Fröhlich, Michael Gallatin, Johanna Griffin, Robert Gundel, Rupert Hallmann, Rudy Hammer, Dorian Haskard, Herbert Hechtman, Martin Hemler, Martin Hofmann, Nancy Hogg, Larry Husten, Mark Jutila, Geoffrey Kansas, Joachim Kaysser, Marcus Kehrli, Marilyn Kehry, Kei Kishimoto, Gordon Letts, Peter Lipsky, Steven Marlin, Stephan Martin, Rodger McEver, Ira Mellman, James Neuberger, Leah Neumaier, Nancy Oppenheimer-Marks, Laurie Phillips, Louis Picker, Martyn Robinson, Steven Rosen, Robert Rothlein, Francisco Sánchez-Madrid, Linda Scharschmidt, Michael Schaude, Elke Seewaldt-Becker, C. Wayne Smith, Roger Snow, Donald Staunton, Lloyd Stoolman, Peter Swetly, Robert Wallace, Ingeborg Wernicke, and Robert Winn.

Contents

Contributors

Omid Abbassi, M.D., Ph.D., Baylor College of Medicine, One Baylor Plaza, Houston, TX 77030 USA

David Adams, M.D., M.R.C.P., The Queen Elizabeth Hospital, Edgbaston, Birmingham, B15 2TH, England

Mark Ackermann, D.V.M., Ph.D., Immunology of Ruminant Perinatal Diseases, Metabolic Diseases and Immunology Research Unit, National Animal Disease Center, Box 70, 2300 Dayton Avenue, Ames, IA 50010 USA

F. Amblard, Ph.D., Institut d'Embryologie du CNRS et du Collège de France, 49Bis, Avenue de la Belle Gabrielle, 94736 Nogent Sur Marne, France

Donald C. Anderson, M.D., Departments of Pediatrics, Cell Biology and Microbiology and Immunology; Section Leukocyte Biology, Baylor College of Medicine, Texas Children's Hospital, Clinical Care Center, Suite 1130, 6621 Fannin, MC 3-3272, Houston, TX 77030-2399 USA

David Andrew, Ph.D., Department of Pathology, Stanford University School of Medicine, Stanford, CA 94305 USA

Karl-E. Arfors, La Jolla Institute for Experimental Medicine, 11099 North Torrey Pines Road, #130, La Jolla, CA 92037 USA

Lawrence W. Argenbright, Boehringer Ingelheim Pharmaceuticals, Inc., 900 Ridgebury Road, Box 368, Ridgeflied, CT 06877-0368 USA

M. Amin Arnaout, M.D., Renal Unit, MGH East, Harvard Medical School, Massachusetts General Hospital, 149 13th Street, Charlestown, MA 02129 USA

Alicia G. Arroyo, M.D., Department of Immunology, Hospital de la Princesa, Universidad Autónoma de Madrid, Madrid, Spain

Hugh Auchincloss, Jr., M.D., Department of Surgery, Massachusetts General Hospital, Fruit Street, Boston, MA 02114 USA

C. Barbat, INSERM, U 132, d'Hematologie, Departement de Pédiatrie, Hôpital Necker, Enfants Malades, 149, rue de Sèvres, 75743 Paris Cédex 15, France

Randall W. Barton, Ph.D., Department of Immunology, Boehringer Ingelheim Pharmaceuticals, Inc., 900 Ridgebury Road, Box 368, Ridgefield, CT 06877-0368 USA

Paul A. Bates, Ph.D., Biomolecular Modelling Laboratory, Imperial Cancer Research Fund, P.O. Box 123, Lincoln's Inn Fields, London WC2A 3PX, United Kingdom

Anthony R. Berendt, B.M., B.C.H., Institute of Molecular Medicine, The Molecular Parasitology Group, John Radcliffe Hospital, Headington, Oxford OX3 9DU, United Kingdom

Derek Brown, Ph.D., Department of Pathology, Stanford University, Stanford, CA 94305 USA

Eric Brown, M.D., Department of Medicine, Washington University School of Medicine, 660 S. Euclid, St. Louis, MO 63110 USA

Alan R. Burns, Pulmonary Research Laboratory, St. Paul's Hospital, University of British Columbia, Vancouver, B.C. V6Z 1T6, Canada

Eugene C. Butcher, M.D., Department of Pathology, Stanford University School of Medicine, Stanford, CA 94305-5324 USA

Carols Cabañas, Ph.D., Macrophage Laboratory, Imperial Cancer Research Fund, Lincoln's Inn Fields, London WC2A 3PX, United Kingdom

Miquel R. Campanero, Ph.D., Department of Immunology, Hospital de la Princesa, Universidad Autónoma de Madrid, Madrid, Spain

Bosco M.C. Chan, Ph.D., Department of Pathology, Harvard Medical School, Dana Farber Cancer Institute, Room #M613, 44 Binney Street, Boston, MA 02115 USA

Robert B. Colvin, M.D., Department of Pathology, Harvard Medical School, Massachusetts General Hospital, 100 Blossom Street, The Cox-5 Building, Boston, MA 02114 USA

A. Benedict Cosimi, M.D., Department of Clinical Transplant Surgery, Harvard Medical School, Massachusetts General Hospital, MGH White 4, Room #5, Boston, MA 02114 USA

Alister G. Craig, Ph.D., Institute of Molecular Medicine, The Molecular Parasitology Group, John Radcliffe Hospital, Headington, Oxford OX3 9DU, United Kingdom

John J. Cush, Ph.D., Harold C. Simmons Arthritis Research Center, The University of Texas, Southwestern Medical School, Department of Internal Medicine, 5323 Harry Hines Blvd., Dallas, TX 75235 USA

Nava Dana, M.D., The Leukocyte Biology and Inflammmation Program, Renal Unit and Department of Medicine, Massachusetts General Hospital and Harvard Medical School, Boston, MA 02124 USA

Laurie S. Davis, Ph.D., Department of Medicine, Rheumatic Diseases Division, University of Texas Southwestern Medical Center at Dallas, 5323 Harry Hines Blvd., Dallas, TX 75235-8884 USA

Manuel O. de Landázuri, M.D., Department of Immunology, Hospital de la Princesa, Universidad Autónoma de Madrid, Madrid, Spain

Francis L. Delmonico, M.D., Harvard Medical School, Massachusetts General Hospital, Boston, MA 02114 USA

Claire M. Doerschuk, M.D., Department of Pediatrics, Pathology and Anatomy, Indiana University, Riley Hospital for Children, Room 293, 702 Barnhill Drive, Indianapolis, IN 46202-5225 USA

Monique Doré, D.V.M., Ph.D., Leukocyte Biology Section, Baylor College of Medicine, Texas Children's Hospital, Clinical Care Center, Suite 1130, 6621 Fannin, MC 3-3272, Houston, TX 77030-2399 USA

Ian Dransfield, Ph.D., Respiratory Medicine Unit, City Hospital, Greenbank Drive, Edinburgh EH10 5SB, United Kingdom

Mariano J. Elices, Ph.D., CYTEL Corporation, 3525 John Hopkins Court, San Diego, CA 92121 USA

Dehmani M. Fathallah, Renal Unit, MGH East, Harvard Medical School, Massachusetts General Hospital, 149 13th Street, Charlestown, MA 02129 USA

Dr. Carl G. Figdor, Division of Immunology, The Netherlands Cancer Institute, Plesmanlaan 121, 1066 CX Amsterdam, The Netherlands

Alain Fisher, M.D., Ph.D., INSERM, U 132, d'Hematologie, Departement de Pédiatrie, Hôpital Necker, Enfants Malades, 149, rue de Sèvres, 75743 Paris Cédex 15, France

James D. Fortenberry, M.D., Department of Pediatrics, Sections of Critical Care Medicine and Leukocyte Biology, Baylor College of Medicine, Texas Children's Clinical Care Center, Suite 1130, 6621 Fannin, MC 3-3272, Houston, TX 77030-2399 USA

W. Michael Gallatin, Ph.D., ICOS Corporation, 22021 20th Avenue, S.E., Bothell, WA 98021 USA

Rosario García-Vicuña, M.D., Section of Rheumatology, Hospital de la Princesa, Madrid, Spain

Michael A. Gimbrone, Jr., M.D., Vascular Research Division, Department of Pathology, Harvard Medical School, Brigham and Women's Hospital, 75 Francis Street, Boston, MA 02115 USA

Hattie D. Gresham, Ph.D., Department of Pharmacology, University of Missouri, M517B, #1 Hospital Drive, Columbia, MO 65212 USA

Robert H. Gundel, Ph.D., Department of Pharmacology, Boehringer Ingelheim Pharmaceutics, Inc., 900 Ridgebury Road, Box 368, Ridgefield, CT 06877-0368 USA

John M. Harlan, M.D., University of Washington, Harborview Medical Center, 325 Ninth Avenue, Seattle, WA 98104 USA

Dorian Haskard, D.M. M.R.C.P., Rheumatology Unit, The Royal Post-Graduate Medical School, Hammersmith Hospital, University of London, Du Cane Road, London W12 0NN, United Kingdom

Craig Haug, M.D., Transplant Unit, Harvard Medical School, Massachusetts General Hospital, Boston, MA 02114 USA

P. Hauss, INSERM, U 132, d'Hematologie, Departement de Pédiatrie, Hôpital Necker, Enfants Malades, 149, rue de Sèvres, 75743 Paris Cédex 15, France

Herbert B. Hechtman, M.D., Department of Surgery, Harvard Medical School, Brigham and Women's Hospital, 75 Francis Street, Boston, MA 02115 USA

Martin E. Hemler, Ph.D., Department of Pathology, Harvard Medical School, Dana Farber Cancer Institute, Room #M613, 44 Binney Street, Boston, MA 02115 USA

James Hill, M.B., Ch.B., Harvard Medical School, Brigham and Women's Hospital 75 Francis Street, Boston, MA 02115 USA

C. Hivroz, Ph.D., INSERM, U 132, d'Hematologie, Departement de Pédiatrie, Hôpital Necker, Enfants Malades, 149, rue de Sèvres, 75743 Paris Cédex 15, France

Patricia A. Hoffmann, M.S., Department of Cell Adhesion, ICOS Corporation, 2202 20th Avenue SE, Bothell, WA 98021 USA

Nancy Hogg, Ph.D., Macrophage Laboratory, Imperial Cancer Research Fund, P.O. Box 123, Lincoln's Inn Fields, London WC2A 3PX, United Kingdom

J.C. Hollers, Well Stream, 7600 Chevy Chase Drive, Austin, TX 78752 USA

Kun Huang, M.D., Ph.D., Department of Anatomy and Program in Immunology, University of California, San Francisco School of Medicine, Box 0452, San Francisco, CA 94143-0452 USA

Stefan Hubscher, Pathology Department, University of Birmingham, Medical School, Birmingham, England

Bonnie J. Hughes, B.S., Department of Pediatrics, Leukocyte Biology Section, Texas Children's Hospital, Clinical Care Center, Suite #1130, 6621 Fannin, MC 3-3272, Houston, TX 77030-2399 USA

Hugo E. Jasin, M.D., Division of Rheumatology and Clinical Immunology, Department of Internal Medicine, Mail Slot #509, University of Arkansas for Medical Sciences, College of Medicine, 4301 W. Markham, Little Rock, AR 72205 USA

Mark A. Jutila, Ph.D., Veterinary Molecular Biology, Montana State University, South 19 and Lincoln, Bozeman, MT 59717 USA

Geoffrey Kansas, Ph.D., Department of Pathology, Division of Tumor Immunology, Dana Farber Cancer Institute, Harvard Medical School, 44 Binney Street, Boston, MA 02115 USA

Arthur F. Kavanaugh, Ph.D., Harold C. Simmons Arthritis Center, The University of Texas Southwestern Medical School, Department of Internal Medicine, 5323 Harry Hines Blvd., Dallas, TX 75235 USA

Marcus E. Kehrli Jr., D.V.M., Ph.D., Immunology of Ruminant Perinatal Diseases, Metabolic Diseases and Immunology Research Unit, National Animal Disease Center, Box 70, 2300 Dayton Avenue, Ames, IA 50010 USA

Takashi Kei Kishimoto, Ph.D., Boehringer Ingelheim Pharmaceuticals, Inc., 900 Ridgebury Rd., Box 368, Ridgefield, CT, 06877-0368 USA

Joyce Koenig, M.D., Department of Pediatrics, Sections of Neonatology and Leukocyte Biology, Baylor College of Medicine, Texas Children's Clinical Care Center, Suite 1130, 6621 Fannin, MC 3-3272, Houston, TX 77030-2399 USA

Armando Laffón, M.D., Section of Rheumatology, Hospital de la Princesa, Madrid, Spain

O. Lecomte, Ph.D., INSERM, U 132, d'Hematologie, Departement de Pédiatrie, Hôpital Necker, Enfants Malades, 149, rue de Sèvres, 75743 Paris Cédex 15, France

L. Gordon Letts, Ph.D., Department of Pharmacology, Boehringer Ingelheim Pharmaceuticals, Inc., 900 Ridgebury Road, Box 368, Ridgefield, CT 06877-0368 USA

Ellis Lightfoot, Internal Medicine and Microbiology, University of Texas Southwestern Medical Center at Dallas, 5323 Harry Hines Blvd., Dallas, TX 75235-8884 USA

Thomas F. Lindsay, M.D., Harvard Medical School, Brigham and Women's Hospital 75 Francis Street, Boston, MA 02115 USA

Peter E. Lipsky, M.D., Internal Medicine and Microbiology, University of Texas Southwestern Medical Center at Dallas, 5323 Harry Hines Blvd., Dallas, TX 75235-8884 USA

S. Loche, Ph.D., Centre de Biochimie, Université de Nice, Parc Valrose, 06034, Nice Cedex, France

Francis W. Luscinskas, M.D., Department of Pathology, Harvard Medical School, Brigham and Women's Hospital, 75 Francis Street, Boston, MA 02115 USA

Elizabeth Mainolfi, Department of Immunology, Boehringer Ingelheim Pharmcaeuticals, Inc., 900 Ridgebury Road, Box 368, Ridgefield, CT 06877-0368 USA

Steven D. Marlin, Ph.D., Department of Immunology, Boehringer Ingelheim Pharmaceuticals, Inc., 900 Ridgebury Road, Box 368, Ridgefield, CT 06877-0368 USA

Fabienne Mazerolles, Ph.D., INSERM, U 132, d'Hematologie, Department de Pédiatrie, Hôpital Necker, Enfants Malades, 149, rue de Sèvres, 75743 Paris Cédex 15, France

Alison McDowall, Macrophage Laboratory, Imperial Cancer Research Fund, P.O. Box 123, Lincoln's Inn Fields, London WC2A 3PX, United Kingdom

Rodger P. McEver, M.D., Department of Medicine, University of Oklahoma, Health Science Center, Oklahoma Medical Research Foundation, 825 N.E. 13th Street, Oklahoma City, OK 73104 USA

Ira Mellman, Ph.D., Department of Cell Biology, Yale University School of Medicine, 333 Cedar Street, P.O. Box 3333, New Haven, CT 06510 USA

S. Meloche, Centre de Biochimie, Université de Nice, Parc Valrose, 06034 Nice Cedex, France

William J. Mileski, M.D., Department of Surgery, University of Texas Southwestern Medical Center at Dallas, 5323 Harry Hines Blvd., Dallas, TX 75235-9031 USA

James Neuberger, M.D., F.R.C.P., The Liver Unit, Queen Elizabeth Hospital, Queen Elizabeth Medical Center, Edgbaston, Birmingham B15 2TH, United Kingdom

Christopher I. Newbold, D. Phil., Institute of Molecular Medicine, The Molecular Parasitology Group, John Radcliffe Hospital, Headington, Oxford OX3 9DU, United Kingdom

Stephen H. Norris, Ph.D., Department of Drug Metabolism, Boehringer Ingelheim Pharmaceuticals, Inc., 900 Ridgebury Road, Box 368, Ridgefield, CT 06877 USA

Nancy Oppenheimer-Marks, Ph.D., Harold C. Simmons Arthritis Center, The University of Texas Southwestern Medical School, Department of Internal Medicine, 5323 Harry Hines Blvd., Dallas, TX 75235 USA

Susan Ortlepp, M.Sc., Department of Pathology, Stanford University, Stanford, CA 94305

Louis J. Picker, M.D., Department of Pathology, Laboratory of Molecular Pathology, University of Texas, Southwestern Medical Center at Dallas, 5323 Harry Hines Blvd., Dallas, TX 75235-9072 USA

Peter Pietschmann, Ph.D., Harold C. Simmons Arthritis Center, The University of Texas Southwestern Medical School, Department of Internal Medicine, 5323 Harry Hines Blvd., Dallas, TX 75235 USA

Antonio A. Postigo, M.D., Department of Immunology, Hospital de la Princesa, Universidad Autónoma de Madrid, Madrid, Spain

Rafael Pulido, Ph.D., Department of Immunology, Hospital de la Princesa, Universidad Autónoma de Madrid, Madrid, Spain

Charles L. Rice, M.D., Department of Surgery, University of Washington School of Medicine, Harborview Medical Center, 325 9th Avenue, ZA-16, Seattle, WA 98104 USA

Martyn K. Robinson, Ph.D., Inflammation Biology Group, Celltech Limited, 216 Bath Road, Slough, Berkshire SL1 4EN, United Kingdom

Robert Rothlein, Ph.D., Cellular Adhesion/Immunology, Boehringer Ingelheim Pharmaceuticals, Inc., 900 Ridgebury Road, Box 368, Ridgefield, CT 06877-0368 USA

Hugh Rosen, M.C. C.H.B.D., Merck Sharp & Dome Research Laboratory, P.O. Box 2000, Rahway, NJ 07065 USA

Steven D. Rosen, Ph.D., Department of Anatomy & Program in Immunology, University of California, San Francisco School of Medicine, Box 0452, San Francisco, CA 94143-0452 USA

Stephen J. Rosenman, Ph.D., Department of Cell Adhesion, ICOS Corporation, 22021 20th Avenue SE, Bothell, WA 98021 USA

Francisco Sánchez-Madrid, Ph.D., Department of Immunology, Hospital de la Princesa, Universidad Autónoma de Madrid, Madrid, Spain 28006

Linda Scharschmidt, M.D., Clinical Research, Boehringer Ingelheim Pharmaceuticals, Inc., 900 Ridgebury Road, Box 368, Ridgefield, CT 06877-0368 USA

R. Sekaly, Ph.D., IRCM, 110. Avenue des Pins Ouest, Montreal, Canada

Sam R. Sharar, M.D., Department of Anesthesia, University of Washington School of Medicine, Harborview Medical Center, 325 9th Avenue, ZA-16, Seattle, WA 98014 USA

David Shepro, Ph.D., The Biological Science Center, Boston University, 5 Cummington Street, Boston, MA 02215 USA

Dale E. Shuster, M.S., Ph.D., Immunology of Ruminant Perinatal Diseases, Metabolic Diseases and Immunology Research Unit, National Animal Disease Center, Box 70, 2300 Dayton Avenue, Ames, IA 50010 USA

C. Wayne Smith, M.D., Department of Pediatrics, Leukocyte Biology Section, Baylor College of Medicine, Texas Children's Hospital, Clinical Care Center, Suite 1130, 6621 Fannin, MC 3-3272, Houston, TX 77030-2399 USA

Olivier Spertini, M.D., Division of Hematology (18), CHUV, 1011 Luissane, Switzerland

Paul Stevens, Ph.D., Department of Pathology, Stanford University, Stanford, CA 94305

Lloyd M. Stoolman M.D., University of Michigan, Department of Pathology, M4224, Medical Services Bldg. 1, Ann Arbor, MI 48109-0602 USA

Thomas F. Tedder, Ph.D., Department of Pathology, Dana Farber Cancer Institute, Harvard Medical School, Division of Tumor Immunology, 44 Binney Street, Boston, MA 02115 USA

Martin H. Thornhill, M.B., F.D.S.R.C.S., Ph.D., Department of Oral Medicine and Periodontology, The London Hospital Medical College, Turner Street, London E1 2AD, United Kingdom

Nina Tolkoff-Rubin, M.D., Dialysis Unit, Harvard Medical School, Massachusetts General Hospital, Boston, MA 02114 USA

C. Robert Valeri, M.D., Naval Blood Research Laboratory, 615 Albany Street, Boston, MA 02118 USA

Dr. Yvette van Kooyk, Division of Immunology, The Netherlands Cancer Institute, Plesmanlaan 121, 1066 CX Amsterdam, The Netherlands

Ulrich H. von Andrian, La Jolla Institute for Experimental Medicine, 11099 North Torrey Pines Road, #130, La Jolla, CA 92037 USA

Bruce Walcheck, Veterinary Molecular Biology, Montana State University, South 19 and Lincoln, Bozeman, MT 59717 USA

Craig D. Wegner, Ph.D., Department of Pharmacology, Boehringer Ingelheim Pharmacentrical Inc., 900 Ridgebury Road, Box 368, Ridgefield, CT 06877-0368 USA

Robert K. Winn, Ph.D., Department of Surgery, University of Washington School of Medicine, Harborview Medical Center, 325 9th Avenue, ZA-16, Seattle, WA 98104 USA

Introduction

There has been dramatic progress in the knowledge and understanding of leukocyte adherence over the last three years. When the first meeting on *"Structure and Function of Molecules Involved in Leukocyte Adhesion"* was held in Titisee in 1988, the program was focused on the newly defined structures of the leukocyte integrins and ICAM-1. These were thought to be primarily involved in cell-to-cell contact. Only a modest amount of information was available concerning the potential anti-inflammatory effects of interfering with leukocyte adhesion molecules *in vivo*. At that time, it was informally agreed by many of the participants that a meeting would be held three years later which would concentrate on *in vivo* effects of antibodies to leukocyte adhesion molecules, including CD18, CD11, and CD54, anticipating that the scant data using these antibodies *in vivo* would be expanded. This is what has happened. Inhibition of leukocyte adhesion *in vivo* has defined new processes whereby leukocytes contribute to the pathophysiology of disease in circumstances in which this had not previously been considered. Extension of burn sites and delayed graft function following transplantation are two examples of this. Moreover, preliminary data on the use of anti–ICAM-1 in man have been generated, suggesting efficacy in disease states as disparate as graft rejection and rheumatoid arthritis. Thus, the predicted therapeutic potential of adhesion antagonists in human disease appears to be validated, with new potential clinical indications rapidly emerging as more is learned about the role of adherence interactions in various pathologic states.

New functions for adhesion molecules have also become apparent. Information developing over the past three years had clearly documented that adhesion molecules function as more than structures that mediate cell-to-cell contact. Emerging information has clearly indicated that ligation of leukocyte adhesion molecules can generate intracellular signals that co-stimulate functional activation of leukocytes. Also, within the past three years, the selectin family of adhesion molecules has been defined, carbohydrate ligands for the selectins have been identified, and a role for selectins in mediating the phenomenon of neutrophil rolling along the vascular endothelium has been described. In addition, new ligands for integrins have been described: the importance of VLA-4 interactions with VCAM-1 and LFA-1 interaction with ICAM-2 and ICAM-3 is becoming apparent. Finally, the role of adhesion molecules as receptors for pathogens as different as malaria and rhinovirus has also been documented.

These new themes and directions provided the focus of the second Titisee symposium on *The Structure and Function of Leukocyte Adhesion* held on October 2–6, 1991. It became clear that the amount of new information concerning leukocyte adhesion had grown exponentially over the last three years, as the roles that these various receptor–counter-receptor pairs play in normal physiology, as well as pathobiologic events, have been delineated. Based on past experience, it will be difficult to predict the evolution of this rapidly expanding field of biology, but it is clearly apparent that new and important understandings of physiologic processes, as well as opportunities for therapeutic interventions, will continue to develop.

THE EDITORS

Part 1
Structure and Function of the Integrins

1
Leukocyte Integrin Activation

Nancy Hogg, Carlos Cabañas, and Ian Dransfield

Introduction

The three leukocyte integrins LFA-1, CR3/Mac-1 and p150,95 are heterodimeric molecules each comprising a unique α subunit noncovalently associated with a common β_2 subunit (1). A feature of the N terminus of the α subunit is the seven domains of ~60 amino acids (domains I–VII) of which the last three or four are homologous to divalent cation binding or EF-hand-type sites. Integrins are known to require bound divalent cations such as Mg^{2+} and/or Ca^{2+} in order to function and it is to this region that they bind (2,3). The roles of these molecules in immune functions are best characterized for LFA-1 (CD11a/CD18) and CR3(CD11b/CD18). Antibody blocking studies have demonstrated a key role for LFA-1 in many cellular interactions of leukocytes and CR3 is important in functions such as myeloid cell phagocytosis, adherence to activated endothelium, and chemotaxis (4–6). These activities are thought to represent *in vitro* correlates of the response to tissue inflammation.

Binding of leukocyte integrins to their counter-receptors or ligands must be carefully controlled in order to avoid disadvantageous reactions. Thus, the avidity of CR3 for a ligand on endothelium can be regulated by cellular activation as shown by studies using phorbol esters and other stimulants such as f-Met-Leu-Phe (7–9). In addition, the interaction of LFA-1 with its ligand, termed intercellular adhesion molecule–1 (ICAM-1, CD54) (10), is transiently induced in response to T-cell–receptor cross-linking using CD3 mAbs, providing a mechanism for strengthening initial antigen-specific T-cell–target interactions (11,12). Interestingly, activation with phorbol ester or through other T-cell–surface molecules and their ligands (e.g., CD2/LFA-3) results in a more prolonged LFA-1/ ICAM-1 interaction (12). Cross-linking other cell-surface receptors such as CD14 (13), CD43 (14), CD44 (15), and MHC class II (16) molecules also causes enhanced LFA-1/ICAM-1 binding, although the kinetics of these reactions have not been determined. Such results suggest that a common intracellular pathway of LFA-1 activation must exist. That this

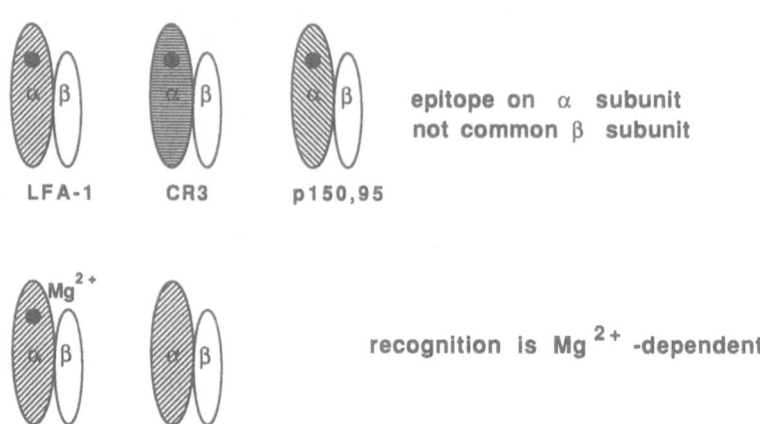

FIGURE 1.1. Summary of characteristic features of mAb 24 specificity.

activation is not confined to LFA-1 is demonstrated by similar results obtained after cross-linking of CD3, for VLA-4, VLA-5, and VLA-6, i.e., enhanced matrix protein binding (17). Finally such activation may also extend to other types of molecules such as selectins. Cross-linking either CD2 or CD3 caused a threefold enhancement in the ability of L-selectin (Mel-14, LAM-1) to bind to ligand (18). This suggests that the memory T cell has a "master-switch" triggered by antigen-specific activation which controls many receptors used by the T cell to adhere to other cells and to the matrix on which T cells move.

We have described mAb 24 that binds to an epitope present on the α subunits of all three leukocyte integrins (19,20) (Fig. 1.1). Recognition of this epitope in the intact heterodimer is Mg^{2+}-dependent and for intact cells, expression parallels receptor activity. Therefore it can be considered that the 24 epitope acts as a "reporter" of leukocyte integrin activation.

Results

We were initially interested to discover any effect of mAb 24 upon functions known to be dependent on active leukocyte integrins. We therefore examined antigen presentation by monocytes to T cells (21) and lymphokine-activated killer (LAK) cell cytotoxicity assays (22), which are

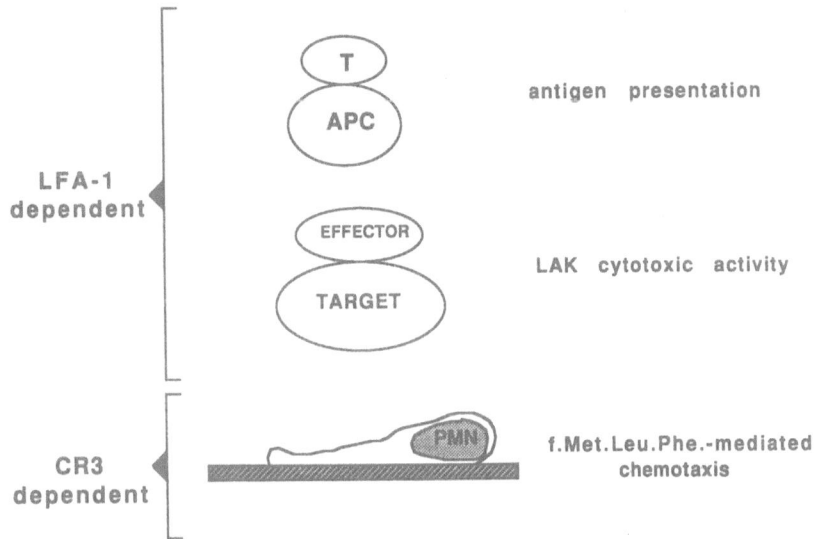

FIGURE 1.2. MAb 24 inhibits antigen-specific T-cell proliferation and LAK cell activity, which are both LFA-1-dependent, and f-Met-Leu-Phe–stimulated neutrophil chemotaxis, which is CR3-dependent.

both LFA-1–dependent, and also f-Met-Leu-Phe–stimulated chemotaxis of neutrophils, a response which has been found to be CR3-dependent (23) (Fig. 1.2). The result was that all three of these functions were inhibited by mAb 24, implicating the recognized epitope as a functionally important region of the leukocyte integrins (24).

If the mechanism by which mAb 24 inhibited function could be elucidated then we thought that further insight might be gained into the process of leukocyte integrin activation. One possibility was that the inhibitory effects of mAb 24 on function might occur as a result of blockade of receptor/ligand binding. Peptide ligands, RGD, and fibrinogen γ-chain sequences, can be cross-linked to the putative cation-binding domains of the β_3 integrins, VNR, and IIb/IIIa (2,25,26). These observations have suggested that for integrins, divalent cation binding and interaction with ligand are intimately associated. However, we found that mAb 24 had no effect on LFA-1– or CR3-dependent binding assays, suggesting that the leukocyte integrins may interact with ligand in a different manner than RGD-specific integrins (Fig. 1.3). Another possibility is that the 24 epitope, which behaves as an indicator of Mg^{2+} binding, is not physically located within the putative cation-binding domains. The localization of this epitope is presently under investigation.

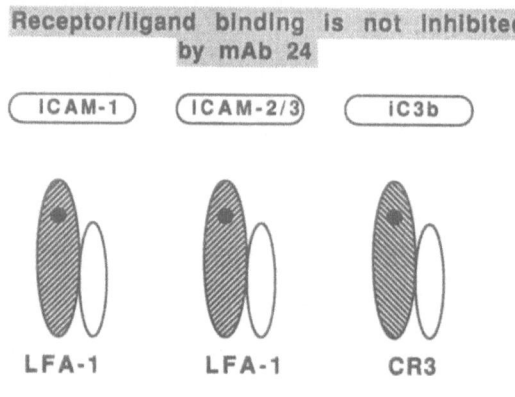

FIGURE 1.3. Effect of mAb 24 on LFA-1/ICAM-1,2,3–dependent aggregation and on macrophage/iC3b-coated erythrocyte CR3-dependent rosetting. Phorbol ester–induced LFA-1/ICAM-1–dependent aggregation of JY lymphoblastoid cell line and LFA-1/ICAM-2,3–dependent aggregation of SKW3 T-cell lymphoma cell line are not inhibited by mAb 24. Binding of iC3b-E by monocyte-derived macrophages (CR3-dependent) is also not inhibited by mAb 24.

Although the inhibitory effect upon function of mAb 24 does not occur as a result of blocking receptor/ligand binding, it was of interest to know whether the mAb promoted this type of interaction. When mAb 24 was added to JY cells undergoing spontaneous aggregation or phorbolesten-stimulated T-cell aggregation, it was able to enhance LFA-1–dependent interactions which were already underway. These observations, coupled with previous work showing that the 24 epitope is expressed only when LFA-1 is functionally active, suggest that mAb 24 is able to promote an increase in LFA-1–dependent binding to ligand only following prior activation of LFA-1. One explanation might be that mAb 24 induces intracellular signals, enabling LFA-1 to combine with ligand with increased avidity. Such activity has been suggested for other LFA-1 α-subunit mAbs which enhance proliferation (27–30). However, inhibitory effects on function might occur if mAb 24 acts extracellularly by blocking conformational alterations resulting from activation.

In order to test these possibilities, we analyzed the effect of mAb 24 in an assay which allows analysis of activation and subsequent deactivation of LFA-1 binding activity to ICAM-1 (11,12) (Fig 1.4). MAb 24 had no effect on the rate of initial adhesion of T cells. However, the resulting prolongation of the initial adhesion phase by mAb 24 suggests that, by interacting with its epitope, mAb 24 prevents de-adhesion of LFA-1 and ICAM-1. It can be speculated that LFA-1 is unable to dissociate from ICAM-1 because an essential conformational alteration of the α and β subunits is prevented. Further evidence for this suggestion comes from the formation of small aggregates in antigen-stimulated cultures of

FIGURE 1.4. The de-adhesion of the transient LFA-1-dependent T cell binding to ICAM-1 (induced by CD3 cross-linking) is prevented by mAb 24.

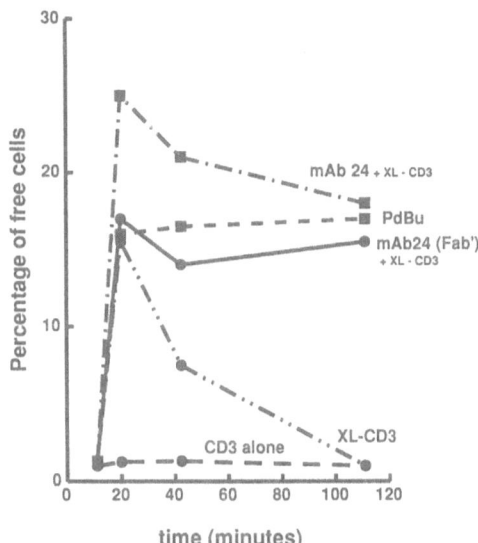

peripheral blood cells in the presence of mAb 24. These aggregates form after a few hours exposure to antigen and subsequently appear to remain stable in size. Such a situation would prevent access of other T cells to adherent antigen-presenting cells, precluding further stimulatory events. Similarly the ability of mAb 24 to prevent LAK cytotoxicity increased as the effector:target-cell ratio decreased, which could be explained by an increased dependence on effector-cell recycling. Finally, it can be speculated that neutrophils moving on a substrate in a chemotactic gradient might accomplish this end at least partly via an attachment-release mechanism involving CR3, as has been demonstrated for neutrophil binding to endothelium (7,9). Thus it may be that mAb 24 causes leukocyte integrins to retain an "activated" conformation, preventing a return to the "nonactivated" state. As a consequence, leukocyte integrin-dependent functions are inhibited, implying a requirement for dynamic receptor/ligand binding in these processes.

We have proposed that the activation step causes an alteration in α and β subunits such that the 24 epitope and Mg^{2+} binding sites are all revealed (19). This has suggested that control of affinity of Mg^{2+} binding represents a potential mechanism for positive regulation of receptor function. Such speculation has been modified recently as a result of experiments which have revealed a role for a second divalent cation, Ca^{2+} in the regulation of LFA-1 activation (31). Identifying a role for Ca^{2+} has come about as a result of a series of experiments in which Mg^{2+} failed to induce expression of the 24 epitope, as had been found previously. This

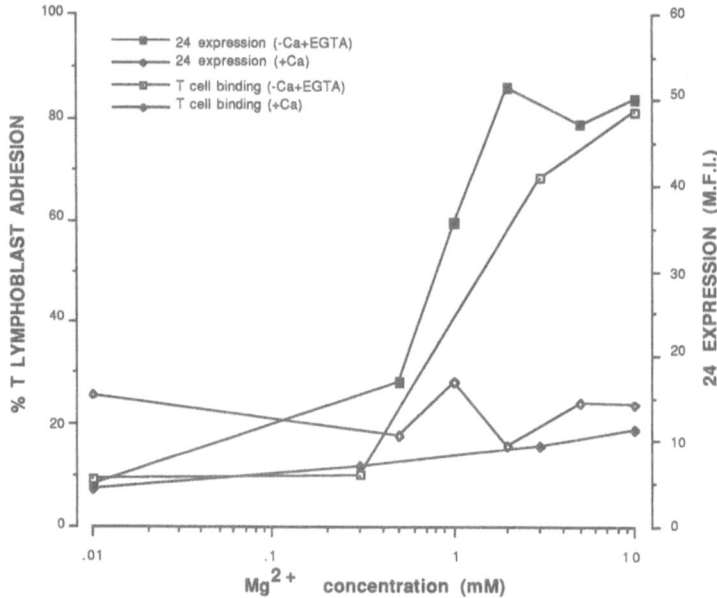

FIGURE 1.5. Inhibitory effect of Ca^{2+} on Mg^{2+}-induced 24 epitope expression and LFA-1–mediated binding of T cells to ICAM-1. Fc chimeric protein. M.F.I. = mean fluorescence intensity.

seeming anomaly was explained by a change in the protocol for preparation of T cells in which a short EDTA wash step had been omitted. That Ca^{2+} must be bound to leukocyte integrins with sufficient affinity not to be removed simply by washing the cells was demonstrated in the following way. Using cells that were washed only in Ca^{2+}/Mg^{2+}-free buffer, low levels of binding of mAb 24 in the presence of Mg^{2+} and low levels of LFA-1–mediated T-cell binding to ICAM-1 were observed. However, when these T cells were pretreated with EGTA, binding of mAb 24 in the presence of Mg^{2+} was increased nearly tenfold and the majority T lymphoblasts adhered to ICAM-1. This situation could be reversed by adding 1 mM Ca^{2+} to the T cells (Fig. 1.5). It was therefore concluded that bound Ca^{2+} imposes a conformation of LFA-1 that is neither recognized by mAb 24 nor conducive to avid ICAM-1 binding. These observations suggest that Ca^{2+} binding to leukocyte integrins (LFA-1) acts as a negative regulator of the functional activity of these molecules. Maintenance of integrin in such an inactive state would have the practical importance of preventing leukocytes from randomly adhering to one another in the circulation until an appropriate encounter caused the stimulation necessary for the release of Ca^{2+} and acquisition of "active" conformation.

The negative effects of Ca^{2+} on LFA-1 function are seemingly at

FIGURE 1.6. Diagrammatic representation of a view of events leading to the activation of a leukocyte integrin

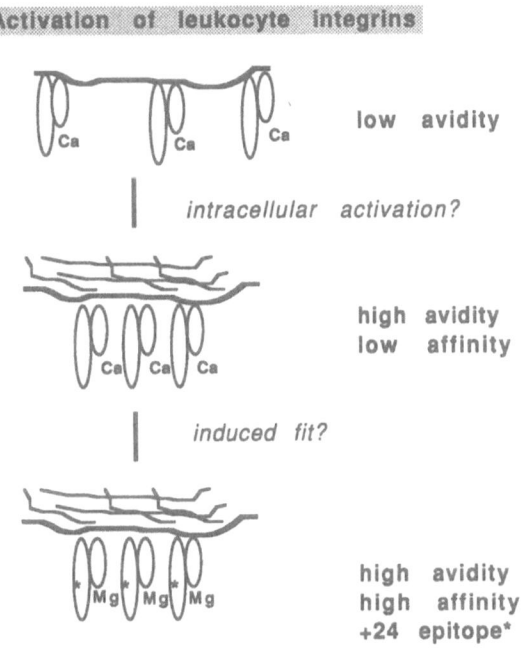

Activation of leukocyte integrins

low avidity

intracellular activation?

high avidity
low affinity

induced fit?

high avidity
high affinity
+24 epitope*

variance with reports from several groups suggesting that Ca^{2+} has synergistic effects in restoration of LFA-1 functional activity at suboptimal concentrations of Mg^{2+}. This apparent discrepancy with our findings may be explained by differences in protocol. As stated, Ca^{2+} already bound to LFA-1 on primary T cells exerts a negative regulatory role upon LFA-1 function. Thus in our studies pretreatment with EGTA in order to remove bound Ca^{2+} increases the functional activity of LFA-1 which can be induced with Mg^{2+} (31). However, in the several studies in which Ca^{2+} was found to synergize with Mg^{2+} both activated cells or phorbol ester–treated cells were employed, suggesting that under these conditions there are different requirements for divalent cations (32–34).

Figure 1.6 illustrates how the foregoing information can be brought together to give a scheme for integrin activation. In the resting T cell, monomeric LFA-1 heterodimers are able to engage ICAM-1 only with low avidity because Ca^{2+} has imposed an unfavorable conformation for such ligation. Activation is initiated by an intracellular signal following on from membrane cross-linking events. This signal is postulated to cause as its end effect some altered interaction of cytoplasmic "tails" of β subunits (35) with cytoskeletal elements, drawn to show clustering of integrins in the membrane. Both talin (36) and α-actinin (37) may associate with activated integrin potentially after a phosphorylation event either on the integrin β subunit or cytoplasmic protein. Clustering of receptors would by itself bring about an increase in receptor avidity, without requirement for change in individual receptor affinity simply by increasing the chances for receptor/ligand interaction. A second set of alterations may then produce high-affinity receptors with an overall increase in the affinity of binding. It is in this form that the $\alpha\beta$ heterodimer binds Mg^{2+} and expresses the 24 epitope. Interestingly, binding of mAb NKI-L16 to LFA-1 is a Ca^{2+}-dependent process which causes leukocyte aggregation and has been suggested to be a reflection of LFA-1 membrane distribution (38,39). By interacting with the Ca^{2+} bound form of integrin, this mAb may induce an allosteric alteration leading to the active conformation of LFA-1. For the integrin IIbIIIa, it is suggested that occupancy of receptor by peptide RGD, a ligand mimetic, can bring about the conformational alterations which increase binding affinity of the receptor (40). Whether this is a general rule of integrin activation remains to be tested.

Conclusions

We have analyzed selected features of leukocyte integrin activation through the use of mAb 24 which binds to leukocyte integrin α subunits and is dependent upon Mg^{2+} binding to integrin for detection of its epitope. The correlation of 24 expression with LFA-1/ICAM-1 ligation has led us to suggest that Mg^{2+} is required for LFA-1 function but that cellular regulation of LFA-1 avidity may be negative and mediated by bound Ca^{2+} which maintains a nonavid conformation of this integrin. For the LFA-1 and CR3, a dynamic alteration between active and inactive receptors is required for effective integrin function. This may be necessary for cell "cycling" but raises the possibility that integrin signalling influences the cellular processes following on from the adhesion phase.

References

1. Sanchez-Madrid, F., J.A. Nagy, E. Robbins, P. Simon, and T.A. Springer. 1983. A human leukocyte differentiation antigen family with distinct α-subunits and a common β-subunit: the lymphocyte function-associated antigen (LFA-1), the C3bi complement receptor (OKM1/Mac-1), and the p150,95 molecule. *J. Exp. Med.* 158:1785.
2. Smith, J.W. and D.A. Cheresh. 1991. Labelling of integrin $\alpha_v\beta_3$ with $^{58}Co(III)$. Evidence of metal ion coordination sphere involvement in ligand binding. *J. Biol. Chem.* 266:11429.
3. Rivas, G.A. and J. Gonzàlez-Rodríguez. 1991. Calcium binding to human platelet integrin GPIIb/IIIa and to its constituent glycoproteins. *Biochem. J.* 176:35.
4. Arnaout, M.A. 1990. Structure and function of the leukocyte adhesion molecules CD11/CD18. *Blood* 75:1037.

5. Springer, T.A. 1990. Adhesion receptors of the immune system. *Nature* 346:425.
6. Smith, C.W., S.D. Marlin, R. Rothlein, C. Toman, and D.C. Anderson. 1989. Cooperative interactions of LFA-1 and Mac-1 with intercellular adhesion molecule-1 in facilitating adherence and transendothelial migration of human neutrophils in vitro. *J. Clin. Invest.* 83:2008.
7. Wright, S.D. and B.C. Meyer. 1986. Phorbol esters cause sequential activation and deactivation of complement receptors on polymorphonuclear leukocytes. *J. Immunol.* 136:1759.
8. Buyon, J.P., S.B. Abramson, M.R. Philips, S.G. Slade, G.D. Ross, G. Weissmann, and R.J. Winchester. 1988. Dissociation between increased surface expression of Gp165/95 and homotypic neutrophil aggregation. *J. Immunol.* 140:3156.
9. Lo, S.K., G.A. Van Seventer, S.M. Levin, and S.D. Wright. 1989. Two leukocyte receptors (CD11a/CD18 and CD11b/CD18) mediate transient adhesion to endothelium by binding to different ligands. *J. Immunol.* 143:3325.
10. Rothlein, R., M.L. Dustin, S.D. Marlin, and T.A. Springer. 1986. A human intercellular adhesion molecule (ICAM-1) distinct from LFA-1. *J. Immunol.* 137:1270.
11. Dustin, M.L., and T.A. Springer. 1989. T-cell receptor cross-linking transiently stimulates adhesiveness through LFA-1. *Nature* 341:619.
12. van Kooyk, Y., P. van de Wiel-van Kemenade, P. Weder, T.W. Kuijpers, and C.G. Figdor. 1989. Enhancement of LFA-1-mediated cell adhesion by triggering through CD2 or CD3 on T lymphocytes. *Nature* 342:811.
13. Lauener, R.P., R.S. Geha, and D. Vercelli. 1990. Engagement of the monocyte surface antigen CD14 induces lymphocyte function-associated antigen-1/intercellular adhesion molecule-1-dependent homotypic adhesion. *J. Immunol.* 145:1390.
14. Nong, Y.-H., E. Remold-O'Donnell, T.W. LeBien, and H.G. Remold. 1989. A monoclonal antibody to sialophorin (CD43) induces homotypic adhesion and activation of human monocytes. *J. Exp. Med.* 170:259.
15. Koopman, G., Y. van Kooyk, M. de Graaff, C.J.L.M. Meyer, C.G. Figdor, and S.T. Pals. 1990. Triggering of the CD44 antigen on T lymphocytes promotes T cell adhesion through the LFA-1 pathway. *J. Immunol.* 145:3589.
16. Mourad, W., R.S. Geha, and T. Chatila, 1990. Engagement of major histocompatibility complex class II molecules induces sustained, lymphocyte function-associated molecule 1-dependent adhesion. *J. Exp. Med.* 172:1513.
17. Shimizu, Y., G.A. Seventer, K.J. Horgan, and S. Shaw. 1990. Regulated expression and binding of three VLA (β1) integrin receptors on T cells. *Nature* 345:250.
18. Spertini, O., G.S. Kansas, J.M. Munro, J.D. Griffin and T.F. Tedder. 1991. Regulation of leukocyte migration by activation of the leukocyte adhesion molecule-1 (LAM-1) selectin. *Nature* 349:691.
19. Dransfield, I. and N. Hogg. 1989. Regulated expression of Mg^{2+} binding epitope on leukocyte integrin α subunits. *EMBO J.* 8:3759.
20. Dransfield, I., A.-M. Buckle, and N. Hogg. 1990. Early events of the immune response mediated by leukocyte integrins. *Immunol. Rev.* 114:29.

21. Dougherty, G.J. and N. Hogg. 1987. The role of monocyte lymphocyte function-associated antigen 1 (LFA-1) in accessory cell function. *Eur. J. Immunol.* 17:943.

22. Robertson, M.J., M.A. Caligiuri, T.J. Manley, H. Levine, and J. Ritz. 1990. Human natural killer cell adhesion molecules. Differential expression after activation and participation in cytolysis. 1990. *J. Immunol.* 145:3194.

23. Anderson, D.C., L.J. Miller, F.C. Schmalstieg, R. Rothlein, and T.A. Springer. 1986. Contributions of the Mac-1 family to adherence-dependent granulocyte functions: structure-function assessments employing subunit-specific monoclonal antibodies. *J. Immunol.* 137:15.

24. Dransfield, I., C. Cabañas, J. Barrett, and N. Hogg. 1992. Interaction of leukocyte integrins with ligand is necessary but not sufficient for function. *J. Cell Biol.* 116:1527.

25. D'Souza, S.E., M.H. Ginsberg, T.A. Burke, and E.F. Plow. 1990. The ligand binding site of the platelet integrin receptor GPIIb-IIIa is proximal to the second calcium binding domain of its α subunit. *J. Biol. Chem.* 265:3440.

26. Smith, J.W. and D.A. Cheresh. 1990. Integrin ($\alpha_v\beta_3$)-ligand interaction. Identification of a heterodimeric RGD binding site on the vitronectin receptor. *J. Biol. Chem.* 265:2168.

27. van Noesel, C., F. Miedema, M. Brouwer, M.A. de Rie, L.A. Aarden, and R.A.W. van Lier. 1988. Regulatory properties of LFA-1 α and β chains in human T-lymphocyte activation. *Nature* 333:850.

28. Carrera, A.C., M. Rincón, F. Sánchez-Madrid, M. López-Botet, and M.O. de Landazuri. 1988. Triggering of co-mitogenic signals in T cell proliferation by anti-LFA-1 (CD18, CD11a), LFA-3, and CD7 monoclonal antibodies. *J. Immunol.* 141:1919.

29. Wacholtz, M.C., S.S. Patel, and P.E. Lipsky. 1989. Leukocyte function-associated antigen 1 is an activation molecule for human T cells. *J. Exp. Med.* 170:431.

30. Pardi, R., J.R. Bender, C. Dettori, E. Giannazza, and E.G. Engleman. 1989. Heterogeneous distribution and transmembrane signaling properties of lymphocyte function-associated antigen (LFA-1) in human lymphocyte subsets. *J. Immunol.* 143:3157.

31. Dransfield, I., C. Cabañas, A. Craig, and N. Hogg. 1992. Divalent cation regulation of the function of the leukocyte integrin LFA-1. *J. Cell Biol.* 116:219.

32. Martz E. 1980. Immune lymphocyte to tumor cell adhesion. Magnesium sufficient, calcium insufficient. *J. Cell Biol.* 84:585.

33. Rothlein, R. and T.A. Springer. 1986. The requirement for lymphocyte function-associated antigen 1 in homotypic leukocyte adhesion stimulated by phorbol ester. *J. Exp. Med.* 163:1132.

34. Cabañas, C. and N. Hogg. 1991. Lymphocyte-fibroblast. A useful model for analysis of the interaction of the leukocyte integrin LFA-1 with ICAM-1. *FEBS Letters* 292:284.

35. Hibbs, M.L., S. Jakes, S.A. Stacker, R.W. Wallace, and T.A. Springer. 1991. The cytoplasmic domain of the integrin lymphocyte function-associated antigen 1 β subunit: sites required for binding to intercellular adhesion molecule 1 and the phorbol ester-stimulated phosphorylation site. *J. Exp. Med.* 174:1227.

36. Burn, P., A. Kupfer, and S.J. Singer. 1988. Dynamic membrane-cytoskeletal interactions: specific association of integrin and talin arises in vivo after phorbol ester treatment of peripheral blood lymphocytes *Proc. Natl. Acad. Sci. USA* 85:497.

37. Otey, C.A., F.M. Pavalko, and K. Burridge. 1990. An interaction between α-actinin and the β_1 subunit in vitro *J. Cell Biol.* 111:721.

38. Keizer, G.D., W. Visser, M. Vliem, and C.G. Figdor. 1988. A monoclonal antibody (NKI-L16) directed against a unique epitope on the α-chain of human leukocyte function-associated antigen 1 induces homotypic cell-cell interactions. *J. Immunol.* 140:1393.

39. van Kooyk, Y., P. Weder, F. Hogervorst, A.J. Verhoeven, G. van Seventer, A. A. te Velde, J. Borst, G.D. Keizer, and C.G. Figdor. 1991. Activation of LFA-1 through a Ca^{2+}-dependent epitope stimulates lymphocyte adhesion. *J. Cell Biol.* 112:345.

40. Du, X., E.F. Plow, A.L Frelinger, T.E. O'Toole, J.C. Loftus, and M.H. Ginsberg. 1991. Ligands "activate" integrin $\alpha_{IIb}\beta_3$ (platelet GPIIb-IIIa). *Cell* 65:409.

2
Activation of LFA-1: Role of Cations

CARL G. FIGDOR AND YVETTE VAN KOOYK

Introduction

Several observations indicate that the affinity of integrin receptors for their ligands can be modulated. Resting leukocytes or platelets do not adhere spontaneously, but a variety of stimuli can induce β1- (VLA-4, VLA-5, VLA-6), β2- (LFA-1, CR3), and β3-integrin (IIb/IIIa)-mediated cell–cell interactions. Exposure of lymphocytes, myeloid cells, or platelets to phorbol ester (PMA) strongly induces cell aggregation (1,2). Similarly, FMLP can stimulate CR3-mediated adhesion of granulocytes to endothelial cells (3,4) and activation of platelets by thrombin or ADP causes IIb/IIIa-mediated aggregation (5). A prominent characteristic in all these observations is that adhesion is induced without an apparent increase in receptor expression. This suggests that changes in affinity of the receptor for its ligand, or changes in the avidity (for instance, by alteration of the organization of the adhesion receptors at the cell surface), directly affect cell adhesion. Second messengers play a pivotal role in integrin activation, although at present the precise intracellular circuits that regulate integrin-mediated cell adhesion are not completely understood.

Activation of LFA-1

Recently we (6), and others (7) found that, except from PMA, monoclonal antibodies directed against the cell-surface molecules expressed by T cells, such as CD2 and CD3, can also stimulate homotypic cell aggregation (Table 2.1). In addition, signals through CD43, CD44, and MHC class II but not MHC class I also can induce LFA-1–mediated adhesion of cloned T lymphocytes (Table 2.1). These observations indicate that a number of surface molecules expressed by T cells can transduce signals and activate LFA-1 via intracellular signaling pathways. These findings imply the existence of at least two forms of LFA-1: an inactive and an active form of LFA-1. A strong argument in favor of the existence of these

14

TABLE 2.1. Homotypic aggregation of HY T cells.

Treatment	% aggregation
Medium	10
PMA	80
CD2	80
CD3	60
MHC class I	10
MHC class II	70
CD43	70
CD44	80

Homotypic aggregation of HY cells induced by PMA (10 ng/ml) or purified mAb (10 μg/ml) directed against surface molecules expressed by T cells: CD3, MHC class I, MHC class II, CD43, and CD44. Aggregation was scored after 40-min incubation of cells at 37°C. Aggregation was determined using a light microscope. Percentage of aggregation was determined by counting the number of free cells, using the equation % aggregation = 100 × [1 − (no free cells/no input cells)].

two states of LFA-1 is that if only one form of LFA-1 existed, spontaneous aggregation of peripheral blood leukocytes would cause injury of the microvascular network, induce micro-embolization, and might thereby compromise normal leukocyte circulation. Resting leukocytes do not tend to aggregate to each other, although they express significant levels of LFA-1 and ICAM-1, indicating that LFA-1 must be activated for high-affinity ligand binding. Similarly, cloned cytotoxic T lymphocytes (CTL) and NK cells, which express extremely high levels of LFA-1 and ICAM-1, do not aggregate (6). Only upon stimulation with antigen, PMA, or via CD2, CD3, CD43, CD44, or MHC class II by mAbs, can rapid cell aggregation (<20 min) of CTL be observed. This cell clustering is LFA-1–dependent since it is abrogated by anti-LFA-1 antibodies. In addition, surface expression of LFA-1 and ICAM-1 does not change during activation of T-cell or NK-cell clones (6), demonstrating that cell aggregation is not caused just by augmented surface expression. These observations support the hypothesis that activation of LFA-1 can be induced through different surface receptors, indicating that the LFA-1/ICAM-1 is a common adhesion pathway which may be used by leukocytes under quite different physiological conditions.

Interestingly, we observed (Fig. 2.1A) that, depending on the stimulus used, activation of LFA-1 was either transient (MHC class II, CD3) or sustained (CD2) for a prolonged period of time ($\geq$2 h). These results indicate that different and complex signaling pathways may be employed to regulate activation of LFA-1 to ensure the required stable or transient interactions in the myriad of functions mediated by this molecule. A good example to demonstrate the mode of action of LFA-1 is the cytotoxic T-cell (CTL)–target-cell interaction, the effector phase of an immune re-

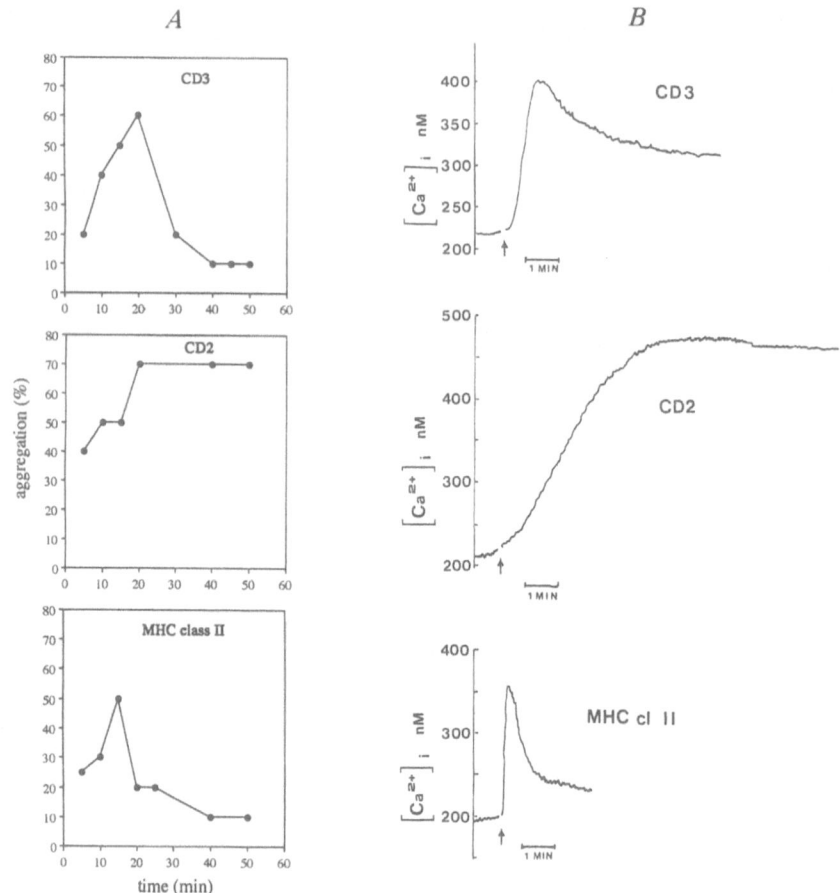

FIGURE 2.1. *A.* Kinetics of homotypic aggregation of JS136 T cells induced by mAbs (10 μg/ml) directed against CD2, CD3, and MHC class II antigens in the presence of 1 mM Mg^{2+} and Ca^{2+}. Aggregate formation was induced as described in Table 2.1 and followed in time. At different time points aggregates were disrupted by vigorous pipetting to form single-cell suspensions. Subsequently the capacity of the cells to re-aggregate was measured in a second aggregation period of 20 min at 37°C, without further additional stimuli. Formation of new clusters indicates that the receptor is still active, whereas unresponsiveness of the cells to form clusters indicates that the receptor is inactive. *B.* Measurements of intracellular Ca^{2+} levels ($[Ca^{2+}]_i$) in JS136 T cells. Cells were loaded with the fluorescent calcium-sensitive dye Indo-1/AM for 40 min at 37°C. Cells were washed and $[Ca^{2+}]_i$ changes were measured in Hepes buffer containing 1 mM $CaCl_2$ and 1 mM $MgCl_2$ upon addition of anti-CD3, anti-CD2, or anti-MHC class II antibodies (10 μg/ml) in a spectrofluorimeter. (↑) indicates moment of addition of the antibodies.

sponse. When a CTL encounters a target cell initial cell–cell contact is established through LFA-1–ICAM interaction. Cell binding is nonspecific and of low avidity and affinity, since it does not implicate antigen recognition by the T-cell receptor. Nevertheless, target-cell binding is strong enough to facilitate recognition of antigen by the T-cell receptor/CD3 complex. If no antigen is recognized, the CTL detaches from its target cell. If, however, antigen is recognized by the T-cell receptor, CD3 creates a signal leading to activation of LFA-1. Activated LFA-1 then mediates high-avidity binding between CTL and target cell, thereby strengthening the adhesion between both cells. This facilitates the formation of intercellular clefts, enabling efficient delivery of cytotoxic molecules into the target cell. In addition activated LFA-1 may interact with cytoskeletal elements, thereby directing the migration of cytotoxic granules. Modulation of the TCR/CD3 complex from the cell surface after peptide MHC binding, possibly by phosphorylation of the CD3 γ component (8), might abrogate signaling, thus acting as a negative feedback signal reversing LFA-1 into its inactive state and providing the CTL of a mechanism to detach from a target cell. Similarly, in the induction phase of an immune response, when antigen is presented by MHC to the T-cell receptor, LFA-1, both on the antigen-presenting cell (B cell, macrophage) and on the responder cell, may be engaged in cell–cell contact, since CD3 may activate LFA-1 on the T cell and MHC class II may stimulate LFA-1 on the antigen-presenting cell, as also suggested by Mourad et al. (9). Activation and deactivation of LFA-1 provide leukocytes with a general mechanism with which to attach and detach to cells and to migrate throughout the body. Since ligands of LFA-1 are widely spread, lymphocytes can migrate to essentially every organ in our body.

Intracellular Messengers and Affinity Modulation of LFA-1

PMA has already early been recognized as a potent activator of LFA-1–mediated adhesion. Whereas PMA directly activates PKC, activation through CD2 and CD3 is thought to stimulate inositol phospholipid metabolism, thereby giving rise to activation of PKC and a rise in $[Ca^{2+}]_i$ levels (10,11). Interestingly, we observed that a rise in the $[Ca^{2+}]_i$ level of cloned T cells by the addition of ionomycin also stimulates LFA-1–mediated adhesion (not shown). This is in line with other reports showing a direct correlation between a rise in $[Ca^{2+}]_i$ levels and the high-affinity state of the leukocyte integrin receptors (12,13), indicating that $[Ca^{2+}]_i$ plays an important role in activation of LFA-1. Furthermore, we also observed that the anti-CD2–, anti-CD3–, and anti–MHC class II–induced LFA-1–mediated adhesion is associated with enhanced $[Ca^{2+}]_i$ levels.

Moreover, we found a direct correlation between the length during which the LFA-1 molecule was activated and the duration of the increased $[Ca^{2+}]_i$ levels (Fig. 2.1B). This further emphasizes the role of $[Ca^{2+}]_i$ in the intracellular processes that lead to activation of the LFA-1 molecule. These findings indicate that at least two distinct signaling pathways are involved in the activation of LFA-1, in which either PKC and/or $[Ca^{2+}]_i$ is involved. PMA-mediated stimulation of PKC stimulates LFA-1-mediated adhesion without a rise in $[Ca^{2+}]_i$, and ionophore-mediated stimulation of LFA-1 does not implicate activation of PKC.

Activation of LFA-1 by NKI-L16

We previously described (14,15) an anti–LFA-1 antibody, designated NKI-L16 (further called L16), that in contrast to other anti–LFA-1 antibodies, stimulates cell adhesion rather than inhibiting LFA-1–dependent cell interactions. All its known characteristics are summarized in Table 2.2. L16-induced cell aggregation does not implicate "outside-in" signaling but is merely thought to act by modulating the conformation of LFA-1 so that the affinity for ligand binding is greatly enhanced (15). Time-course studies measuring cell aggregation induced by L16 or by F(ab)' fragments showed kinetics strikingly similar to that observed when cells were stimulated with PMA (14). This observation led us to suggest that stimulation of cells with PMA or through the CD3, CD2, or MHC class II receptor, although by an entirely different mechanism, might ultimately also result in a conformational change of LFA-1, thus increasing the affinity for its ligands.

TABLE 2.2. Characteristics of the L16 epitope.

Antibody induces: (IgG, Fab')	—cell aggregation, homotypic (T, B, NK, Mo, Gr) heterotypic
	—Ca^{2+} and/or Mg^{2+} is required and sufficient for cell aggregation
	—binding to purified ICAM-1
	—binding to cells transfected with ICAM-1
	—binding to endothelial cells
Signal transduction:	—not clear, probably not
	—no Ca^{2+} flux
	—no effect of PKC inhibitors
	—no effect of sodium azide
	—no effect deoxyglucose
	—inhibition by sodium azide and deoxyglucose
Characteristics epitope:	—Ca^{2+} dependent
	—Ca^{2+} can be replaced by Sr^{2+} (FACS + immunoprecipitation) or by Mn^{2+} (immunoprecipitation)
	—increased expression induced during cell maturation

The L16 epitope is not within the ligand-binding domain, since addition of the L16 antibody in itself induces LFA-1–dependent adhesion (6). This notion is supported by the observation that L16-induced adhesion is completely blocked by other anti–LFA-1α or -β antibodies (6,15). Since also F(ab)′ fragments of L16 were capable of inducing adhesion, we can exclude that crosslinking of receptors or Fc receptor mediated phenomena are involved. The observation that PKC inhibitors (staurosporin, AMG) were unable to inhibit L16-induced adhesion, but unequivocally inhibited PMA-stimulated adhesion, suggests that L16 does not induce signaling into the cell (15), although the L16-induced adhesion seems to depend on metabolic energy (Table 2.2). In addition, L16-induced adhesion is not associated with a rise in $[Ca^{2+}]_i$ levels (Table 2.2). Together, these results suggest that L16 induces a change in the tertiary structure of LFA-1 which may result in modulation of the ligand-binding affinity.

Recently, we found that the epitope on LFA-1 recognized by the L16 antibody depends on the presence of Ca^{2+} (15). Treatment of cloned T cells, which express L16 abundantly, with a metal-chelating agent (EDTA or EGTA) results in a complete loss of this epitope (Table 2.3) and, more importantly, in loss of capacity to bind to other cells in a LFA-1–dependent manner. Loss of the L16 epitope is not associated with a reduction of other epitopes of LFA-1, showing that the molecule is still pre-

TABLE 2.3. Expression of the L16 epitope is required for LFA-1–mediated adhesion.

| | JS-136 | | | |
| | | Homotypic aggregation induced by | | |
	L16 expression	L16	CD3	PMA
Medium	+	+	+	+
dPBS	+	−	−	−
dPBS + Mg^{2+}	+	+	−	+
dPBS + Ca^{2+}	+	+	−	−
dPBS + Ca^{2+}/Mg^{2+}	+	+	+	+
EGTA	−	−	−	−
EDTA	−	−	−	−

The effect of Mg^{2+} and Ca^{2+} cations on L16 epitope expression and homotypic aggregation induced by L16, CD3 antibodies (10 μg/ml) or PMA (10 ng/ml). Medium was depleted from extracellular cations (dPBS) with 1% wt/vol Chelex 100 microspheres by rotary mixing for 2 h at room temperature. Cation concentrations were restored by the addition of 1 mM CaCl$_2$ or/and 1 mM MgCl$_2$. Extracellular and receptor-bound cations were removed by treatment of the cells with 5 mM EGTA or 5 mM EDTA for 10 min at 4°C. L16 expression on JS136 T cells as determined by using FACScan analysis. Homotypic aggregation was performed as described in Table 2.1 and scored after 50-min incubation of T cells at 37°C. Note that suspension of cells in cation-depleted medium (dPBS) does not affect the L16 expression whereas the epitope is lost after treatment of the cells with EDTA or EGTA.

sent at the cell surface. In contrast to cloned T lymphocytes, resting PBLs express LFA-1 on their cell surface but they generally lack L16 expression or exhibit only low levels of this epitope at the cell surface (donor-dependent). We observed that the ability of lymphocytes to aggregate upon stimulation directly correlates with expression of this L16 epitope by LFA-1, suggesting that expression of the L16 epitope is a prerequisite for LFA-1–dependent adhesion. This notion is supported by the finding that stimulation of resting lymphocytes with IL-2 or PMA (3–12 h) results in a significant increase in expression of the L16 epitope and correlates with their capacity to aggregate. This is not due to an increased number of LFA-1 molecules expressed on the cell membrane, since it is not accompanied by a concomitant rise in expression of LFA-1 epitopes other than L16 (15). These observations show that the majority of LFA-1 molecules expressed by resting lymphocytes lack the L16 epitope, but that this epitope becomes readily expressed upon *in vitro* culture of the cells. T-cell or NK-cell clones express high levels of LFA-1 molecules, all of which expose the L16 epitope. Recently we observed that several T-cell leukemia cell lines, such as CEM, lack L16 expression (unpublished). In accordance to the hypothesis that L16 is required for LFA-1–mediated function, these cells fail to aggregate when stimulated despite a normal expression of LFA-1.

The Role of Cations in LFA-1–Mediated Adhesion

Integrin-mediated cell adhesion is a temperature- and energy-dependent process which requires an intact cytoskeleton and the presence of divalent cations, notably Mg^{2+} and/or Ca^{2+}. In most cases Ca^{2+} or Mg^{2+} can be substituted by other divalent cations. We observed that Mn^{2+} causes spontaneous activation of LFA-1 resulting in homotypic aggregation of T cells or binding of T cells to L cells that express ICAM-1 (unpublished). The underlying mechanism and its physiological relevance are unknown, although recent observations by Altieri et al. (16) indicate that the high-affinity state of the CR3 receptor induced by Mn^{2+} on monocytes may be due to direct binding of Mn^{2+} to CR3 and cause a conformational change in the molecule, which results in an increased affinity for its ligand. These observations underscore the important role of metal ions in integrin receptor activation.

When we compare the cation requirements for activation of LFA-1 using different stimuli, two main categories can be distinguished (Fig. 2.2). Stimuli (anti-CD3, anti-CD43, anti–MHC class II) require the presence of both Ca^{2+} and Mg^{2+}, and stimuli (CD2, CD44) that induce a prolonged activation of LFA-1 require only Mg^{2+}. Interestingly, this cation dependency correlates with the type of response that is observed. The

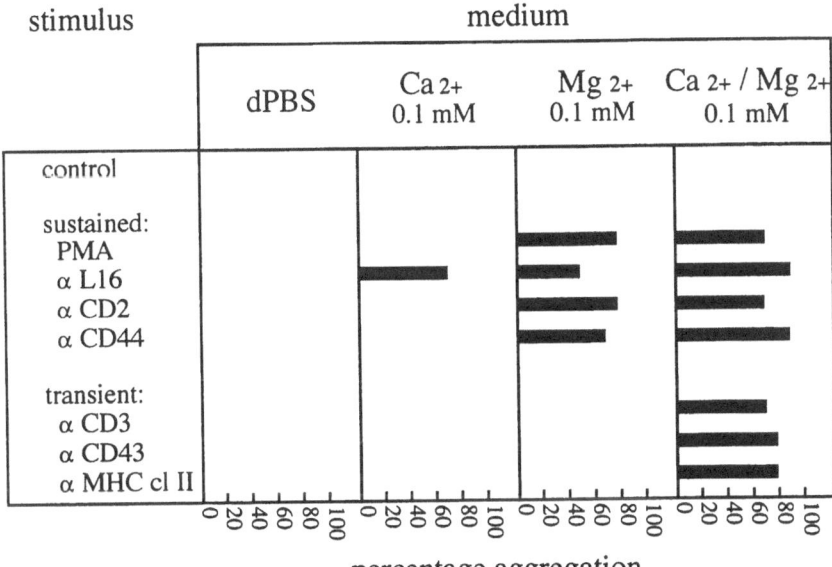

FIGURE 2.2. Cation dependence of homotypic aggregation of HY T cells induced by PMA (10 ng/ml) or antibodies directed against CD11a (L16), CD2, CD3, CD43, CD44, and MHC class II (10 μg/ml). Extracellular cations were depleted by rotary mixing all solutions (PBS and stimuli) with Chelex 100 as mentioned in Table 2.3. Divalent cation concentrations were restored by 0.1 mM Ca^{2+} and/or Mg^{2+}. Aggregation was scored as mentioned in Figure 2.1, after 50-min incubation of cells at 37°C.

stimuli that require both cations generate a transient response. Furthermore, it should be noted that the L16 antibody (Fig. 2.2) is the only agent that is able to induce LFA-1–mediated adhesion in the absence of Mg^{2+}. The presence of only Ca^{2+} is sufficient to induce adhesion (Fig. 2.2).

Distinct Forms of LFA-1

Several conclusions can be drawn from the results shown in this paper. First, the data demonstrate that several stimuli are able to regulate LFA-1–mediated adhesion by activating and de-activating the molecule. Inactive LFA-1, as depicted in the model (Fig. 2.3a), is not able to bind its ligands with high avidity and affinity. Second, our results show that LFA-1 cannot be stimulated into its active state unless it expresses an epitope recognized by the L16 antibody (Fig. 2.3b). We termed this form of LFA-1 "potentially active LFA-1," since upon stimulation a rapid transition

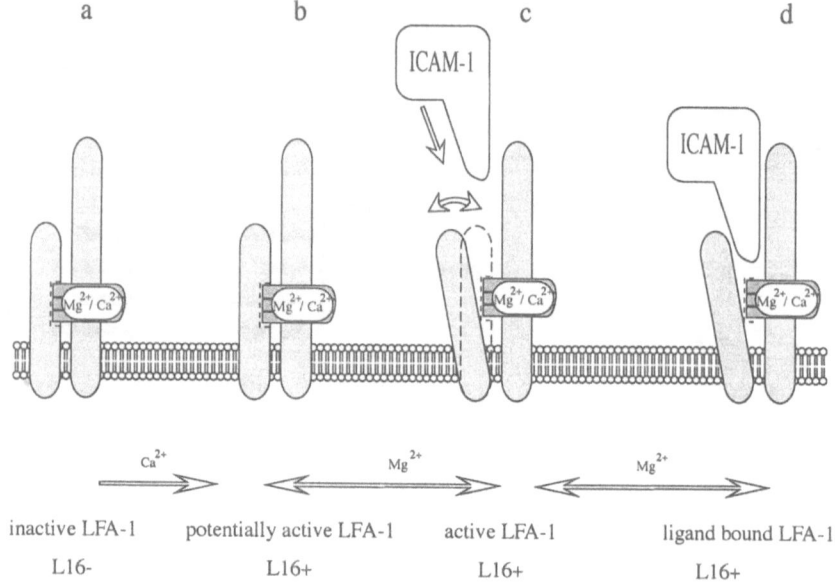

FIGURE 2.3. A model showing four distinct forms of LFA-1:

a. inactive LFA-1   (L16−)
b. "potentially active" LFA-1   (L16+)
c. active LFA-1   (L16+)
d. ligand bound LFA-1   (L16+)

Transition from a to b is dependent on the presence of Ca^{2+}, whereas activation of LFA-1 (c) generally requires the presence of Mg^{2+}.

into fully activated LFA-1 (Fig. 2.3c) is observed. It should be noted that expression of the L16 epitope is required, but not sufficient to induce cell adhesion. Activation of LFA-1 (Fig. 2.3c) requires a strong stimulatory signal, resulting in binding of its ligand (Fig. 2.3d). Currently what causes the Ca^{2+}-dependent L16 epitope is unknown, but it cannot be excluded that reorganization of LFA-1 at the cell surface is also important, thereby facilitating high-avidity ligand binding. Third, extracellular divalent cations play an extremely important role in LFA-1–mediated cell adhesion, and can modulate the response. This notion is not only supported by our results but also by those of Dransfield et al. (17), who recognized an Mg^{2+}-dependent epitope on the α chains of the $\beta2$ integrins that seems to be expressed only when LFA-1 is in its active (Fig. 2.3c) state. In addition, data obtained with Mn^{2+} (unpublished; 16) underscore the important role of cations in modulating integrin activity.

The leukocyte integrins express three cation-binding domains (18). This allows the possibility that occupancy by either Ca^{2+} or Mg^{2+} alone or by a combination of both may have dramatic consequences for the functional status of the receptor. Stimulation of cell aggregation with PMA absolutely requires the presence of Mg^{2+}, and exposure of cloned T lymphocytes to EDTA completely abrogates the capacity of these cells to aggregate. However, reconstitution of the cells with Mg^{2+} but not Ca^{2+} causes a dramatic spontaneous aggregation which may be due to occupancy of all three metal-binding sites by Mg^{2+} (unpublished), supporting the hypothesis that Mg^{2+} may induce an activated state of the receptor under physiological circumstances. On the other hand, we found that in the absence of Mg^{2+}, but in the presence of Ca^{2+}, L16 is still able to induce LFA-1–mediated cell adhesion (Fig. 2.2), demonstrating that LFA-1–mediated adhesion can also occur in the complete absence of Mg^{2+}. In addition, we observed that the addition of Mn^{2+} to the medium immediately results in induction of LFA-1–dependent adhesion of T cells (unpublished). This may indicate that Mn^{2+} can substitute Mg^{2+} or Ca^{2+} and is able to activate LFA-1. The observation that L16 induces LFA-1–mediated adhesion also in the presence of only Ca^{2+} suggests that this metal ion is not only required to express the epitope (Table 2.3), but is also sufficient to allow LFA-1–mediated function. In addition, it may indicate that both Ca^{2+} and Mg^{2+} ions bind to LFA-1, even when the molecule is active.

References

1. Rothlein, R. and T.A. Springer. 1986. The requirement for lymphocyte function-associated antigen in homotypic leukocyte adhesion stimulated by phorbol ester. *J. Exp. Med.* 163:1132.
2. Patarroyo, M., P.G. Beatty, J.W. Fabro, and C.G. Gahmberg. 1985. Identification of a cell surface protein complex mediating phorbol ester induced adhesion (binding) among human mononuclear leukocytes. *Scand. J. Immunol.* 22:171.
3. Buyon, J.P., S.B. Abramson, M.R. Philips, S.G. Slade, G.D. Ross, G. Weissman, and R.J. Winchester. 1988. Dissociation between increased surface expression of Gp165/95 and homotypic neutrophil aggregation. *J. Immunol.* 140:3156.
4. Vedder, N.B., and J.M. Harlan. 1988. Increased surface expression of CD11b/CD18 (MAC-1) is not required for stimulated neutrophil adherence to cultured endothelium. *J. Clin. Invest.* 81:676.
5. Marguerie, G.A., E.F. Plow, and T.S. Edgington. 1979. Human platelets possess an inducible and saturable receptor specific for fibrinogen. *J. Biol. Chem.* 254:5357.
6. Van Kooyk, Y., P. van de Wiel-van Kemenade, P. Weder, T.W. Kuijpers, and C.G. Figdor. 1989. Enhancement of LFA-1 mediated cell adhesion by triggering through CD2 or CD3 on T lymphocytes. *Nature* 342:811.
7. Dustin, M.L., and T.A. Springer. 1989. T cell receptor cross-linking transiently stimulates adhesiveness through LFA-1. *Nature* 341:619.
8. Alexander, D.R., and D.A. Cantrell. 1989. Kinases and phosphatases in T cell activation. *Immunol. Today* 10:200.

9. Mourad, W., R.S. Geha, and T. Chatila. 1990. Engagement of major histo-compatibility complex class II molecules induces sustained, lymphocyte function-associated molecule-1-dependent cell adhesion. *J. Exp. Med.* 172: 1513.

10. Imboden, J.B., and J.D. Stobo. 1985. Transmembrane signalling by the T cell antigen receptor. Perturbation of the T3-antigen receptor complex generates inositol phosphates and releases calcium ions from intracellular stores. *J. Exp. Med.* 161:446.

11. Pantaleo, G., D. Olive, A. Poggi, W.J. Kozumbo, L. Moretta, and A. Moretta. 1987. Transmembrane signalling via the T11-dependent pathway of human T cell activation. Evidence for the involvement of 1,2-diacylglycerol and inositol phosphates. *Eur. J. Immunol.* 17:55.

12. Jaconi, J.E.E., J.M. Theler, W. Schlegel, R.D. Appel, S.D. Wright, and P.D. Lew. 1991. Multiple elevation of cytosolic free Ca^{2+} in human neutrophils: initiation by adherence receptors of the integrin family. *J. Cell Biol.* 112:1249.

13. Altieri, D.C., W.L. Wiltse, and T.S. Edgington. 1990. Signal transduction initiated by extracellular nucleotides regulates the high affinity ligand recognition of the adhesive receptor CD11b/CD18. *J. Immunol.* 145:662.

14. Keizer, G.D., W. Visser, M. Vliem, and C.G. Figdor. 1988. A monoclonal antibody (NKI-L16) directed against a unique epitope on the alpha chain of LFA-1 induces homotypic cell-cell interaction. *J. Immunol.* 140:1393.

15. Van Kooyk, Y., P. Weder, F. Hogervorst, A.J. Verhoeven, G. van Seventer, A.A. te Velde, J. Borst, G.D. Keizer, and C.G. Figdor. 1991. Activation of LFA-1 through a Ca^{2+} dependent epitope stimulates lymphocyte adhesion. *J. Cell Biol.* 112:345.

16. Altieri, D.C. 1991. Occupancy of CD11b/CD18 (Mac-1) divalent ion binding site(s) induces leukocyte adhesion. *J. Immunol.* 147:1891.

17. Dransfield, I., A.-M. Buckle, and N. Hogg. 1990. Early events of the immune response mediated by leukocyte integrins. *Immunol. Rev.* 114:29.

18. Larson, R.S., and T.A. Springer. 1990. Structure and function of leukocyte integrins. *Immunol. Rev.* 114:181.

3
An Antibody That Promotes Adhesion Events Mediated by Both LFA-1 and CR3

Martyn K. Robinson, David Andrew, Hugh Rosen, Derek Brown, Susan Ortlepp, P. Stephens, and Eugene C. Butcher

Introduction

Members of the Leu-CAM (β^2 integrin) family of adhesion molecules are widely distributed on circulating lecukocytes, where they play an important role in many cell–cell interactions (reviewed in 1). This family forms a subset of the integrin supergene family and its members show the typical integrin noncovalently associated α/β heterodimer structure (2). There are three members of the family; LFA-1 (CD11a/CD18), CR3 (CD11b/CD18), and p150.95 (CD11c/CD18) with α-chain molecular weights of 170, 165, and 150 kDa, respectively (2). These glycoproteins exist on the surface of leukocytes as noncovalently linked heterodimers; all share a common β chain (CD18) with a molecular weight of 95 kDa. The different α chains which characterize the individual members of the group share extensive homology (3).

The control of the Leu-CAM family of adhesion molecules appears to be exerted at several different levels. In the case of CR3 and p150.95 a mechanism termed "prevalence modulation" may operate. This involves the rapid mobilization of presynthesized molecules from intracellular granules to the cell surface following cell activation (4–7). There is increasing evidence that the cell can also control its adhesive state by regulating the affinity of members of the Leu-CAM family for their ligands (8–10). The mechanism by which the changes of affinity are controlled is incompletely understood but may involve a conformation change associated with phosphorylation of residues in the cytoplasmic domain of CD18.

Here we describe an anti-CD18 antibody (KIM 127) that promotes both LFA-1– and CR3-dependent aggregation. It is believed that the binding of this antibody provokes a conformational change that increases the affinity of LFA-1 and CR3 for their ligands.

Results

The KIM 127 hybridoma line was produced by fusing NSO cells with the spleen of a mouse immunized with purified human CR3. The CR3 was immunopurified from a human leukocyte lysate on Sepharose beads coated with a rat monoclonal antibody (M1/70) that weakly binds to human CR3.

Cell Aggregation

The human B lymphoblastoid cell line JY grows as a single-cell suspension or as loose aggregates. The addition of hybridoma supernatant from KIM 127 caused a very dramatic increase in the aggregation of the JY cells (Fig. 3.1). The addition of an isotype-matched control antibody that binds to the MHC class I marker on JY cells did not cause any increase in the background level of JY aggregation. KIM 127 supernatant also markedly changed the aggregation of the human T-cell-line MOLT 4.

The aggregating activity associated with KIM 127 supernatant was removed by pretreatment of the supernatant with Sepharose beads coated with goat antimouse immunoglobulin but not by beads coated with nonspecific goat immunoglobulin. Protein G–purified immunoglobulin also promoted the same aggregation, as did monovalent Fab fragments

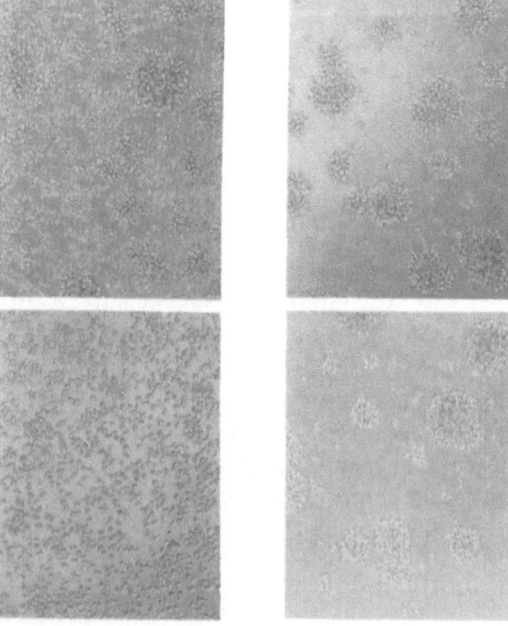

FIGURE 3.1. Shown are the effects of KIM 127 antibody on the human B lymphoblastoid cell line JY. Antibodies were used at 20 μg/ml in RPM1 containing 10% bovine calf serum. The JY cells were incubated with the antibody for 1 h at 37°C before being photographed. a. JY cells incubated with anti-MHC class I antibody (IgG1). b. JY cells incubated with KIM 127 (IgG1). c. JY cells incubated with KIM 127 and R3.1 (anti-CD11a). d. JY cells incubated with KIM 127 F(ab) fragments. (×100).

TABLE 3.1. Aggregation of MOLT-4 Cells in the presence of KIM 127.

	Antibody concentration (μg/ml)					
	20	10	5	2.5	1.25	0.625
KIM 127 whole antibody	+ + + + +	+ + + +	+ + +	+ +	$\frac{1}{2}$+	$\frac{1}{2}$+
KIM 127 Fab	+ + + +	+ +	$\frac{1}{2}$+	$\frac{1}{2}$+	$\frac{1}{2}$+	$\frac{1}{2}$+

of the antibody. Table 3.1 shows that a somewhat higher concentration of Fab fragment was required, which may reflect the reduced avidity of the monovalent fragment.

In a static assay aggregation of MOLT-4 cells in the presence of KIM 127 normally took several hours at 37°C. However, Figure 3.2 shows that if the cells were subjected to mechanical agitation, aggregates could be detected between 30 and 60 min. Figure 3.2 also shows that the aggrega-

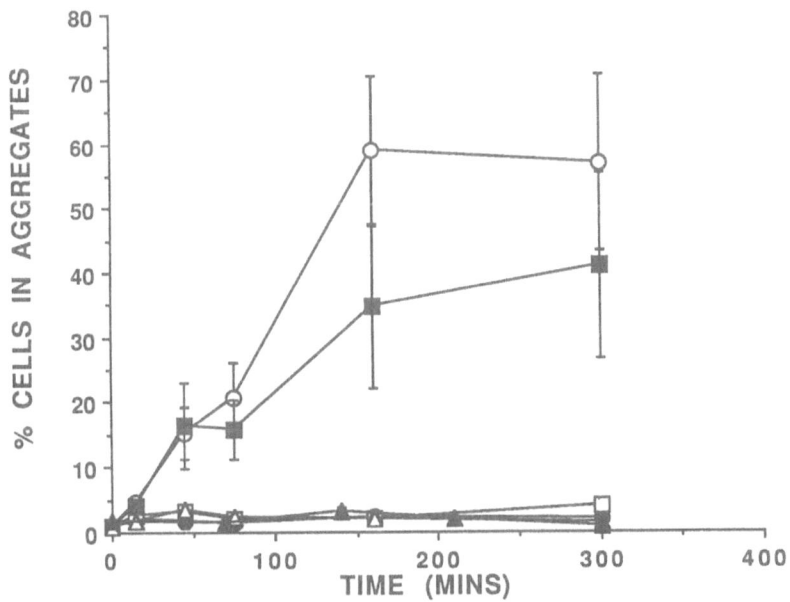

FIGURE 3.2. Figure 3.2 shows a time course of MOLT-4 aggregate formation in the presence of a variety of antibodies. The results represent the mean (±SE) of four experiments. KIM 127 (anti-CD18), R3.1 (anti–LFA-1), and BB7.5 (anti–MHC class 1) were used at a final concentration of 10 μg/ml. LM2/1 (anti-CR3) was used at 1-in-100 dilution of ascites. Where shown EDTA was added to a final concentration of 10 mM. ○ KIM 127, ● KIM 127 + EDTA, □ KIM 127 + R3.1, ■ KIM 127 + LM2/1, △ BB7.5, ▲ KIM 127 AT 4°C.

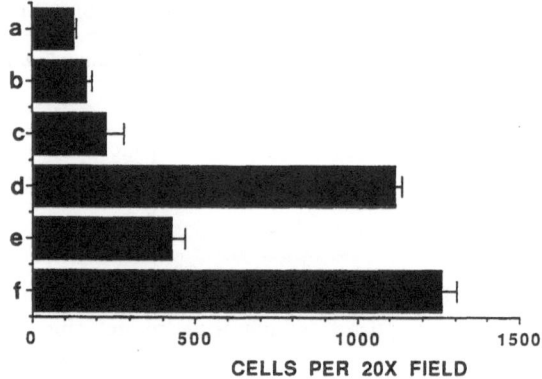

FIGURE 3.3. Figure 3.3 shows that the activity in KIM 127 hybridoma supernatant is associated with the immunoglobulin fraction. Human polymorphonuclear cells isolated on Ficoll-Hypaque (Pharmacia) were resuspended at 3×10^6/ml in RPM1 containing 10% bovine calf serum; 0.3 ml of suspension was added per well in a Lab-tek slide that had been pre-incubated for 2 h at 37°C with growth medium containing 10% bovine calf serum. Finally 0.1 ml of overgrown hybridoma supernatant was added to each well; prior to addition some of the supernatants were pre-incubated with either goat antimouse agarose beads (Sigma) or agarose beads coated with normal goat immunoglobulin (Sigma). Cells were incubated with hybridoma supernatant for 20 min at 37°C and then a further 40 min at room temperature. The nonadherent cells were removed by washing, the adherent cells were fixed, and the number of cells present per $20 \times$ field was recorded. The figure shows the mean and standard deviation from five fields per well. a. Anti–MHC class I hybridoma supernatant pre-incubated with goat immunoglobulin beads. b. Anti–MHC class 1 hybridoma supernatant pre-incubated with goat antimouse mouse immunoglobulin beads. c. Anti–MHC class 1 supernatant. d. KIM 127 hybridoma supernatant pre-incubated with goat immunoglobulin beads. e. KIM 127 hybridoma supernatant pre-incubated with goat antimouse agarose beads. f. KIM 127 hybridoma supernatant.

tion was sensitive to both temperature and the presence of EDTA. The aggregation could be blocked by antibodies against LFA-1. One of the anti–LFA-1 antibodies (DA36) has been shown to not interfere with KIM 127 binding (D. Andrew, unpublished data).

Supernatant from the KIM 127 hybridoma was also found to cause a very large increase in the adhesion of neutrophils to a protein-coated glass. This effect could also be removed by pretreating the supernatant with goat antimouse immunoglobulin-coated beads but not by pretreatment with normal goat immunoglobulin-coated beads (Fig. 3.3). A similar increase in neutrophil adhesion was promoted by purified KIM 127 antibody but not by an isotype-matched cell-binding control antibody

FIGURE 3.4. Figure 3.4 shows the effects of antibodies on neutrophil binding to protein-coated glass. The results represent the mean ± SE of three experiments. KIM 127, R3.1 (anti–LFA-1), and BB7.5 (anti–MHC class 1) were used at a final concentration of 6 μg/ml. LM2/1 (anti-CR3) was used as a 1-in-1,500 dilution of ascites.

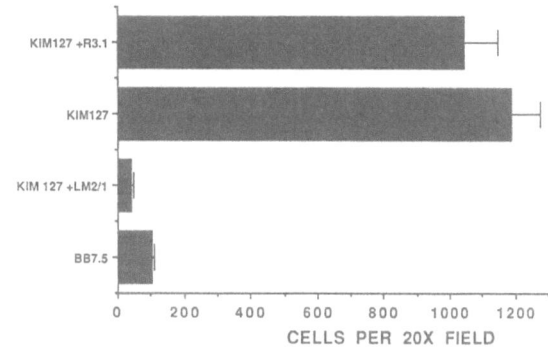

(Fig. 3.4). This adhesion could be blocked by an anti-CR3 monoclonal but not an anti–LFA-1 antibody (Fig. 3.4).

KIM 127 Binds to CD18

A Western blotting experiment showed that KIM 127 recognized a protein that comigrated with radiolabeled CD18 (Fig. 3.5). In order to confirm that CD18 was actually the ligand for KIM 127, the gene encoding human CD18 was cloned by polymerase chain-reaction amplification from U937 cells. The gene was expressed in the mouse myeloma line

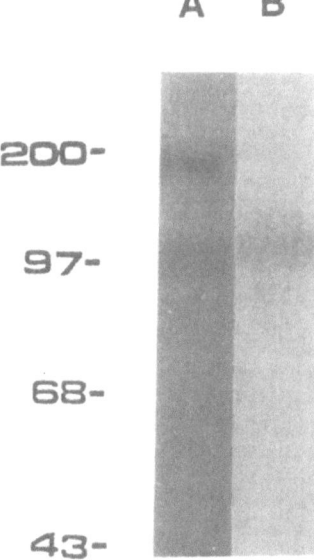

FIGURE 3.5. Figure 3.5 shows an autoradiograph of material immune precipitated with an anti-CD18 antibody (TS1/18) from a lysate of I^{125} surface-labeled JY cells (lane A). The precipitated material was run on nonreducing SDS-PAGE and transferred to nitrocellulose. A sample of a JY cell lysate was also run on the same gel under the same conditions, transferred to nitrocellulose, and probed with KIM 127 (lane B).

JY CELLS

NEUTROPHILS

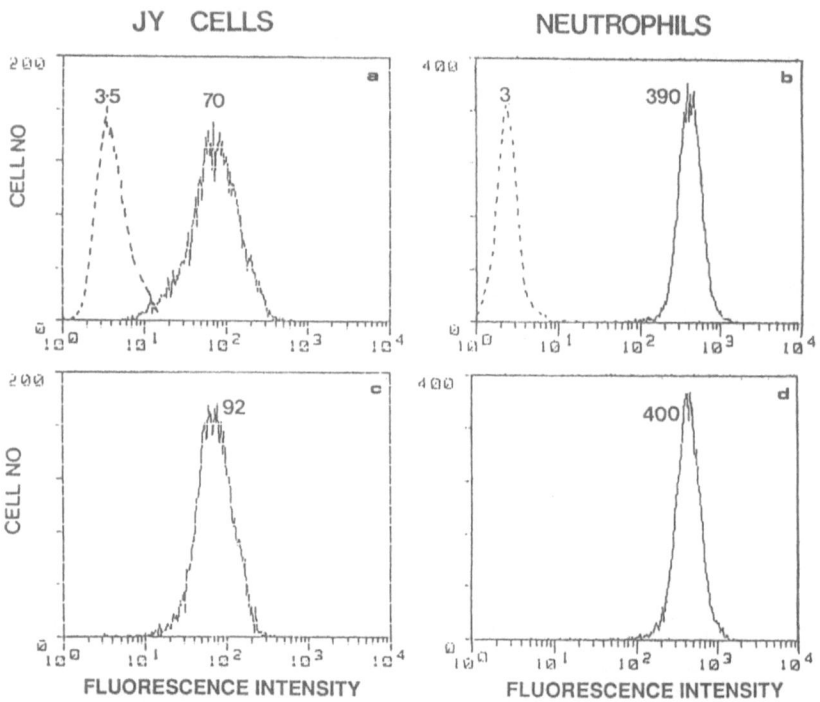

FIGURE 3.6. Figure 3.6 shows the effects on LFA-1 or CR3 expression of pre-incubating JY (panels a and c) or human PMN (panels b and c) with Fab'2 fragments of KIM 127. Fab'2 fragments of KIM 127 (20 μg/ml) were incubated with either JY cells or human PMN for 30 min at 37°C. JY cells were then stained at 37°C with DA36 (anti–LFA-1) and a goat antimouse IgG Fc FITC conjugate. After incubation with KIM 127 Fab'2 fragments the human PMNs were stained with LM2/1 (anti-CR3) and the same FITC conjugate. Panel a shows the fluorescence intensity for DA36 (anti–LFA-1) staining of JY cells after incubation with KIM 127 Fab'2 fragments (solid line). The dotted line shows the background fluorescence intensity observed with KIM 127 Fab'2 and antimouse Fc FITC conjugate. Panel c shows the fluorescence intensity for DA36 staining of JY cells without KIM 127 Fab'2 pretreatment. Similarly, in panel b the solid line shows the staining of PMN with LM2/1 after KIM 127 Fab'2 pretreatment and the dotted line shows the staining with the Fab'2 fragment alone. Panel d shows the fluorescence intensity of L1/2 staining of PMN without KIM 127 pretreatment. The numbers indicate the mode fluorescence for each experiment.

NSO. KIM 127 shows no binding to the parent cell line but shows good binding after transfection with the CD18 gene. As expected, the antihuman LFA-1 chain antibody DA36 does not bind to either the parent NSO line or the transfected line. This clearly shows that KIM 127 recognizes an epitope on CD18. This is consistent with it stimulating both LFA-1– and CR3–dependent adhesion events.

Incubation With KIM 127 Does Not Lead to Increased Surface
Expression of LFA-1 or CR3

The increase in LFA-1– or CR3-mediated adherence promoted by KIM
127 was not associated with increased expression of LFA-1 or CR3. Figure
3.6 shows that the incubation of JY cells or human PMN with KIM 127
F(ab')$_2$ fragments under conditions previously shown to stimulate both
JY-cell aggregation and PMN adhesion to protein-coated glass (D.
Andrew, unpublished results) does not lead to an increase in JY expression of LFA-1 or PMN expression of CR3.

Characteristics of the Epitope Recognised By KIM 127

FACS analysis showed that binding of KIM 127 to JY cells was temperature sensitive (Fig. 3.7); a similar temperature sensitivity was found for
the KIM 127 epitope on both MOLT-4 and human PMN. Dransfield and
Hogg (11) have previously reported another antibody which bound
to members of the Leu-CAM family of adhesion molecules in a
temperature-sensitive manner. The antibody reported by these workers

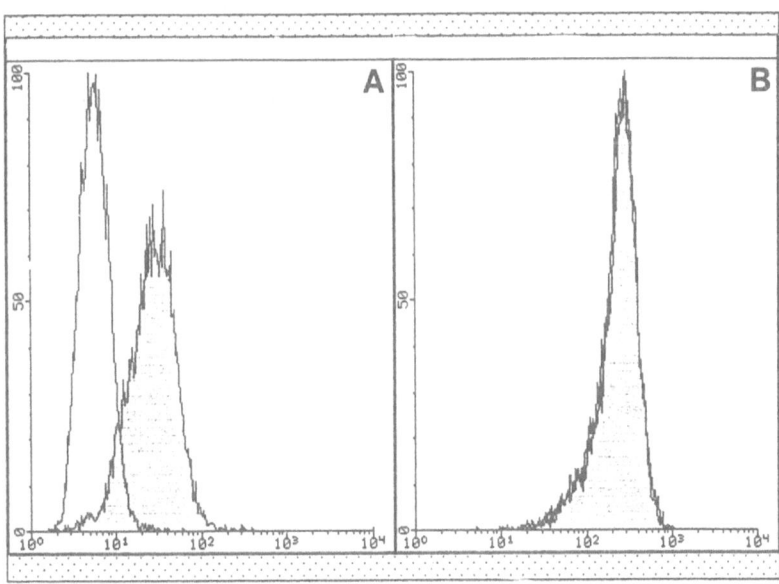

FIGURE 3.7. Figure 3.7 shows a plot of cell number against relative fluorescence intensity (log scale) for MOLT-4 cells stained with KIM 127 (panel A) or R3.1
(panel B). The shaded histogram shows the results of staining at 37°C while the
unshaded historgraph shows the results of staining at 4°C. The mode fluorescence
intensity for MOLT-4 cells stained with an isotype-matched nonbinding control
antibody (MOPC21) was 3.

(Mab 24) bound to an epitope on the α chains of all three members of this family and binding abolished by compounds which chelated divalent cations. However, the binding of KIM 127 to JY cells was not reduced by incubation with either EDTA or EGTA.

Discussion

Members of the Leu-CAM family of adhesion molecules play a key role in leukocyte adhesion. It seems likely that leukocyte extravasation involves an initial interaction with the endothelium to slow the progress of blood-borne leukocytes. This is probably mediated by members of the selectin family binding to carbohydrate structures present either as glycoproteins or glycolipids. This initial contact or agents such as FMLP, C5a, or IL8 diffusing from the tissues probably trigger a conformational change in members of the Leu-CAM family resulting in a much firmer adhesion and allowing the leukocytes to extravasate.

To date the mechanisms that trigger changes in the Leu-CAM family of adhesion molecules are not understood. It has been suggested that phosphorylation of CD18 may be important in this mechanism.

Here we present data on an antibody that binds to CD18 and appears to mimic the natural trigger mechanism in that the antibody binding even as a monovalent fragment seems to increase the affinity of both LFA-1 and CR3 for their ligands. It is interesting that the antibody should bind to CD18 and stimulate both LFA-1– and CR3-dependent adhesion.

Although LFA-1 and CR3 bind to different ligands or to different parts of the same molecule (in the case of ICAM-1) they share a lot of structural homology. It seems reasonable that although the ligand binding sites may have changed during evolution the mechanism that promotes change from a low-affinity to a high-affinity state should still retain a number of common features.

KIM 127 binding seems to trigger a conformational change common to both CR3 and LFA-1 activation. It will be interesting to determine whether KIM 127 will bind to p150.95 and also cause increased adhesion of this molecule to its ligands.

References

1. Patarroyo, M., and M.W., Makgoba, 1989. Leukocyte adhesion to cells. *Scand. J. Immunol.* 30:129.
2. Sanchez-Madrid, F., J.A. Nagy, E.R. Robbins, P. Simon, and T. Springer, 1983. A human leukocyte differentiation antigen family with distinct α-subunits and a common β-subunit. *J. Exp. Med.* 158:1785.
3. Larson, R.S., A.L. Corbi, L. Berman, and T. Springer, 1989. Primary structure of the leukocyte function-associated molecule-1 α subunit. An integrin

with an embedded domain defining a protein superfamily. *J. Cell Biol.* 108:703.

4. Todd, R.F., III, M.A. Arnaout, R.E. Rosin, C.A. Crowley, W.A. Peters, and B.M. Babior, 1984. Subcellular localization of the large subunit of Mol (Molα: formerly gp110), a surface glycoprotein associated with neutrophil adhesion. *J. Clin. Invest.* 74:1280.

5. Berger, M., J. Oshea, A.S. Cross, T.M. Folks, T.M. Chused, E.J. Brown, and M.M. Frank, 1984. Human neutrophils increase expression of C3bi as well as C3b receptors upon activation. *J. Clin. Invest.* 74:1566.

6. Arnaout, M.A., H. Spits, C. Terhorst, J. Pitt, and R.F. Todd III, 1984. Deficiency of a leukocyte surface glycoprotein (LFA-1) in two patients with Mol. deficiency: effects of cell activation on Mol/LFA-1 surface expression in normal and deficient leukocytes. *J. Clin. Invest.*, 74:1291.

7. O'Shea, J.J., E.J. Brown, B.E. Seligmann, J.A. Metcalf, M.M. Frank, and J.I. Gallin, 1984. Evidence for distinct intracellular pools of receptors for C3b and C3bi in human neutrophils. *J. Immunol.* 134:2580.

8. Buyon, J.P., S.B. Abramson, M.R. Philips, S.G. Slade, G.D. Ross, G. Weissmann, and R.J. Winchester, 1988. Dissociation between increased surface expression of Gp 165/95 and homotypic neutrophil aggregation. *J. Immunol.* 140:3156.

9. Philips, M.R., J.P. Buyon, R.J. Winchester, g. Weissmann, and S.B. Abramson, 1988. Upregulation of the ic3b receptor (CR3) is neither necessary nor sufficient to promote neutrophil aggregation. *J. Clin. Invest.* 82:495.

10. Dustin, M.L. and T.A. Springer, 1989. T-cell receptor cross-linking transiently stimulates adhesiveness through LFA-1. *Nature.* 341:619.

11. Dransfield, I. and N. Hogg, 1989. Regulated expression of Mg^{2+} binding epitope on leukocyte integrin α subunits. *EMBO.* 8:3759.

4
Function of a Soluble Human $\beta2$ Integrin CD11b/CD18

NAVA DANA, DEHMANI M. FATHALLAH, AND M. AMIN ARNAOUT

Introduction

Phagocytic cells, ploymorphonuclear cells (PMNs), and monocytes play a critical role in tissue destruction in many clinical disorders, including ischemia-reperfusion injury, allograft rejection, and many inflammatory and autoimmune disorders (1). Interventions aimed at attenuating this harmful role may have great therapeutic benefit. However, given the myriad of proteinases, oxidants, and cationic proteins produced by phagocytes after activation with chemoattractants and cytokines and during phagocytosis (2), this goal has proven elusive.

A new approach to prevention of tissue injury has been suggested by unraveling the molecular basis of an inherited immune-deficiency disease (leukocyte adhesion molecule deficiency) manifested clinically by life-threatening recurrent bacterial infections in which affected individuals have a global deficiency in phagocyte-mediated acute inflammatory responses due to lack of expression of CD11/CD18 ($\beta2$ integrins) (3,4). CD11/CD18 are adhesion receptors that promote interaction of leukocytes with each other (5,6), with endothelial cells (during transmigration) (7–9), and with specific opsonins deposited on invading bacteria and rejected or hypoxic tissues (10,11). The CD11/CD18 family consists of three heterodimeric surface-membrane glycoproteins, each with a distinct α subunit (CD11a, -b, or -c) noncovalently associated with an identical β subunit (CD18) (12,13). As in other integrins, association of the CD11 and CD18 subunits is required for normal surface-membrane expression and function of these receptors (14,15). The CD11b/CD18 (CR3) heterodimer is a major $\beta2$ integrin on PMNs and monocytes and mediates many of the inflammatory functions in these cells (3,16–18). Murine monoclonal antibodies (mAbs) directed against CD11b/CD18 reduce the degree of ischemia-reperfusion injury by 50–80% in several animal models of phagocyte-dependent acute tissue injury (19–21) and prevent development of insulin-dependent diabetes mellitus in susceptible mouse strains (22) through prevention of phagocyte accumulation in damaged tissues and their interaction with complement iC3b. Murine antibodies, however, usually elicit an antiglobulin response in humans and are thus of limited therapeutic usefulness. We therefore considered that antagonism of phagocyte interactions with inflamed endothelium and with iC3b might also be achieved by using a soluble and functional form of human CD11b/CD18.

In this paper we describe the expression of a recombinant soluble form of CD11b/CD18 (sCD11b/CD18) which appears to maintain functional integrity. The soluble receptor remains complexed as an α/β heterodimer and binds specifically to iC3b in a divalent cation-dependent manner similar to that of the wild-type receptor (mCD11b/CD18) and inhibits binding of PMN to recombinant interleukin-1 (rIL-1)-activated endothelium.

Results

Expression of Soluble CD11b/CD18 Heterodimer in COS Cells

Using site-directed mutagenesis to insert in frame stop codons at the predicted boundaries between extracellular and transmembrane regions in wild-type CD11b and CD18, two mutant cDNAs were generated. The mutant constructs are shown in Figure 4.1. To determine whether COS cells can express a secreted protein heterodimer, we cotransfected the mutated human CD11b and CD18 cDNA constructs into COS M6 monkey kidney cells. After metabolic labeling, harvested supernatants were used for immunoprecipitation with anti-CD11b (44a) and anti-CD18 (TS18) monoclonal antibodies (generous gifts from R.F. Todd III and S. Burakoff, respectively). Figure 4.2 shows that both subunits of the receptor were immunoprecipitated from mutated CD11b/CD18-transfected but not from sham-transfected COS cell supernatants. The detergent-soluble fraction of the pulsed COS cell contained little or no radiolabeled recombinant CD11b/CD18 (data not shown), indicating that the bulk of the synthesized receptor was secreted. The sCD11b/CD18 heterodimer had an apparent molecular mass of 149 kD and 84 kD for its CD11b and CD18 subunits, respectively (Fig. 4.2), compared with 155 kD and 94 kD for the membrane-associated forms (10,24), in agreement with an expected loss of 45 and 69 amino acids from CD11b and CD18, respectively. Monoclonal antibodies directed against either the sCD11b or sCD18 subunits immunoprecipitated both subunits (Fig. 4.2). Quantitative analysis of the immunoprecipitated bands by use of densitometry indicated that ~69% of the secreted material is present in an α/β heterodimer complex. In two separate transfections, the total amount of sCD11b detected in 1 ml of culture supernatant was 20 +/− 1 ng.

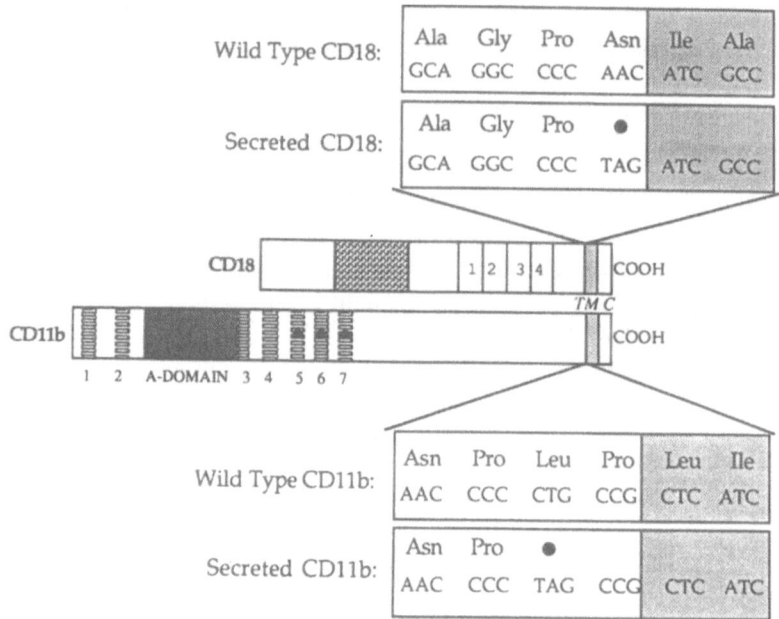

FIGURE 4.1. Schematic diagram showing the structure of normal CD11b/CD18 heterodimer and position in the immediate extracellular region where stop codons were introduced. The extracellular region of CD11b contains seven short repeats, three of which contain putative metal-binding regions (▲), and a large domain (A-domain) containing several epitopes for functionally active mAbs. The transmembrane (TM) region and cytoplasmic tail (C) are shown. The extracellular region of CD18 contains four cysteine-rich repeats and a more N-terminal region (outlined) involved in heterodimer formation (23).

sCD11b/CD18 Binds to iC3b

To determine whether this soluble heterodimer receptor was functional we first measured the ability of sCD11b/CD18 to bind to its well-characterized ligand iC3b, the major opsonic fragment of complement C3, using mCD11b/CD18 as a positive control. The ^{125}I-labeled mCD11b/CD18 bound specifically to iC3b (Fig. 4.3A, lanes b and d) but not to C3b (Fig. 4.3A, lanes a and c), confirming previous reports (25,26). Under the same conditions, metabolically labeled supernatants containing the sCD11b/CD18 heterodimer bound to ATS-iC3b (Fig. 4.3B, lane b). Binding was specific in that no binding was observed to ATS-C3b (Fig. 4.3B, lane c). No comparable binding activity to iC3b was detected in supernatants from the sham-transfected COS cells (Fig. 4.3B, lane a). Binding of mCD11b/CD18 and sCD11b/CD18 to iC3b was inhibited in the presence of EDTA (data not shown). Binding of sCD11b/CD18 to iC3b was also critically dependent on the presence of the recombinant receptor as a

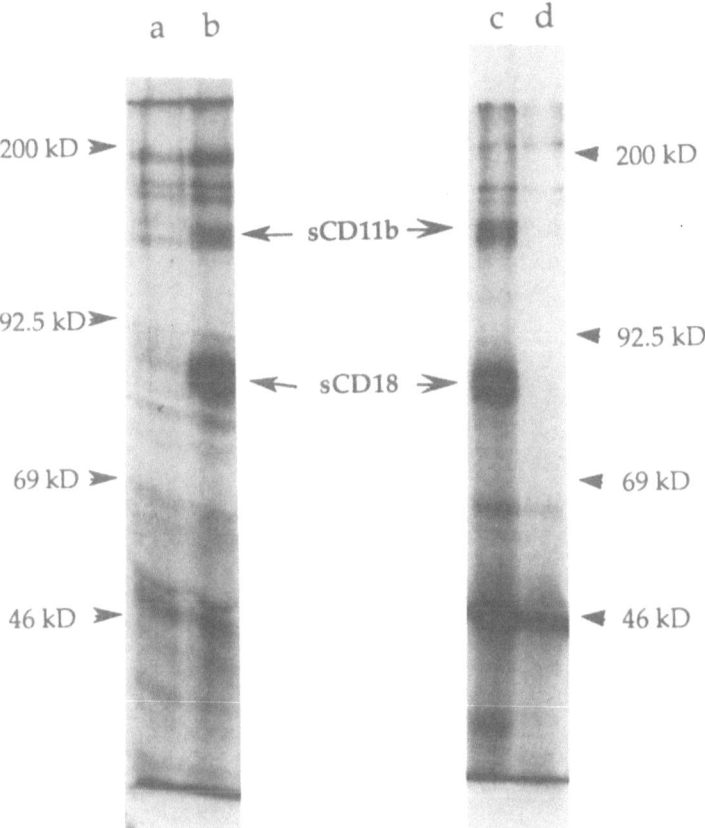

FIGURE 4.2. Autoradiograph of a SDS/7.0% PAGE showing immunoprecipitates from [^{35}S]methionine-labeled COS cell supernatants after cotransfection of COS cells with the mutated CD11b and CD18 cDNA constructs. A negative control ascites (NS1) (lanes a and d), an anti-CD18 mAb (TS18) (lane b), and an anti-CD11b mAb (44a) (lane c) were used. Arrowheads represent migration of molecular mass standards myosin, phosphorylase b, bovine serum albumin, and ovalbumin.

heterodimer because neither sCD11b nor sCD18 subunits, expressed separately, bound to iC3b-Sepharose (Fig. 4.4).

sCD11b/CD18 Inhibits PMN Binding to Inflamed Endothelium

The ability of sCD11b/CD18 to inhibit binding of human PMNs to inflamed endothelium was also examined and compared with levels of inhibition found using anti-CD11b or anti-CD18 mAbs (8,9); 44a and TS18 mAbs inhibited adhesion of PMN to rIL-1–treated endothelium by 32 +/

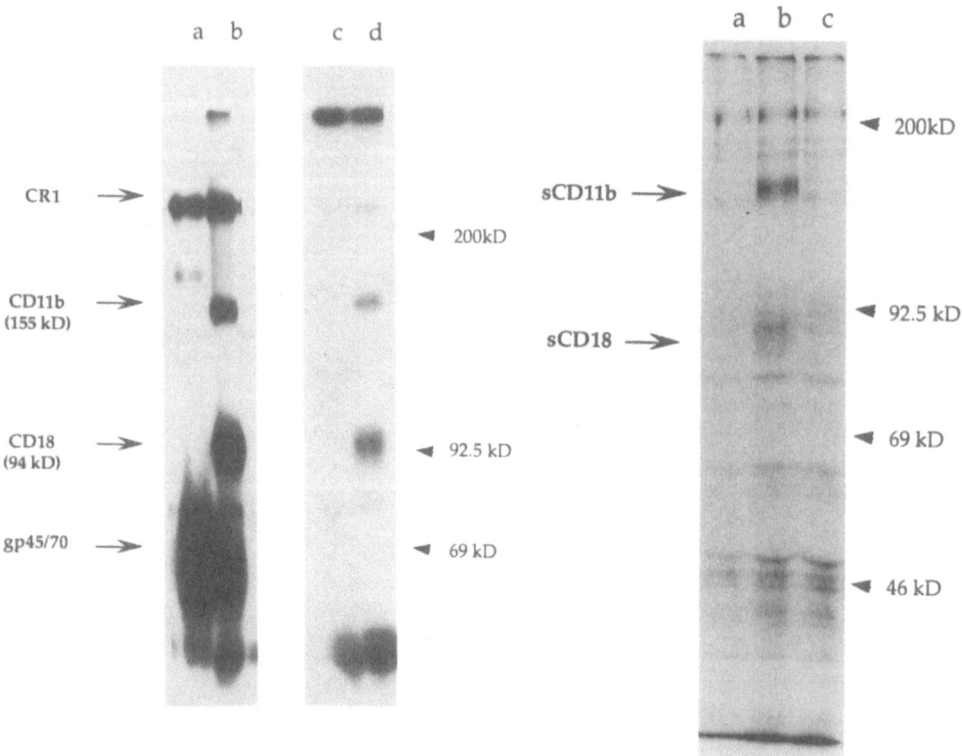

FIGURE 4.3. Binding of membrane (A) and secreted (B) CD11b/CD18 to iC3b. A. Autoradiograph of a SDS/6% PAGE of PMN-derived ^{125}I-surface-labeled glycoproteins eluted from ATS-C3b (lane a) and ATS-iC3b (lane b). Eluants from ATS-C3b (lane a) contained complement receptor type 1 (CR1) (250 kD) and the C3-binding regulatory protein gp45/70 (45–70 kD) (25,26). Eluants from ATS-iC3b (lane b) contained two additional proteins at 155 kD and 94 kD, representing mCD11b/CD18 heterodimer. CD11b/CD18 was immunoprecipitated with 44a mAb from material eluted from ATS-iC3b (lane d) but not from ATS-C3b (lane c). Molecular mass standards (arrowheads) were myosin, phosphorylase b, and bovine serum albumin. B. Autoradiograph of an SDS/8% PAGE showing specific binding of metabolically labeled sCD11b/CD18 heterodimer to iC3b. sCD11b/CD18 was eluted from ATS-iC3b (lane b) but not from ATS-C3b (lane c). No specific radiolabeled material was present in eluant of ATS-iC3b exposed to sham-transfected COS cells (lane a). Molecular mass standards (arrowheads) are as in Figure 4.2. Gel exposure to X-ray film (with an intensifying screen) was at −80°C for 5 days. Similar results were seen with supernatants from two other transfections.

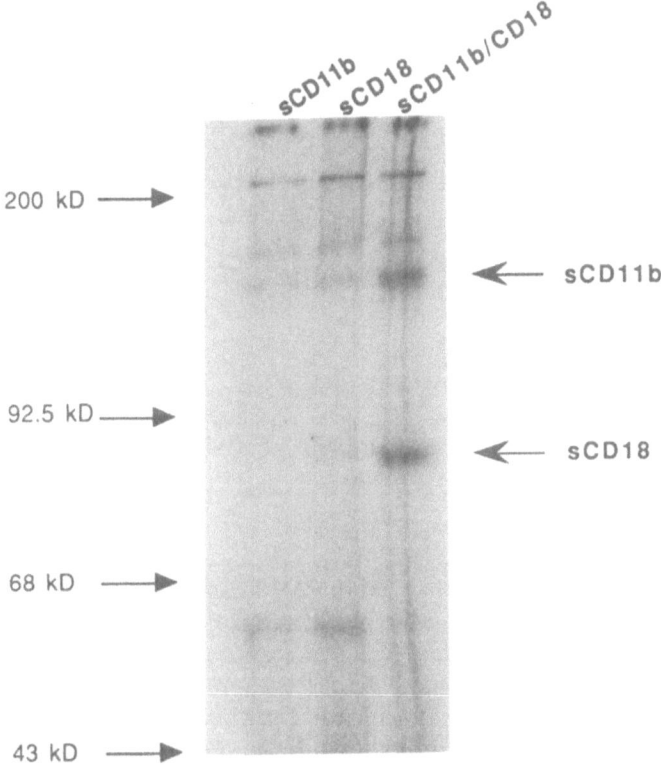

FIGURE 4.4. Autoradiograph of a SDS/10% PAGE evaluating binding of COS cell supernatants containing metabolically labeled sCD11b, sCD18, or sCD11b/CD18 heterodimer to iC3b. Binding activity was observed only in supernatants from COS cells expressing sCD11b/CD18 but not in those expressing either sCD11b or sCD18 alone. Culture supernatants contained equivalent amounts of sCD11b, sCD18, or sCD11b/CD18, as determined by immunoprecipitation studies (data not shown). Molecular mass standards (arrows) are the same as for Figure 4.2.

−9.3% and 72 +/− 6.4% (mean +/− SEM, n = 4), respectively (Fig. 4.5), in agreement with previous data (8,9). Supernatants containing sCD11b/ CD18 (∼7 ng of sCD11b added to each well containing 2.7 ng of surface-expressed receptor, or ∼3-fold molar excess of secreted to membrane-bound receptor) were also effective in blocking PMN adhesion to rIL-1–induced endothelium (58 +/− 7.2% inhibition, p < 0.001 compared to su-pernatants from sham-transfected cells) (Fig. 4.5).

Conclusion

This report demonstrates the synthesis and expression of a functional and soluble form of a recombinant integrin heterodimer. The evidence that

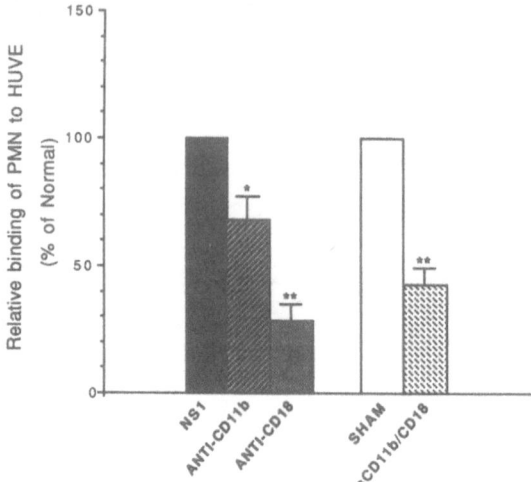

FIGURE 4.5. Histogram showing relative binding of PMN to rIL-1–treated HUVE in the presence of sCD11b/CD18 and mAbs 44a and TS18. Data points are mean ± SEM of triplicate determinations from four independent experiments. * and **, $p < 0.05$ and $p < 0.001$, respectively (when compared with adhesion in the presence of NS1 or sham-transfected cell supernatants).

the recombinant $\beta2$ integrin CD11b/CD18 can be expressed in a water-soluble form is supported by the observations that a truncated form of the receptor of the expected molecular size is present in supernatants from the transfected cells. That the bulk of the expressed receptor is in a heterodimeric form is shown by the ability of mAbs directed against either the CD11b or CD18 subunits to immunoprecipitate both subunits (Fig. 4.2). These findings also suggest that neither the cytoplasmic end nor the transmembrane regions of the CD11b or CD18 subunits are required for heterodimer formation. We cannot exclude the possibility, however, that these regions may improve the stability of this complex and also increase receptor-ligand affinity and resultant functional activity.

Our ability to express a recombinant sCD11b/CD18 heterodimer contrasts with the significant difficulties being encountered in expressing heterodimeric forms of other receptors (27) and, perhaps, reflects special features in the requirements for formation of an α/β complex in integrins. The present findings agree with recent molecular cloning studies (23,28,29) and EM studies of integrins (30) that suggest that α/β association requires direct binding of the N-terminal halves of the two subunits, forming a mushroomlike head with the remainder of each subunit forming a free tail embedded in the plasma membrane.

The secreted receptor reacted with mAbs directed against functional epitopes or domains (13,16) suggesting that the overall conformation of the native form of the receptor is maintained. This suggestion was directly confirmed by examination of the ligand-binding properties of the recombinant receptor. sCD11b/CD18 bound to its ligand, iC3b, under similar conditions to those previously used to show direct and specific binding of mCD11b/CD18 in that no binding was detected to C3b. Binding also required expression of the secreted receptor in its heterodimeric form because no binding activity was displayed by its secreted subunits when expressed separately (Fig. 4.4). The secreted heterodimer also significantly inhibited binding of PMNs to rIL-1–activated endothelium.

β2 integrins have been suggested to require the cytoplasmic portion for enhanced adhesion to various ligands. Stimulus-induced phosphorylation of CD18 has been directly demonstrated (31) and occurs with kinetics similar to those of stimulus-induced adhesive interactions (32). The ability of sCD11b/CD18 to block β2-integrin–mediated cell adhesion indicates that the ligand-binding properties themselves are intact at least with regard to interaction with iC3b and rIL-1–activated endothelium. Similar findings have also been observed in other adhesion receptor families such as the cadherins (33). The observed posttranslational modifications in mCD11/CD18 receptors may enhance their affinity to ligands through interactions with cytoskeleton and/or expression of conformational epitopes or binding sites that may also become expressed in the recombinant receptor. Comparison of the binding affinities of secreted vs. membrane-bound integrins will now directly address this question.

The purified sCD11b/CD18 described in this report should be useful in identifying other putative ligands for this receptor that appear to mediate binding of phagocytes to endothelium (8,9) and to a variety of microorganisms (34). It should also help in better delineation of the areas involved in heterodimer formation. Generation of larger quantities of this receptor from stably transfected cell lines should also permit a detailed evaluation of its functional activity in a large number of adhesive interactions mediated by CD11b/CD18 (e.g., phagocytosis, chemotaxis, binding to clotting factors, opsonized-zymosan–induced oxygen free radicals and proteolytic-enzyme release) (3). The antiadhesive and complement iC3b-binding activities of sCD11b/CD18 demonstrated in this report make it (or a derivative) an attractive antiinflammatory candidate, which could effectively block phagocyte emigration into inflamed organs as well as complement-dependent tissue injury. These effects may be additive when used in combination with the secreted form of the monomeric receptor CR1 (35). This approach should also apply to other integrins important in inflammatory and thromboocclusive disorders (36–38).

Acknowledgments. We thank Dr. Vibeke Videm, Ms. Maria Giovino, and Mr. Craig Nelson for expert technical assistance. This work was supported by the National Institutes of Health, the Arthritis Foundation, and the March of Dimes Birth Defects Foundation.

References

1. Malech, H.L. and Gallin, J.I. 1987. Neutrophils in human diseases. *N. Engl. J. Med.* 317:687.

2. Henson, P.M. and Johnston, R.B., Jr. 1987. Tissue injury in inflammation: Oxidants, proteinases, and cationic proteins. *J. Clin. Invest.* 79:669.
3. Arnaout, M.A. 1990. Stucture and function of the leukocyte adhesion molecules CD11/CD18. *Blood* 75:1037.
4. Arnaout, M.A. 1990. Leukocyte adhesion molecule deficiency: its structural basis, pathophysiology and implications for modulating the inflammatory response. *Immunol. Rev.* 114:145.
5. Arnaout, M.A., Hakim, R.M., Todd, R.F., III, Dana, N. and Colten, H.R. 1985. Increased expression of an adhesion-promoting surface glycoprotein in the granulocytopenia of hemodialysis. *N. Engl. J. Med.* 312:457.
6. Schwartz, B.R., Ochs, H.D., Beatty, P.G. and Harlan, J.M. 1985. A monoclonal antibody-defined membrane antigen complex is required for neutrophil-neutrophil aggregation. *Blood.* 65:1553.
7. Diener, A.M., Beatty, P.G., Ochs, H.D., and Harlan, J.M. 1985. The role of the neutrophil membrane glycoprotein 150 (GP-150) in neutrophil-mediated endothelial cell injury in vitro. *J. Immunol.* 135:537.
8. Arnaout, M.A., Lanier, L.L. and Faller D.V. 1988. The relative contribution of the leukocyte molecules Mol, LFA-1, p150,95 (LeuM5) in adhesion of granulocytes and monocytes to vascular endothelium is tissue- and stimulus-specific. *J. Cell. Physiol.* 137:305.
9. Luscinskas, F.W., Brock, A.F., Arnaout, M.A. and Gimbrone, M.A. 1989. Endothelial-leukocyte adhesion molecule-1-dependent and leukocyte (CD11/CD18)-dependent mechanisms contribute to polymorphonuclear leukocyte adhesion to cytokine-activated human vascular endothelium. *J. Immunol.* 142:2257.
10. Arnaout, M.A., Todd, R.F., III, Dana, N., Melamed, J., Schlossman, S.F. and Colten, H.R. 1983. Inhibition of phagocytosis of complement C3- or immunoglobulin-G-coated particles and of C3bi binding by monoclonal antibodies to a monocyte-granulocyte membrane glycoprotein (Mol). *J. Clin. Invest.* 72:171.
11. Klebanoff, J.S., Beatty, P.G., Schreiber, R.D., Ochs, H. and Waltersdorph, A.M. 1985. Effect of antibodies directed against complement receptors on phagocytosis by polymorphonuclear leukocytes: use of iodination as a convenient measure of phagocytosis. *J. Immunol.* 134:1153.
12. Trowbridge, I.S. and Omary, M.B. 1981. Molecular complexity of surface glycoproteins related to the macrophage differentiation antigen Mac-1. *J. Exp. Med.* 154:1517.
13. Sanchez-Madrid, F., Nagy, J.A., Robbins, E., Simon, P.A., and Springer, T.A. 1983. A human leukocyte differentiation antigen family with distinct alpha subunits and a common beta subunit: the lymphocyte function associated antigen (LFA-1), the C3bi complement receptor (OKM1/Mac-1), and the p150/95 molecule. *J. Exp. Med.* 158:1785.
14. Ruoslahti, E. and Pierschbacher, M.D. 1987. New perspectives in cell adhesion: RGD and integrins. *Science* 238:491.
15. Hynes, R.O. 1987. Integrins: a family of cell surface receptors. *Cell* 48:549.
16. Dana, N., Styrt, B., Griffin, J.D., Todd, R.F., III, Klempner, M.S. and Arnaout, M.A. 1986. Two functional domains in the phagocyte membrane glycoprotein Mol identified with monoclonal antibodies. *J. Immunol.* 137:3259.

17. Ross, G.D., Cain, J.A. and Lachmann, P.J. 1985. Membrane complement receptor type 3 has lectin-like properties analagous to bovine conglutinin and functions as a receptor for zymosan and rabbit erythrocytes as well as a receptor for iC3b. *J. Immunol.* 134:3307.

18. Brown, E.J., Bohnsack, J.F. and Gresham, H.D. 1988. Mechanism of inhibition of immunoglobulin G-mediated phagocytosis by monoclonal antibodies that recognize the Mac-1 antigen. *J. Clin. Invest.* 81:365.

19. Hernandez, L.A., Grisham, M.B., Twohig, B., Arfors, K.-E., Harlan, J.M. and Granger, D.N. 1987. Role of neutrophils in ischemia-reperfusion-induced microvascular injury. *Am. J. Physiol.* 253:H699.

20. Simpson, P.J., Todd, R.F., III, Fantone, J.C., Mickelson, J.K., Griffin, J.D., and Lucchesi, B.R. 1988. Reduction of experimental canine myocardial reperfusion injury by a monoclonal antibody (anti-Mo1, anti-CD11b) that inhibits leukocyte adhesion. *J. Clin. Invest.* 81:624.

21. Vedder, N.B., Winn, R.K., Rice, C.L., Chi, E.Y., Arfors, K.E. and Harlan, J.M. 1988. A monoclonal antibody to the adherence-promoting leukocyte glycoprotein, CD18, reduces organ injury and improves survival from hemorrhagic shock and resuscitation in rabbits. *J. Clin. Invest.* 81:939.

22. Hutchings, P., Rosen, H., O'Reilly, L., Simpson, E., Gordon, S. and Cooke, A. 1990. Transfer of diabetes in mice prevented by blockade of adhesion-promoting receptor in macrophages. *Nature* 348:639.

23. Arnaout, M.A., Dana, N., Gupta, S.K., Tenen, D.G. and Fathallah, D.F. 1990. Point mutations impairing cell surface expression of the common β subunit (CD18) in a patient with leukocyte adhesion molecule (Leu-CAM) deficiency. *J. Clin. Invest.* 85:977.

24. Todd, R.F., III, Nadler, L.M. and Schlossman, S.F. 1981. Antigens on human monocytes identified with monoclonal antibodies. *J. Immunol.* 126:1442.

25. Malhotra, V., Hogg, N. and Sim, R.B. 1986. Ligand binding by the p150/95 antigen of U937 monocytic cells: properties in common with complement receptor type 3. *Eur. J. Immunol.* 16:1117.

26. Cole, J.L., Housley, G.A., Jr., Dykman, T.R., MacDermot, R.P. and Atkinson, J.P. 1985. Identification of an additional class of C3-binding membrane proteins of human peripheral blood leukocytes and cell lines. *Proc. Natl. Acad. Sci., USA* 82:859.

27. Traunecker, A., Dodler, B., Oleveri, F. and Karjalainen, K. 1989. Solubilizing the T-cell receptor—problems in solution. *Immunol. Today* 10:29.

28. Newman, P.J., Seligsohn, U., Lyman, S., Poncz, M. and Coller, B.S. 1990. The molecular genetic basis of Glanzmann thrombasthenia in the Iraqi-Jewish and Arab populations in Israel. *Clin. Res.* 38:467A.

29. Loftus, J.C., O'Toole, T.E., Plow, E.F., Glass, A., Frelinger, A.L., III, and Ginsberg, M.H. 1990. A β_3 integrin mutation abolishes ligand binding and alters divalent cation-dependent conformation. *Science* 249:915.

30. Nermut, M.V., Green, N.M., Eason, P., Yamada, S.S. and Yamada, K.M. 1988. Electron microscopy and structural model of human fibronectin receptor. *EMBO J.* 7:4093.

31. Chatila, T., Geha, R.S. and Arnaout, M.A. 1989. Constitutive and stimulus-induced phosphorylation of CD11/CD18 leukocyte adhesion molecules. *J. Cell Biol.* 109:3435.

32. Buyon, J.P., Slade, S.G., Reibman, J., Abramson, S.B., Philips, M.R.,

Weismann, G. and Winchester, R. 1990. Constituitive and induced phosphorylation of the α and β chains of the CD11/CD18 leukocyte integrin family: relationship to adhesion-dependent functions. *J. Immunol.* 144:191.

33. Takeichi, M. 1990. Cadherins: a molecular family important in selective cell-cell adhesion. *Annu. Rev. Biochem.* 59:237.

34. Bullock, W.E. and Wright, S.D. 1987. Role of adherence-promoting receptors, CR3, LFA-1 and p150/95, in binding of Histoplasma capsulatum by human macrophages. *J. Exp. Med.* 165:195.

35. Weisman, H.F., Bartow, T., Leppo, M.K., Marsh, H.C., Carson, G.R., Concino, M.F., Boyle, M.P., Roux, K.H., Weisfeldt, M.L. and Fearon, D.T. 1990. Soluble human complement receptor type 1: In vivo inhibitor of complement suppressing post-ischemic myocardial inflammation and necrosis. *Science* 249:146.

36. Fitzgerald, L.A., Steiner, B., Rall, S.C., Lo, S.S. and Philips, D.R. 1987. Protein sequence of endothelial glycoprotein IIIa derived from a cDNA clone. *J. Biol. Chem.* 262:3936.

37. Poncz, M., Eisman, R., Heidenreich, R., Silver, S.M., Vilaire, G., Surrey, S., Schwartz, E. and Bennet, J.S. 1987. Structure of the platelet membrane glycoprotein IIb. *J. Biol. Chem.* 262:8476.

38. Coller, B.S. 1990. Platelets and thrombolytic therapy. *N. Engl. J. Med.* 322:33.

5
Mac-1 (CD11b/CD18) and Adherence-Dependent Neutrophil Locomotion

B.J. HUGHES, J.C. HOLLERS, AND C. WAYNE SMITH

Introduction

CD18 integrins are necessary for adherence-dependent neutrophil locomotion *in vitro* as shown by the fact that neutrophils from patients with CD18 deficiency exhibit markedly reduced migration on protein coated glass or plastic (1), and on monolayers of endothelial cells (2,3). Adhesion to the these surfaces apparently requires stimulation of the neutrophil (e.g., with chemotactic factors), and involves CD11b/CD18 (Mac-1) for the attachment to protein-coated surfaces (4–7), and both Mac-1 and CD11a/CD18 (LFA-1) for attachment to endothelial cells (3). A well known effect of chemotactic stimulation is the mobilization of Mac-1 from secondary granules in the neutrophil to the cell surface (4,8,9–14). The functional significance of this event is uncertain, and recent evidence suggests that constitutively expressed Mac-1 is sufficient for the increased adhesion induced by a single chemotactic stimulus (15–19). In this report, we propose the hypothesis that the newly arrived Mac-1 can participate in adhesion, but our evidence indicates that in order to do so, the neutrophil must experience a further increase in the concentration of the chemotactic stimulus.

Results

Stepwise Increases in Concentrations of fMLP Result in Stepwise Increases in Surface Mac-1

In these studies, neutrophils were exposed to a given concentration of the chemotactic peptide f-Met-Leu-Phe [fMLP] for 15 min in the presence of anti–Mac-1 monoclonal antibody (MAb). Cells were then washed in PBS containing this [fMLP] in order to remove the unbound MAb without changing the [fMLP]. These cells were then exposed to an increase in [fMLP] for 15 min. The newly upregulated Mac-1 was detected by flow

45

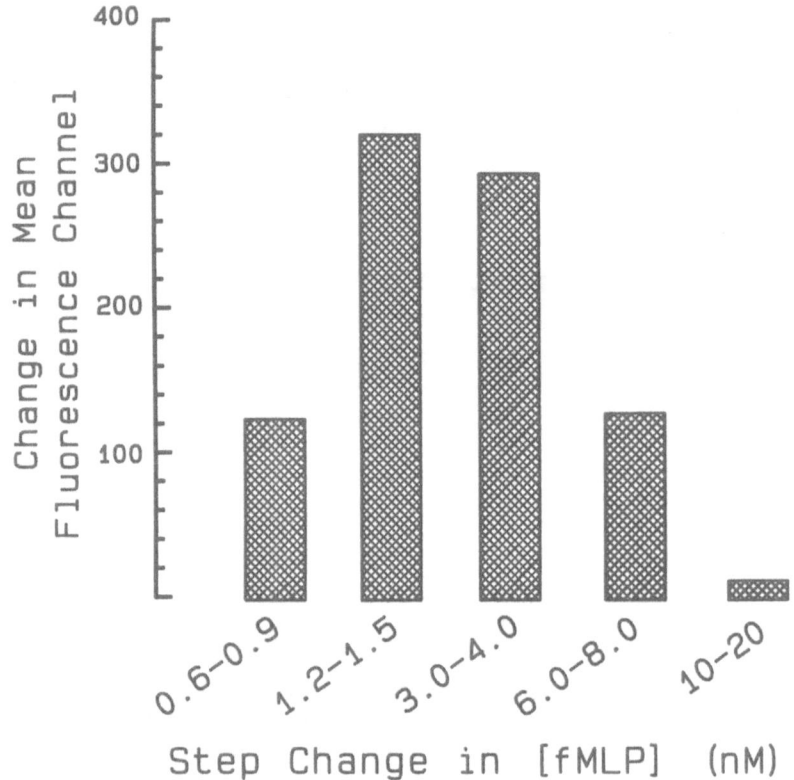

FIGURE 5.1. Effect of a single step change in [fMLP] on the binding of MAb 904 to isolated human neutrophils. Isolated neutrophils were suspended in PBS containing the concentration of fMLP shown as the initial value in the pair of concentrations under each bar for 15 min at room temperature in the presence of MAb 904. The cells were then washed in PBS containing this concentration of fMLP in order to remove the unbound MAb. Cells were then resuspended in the second concentration of fMLP and incubated for an additional 10 min. The ability of this second stimulus to increase surface binding sites for 904 was evaluated using biotinylated 904 and streptavidin-labeled PE. The mean fluorescence for the cell-associated PE is plotted. These data are the mean of duplicate determinations from a single experiment and are representative of results from three separate experiments.

cytometry using the same MAb biotin labeled, followed by streptavidin phycoerythrin (PE). In order to distinguish only the Mac-1 up-regulated by the new stimulus, parallel controls were prepared without increasing the [fMLP]. The results of such studies over a wide concentration range are shown in Fig. 5.1. The maximum change in cell-surface Mac-1 was found in the range of 1 to 8 nM fMLP, a range shown by numerous inves-

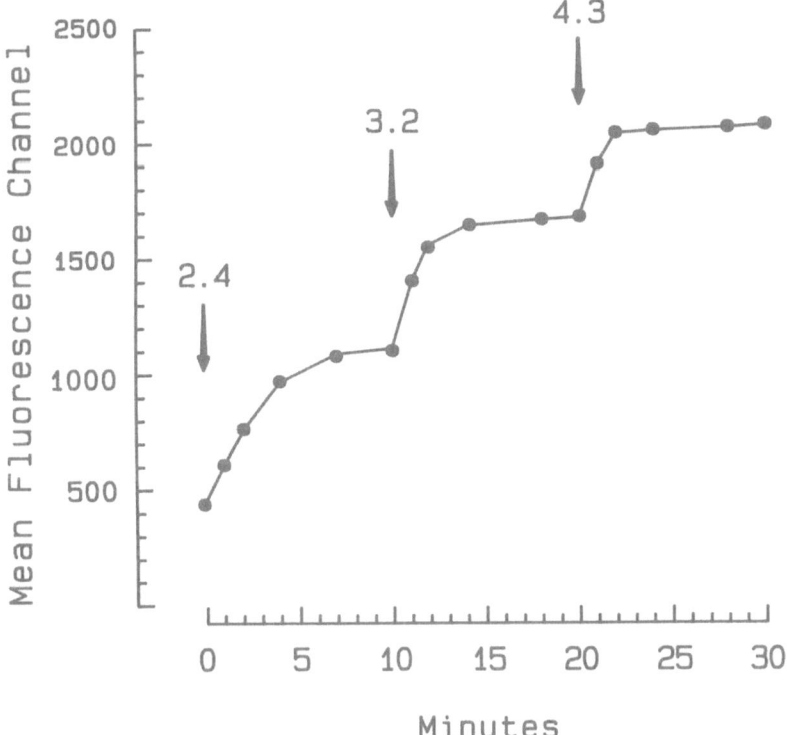

FIGURE 5.2. Kinetic study of the increases in surface Mac-1 following increases in [fMLP]. Isolated neutrophils were suspended in PBS containing 10 μg/ml anti-CD11b (Leu-15 labeled with PE) and incubated for 5 min. Sufficient fMLP was added to the cell suspension to attain the final concentration (nM) shown, and cells (3,000) were repeatedly sampled directly for flow cytometry. At 10 min and 20 min the [fMLP] was increased as shown. These data are from a single donor, and are representative of results with cells from three separate donors.

tigators to be optimal for chemotaxis (20). Little response to the change in [fMLP] was seen beyond this range.

Kinetics of Increases in Surface Mac-1

To study the kinetics of sequential stepwise increases in the [fMLP] on the surface expression of Mac-1, we used directly labeled anti-CD11b MAb Leu-15. As shown in Figure 5.2, cells were exposed to 2.4 nM fMLP in the presence of excess Leu-15-PE, and flow cytometry readings were taken at 1-min intervals for 10 min. The [fMLP] was then increased to 3.2 nM, and readings were again taken at 1-min intervals for 10 min.

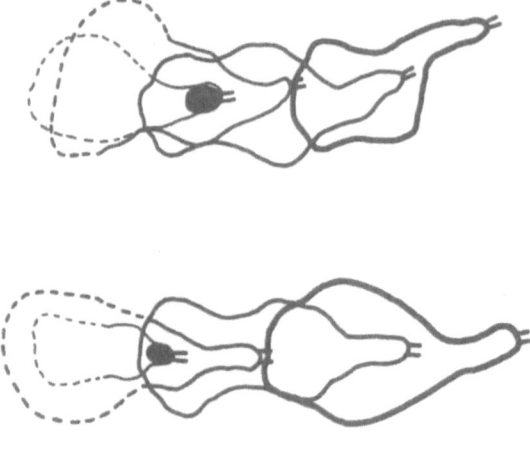

FIGURE 5.3. Effects of stepwise increases in [fMLP] on adherence-dependent migration of neutrophils. Isolated neutrophils hanging from the upper wall of a parallel plate flow chamber were exposed to a single step in [fMLP] from 0.3 to 1 nM. The fluid in the chamber was stationary after the new [fMLP] was attained. A frequently observed behavior is depicted by tracings of the outlines of cells moving over a 15-min observation period. For this illustration, cell movement was oriented from right to left. Observations were made under phase-contrast microscopy (100× objective, focal plane was that of the tip of the uropod). The dotted line indicates portions of the cell that were out of focus and the dark spot indicates the position of the attached uropod when the cell was hanging from the surface.

This was followed by another increase to 4.3 nM for 10 min. Mac-1 mobilization was rapid after each step in [fMLP] and appeared to plateau after approximately 2–4 min.

Evaluation of Mac-1–Dependent Cell Locomotion

Neutrophils were allowed to attach to the upper wall (i.e., ceiling) in a parallel plate flow chamber. This allowed us to evaluate adherence-dependent locomotion directly, since neutrophils hanging from a planar surface will be able to migrate only if they repeatedly attach as they translocate. We assured that this adhesion was Mac-1-dependent by coating the glass with 10% serum for 2 min prior to the introduction of the neutrophils. In previous studies we have shown that adherence of human neutrophils to coated glass is blocked by anti-CD11b monoclonal anti-

FIGURE 5.4. Neutrophils were treated exactly as in Figure 5.3 with one exception: They were exposed to repeated step-wise increases in [fMLP] at 200-s intervals. Arrows indicate the points at which the [fMLP] was changed. From right to left, 0.3 to 1 nM; 2.4 to 3.2 nM; 3.2 to 4.3 nM. This is representative of the behavior of numerous cells evaluated in this manner.

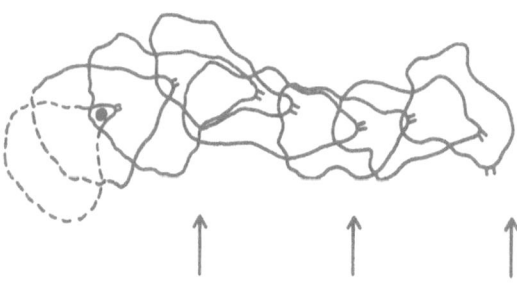

bodies (4). The attached cells were observed under phase-contrast optics, and their behavior was recorded on videotape for subsequent analysis. To expose the cells to changes in [fMLP], the chamber volume was replaced with PBS containing the specified [fMLP]. As seen in previous studies (21), a single step in [fMLP] to a level that stimulates shape change (e.g., 1 nM fMLP) induced migration for a short distance followed by progressive detachment until most cells were simply hanging by the tip of the uropod (Fig. 5.3). Thus, a single increment in the chemotactic stimulus failed to induce the repeated attachment needed for continued migration. The balance between adherence and detachment was apparently shifted in favor of detachment. If, however, the attached neutrophils were exposed to progressive steps in the [fMLP], they remained attached and continued to migrate (Fig. 5.4). If no further increases in the stimulus were introduced, cells detached as shown in Figure 5.3. Step increases of 10% in [fMLP] when introduced at 200-s intervals significantly increased the percentage of neutrophils that remained attached (Fig. 5.5) and significantly increased the distance migrated from a mean of 20 ± 10 μm to 71 ± 12 μm over the observation period ($p < 0.01$, n = 5). Steps of 33% at 200-s intervals almost completely inhibited detachment (Figs. 5.5 and 5.6), and increased migration to a mean of 253 ± 43 μm ($p < 0.01$, n = 5). If the interval between steps in [fMLP] was lengthened to 400s, many cells detached before the next [fMLP] could be introduced (data not shown).

Newly Expressed Adhesion Sites Are Mac-1–dependent

Albumin-coated latex beads (ACLB) (21) were used to assess the distribution of surface adhesion sites on neutrophils. As shown in previous studies, unstimulated neutrophils in suspension bind few ACLB but when exposed to a single stimulus of fMLP demonstrate bipolar morphology,

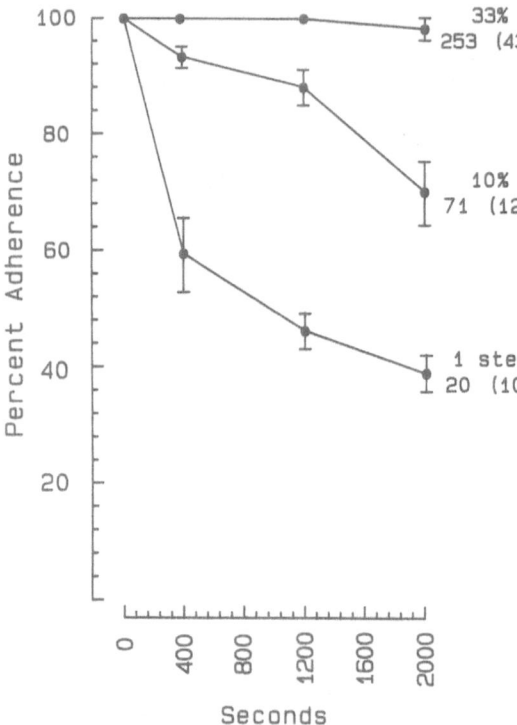

FIGURE 5.5. Effects of stepwise increases in [fMLP] on detachment of neutrophils from a protein-coated glass surface. Isolated neutrophils hanging from the upper wall of a parallel plate flow chamber were exposed to a single step or to multiple steps in [fMLP]. The chamber volume was exchanged every 200 s, but the fluid in the chamber was stationary between each exchange. The percentage of cells remaining attached at the times indicated was determined during three experimental conditions: (1) A single-step increase from 1 to 10 nM fMLP occurring at the first exchange, with subsequent exchanges maintaining the 10 nM concentration. (2) Step increases in [fMLP] of 10%. (3) Step increases in [fMLP] of 33%. Results are given as mean ± SD for five separate experiments. The values given beside each line are the mean distances migrated in μm (±SD).

and avidly bind ACLB in a generalized distribution. These cells are also highly adherent to protein-coated glass. With a second and higher stimulus (e.g., 1–5nM fMLP) the ACLB move to the uropod (Fig. 5.7). Cells treated this way have been shown to exhibit a low adherence to protein-coated glass (21). A third and higher stimulus (e.g. 10 nM fMLP) creates new binding sites at the anterior region of the cell as evidenced by bead binding to this area of membrane ruffling (Fig. 5.7), and cells are once again highly adherent to protein-coated glass. Over a period of 5 min these ACLB move to the uropod, and cell adherence again drops. In previously published studies (4) we have shown that with a single stimulus of fMLP in the presence of either anti-CD18 or anti-CD11b, the percentage

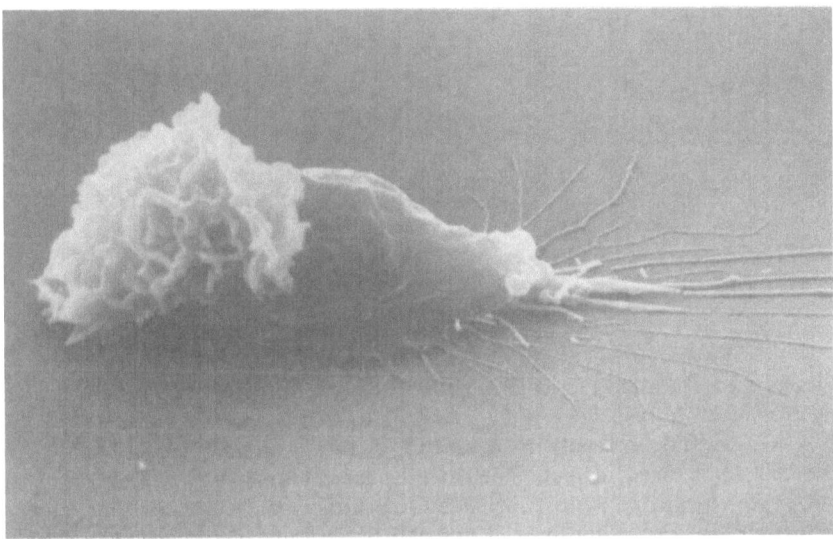

FIGURE 5.6. Example of adherent neutrophils exposed to repeated increases in [fMLP]. This neutrophil was from the experimental condition described in Figure 5.5 with 33% increments in [fMLP]. The cells were fixed by infusion of 1% glutaraldehyde 20 s after the fourth step in the stimulus level. Note the markedly ruffled anterior region and the retraction fibers indicating previous sites of adhesion. (Scanning electron microscopy, 7,000×)

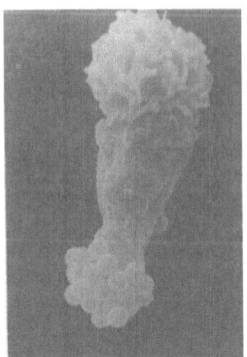
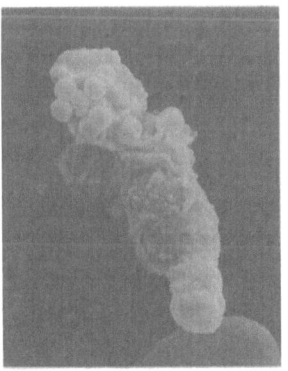
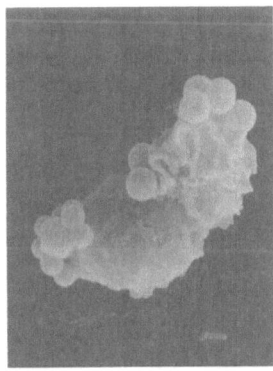

FIGURE 5.7. Binding of albumin-coated latex beads to stimulated neutrophils. Left panel. Neutrophil 5 min after [fMLP] was increased from 1 to 5 nM showing the migration of ACLB toward the uropod (lower pole of the cell in this photograph). Middle panel. Neutrophil 20 s after [fMLP] was increased from 1 to 5 nM showing the attachment of ACLB to the ruffled end of the cell (upper pole of the cell in this photograph). Right panel. Neutrophil exposed to three levels of FMLP. The smaller ACLB moved to the uropod (lower left) 5 min after the second stimulus (5 nM) and the largeri ACLB attached 20s after the third stimulus (10 nM). Scanning electron microscopy, 4,000×.

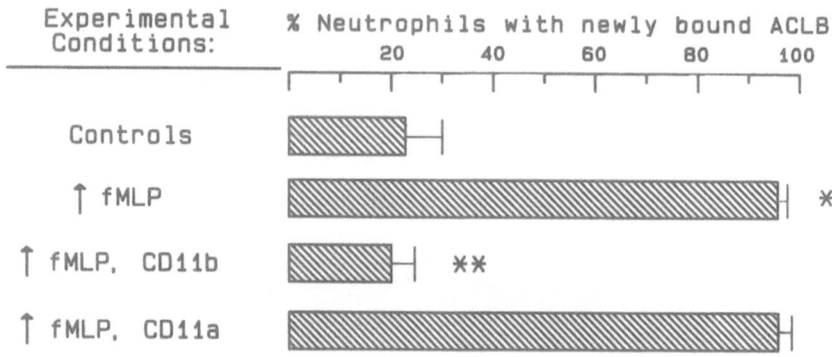

FIGURE 5.8. The Mac-1 dependence of new adhesion sites for ACLB. Isolated neutrophils were suspended in PBS containing ACLB and 1 nM fMLP for 5 min at 37°C. The [fMLP] was then increased to 5 nM for an additional 15 min before the cells were fixed in glutaraldehyde. Under differential interference-contrast (DIC) microscopy cells were examined for surface-bound ACLB. Greater than 90% were polarized with most of the ACLB clustered on the uropod. The percentage of cells with five or more ACLB at sites on the cell surface other than the uropod was determined. This value is shown as control. In the remaining experimental conditions cells were treated as the controls with the exception that a third concentration of fMLP (10 nM) was introduced for 20s priors to fixation in 1% glutaraldehyde. Note that greater than 90% of these cells had five or more beads bound to sites other than the uropod. The effect of adding MAb (10 μg/ml against the indicated β2 integrin subunit) to such cell preparations is shown. The results are derived from four separate experiments with >100 cells being evaluated for each experimental condition in each experiment, and results are plotted as mean ± SD. *, p < 0.01 compared with control; **, p < 0.01 compared with cells receiving the third fMLP stimulus without added MAb.

of cells binding ACLB and the numbers of beads bound per cell were greatly inhibited. Anti-CD11a had no affect on binding, indicating that the adhesion is Mac-1–dependent. In the present study, MAbs 904 (anti-CD11b) or R3.1 (anti-CD11a) was introduced coincident with the second fMLP stimulus (5 nM) in the protocol given above and retained with the cells and ACLB through the third stimulus (10 nM). Thus, the antibodies were available to block further adhesion after ACLB binding was induced by the first (1 nM) fMLP stimulus, the objective being to determine if these MAb would block subsequent ACLB binding induced by the second and third levels of fMLP stimulation. Since the ACLB that bound after the first (1 nM) stimulus move to the uropod of the cell, only ACLB on the body or lamella of the cell (examples of such binding in Fig. 5.7) were enumerated. These would reflect beads bound as a result of the steps in the chemotactic stimulus. As shown in Figure 5.8, the anti–Mac-1 MAb almost completely inhibited this ACLB binding. In contrast, MAb R3.1 (anti-CD11) was without effect.

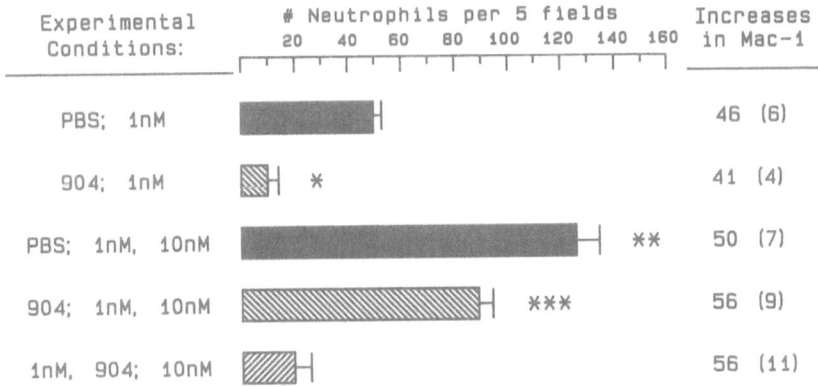

Experimental Conditions:	# Neutrophils per 5 fields	Increases in Mac-1
PBS; 1nM		46 (6)
904; 1nM	*	41 (4)
PBS; 1nM, 10nM	**	50 (7)
904; 1nM, 10nM	***	56 (9)
1nM, 904; 10nM		56 (11)

FIGURE 5.9. Newly arrived surface Mac-1 and the adhesion of neutrophils to KLH-coated glass. The results of five experimental conditions are shown. In the first, isolated neutrophils were suspended in PBS for 15 min, washed in PBS, and then exposed to 1 nM fMLP 15 min before being injected into the adherence chamber. Adhesion to KLH-coated glass is shown (mean ± SEM, n = 8 for all adhesion studies), as well as the number of new binding sites for monoclonal antibody 904 induced by the 1 nM fMLP stimulus (mean in thousands with SEM in parenthesis, n = 3 for all binding site studies). In the second condition, neutrophils were incubated with 904 (10 μg/ml) for 15 min, washed in PBS to remove unbound MAb; adherence and increase in 904 binding sites were determined. In the third, neutrophils were incubated in PBS for 15 min, washed, and then exposed to 1 nM fMLP for 15 min. This was followed by another increase in [fMLP] to 10 nM 5 min prior to determining adhesion and the increase in binding sites for 904 resulting from the 1–10 nM step in [fMLP]. This level of adhesion represented attachment of 49% of the neutrophils. In the fourth experimental condition, unstimulated neutrophils were incubated with 904 (10 μg/ml) for 15 min, washed to remove unbound MAb, and then exposed to 1 nM fMLP for 15 min. This was followed by an increase in [fMLP] to 10 nM 5 min prior to determining adherence and new binding sites for 904. In the fifth condition, neutrophils were exposed to 1 nM fMLP in the presence of 904 (10 μg/ml), washed in PBS containing 1 nM fMLP to remove unbound MAb without changing the [fMLP], and then exposed to 10 nM fMLP before determining adherence and new binding sites for 904. *, $p < 0.01$ compared with the results in the first condition; **, $p < 0.01$ compared with the results in the second condition; ***, $p > 0.01$ compared with the results in the second condition. The adhesion of unstimulated control neutrophils in this assay averaged less than 10 cells/5 fields, n = 45.

Evaluation of Newly Arrived Surface Mac-1 in Cell Adhesion

Though these studies show that increases in surface Mac-1 occur with each step in [fMLP], and the increases appear to be coincident with Mac-1–dependent adhesive events such as attachment of ACLB or adherence-dependent locomotion, they do not address the question of whether the newly arrived surface Mac-1 participates in these adhesive events. The work of other investigators suggests that it may not (15–19).

Monoclonal antibodies were used to address the question of whether newly arrived surface Mac-1 can participate in adhesion to protein-coated glass. KLH-coated glass was chosen as a substrate since we have previously found that adhesion of neutrophils to this surface requires chemotactic stimulation and is Mac-1–dependent (22), and contact with this surface does not appear to directly activate the neutrophil (22). Unstimulated neutrophils were incubated with anti-CD11b (904) for 15 min and then washed to remove the unbound MAb prior to the initial stimulation with the chemotactic factor. As shown in Figure 5.9, the antibody-pretreated cells, in contrast to control cells, exhibited a level of adherence that was not significantly different from unstimulated neutrophils. This failure to increase adherence occurred even though both control and MAb-pretreated cells had the same increase in binding sites for 904 following the initial chemotactic stimulus. Thus, as other investigators have concluded, it appeared that the newly arrived Mac-1 does not participate in adhesion. However, the data in Figure 5.9 also show that cells on which constitutive Mac-1 had been blocked by MAb 904 increased their adhesiveness as well as normal cells when exposed to another step in the [fMLP]. Over the eight replicates of this experiment, a mean of 76 cells/5 fields were recruited into the adhesive subpopulation when normal cells were exposed to a step of 1 to 10 nM fMLP, and a mean of 78 cells/5 fields were recruited into the adhesive subpopulation from cells on which constitutive Mac-1 was blocked by 904. These results are consistent with the interpretation that Mac-1 arriving at the cell surface during the first (1 nM) fMLP stimulus period was stimulated to participate in the adhesion induced by the second (10 nM) fMLP stimulus. This interpretation is further supported by the finding that neutrophils exposed to 904 after the initial 1 nM stimulus, and then washed to remove unbound MAb, exhibited only a small increase in adhesion following the 10 nM stimulus. The number of new binding sites for 904 that were brought to the surface after the step from 1 to 10 nM fMLP was equivalent for both control cells and MAb-pretreated cells.

Conclusions

The studies in this report address the hypothesis that increases in surface Mac-1 following chemotactic stimulation contribute to cell adhesion. Previous studies have indicated that Mac-1 translocated to the surface by a chemotactic or degranulating stimulus does not participate in the increased adhesion induced by the stimulus. Our results are consistent with this conclusion. We found that when MAb 904 was used to block the constitutive Mac-1, and then was removed by washing, a chemotactic stimulus failed to increase Mac-1–dependent adhesion. However, we evaluated the possibility that the newly arrived Mac-1 could function in adhesive events at some time later if the neutrophil was exposed to an additional stimulus. The experimental protocol involved the assessment of neutrophil responsiveness to stepwise increases in the concentration of a chemotactic factor. We evaluated (1) increases in surface Mac-1, (2) increases in adhesion to and detachment from protein-coated artificial surfaces, (3) adherence-dependent migration on protein-coated surfaces, and (4) the contribution of Mac-1 to adherence and migration using MAbs that block adhesion. Our results indicate that Mac-1 mobilization is demonstrable over a concentration range of fMLP that is optimal for inducing chemokinetic and chemotactic migration, and adherence-dependent locomotion is modulated by the same range of stimulus concentrations. Our results also indicate that newly upregulated Mac-1 can contribute to adhesion if the neutrophils are exposed to subsequent increases in the chemotactic stimulus. This was evidenced by the finding that when constitutive Mac-1 was blocked by MAb 904, subsequent levels of chemotactic stimulation both upregulated Mac-1, and in the absence of additional blocking antibody, promoted Mac-1–dependent adhesion.

Acknowledgements. The work was supported by NIH grants AI23521 and AI19031. The expert secretarial assistance of Michelle Swarthout and Carol McGary is acknowledged.

References

1. Schmalstieg, F.C., H.E. Rudloff, G.R. Hillman, and D.C. Anderson. 1986. Two dimensional and three dimensional movement of human polymorphonuclear leukocytes: Two fundamentally different mechanisms of location. *J. Leuk. Biol.* 40:677–691.
2. Smith, C.W., R. Rothlein, B.J. Hughes, M.M. Mariscalco, F.C. Schmalstieg, and D.C. Anderson. 1988. Recognition of an endothelial determinant for CD18–dependent human neutrophil adherence and transendothelial migration. *J. Clin. Invest.* 82:1746–1756.
3. Smith, C.W., S.D. Marlin, R. Rothlein, C. Toman, and D.C. Anderson. 1989. Cooperative interactions of LFA-1 and Mac-1 with intercellular adhesion molecule-1 in facilitating adherence and transendothelial migration of human neutrophils in vitro. *J. Clin. Invest.* 83:2008–2017.
4. Anderson, D.C., L.J. Miller, F.C. Schmalstieg, R. Rothlein, and T.A. Springer. 1986. Contributions of the Mac-1 glycoprotein family to adherence-dependent granulocyte functions: Structure-function assessments employing subunit-specific monoclonal antibodies. *J. Immunol.* 137:15–27.
5. Wright, S.D., and M.T.C. Jong. 1986. Adhesion-promoting receptors on human macrophages recognize Escherichia coli by binding to lipopolysaccharide. *J. Exp. Med.* 164:1876–1888.

6. Wright, S.D., S.M. Levin, M.T.C. Jong, Z. Chad, and L.G. Kabbash. 1989. CR3 (CD11b/CD18) expresses one binding site for Arg-Gly-Asp-containing peptides and a second site for bacterial lipopolysaccharide. *J. Exp. Med.* 169:175–183.

7. Wright, S.D., S.K. Lo, and P.A. Detmers. 1989. Specificity and regulation of CD18-dependent adhesion. In Leukocyte Adhesion Molecules: Structure, Function and Regulation. T.A. Springer, D.C. Anderson, R. Rothlein, and A.S. Rosenthal, editors. Springer-Verlag, New York. 190–207.

8. Berger, M., J.J. O'Shea, A.S. Cross, T.M. Folks, T.M. Chused, E.J. Brown, and M.M. Frank. 1984. Human neutrophils increase expression of C3bi as well as C3b receptors upon activation. *J. Clin. Invest.* 74:1566–1571.

9. Todd, III, R.F., M.A. Arnaout, R.E. Rosin, C.A. Crowley, W.A. Peters, and B.M. Babior. 1984. Subcellular localization of the large subunit of Mol (Mol$_1$; formerly gp110), a surface glycoprotein associated with neutrophil adhesion. *J. Clin. Invest.* 74:1280–1290.

10. Jones, D.H., F.C. Schmalstieg, H.K. Hawkins, B.L. Burr, H.E. Rudloff, S.S. Krater, C.W. Smith, and D.C. Anderson. 1989. Characterization of a new mobilizable Mac-1 (CD11b/CD18) pool that co-localizes with gelatinase in human neutrophils. In Leukocyte Adhesion Molecules: Structure, Function, and Regulation. T.A. Springer, D.C. Anderson, A.S. Rosenthal, and R. Rothlein, editors. Springer-Verlag, New York. 106–124.

11. Petty, H.R., J.W. Francis, III, R.F. Todd, P.R. Petrequin, and L.A. Boxer. 1987. Neutrophil C3bi receptors: Formation of membrane clusters during cell triggering requires intracellular granules. *J. Cell. Physiol.* 133:235–242.

12. Boxer, L.A., R.A. Haak, H.-H. Yang, J.B. Wolach, J.A. Whitcomb, C.H. Butterick, and R.L. Baehner. 1982. Membrane-bound lactoferrin alters the surface properties of polymorphonuclear leukocytes. *J. Clin. Invest.* 70:1049–1057.

13. Oseas, R., H.-H. Yang, R.L. Baehner, and L.A. Boxer. 1981. Lactoferrin: a promoter of polymorphonuclear leukocyte adhesiveness. *Blood* 57(5):939–945.

14. Bainton, D.F., L.J. Miller, T.K. Kishimoto, and T.A. Springer. 1987. Leukocyte adhesion receptors are stored in peroxidase-negative granules of human neutrophils. *J. Exp. Med.* 166:1641–1653.

15. Buyon, J.P., M.R. Philips, S.B. Abramson, S.G. Slade, G. Weissmann, and R. Winchester. 1988. Mechanism regulating recruitment of CD11b/CD18 to the cell surface is distinct from that which induces adhesion in homotypic neutrophil aggregation. In Leukocyte Adhesion Molecules: Structure, Function, and Regulation. T.A. Springer, D.C. Anderson, A.S. Rosenthal, and R. Rothlein, editors. Springer-Verlag, New York. 72–83.

16. Buyon, J.P., S.B. Abramson, M.R. Philips, S.G. Slade, G.D. Ross, G. Weissmann, and R.J. Winchester. 1988. Dissociation between increased surface expression of Gp165/95 and homotypic neutrophil aggregation. *J. Immunol.* 140:3156–3160.

17. Buyon, J.P., S.G. Slade, J. Reibman, S.B. Abramson, M.R. Philips, G. Weissman, and R. Winchester. 1990. Constitutive and induced phosphorylation of the alpha and beta-chains of the CD11/CD18 leukocyte integrin family. *J. Immunology* 144:191–197.

18. Detmers, P.A., S.D. Wright, E. Olsen, B. Kimball, and Z.A. Cohn. 1987.

Aggregation of complement receptors on human neutrophils in the absence of ligand. *J. Cell Biol.* 105:1137–1145.

19. Vedder, N.B., and J.M. Harlan. 1988. Increased surface expression of CD11b/CD18 (Mac-1) is not required for stimulated neutrophil adherence to cultured endothelium. *J. Clin. Invest.* 81:676–682.

20. Zigmond, S.H. 1978. Chemotaxis by polymorphonuclear leukocytes. *J. Cell Biol.* 77:269–287.

21. Smith, C.W., and J.C. Hollers. 1980. Motility and adhesiveness in human neutrophils. Redistribution of chemotactic factor induced adhesion sites. *J. Clin. Invest.* 65:804–812.

22. Shappell, S.B., C. Toman, D.C. Anderson, A.A. Taylor, M.L. Entman, and C.W. Smith. 1990. Mac-1 (CD11b/CD18) mediates adherence-dependent hydrogen peroxide production by human and canine neutrophils. *J. Immunol.* 144:2702–2711.

6
Functional Comparison of the Integrins $\alpha^4\beta_1$ (VLA-4) and $\alpha^4\beta_7$

Bosco M.C. Chan, Mariano J. Elices, and Martin E. Hemler

Introduction

VLA-4—a Functionally Versatile Integrin

Many adhesive interactions between cells and their surrounding extracellular matrix are mediated by receptors in the integrin family. In addition, several cell–cell interactions are mediated by integrins. Ligands specifically recognized by integrins include ECM components (collagen, laminin, fibronectin, vitronectin), cell-surface proteins in the Ig superfamily (ICAM-1, ICAM-2, ICAM-3, VCAM-1), and serum proteins (fibrinogen, von Willebrand factor, complement fragment C3bi). At present the integrin family includes at least 14 α subunits, and at least seven β subunits, arranged to form at least 19 distinct heterodimers as shown schematically in Figure 6.1.

One of the more functionally versatile integrins is VLA-4, which is expressed on monocytes, thymocytes, normal peripheral blood B and T cells [1], and on some melanoma cells [2]. The VLA-4 integrin not only binds to fibronectin [3–5] and VCAM-1 (vascular cell adhesion molecule-1, found on the surface of activated endothelial cells) [6–9] but also may participate in homotypic aggregation [10,11] and other cell–cell interactions [12,13]. Also, it has become increasingly obvious that this integrin may be important in a number of diverse contexts in which mononuclear leukocytes [1,7–9,14–16] or eosinophils [17,18] are recruited to inflammatory sites. In addition, VLA-4 interaction with VCAM-1 is postulated to be important during metastasis of some melanoma cells [2,19].

Functions of VLA-4 Compared to Other α^4-Containing Heterodimers

During studies of the functions of VLA-4 ($\alpha^4\beta_1$ heterodimer), antibodies directed toward the α^4 subunit have often been utilized. However, a possible complication arises because α^4 can associate with at least two distinct β subunits, β_1 in human, and both β_1 and β_p in mouse [20], as indi-

FIGURE 6.1. Organization of seven β subunits and 14 α subunits in the integrin family of adhesion receptors.

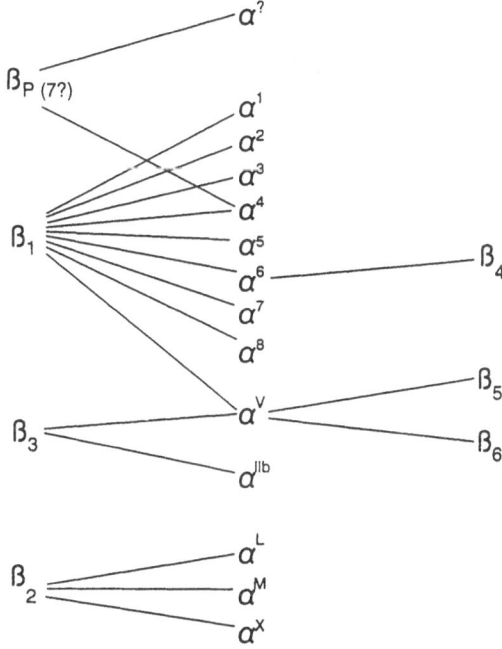

cated in Figure 6.1. A possible human homologue of the murine β_p is β_7 (21,22), but little evidence has yet been obtained to determine whether human α^4 might associate with the β_7 subunit. Also, it has yet to be determined whether other α^4 heterodimers (i.e., those not containing β_1) might mediate adhesion functions similar to those mediated by VLA-4 ($\alpha^4\beta_1$).

Some preliminary evidence has indicated that human α^4 might associate with a β subunit other than β_1. For example, a human B cell line, called JY, has been described which contains a high level of α^4, but little or no detectable β_1 subunit (1). In addition, freshly isolated human B lymphocytes were observed to contain approximately two-fold more α^4 than β_1 (23). It is assumed that this previously observed "excess" α^4 is associated with another β subunit, other than β_1, because integrin heterodimer formation has been established as being an obligatory prerequisite for cell-surface expression (24–27).

Results and Discussion

Selection of β_1-Positive JY Cells

Here we have investigated the biochemical and functional properties of a previously described B lymphoblastoid cell line called JY, which ex-

TABLE 6.1. Flow cytometric analysis of JY and β_1-JY cells.

Cell	Fluorescence Intensity (MFI)	
	β_1	α^4
JY	3	50
β_1-JY	36	45

Flow cytometry was carried out using a FACScan (Becton Dickenson Co.). In each case, a single homogeneous peak was observed, and median flourescence intensity (MFI) was obtained by subtracting background autofluorescence. Expression of α^4 was assessed using MAb B-5G10 (43), and β_1 was measured using MAb A-1A5 (44).

pressed substantial levels of α^4 but minimal β_1 (1). For comparison with native JY cells, β_1-positive JY cells were selected. Full-length β_1 cDNA (28), in the pFneo expression vector (29–31), was transfected into JY cells as previously described (32), and after selection for β_1-positive cells, flow cytometry was carried out. As shown in Table 6.1, β_1 expression was at a level of 36 mean fluorescence intensity (MFI) units in the β_1-JY cells, but was only 3 MFI units in JY cells. Notably, α^4 expression did not increase in the β_1-JY cells, and, in fact, it decreased slightly (from 50 to 45 MFI). Although it initially seemed that β_1 had been transfected into JY cells, vector control experiments also led to β_1 expression (not shown), and it became clear that the transfection procedure itself led to selection and expansion of an endogenous β_1-positive JY cell.

Immunoprecipitation Experiments

To test the hypothesis that the selected β_1 subunit was replacing another β subunit, and to identify the other β subunit, immunoprecipitation experiments were carried out. Immunoprecipitation experiments from JY cells, using anti-α^4 MAb, yielded a trace of a typical 150-kD α^4 band, in addition to substantial amounts of the 80-kD and 70-kD cleavage products often derived from α^4 (33). In addition, an associated β subunit of approximately 100 kD was coprecipitated with the α^4 proteins. Immunoprecipitation from β_1-JY cells yielded an almost identical pattern, except that the coprecipitated β subunit had an apparent Mr that was slightly larger than that from JY cells. As expected, the use of an anti-β_1 MAb immunoprecipitated no β_1 (or any other proteins) from JY cells, but yielded a moderately strong $\alpha^4\beta_1$ complex from β_1-JY cells. This result together with the flow cytometry results confirmed that the β_1 present in β_1-JY cells formed a cell-surface heterodimer with α^4.

In further immunoprecipitation experiments, we utilized a polyclonal rabbit anti-β_7 reagent directed against a synthetic peptide from the β_7 cytoplasmic tail sequence (21,22) to demonstrate that an $\alpha^4\beta_7$ complex

TABLE 6.2. Immunoprecipitation analysis of JY and β_1-JY cells.

Cell	Anti-α^4	Anti-β_1	Anti-β_7
JY	Moderate	Absent	Strong
β_1-JY	Moderate	Moderate	Moderate

Immunoprecipitation was carried out using the anti-α^4 MAb B-5G10, the anti-β_1 MAb A-1A5, or a rabbit anti-β_7 polyclonal serum made against a synthetic peptide (RREYSR-FEKEQQQLNWKQDS) derived from a sequence near the C-terminus of the published β_7 sequence (21,22). In each experiment, a complex of α^4 (weak 150 kD, strong 80 and 70 kD) and either β_1 or β_7 (100–110 kD) was observed. In a control experiment using the P3 MAb, no protein bands were obtained.

could be readily immunoprecipitated from unselected JY cells. Furthermore, the level of α^4 coprecipitated with β_7 was comparable to the level of α^4 obtained using anti-α^4 MAb. Also, $\alpha^4\beta_7$ heterodimer was readily obtained from β_1-JY cells, but at somewhat lower amounts than obtained from the JY cells. The immunoprecipitation results are summarized in Table 6.2. From these results we conclude (1) that moderately high levels of $\alpha^4\beta_7$ heterodimer are expressed in JY cells, (2) that β_1 could partially replace the β_7 subunit, and (3) that $\alpha^4\beta_1$ heterodimer association is somewhat more favored than $\alpha^4\beta_7$ heterodimer formation. Our immunochemical evidence for the presence of β_7 on the surface of a B lymphoblastoid cell line (JY) is consistent with previous northern blot experiments showing β_7 expression on lymphoblastoid cells (21,22).

Contribution of the β_1 Subunit to VLA-4 Function

In cell adhesion assays, we determined that JY cells adhered rather poorly to known ligands for VLA-4 (Table 6.3). In contrast, β_1-JY cells consistently showed 5–40-fold greater adhesivity to the FN-40 (40-kD chymotryptic) fragment of fibronectin, to the 25-amino-acid CS1 peptide derived from fibronectin, and to sVCAM-1 (Table 6.3). To confirm that adhesion to these substrates was mediated by an α^4-containing heterodimer, antibody inhibition experiments were performed. As shown in Table 6.4, attachment of β_1-JY cells to FN-40, CSI peptide, and sVCAM-1 was almost completely inhibited by the anti-α^4 MAb HP2/1, but not by the anti-α^5 MAb PlD6. In contrast, a control experiment showed that adhesion to the FN-120 fragment of fibronectin was minimally inhibited by the anti-α^4 MAb but substantially inhibited by the anti-α^5 MAb. These results are consistent with the presence of a VLA-5 binding site in FN-120, and a VLA-4 binding site in FN-40, as previously established (3–5,34).

Previous studies had directly demonstrated that the integrin α^4 subunit was critically involved in adhesion to both fibronectin and to

TABLE 6.3. Functional analyses of JY and β_1-JY adhesion properties.

		Adhesion (cells bound/mm²)	
Substrate (μg/well)		JY cells	β_1-JY cells
FN-40	0.63	18	97
	2.5	54	235
CS1	1.0	17	100
	4.0	5	125
sVCAM-1	1.25	5	26
	0.5	5	198

Cell attachment assays were carried out in triplicate, in 96-well microtiter plates as previously described (8). Substrates utilized for coating plastic are the 40-kD Hep II chymotryptic fragment of fibronectin (FN-40), a 25-amino-acid synthetic peptide derived from the FN-40 fragment (CSI), and recombinant soluble VCAM-1 (45).

TABLE 6.4. Demonstration of the α^4 dependence of β_1-JY cell adhesion.

	Adhesion (% relative to the control)	
Substrate	Anti-α^4	Anti-α^5
FN-40 (1 μg/ml)	2	89
FN-120 (1 μg/ml)	85	15
CS1 (2 μg/ml)	3	96
sVCAM-1 (0.25 μg/ml)	3	99

Cell adhesion assays were carried out in the presence of inhibitory antibodies HP2/1 (anti-α^4) and P1D6 (anti-α^5). In the presence of MAb P3 (control antibody) 30–50% of the input cells were adherent in each experiment. Concentrations of each substrate used for coating are indicated in parentheses.

VCAM-1 (8), but evidence for β_1 involvement was only indirect. Now we provide additional evidence that the integrin β_1 subunit is also critically involved in these functions. At the same time, we have demonstrated that the $\alpha^4\beta_7$ heterodimer endogenously expressed in JY cells is not an effective adhesion receptor for known VLA-4 ligands, despite being expressed at high levels. It is unlikely that this heterodimer is simply "turned off" in the JY cell, because the same JY cellular environment readily supported the function of the VLA-4 ($\alpha^4\beta_1$) heterodimer. Because replacement of β_7 with β_1 resulted in a marked alteration of α^4 heterodimer function, a role for the β subunit in determining ligand-binding specificity is suggested. In agreement with this, the ligand-binding specificity of other integrin heterodimers also varies depending on the identity of the associated β subunits. For example, $\alpha^6\beta_1$ binds to laminin, whereas $\alpha^6\beta_4$ does not (35).

Additional work will be required to determine the ligand-binding properties of the $\alpha^4\beta_7$ heterodimer. The size of β_7 (slightly smaller than β_1), its association with α^4, and its expression on lymphoid cells are all properties resembling those of the murine β_p subunit. If the human β_7 subunit is indeed the human homologue of the murine β_p subunit, then $\alpha^4\beta_7$ could possibly contribute to lymphocyte "homing" to Peyer's patches, as shown for $\alpha^4\beta_p$ (36,37). Other recent evidence regarding β_7 (38,39) suggests that it may not only complex with α^4 but also may be part of a novel integrinlike complex found on the surface of rat (40), mouse (39), and human (41,42) intraepithelial lymphocytes. Thus β_7, like the integrin β_1, β_2, and β_3 subunits, defines a subfamily of integrin heterodimers, with each subfamily having multiple distinct α subunits associated with a common β subunit.

In conclusion, we have established that the β_1 subunit is essentially involved in determining the specificity of VLA-4 binding to its ligands and that the $\alpha^4\beta_7$ heterodimer appears to be a functionally distinct integrin. This information is of fundamental importance for sorting out the roles of α^4-containing integrins on leukocytes and other cells.

Acknowledgments. This work was partially supported by NIH grants CA42368 and GM38903 (to M.E.H.) and by a postdoctoral fellowship from the MRC of Canada (to B.M.C.C.). We thank Dr. Christina Parker for providing rabbit serum directed against β_7 synthetic peptide.

References

1. Hemler, M.E., M.J. Elices, C. Parker, and Y. Takada. 1990. Structure of the integrin VLA-4 and its cell-cell and cell-matrix adhesion functions. *Immunol. Rev.* 114:45.

2. Taichman, D.B., M.I. Cybulsky, I. Djaffar, B.M. Longenecker, J. Teixidó, G.E. Rice, A. Aruffo, and M.P. Bevilacqua. 1991. Tumor cell surface $\alpha^4\beta_1$ integrin mediates adhesion to vascular endothelium: demonstration of an interaction with the N-terminal domains of INCAM-110/VCAM-1. *Cell Regul.* 2:347.

3. Wayner, E.A., A. García-Pardo, M.J. Humphries, J.A. McDonald, and W.G. Carter. 1989. Identification and characterization of the lymphocyte adhesion receptor for an alternative cell attachment domain in plasma fibronectin. *J. Cell Biol.* 109:1321.

4. Guan, J.-L. and R.O. Hynes. 1990. Lymphoid cells recognize an alternatively spliced segment of fibronectin via the integrin receptor $\alpha4\beta1$. *Cell* 60:53.

5. García-Pardo, A., E.A. Wayner, W.G. Carter, and O.C. Ferreira. 1990. Human B lymphocytes define an alternative mechanism of adhesion to fibronectin: The interaction of the $\alpha4\beta1$ integrin with the LHGPEILDVPST sequence of the type III connecting segment is sufficient to promote cell attachment. *J. Immunol.* 144:3361.

6. Osborn, L., C. Hession, R. Tizard, C. Vassallo, S. Luhowskyj, G. Chi-Rosso, and R. Lobb. 1989. Direct cloning of vascular cell adhesion molecule 1 (VCAM1), a cytokine-induced endothelial protein that binds to lymphocytes. *Cell* 59:1203.

7. Rice, G.E., J.M. Munro, and M.P. Bevilacqua. 1990. Inducible cell adhesion molecule 110 (INCAM-110) is an endothelial receptor for lymphocytes: A CD11/CD18-independent adhesion mechanism. *J. Exp. Med.* 171:1369.

8. Elices, M.J., L. Osborn, Y. Takada, C. Crouse, S. Luhowskyj, M.E. Hemler, and R.R. Lobb. 1990. VCAM-1 on activated endothelium interacts with the leukocyte integrin VLA-4 at a site distinct from the VLA-4/fibronectin binding site. *Cell* 60:577.

9. Schwartz, B.R., E.A. Wayner, T.M. Carlos, H.D. Ochs, and J.M. Harlan. 1990. Identification of surface proteins mediating adherence of CD11/CD18-deficient lymphoblastoid cells to cultured human endothelium. *J. Clin. Invest.* 85:2019.

10. Bednarczyk, J.L. and B.W. McIntyre. 1990. A monoclonal antibody to VLA-4 α chain (CDw49) induces homotypic lymphocyte aggregation. *J. Immunol.* 144:777.

11. Campanero, M.R., R. Pulido, M.A. Ursa, M. Rodriquez-Moya, M.O. De Landazuri, and F. Sánchez-Madrid. 1990. An alternative leukocyte homotypic adhesion mechanism, LFA-1/ICAM-1 independent, triggered through the human VLA-4 integrin. *J. Cell Biol.* 110:2157.

12. Clayberger, C., A.M. Krensky, B.W. McIntyre, T.D. Koller, P. Parham, F. Brodsky, D.J. Linn, and E.L. Evans. 1987. Identification and characterization of two novel lymphocyte function-associated antigens, L24 and L25. *J. Immunol.* 138:1510.

13. Takada, Y., M.J. Elices, C. Crouse, and M.E. Hemler. 1989. The primary structure of the $\alpha4$ subunit of VLA-4: homology to other integrins and a possible cell-cell adhesion function. *EMBO J.* 8:1361.

14. Cybulsky, M.I. and M.A. Gimbrone. 1991. Endothelial expression of a mononuclear leukocyte adhesion molecule during atherogenesis. *Science* 251:788.

15. Allavena, P., C. Paganin, I. Martin-Padura, G. Peri, M. Gaboli, E. Dejana, P.C. Marchisio, and A. Mantovani. 1991. Molecules and structures involved in the adhesion of natural killer cells to vascular endothelium. *J. Exp. Med.* 173:439.

16. Issekutz, T.B. and A. Wyrkretowicz. 1991. Effect of a new monoclonal antibody, TA-2, that inhibits lymphocyte adherence to cytokine stimulated endothelium in the rat. *J. Immunol.* 147:109.

17. Walsh, G.M., J-J. Mermod, A. Hartnell, A.B. Kay, and A.J. Wardlaw. 1991. Human eosinophil, but not neutrophil, adherence to IL-1-stimulated human umbilical vascular endothelial cells is $\alpha^4\beta_1$ (very late antigen-4) dependent. *J. Immunol.* 146:3419.

18. Bochner, B.S., F.W. Luscinskas, M.A. Gimbrone, W. Newman, S.A. Sterbinsky, C.P. Derse-Anthony, D. Klunk, and R.P. Schleimer. 1991. Adhesion of human basophils, eosinophils, and neutrophils to interleukin 1-activated human vascular endothelial cells: contributions of endothelial cell adhesion molecules. *J. Exp. Med.* 173:1553.

19. Rice, G.E. and M.P. Bevilacqua. 1989. An inducible endothelial cell surface glycoprotein mediates melanoma adhesion. *Science* 246:1303.
20. Holzmann, B. and I.L. Weissman. 1989. Peyer's patch-specific lymphocyte homing receptors consist of a VLA-4-like α chain associated with either of two integrin β chains, one of which is novel. *EMBO J.* 8:1735.
21. Yuan, Q., W.-M. Jiang, G.W. Krissansen, and J.D. Watson. 1990. Cloning and sequence analysis of a novel β2-related integrin transcript from T lymphocytes: Homology of integrin cysteine-rich repeats to domain III of laminin B chains. *Inter. Immunol.* 2:1097.
22. Erle, D.J., C. Rüegg, D. Sheppard, and R. Pytela. 1991. Complete amino acid sequence of an integrin β subunit (β_7) identified in leukocytes. *J. Biol. Chem.* 266:11009.
23. Hemler, M.E. 1990. VLA proteins in the integrin family: Structures, functions, and their role on leukocytes. *Ann. Rev. Immunol.* 8:365.
24. O'Toole, T.E., J.C. Loftus, E.F. Plow, A.A. Glass, J.R. Harper, and M.H. Ginsberg. 1989. Efficient surface expression of platelet GPIIb-IIIa requires both subunits. *Blood 74*:14.
25. Bodary, S.C., M.A. Napier, and J.W. McLean. 1989. Expression of recombinant platelet glycoprotein IIbIIIa results in a functional fibrinogen-binding complex. *J. Biol. Chem.* 264:18859.
26. Springer, T.A., W.S. Thompson, L.J. Miller, F.C. Schmalstieg, and D.C. Anderson. 1984. Inherited deficiency of the Mac-1, LFA-1 and pl50, 95 glycoprotein family and its molecular basis. *J. Exp. Med.* 160:1901.
27. Kishimoto, T.K., N. Hollander, T.M. Roberts, D.C. Anderson, and T.A. Springer. 1987. Heterogeneous mutations in the b subunit common to the LFA-1, Mac-1, and pl50, 95 glycoproteins cause leukocyte adhesion deficiency. *Cell* 50:193.
28. Argraves, W.S., S. Suzuki, H. Arai, K. Thompson, M.D. Pierschbacher, and E. Ruoslahti. 1987. Amino acid sequence of the human fibronectin receptor. *J. Cell Biol.* 105:1183.
29. Ohashi, P., T.W. Mak, P. Van den Elsen, Y. Yanagi, Y. Yoshikai, A.F. Calman, C. Terhorst, J.D. Stobo, and A. Weiss. 1985. Reconstitution of an active surface T3/T-cell antigen receptor by DNA transfer. *Nature* 316:606.
30. Saito, T., A. Weiss, J. Miller, M.A. Norcross, and R.N. Germain. 1987. Specific antigen-Ia activation of transfected human T cells expressing Ti $\alpha\beta$-human T3 receptor complexes. *Nature* 325:125.
31. Band, H., F. Hochstenbach, C.M. Parker, J. McLean, M.S. Krangel, and M.B. Brenner. 1989. Expression of human T cell receptor-$\gamma\delta$ structural forms. *J. Immunol.* 142:3627.
32. Chan, B.M. C., N. Matsuura, Y. Takada, B.R. Zetter, and M.E. Hemler. 1991. In vitro and in vivo consequences of VLA-2 expression on rhabdomyosarcoma cells. *Science* 251:1600.
33. Teixidó, J., C.M. Parker, P.D. Kassner, and M.E. Hemler. 1992. Functional and structural analysis of VLA-4 integrin α^4 subunit cleavage. *J. Biol. Chem.* 267:1786.
34. Ruoslahti, E. 1988. Fibronection and its receptors. *Ann. Rev. Biochem.* 57:375.
35. Sonnenberg, A., C.J.T. Linders, P.W. Modderman, C.H. Damsky, M. Au-

mailley, and R. Timpl. 1990. Integrin recognition of different cell-binding fragments of laminin (P1, E3, E8) and evidence that $\alpha6\beta1$ but not $\alpha6\beta4$ functions as a major receptor for fragment E8. *J. Cell Biol.* 110:2145.

36. Holzmann, B., B.W. McIntyre, and I.L. Weissman. 1989. Identification of a murine peyer's patch-specific lymphocyte homing receptor as an integrin molecule with an α chain homologous to human VLA-4α. *Cell* 56:37.

37. Holzmann, B. and I.L. Weissman. 1989. Integrin molecules involved in lymphocyte homing to Peyer's Patches. *Immunol. Rev.* 108:45.

38. Yuan, Q., W. Jiang, D. Hollander, E. Leung, J.D. Watson, and G.W. Krissansen. 1991. Identity between the novel integrin β_7 subunit and an antigen found highly expressed on intraepithelial lymphocytes in the small intestine. *Biochem. Biophys. Res. Commun.* 176:1443.

39. Kilshaw, P.J. and S.J. Murant. 1990. A new surface antigen on intraepithelial lymphocytes in the intestine. *Eur. J. Immunol.* 20:2201.

40. Cerf-Bensussan, N., D. Guy-Grand, B. Lisowska-Grospierre, C. Griscelli, and A.K. Bhan. 1986. A monoclonal antibody specific for rat intestinal lymphocytes. *J. Immunol.* 136:76.

41. Cerf-Bensussan, N., A. Jarry, N. Brousse, B. Lisowska-Grospierre, D. Guy-Grand, and C. Griscelli. 1987. A monoclonal antibody (HML-1) defining a novel membrane molecule present on human intestinal lymphocytes. *Eur. J. Immunol.* 17:1279.

42. Schieferdecker, H.L., R. Ullrich, A.N. Weiss-Breckwoldt, R. Schwarting, H. Stein, E.-O. Riecken, and M. Zeitz. 1990. The HML-1 antigen of intestinal lymphocytes is an activation antigen. *J. Immunol.* 144:2541.

43. Hemler, M.E., C. Huang, Y. Takada, L. Schwarz, J.L. Strominger, and M.L. Clabby. 1987. Characterization of the cell surface heterodimer VLA-4 and related peptides. *J. Biol. Chem.* 262:11478.

44. Hemler, M.E., C.F. Ware, and J.L. Strominger. 1983. Characterization of a novel differentiation antigen complex recognized by a monoclonal antibody (A-1A5): unique activation-specific molecular forms on stimulated T cells. *J. Immunol.* 131:334.

45. Lobb, R.R., G. Chi-Rosso, D.R. Leone, M.D. Rosa, B.M. Newman, S. Luhowskyj, L. Osborn, S.G. Schiffer, C.D. Benjamin, I.G. Dougas, C. Hession, and E.P. Chow. 1991. Expression and functional characterization of a soluble form of vascular cell adhesion molecule 1 (VCAM1). *Biochem. Biophys. Res. Commun.* 178:1498.

7
Functional Mapping and Regulation of VLA-4 Adhesion Activities

Francisco Sánchez-Madrid, Rafael Pulido, Antonio A. Postigo, Miguel R. Campanero, Alicia G. Arroyo, Rosario García-Vicuña, Armando Laffón, and Manuel O. de Landázuri

Introduction

The integrin family includes receptors for extracellular matrix (ECM) components as well as receptors involved in cell–cell adhesive interactions (1–3). The VLA subfamily of integrins is comprised of at least eight heterodimers, each with a common β subunit (the $\beta 1$ chain) noncovalently associated with one of eight different α subunits (4–6). In addition to $\alpha : \beta 1$ heterodimers, novel associations of α and alternative β subunits have been found, which expand the molecular and functional repertoire of integrins (4,7–9). VLA integrins are involved in cell interaction either with extracellular matrix (ECM) components or with cellular ligands (4,10,11), and these adhesive processes appear to be essential for lymphocyte homing (3,4), embryogenesis and histogenesis (12), leukocyte activation (13–16), cytotoxic T lymphocyte activity (17,18), and lymphohemopoiesis (19).

The VLA-4 ($\alpha 4 : \beta 1$; CD49d/CD29) integrin is expressed on thymocytes, resting and activated lymphocytes, and monocytes, as well as by most T- and B-cell lines (4,20,21). Although most members of the VLA family have been involved in cell ECM interactions (4), only for VLA-4 has a cellular ligand been identified, in addition to its involvement in the interaction with ECM components (21). Thus, VLA-4 has been demonstrated to serve as a receptor for Arg-Gly-Asp–independent sites of plasma fibronectin (FN), namely, CS-1 and CS-5 (22–25), as well as for the vascular cell surface adhesion molecule-1 (VCAM-1), a member of the immunoglobulin superfamily expressed on cytokine-activated endothelial cells (26–28). In addition, murine VLA-4 has also been implicated in lymphocyte adhesion to high endothelial venules on Peyer's patches (29,30). More recently, VLA-4 has been found to mediate aggregation of leukocytes based on the ability of specific anti–VLA-$\alpha 4$ monoclonal antibodies (mAb) to trigger homotypic cell aggregation via an LFA-1/ICAM-1–independent mechanism (31,32). These VLA-4–mediated functions correlate with the topographically distinct antigenic sites defined on the VLA-α4 chain (33).

Cellular activation has been shown to increase the avidity of integrins for their ligands (34,35). Recent studies indicated that the expression and function of some VLA heterodimers, including VLA-4, are also regulated during T- and B-lymphocyte maturation and activation (34,36). Thus, activated B cells display an enhanced binding to VCAM-1+ follicular dendritic cell as compared to resting B cells (37,38). Likewise, activation of CD4+ T cells increases VLA-4–mediated binding to intact FN (34). We have investigated the metabolic pathways involved in this up-regulated function of VLA-4, also providing evidence that the increase in adhesiveness of this receptor to its ligands could be also triggered through the β1 subunit. In order to determine whether this regulation also occurs in vivo, we studied the function of VLA-4 on synovial T cells from rheumatoid arthritis patients. We found an up-regulated expression and function of this receptor on these already in vivo activated T cells, suggesting that VLA-4 could play a role in the pathogenesis of certain inflammatory processes.

Herein, we summarize our recent findings on the mapping and functional regulation of VLA-4 adhesion activities.

Results and Conclusions

Functional Mapping of VLA-4

The human integrin VLA-4 is the leukocyte receptor for both the CS-1 and CS-5 regions of plasma FN and the vascular cell surface adhesion molecule-1 (VCAM-1), and also mediates homotypic aggregation upon triggering with specific anti–VLA-4 mAb (22–28,31,32).

To study structure–function relationships for the integrin VLA-4, different antigenic sites within the VLA-4 chain were mapped by cross-competitive mAb-binding studies. All anti-α4 mAbs studied clustered into three distinct classes defining three topographically separated epitopes (A, B, and C) (Fig. 7.1). The ability of the distinct anti-α4 mAbs to induce cell aggregation was evaluated, and it was found that all mAbs directed to epitope A, as well as a subgroup of mAb to epitope B (called B2), caused cell aggregation. Anti-epitope C mAbs were ineffective in promoting cell aggregation. Moreover, aggregation assay has allowed the distinction of two sites with different behavior within epitope B (33).

The VLA-4–mediated cell aggregation is isotype- and Fc-independent and can be induced by either F(ab')$_2$, as well as by monovalent Fab fragments (31). Similar to phorbol-ester-induced homotypic adhesion, cell aggregation triggered through VLA-4 requires divalent cations, integrity of cytoskeleton, and active metabolism (31).

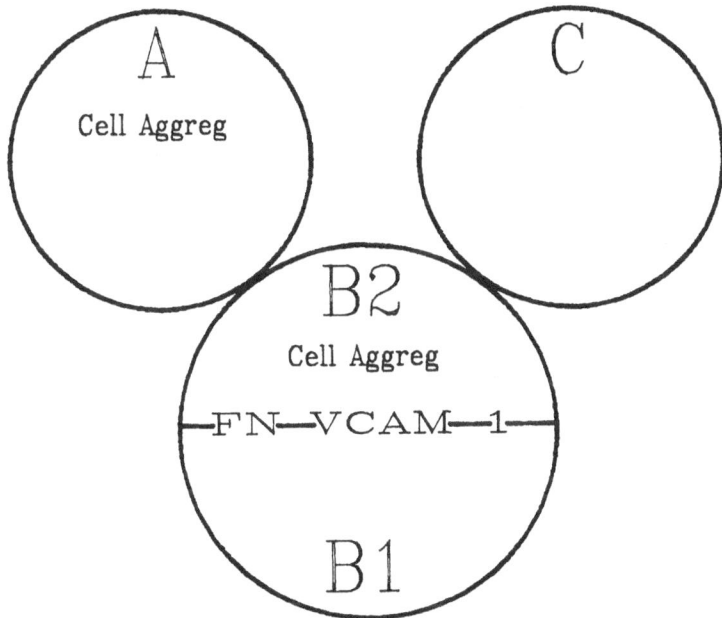

FIGURE 7.1. Model of VLA-4 epitope and functional mapping.

A correlation between VLA-α4 epitopes and VLA-4 interaction with plasma FN and VCAM-1 was also established by studying the effect of anti-α4 mAb on the cell adhesion to a 38-kDa FN proteolytic fragment containing CS-1 (herein called FN-38) or to either VCAM-1 transfected COS cells or TNFα-activated umbilical vein endothelial cells (HUVEC). In conclusion, anti-epitope B mAbs were potent inhibitors of VLA-4–mediated adhesion to both FN-38 and VCAM-1; anti-epitope A mAbs inhibited partially attachment to FN-38, but did not affect cell binding to VCAM-1; and anti-epitope C mAbs were completely ineffective as inhibitors of either interaction with FN-38 or VCAM-1 (33).

The results of the immunochemical and functional epitope mapping studies are summarized in the model proposed in Figure 7.1. As shown, VLA-4–mediated aggregation, attachment to FN38, and adhesion to VCAM-1 can each be inhibited independently, indicating that these are three distinct functions. Each of these epitopes has a different pattern of functional involvement. Thus, epitope C it is not involved in triggering of cell aggregation, nor in adhesion to FN-38 or VCAM-1. In contrast, epitope B2 is involved in all functions, epitope B1 contributes to two functions, and epitope A is involved in cell aggregation as well as having a partial role in adhesion to FN-38.

The specificity of these adhesion processes is demonstrated by the ability of anti–VCAM-1 4B9 mAb to inhibit cell attachment to VCAM-1 but not to FN38, whereas anti-FN sera inhibited attachment to FN38 but not to VCAM-1. On the other hand, VLA-4–mediated aggregation cannot be blocked by anti-FN antisera, soluble FN-38, nor anti–VCAM-1 mAb, thus indicating that VLA-4 interaction with FN or VCAM-1 is not required for triggering of cell aggregation by anti–VLA-α4 mAb. Together these findings indicate that other unknown cell-surface ligand(s) probably exist for VLA-4. However, we cannot discern whether VLA-4–mediated cell aggregation requires a third VLA-4 ligand or whether it could be just explained by intracellular signal transduction mechanisms.

Regulation of VLA-4 Adhesion Activities

Signal Transduction Pathways Regulating T-Cell Binding to FN and VCAM-1

As described for other integrins (3), the constitutive expression of VLA receptors does not necessarily imply that they are functionally active. Thus, the in vitro activation of CD4+ T cells by phorbol esters and other stimuli has been described as inducing VLA-4–mediated attachment to ECM proteins (34).

We have investigated the effect of different stimuli in the binding of peripheral blood T lymphocytes to both VLA-4 ligands. The treatment for 30 min with phorbol esters enhanced T-cell binding to both FN38 and to a recombinant soluble form of VCAM-1 (rs VCAM-1). Other stimuli such as the lectin phytohemagglutinin or the activation by crosslinked CD3 mAb and by a mitogenic combination of CD2 mAb also increased T-cell binding to both VLA-4 ligands (Table 7.1).

In order to clarify the transduction signal pathways implicated in this up-regulated adhesiveness, we investigated the effect of single stimuli. Thus, two specific protein kinase C (PKC) activators, mezerein and dioctanoyl-rac-glycerol (DIC-8), also induced T-cell binding to FN38 and rsVCAM-1. An increase in the intracytoplasmic Ca^{2+} concentration induced by treatment with the Ca^{2+} ionophore A23187 promoted by itself a strong enhancement in T-cell interaction with both ligands (Table 7.1). The increased binding to both ligands triggered by all these stimuli was specificically VLA-4–mediated since it was abrogated by anti-α4 (and anti–VCAM-1 for rsVCAM-1 binding) mAbs but not by other mAbs included as controls.

None of the stimuli that enhance T-cell binding to both VLA-4 ligands altered α4 and β1 cell-surface expression as determined by flow cytometry. Moreover, cell binding to neither FN38 nor VCAM-1 was affected by preincubation with inhibitors of either protein or RNA synthesis and could be observed after PMA or CD3 treatment for only 5 min. Therefore, the data indicate that activation of PKC or rise in the Ca^{2+}

TABLE 7.1. Regulation of VLA-4–mediated T-cell binding to FN38 and VCAM-1 upon activation with different stimuli and treatment with anti-VLA β1 mAbs.

Stimulus	FN38	rsVCAM-1	COL
		% cell attachment	
Control	13	14	0
PMA	73	75	2
PHA	51	65	0
CD3 mAb	44	46	0
CD2 mixture mAb	53	50	0
Mezerein	64	75	0
DIC-8	70	69	0
Ca²⁺ ionophore A23187	52	58	0
Ca²⁺ ionophore A23187*	70	72	0
Anti-β1 TS2/16 mAb	45	36	4
Anti-β1 Alex 1/4 mAb	10	8	3

Binding to FN38 and rsVCAM-1 by T cells treated for 30 min with different stimuli. PMA: 5 ng/ml; PHA: 1%; CD3: 10 μg/ml coated to plates overnight at 4°C; CD2: mitogenic combination: 10 μg/ml of TS2/18 mAb and 1/600 ascites dilution of D66 mAb; mezereine: 100 ng/ml; DIC-8: 20 μg/ml; A23187: 0.2 μM and A23187*: 1 μM; anti-β1 TS2/16 and Alex 1/4 mAb: 1 μg/ml.

levels up-regulate the avidity of T cells for both FN and VCAM-1 through a qualitative change of VLA-4.

Regulation of VLA Integrin Ligand Interaction Through the β1 Chain

The VLA-mediated adhesive interactions can be also regulated through the common β1 chain. In this respect, we have found that certain anti-β1 mAbs can promote binding of distinct VLA heterodimers to both ECM and cellular ligands. These enhancements are not restricted to a particular VLA integrin nor to a single substrate and have been observed to affect at least the VLA-4, VLA-5, and VLA-6 heterodimers with their ligands (FN, LN, and VCAM-1) (Table 7.1 and (39)). This β1-mediated up-regulated cell binding occurs very rapidly, within 5–10 min after anti-β1 mAb addition, is isotype- and Fc-independent, and can be triggered by both divalent and monovalent Fab fragments. Similar regulatory effects have been also described with antibodies specific for some integrin α subunits (31,32,40,41).

Taken together, these observations indicate that conformational changes resulting in an increase of ligand avidity can be induced in different VLA$\alpha\beta$1 heterodimers through the β1 chain. This putative conformational change affects the α chain since the anti-β1–mediated effect could be abrogated by specific anti-α antibodies. However, it is also conceivable that the binding of anti-$\beta 1$ mAbs may trigger intracellular signals resulting in the induction of cell attachment. The increase of intracellular pH triggered by anti-$\beta 1$ mAb has been recently reported (42).

Regulated Expression and Function of VLA-4 Receptors on In Vivo Activated T cells in Rheumatoid Arthritis

The possibility that a regulated expression and function of the VLA-4 integrin could also exist on T cells that are activated in vivo as occurs in certain pathological situations has been investigated. Rheumatoid arthritis (RA) constitutes a human model of chronic articular inflammatory disease in which T cells infiltrating the synovium may play an important role in the pathogenesis of this disease. Previous studies have described the existence of subsets of activated T cells bearing the MHC class II and VLA-1 antigens infiltrating the synovium of these patients (43–47). Recent studies indicated that the expression of three different VLA heterodimers (VLA-4, VLA-5, and VLA-6) is regulated during T-cell maturation, because higher amounts of these integrins are found on memory (CD45RO$^+$) T cells as compared with naive (CD45RO$^-$) T cells (34).

The expression and function of VLA-4 receptors have been studied on T cells localized in the inflamed synovium of patients with RA. A higher proportion of T cells in both synovial membrane (SM) and synovial fluid (SF) expressed VLA-4α and $\beta 1$ chains and the two activation antigens AIM (CD69) (48) and gp95/85 (Ea2) (49) as compared with peripheral blood (PB) T cells from the same patients (50 and Table 7.2).

The functional capacity of VLA-4 receptors expressed by these cells was determined by performing adhesion assays to FN38 and rsVCAM-1 molecules with T cells purified from PB, SM, and SF compartments of RA patients (50,51). The majority of SF T cells were able to adhere to FN 38 and rsVCAM-1, whereas only a reduced percentage of PB T cells

TABLE 7.2. Expression of VLA-4, AIM, and CD45RO antigens and functional activity for binding to FN38 and VCAM-1 of purified T cells from SF, SM, and PB of RA patients.

Purified T cells	Antigen expression (% positive cells)			Cell adhesion (% cell attachment)	
	VLA-4 (CD49d)	AIM(CD69)	CD45RO	FN38	rsVCAM-1
Healthy donors PB	53	2	41	20	19
RA patients PB	43	8	37	30	26
RA patients SM	71	68	65	69	71
RA patients SF	73	50	74	79	75

displayed this capacity. The percentages of PB T cells in RA patients that bound to 38-kDa FN were in same percentage range as lymphocytes bearing the CD45RO T-cell memory marker (Table 7.2).

These results provide evidence for the in vivo regulation of the expression and function of the VLA-4 integrin. Accordingly, we have documented an increased binding of synovial T cells to FN and VCAM-1 in RA patients. SM and SF from RA have been reported to contain high concentrations of FN (52,53). Therefore, T-lymphocyte interaction with FN in the SM could constitute an important event in the infiltration and maintenance of cells perpetuating the tissue injury in this compartment, and also can convey proliferative signals to infiltrating T lymphocytes (13–16). In addition, we have found that the interaction between VLA-4 and VCAM-1 also delivers proliferative signals to synovial T cells (51), as recently described for normal T cells (54).

The increased adhesiveness of VLA-4 from synovial T cells to VCAM-1 ligand can also reflect pathologic phenomena occurring in the inflamed tissue of patients with RA. The high expression of VCAM-1 by the RA lining layer (55) could account for the migration of infiltrating cells to the lining and thus participating in its hyperplasia and in the resulting increased cellularity of SF.

Acknowledgments. We greatly appreciate the critical reading by Dr. A.L. Corbí. This work was supported by grant from INSALUD (FISS 91/0259) and Fundación Ramón Areces (to F.S.-M.).

References

1. Hynes, R.O. 1987. Integrins; a family of cell surface receptors. *Cell.* 48:549–554.
2. Hemler, M.E. 1988. Adhesive proteins receptors on hematopoietic cells. *Immunol Today.* 41:109–513.
3. Springer, T.A. 1990. Adhesion receptors of the immune system. *Nature* (Lond). 346:425–434.
4. Hemler, M.E. 1990. VLA proteins in the integrin family: structures, functions and expression on leukocytes. *Annu. Rev. Immunol.* 8:365–400.
5. Kramer, R.H., K.A. MacDonald, and P.M. Vu. 1989. Human melanoma cells express a novel integrin receptor for laminin. *J. Biol. Chem.* 264:15462–15649.
6. Bossy, B., E. Bossy-Wetzel, and L. Reichardt. 1991. Characterization of the integrin α8 subunit: a new integrin β1-associated subunit, which is prominently expressed on axons and on cells in contact with basal laminae in chick embryos. *EMBO. J.* 10:2375–2385.
7. Bodary, S.C., and J.W. McLean. 1990. The integrin β1 subunit associates with the vitronectin receptor in a human embryonic kidney cell line. *J. Biol. Chem.* 265:5938–5941.

8. Vogel, B.E., G. Tarone, F.G. Giancoti, J. Gailiti, and E. Ruoslahti. 1990. A novel fibronectin receptor with an unexpected subunit composition ($\alpha v\beta 1$). *J. Biol. Chem.* 265:5934–5937.

9. Holzmann, B., and I.L. Weissman. 1989. Peyer's patch-specific lymphocyte homing receptors consist of a VLA-4 like α chain associated with either of two integrin β chains, one of which is novel. *EMBO. J.* 8:1735–1741.

10. Ruoslahti, E. 1991. Integrins. *J. Clin. Invest.* 87:1–6.

11. Carter, W.G., M.C. Ryan, and P.J. Gahr. 1991. Epiligrin, a new cell adhesion ligand for integrin $\alpha 3\beta 1$ in epithelial basement membranes. *Cell* 65:599–610.

12. Darribière, T., K. Guida, H. Larjava, K.E. Johnson, K.M. Yamada, J.P. Thiery, and J.C. Boucaut. 1990. In vivo analyses of integrin $\beta 1$ subunit function in fibronectin matrix assembly. *J. Cell Biol.* 110:1813–1823.

13. Davis, L.S., N. Oppenheimer-Marks, J.L. Bednarcyzk, B.W. McIntyre, and P.E. Lipsky. 1990. Fibronectin promotes proliferation of naive and memory T cells by signaling through both the VLA-4 and VLA-5 integrin molecules. *J. Immunol.* 145:785–793.

14. Matsuyama, T., A. Yamada, J. Kay, K.M. Yamada, S.K. Akiyama, S.F. Schlossman, and C. Morimoto. 1989. Activation of CD4 cells by fibronectin and anti-CD3 antibody. A synergistic effect mediated by the VLA-5 fibronectin receptor complex. *J. Exp. Med.* 170:1133–1148.

15. Nojima, Y., M.J. Humphries, A.P. Mould, A. Komoriya, K.M. Yamada. S.F. Schlossman, and C. Morimoto. 1990. VLA-4 mediated CD3-dependent CD4+ T cell activation via the CS1 alternatively spliced domain of fibronectin. *J. Exp. Med.* 172:1185–1192.

16. Shimizu, Y., G.A. van Seventer, K.J. Horgan, and S. Shaw. 1990. Costimulation of proliferative responses of resting CD4+ T cells by the interaction of VLA-4 and VLA-5 with fibronectin or VLA-6 with laminin. *J. Immunol.* 145:59–67.

17. Clayberger, C., A.M. Krensky, B.W. McIntyre, T.S. Koller, P. Parham, F. Brodsky, D.J. Linn, and E.L. Evans. 1987. Identification and characterization of two novel lymphocyte function-associated antigens, L24 and L25. *J. Immunol.* 138:1510–1514.

18. Takada, Y., M.J. Elices, C. Crouse, and M.E. Hemler. 1989. The primary structure of the $\alpha 4$ subunit of VLA-4: homology to other integrins and a possible cell-cell adhesion function. *EMBO. (Eur. Mol. Biol. Organ). J.* 8:1361–1368.

19. Miyake, K., I.L. Weissman, J.S. Greenberger, and P.W. Kincade. 1991. Evidence for a role of the integrin VLA-4 in lympho-hemopoiesis. *J. Exp. Med.* 173:599–607.

20. Sánchez-Madrid, F., M.O. de Landázuri, G. Morago, M. Cebrián, A. Acevedo, and C. Bernabeu. 1986. VLA-3: a novel polypeptide association within the VLA molecular complex: cell distribution and biochemical characterization. *Eur. J. Immunol.* 16:1343–1349.

21. Hemler, M.E., M.J. Elices, C. Parker, and Y. Takada. 1990. Structure of the integrin VLA-4 and its cell-cell and cell-matrix adhesion functions. *Immunol. Rev.* 114:45–65.

22. Wayner, E.A., A. García-Pardo, M.J. Humphries, J.A. MacDonald, and

W.G. Carter. 1989. Identification and characterization of the T lymphocyte adhesion receptor for an alternative cell attachment domain (CS-1) in plasma fibronectin. *J. Cell. Biol.* 109:1321–1330.

23. García-Pardo, A., E.A. Wayner, W.G. Carter, and O.C. Ferreira. 1990. Human B lymphocytes define an alternative mechanism of adhesion to fibronectin. The interaction of $\alpha 4\beta 1$ integrin with the LHGPEILDVPST sequence of the type III connecting segment is sufficient to promote cell adhesion. *J. Immunol.* 144:3361–3366.

24. Guan, J.L., and R.O. Hynes. 1990. Lymphoid cells recognize an alternatively spliced segment of fibronectin via the integrin receptor $\alpha 4\beta 1$. *Cell.* 60:53–61.

25. Mould, A.P., A. Komoriya, K.M. Yamada, and M.J. Humphries. 1991. The CS5 peptide is a second site in the IIICS region of FN recognized by the integrin $\alpha 4\beta 1$. *J. Biol. Chem.* 266:3579.

26. Osborn, L., R. Hession, C. Tizard, C. Vasallo, S. Luhowskyi, G. Chi-Rosso, and R. Lobb. 1989. Direct expression cloning of vascular cell adhesion molecule I, a cytokine-induced endothelial protein that binds lymphocytes. *Cell.* 59:1203–1211.

27. Elices, M.J., L. Osborn, Y. Takada, C. Crouse, S. Luhowskyi, M.E. Hemler, and R. Lobb. 1990. VCAM-1 on activated endothelium interacts with the leukocyte integrin VLA-4 at site distinct from the VLA-4/fibronectin binding site. *Cell.* 6:577–584.

28. Rice, G., J. Munro, and M. Bevilacqua. 1990. Inducible cell adhesion molecule 110 (INCAM-110) is an endothelial receptor for lymphocytes: A CD11/CD18-independent adhesion mechanism. *J. Exp. Med.* 171:1369–1374.

29. Holzmann, B., and I.L. Weissman. 1989. Integrin molecules involved in lymphocyte homing to Peyer's patches. *Immunol. Rev.* 108:45–62.

30. Holzmann, B., B.W. McIntyre, and I.L. Weissman. Identification of a murine Peyer's patch-specific lymphocyte homing receptor as an integrin molecule with an α chain homologous to human VLA-4 α. *Cell.* 56:37–46.

31. Campanero, M.R., R. Pulido, M. Ursa, M. Rodríguez-Moya, M.O. de Landázuri, and F. Sánchez-Madrid. 1990. An alternative leukocyte homotypic adhesion mechanism, LFA-1/ICAM-1 independent, triggered through the human VLA-4 integrin. *J. Cell. Biol.* 210:2157–2163.

32. Bednarczyk, J.L., and T.W. McIntyre. 1990. A monoclonal antibody to VLA-4 α chain (CDw 49d) induces homotypic lymphocyte aggregation. *J. Immunol.* 144:777–784.

33. Pulido, R., M.J. Elices, M.R. Campanero, L. Osborn, S. Schiffer, A. García-Pardo, R. Lobb, M.E. Hemler, and F. Sánchez-Madrid. 1991. Functional evidence for three distinct and independently inhibitable adhesion activities mediated by the human integrin VLA-4. *J. Biol. Chem.* 266:10241–10245.

34. Shimizu, Y., G.A. Van Seventer, K.J. Horgan, and S. Shaw. 1990. Regulated expression and binding of three VLA ($\beta 1$) integrin receptors on T cells. *Nature* (Lond). 345:250–253.

35. Dustin, M.L., and T.A. Springer. 1989. T-cell receptor cross-linking transiently stimulates adhesiveness through LFA-1. *Nature* (Lond). 341:619–624.

36. Postigo, A., R. Pulido, M.R. Campanero, A. Acevedo, A. García-Pardo, A.L. Corbí, F. Sánchez-Madrid, and M.O. de Landázuri. 1991. Differential expression of VLA-4 integrin by resident and peripheral blood B lympho-

cytes. Acquisition of functionally active $\alpha4\beta1$-fibronectin receptors upon B cell activation. *Eur. J. Immunol.* 21:2437–2445.

37. Freedman, A.S., J.M. Munro, G.G. Rice, M.P. Bevilacqua," C. Morimoto, B.W. McIntyre, K. Rhynhart, J.S. Pober. and L.M. Nadler. 1990. Adhesion of human B cells to germinal centers in vitro involves LVA-4 and INCAM-110. *Science* 249:1030–1033.

38. Koopman, G., Parmentier, H.K., H-J. Schuurman, W. Newman, C.J.L.M. Meijer, and S.T. Pals. 1991. Adhesion of human B cell to follicular dendritic cells involves both the lymphocyte function-associated antigen 1/intercellular adhesion molecule 1 and very late antigen 4/vascular cell adhesion molecule 1 pathway. *J. Exp. Med.* 173:1297–1304.

39. Arroyo, A.G., P. Sánchez-Mateos, M.R. Campanero, J. Martin-Padura, E. Dejana, and F. Sanchez-Madrid. 1992. Regulation of the VLA-integrin-ligand interactions through the $\beta1$ chain. *J. Cell. Biol.* 117:659–670.

40. Gulino, D., J.J. Ryckwaert, A. Andrieux, M.J. Rabiet, and G. Marguerie. 1990. Identification of a monoclonal antibody against platelet gpIIb that interacts with a calcium-binding site and induces aggregation. *J. Biol. Chem.* 265:9575–9581.

41. van-Kooyk, Y., P. Weder, F. Hogervorts, A.J., Verhoeven, G. van Sventer, A.A. te Velde, J. Borst, G.D. Keizer, and C.G. Figdor. 1991. Activation of LFA-1 through a Ca^{2+} dependent epitope stimulates lymphocyte adhesion. *J. Cell Biol.* 112:345–354.

42. Schwartz, M.A., C. Lechene, and D.E. Ingber. 1991. Insoluble fibronectin activates the Na/H antiporter by clustering and immobilizing integrin $\alpha5\beta1$ independent of cell shape. *Proc. Natl. Acad. Sci. USA.* 88:7849–7853.

43. Cavender, P., D. Haskard, C.-L. Yu, P. Miossec, M. Oppenheimer-Marks, and M. Ziff. 1987. Pathways to chronic inflammation in Rheumatoid synovitis. *Fed. Proc.* 46:113–117.

44. Burmester, G.R., D.T.I. Yu, A.A. Irani, H.G. Kunkel, and R.J. Winchester. 1981. Ia+ T cells in synovial fluid and tissues of patients with rheumatoid arthritis. *Arthritis Rheum.* 24:1370–1376.

45. Hemler, M.E., D. Glass, J.S. Coblyn, and J.G. Jacobson. 1986. Very late activation antigens on rheumatoid synovial fluid T lymphocytes: association with stages of T cell activation. *J. Clin. Invest.* 78:696–702.

46. Laffón, A., F. Sánchez-Madrid, M.O. de Landázuri, A. Ariza, C. Ossorio, and P. Sabando. 1989. Very late activation antigen on synovial fluid T cells from patients with rheumatoid arthritis and other rheumatic diseases. *Arthritis Rheum.* 32:386–392.

47. Cush, J.J., and P.E. Lipsky. 1988. Phenotypic analysis of synovial tissue and peripheral blood lymphocytes isolated from patients with rheumatoid arthritis. *Arthritis Rheum.* 10:1230–1238.

48. Cebrián, M., E. Yagüe, M. Rincón, M. López-Botet, M.O. de Landázuri, and F. Sánchez-Madrid. 1988. Triggering of T cell proliferation through AIM an activation inducer molecule on activated human lymphocytes. *J. Exp. Med.* 168:1621–1637.

49. Newman, W., V.A. Fanning, P.E. Rao, E.F. Westberg, and E. Patten. 1986. Early events in lymphocyte activation as defined by three new monoclonal antibodies. *J. Immunol.* 137:370–378.

50. Laffón, A., R. García-Vicuña, A. Humbría, A.A. Postigo, A.L. Corbí, M.O.

de Landázuri, and F. Sánchez-Madrid. 1991. Upregulated expression and function of VLA-4 fibronectin receptor on human activated T cells in rheumatoid arthritis. *J. Clin. Invest.* 88:546–552.

51. Postigo, A.A., R. Garcia-Vicuña, F. Diaz-Gonzalez, A.G. Arroyo, M.O. de Landázuri, G. Chi-Rosso, R.R. Lobb., A. Laffón, and F. Sánchez-Madrid. 1992. Increased binding of synovial T lymphocytes from Rheumatoid Arthritis to endothelial-leukocyte adhesion molecule-1 (ELAM-1)and vascular all adhesion molecule-1 (VCAM-1). *J. Clin. Invest.* 89:1445–1452.

52. Carsons, S., M.W. Mosesson, and H.S. Diamon. 1981. Detection and quantification of fibronectin in synovial fluid from patients with rheumatoid diseases. *Arthritis Rheum.* 124:1261–1267.

53. Herbert, K.W., J.S. Coppock, A.M. Griffith, A. Williams, M.W. Robinson, and D.L. Scott. 1987. Fibronectin and immune complexes in rheumatoid diseases. *Ann Rheum. Dis.* 46:734–740.

54. Damle, N.K., and A. Aruffo. 1991. Vascular cell adhesion molecule 1 induces T-cell antigen receptor-dependent activation of CD4+ T lymphocyte. *Proc. Natl. Acad. Sci. USA* 88:6403–6407.

55. Koch, A.E., J.C. Burrows, G.K. Haynes, T.M. Carlos, J.M. Harlan, and S.J. Leibovich. 1991. Immunolocalization of endothelial and leukocyte adhesion molecules in human rheumatoid and osteoarthritic synovial tissues. *Lab. Invest.* 64:313–320.

8
Signal Transduction Through a Novel Phagocyte Integrin

ERIC J. BROWN AND HATTIE D. GRESHAM

> *There is at bottom only one genuinely scientific treatment for all diseases, and that is to stimulate the phagocytes*—George Bernard Shaw, in *The Doctor's Dilemma* (1911)

Phagocytosis is a highly regulated phenomenon in neutrophils, monocytes, and macrophages, the so-called "professional phagocytes." Resting phagocytes ingest much less avidly than cells stimulated with a variety of mediators found at inflammatory sites. Thus, a neutrophil purified in its resting state from the circulation demonstrates minimal phagocytic capacity; one that has been stimulated with TNF-α, GM-CSF, fMLP, C5a, or other activating agents ingests at a rate and to an extent almost 10-fold greater than normal. Presumably, this is because the activation of the respiratory burst and secretion of lysosomal enzymes often accompanies the process of ingestion by professional phagocytes. Since these products of the activated phenotype can be highly toxic, not only to invading microorganisms, but also to the host, it is reasonable that molecular mechanisms have developed which limit these toxic events to sites where maximal host defense function is required.

Since most host defense is an extravascular event, we reasoned several years ago that interaction with extracellular matrix proteins might provide a signal to enhance phagocytosis. Our original studies with fibronectin and laminin demonstrated that these molecules could indeed stimulate ingestion by all professional phagocytes. More recently, we have begun a detailed study of the properties of the receptor(s) which signal this enhanced phagocytic function. A particular interest of our laboratories has been understanding the molecular mechanisms by which phagocytosis is regulated. In this review, we will detail our current understanding of the nature of the receptors involved in regulation of phagocytosis by extracellular matrix proteins and the molecular nature of the signals involved in the regulation.

Results

The Arg-Gly-Asp Domain of Fibronectin (Fn) is Involved in Phagocytosis Enhancement

Fn is a paradigmatic adhesion protein. It is found in the extracellular matrix in essentially all tissues and is part of the pericellular matrix of adherent cells in tissue culture. Elegant and ground-breaking work from Pierschbacher and Ruoslahti demonstrated that a tripeptide within the sequence of Fn, Arg-Gly-Asp, could mimic the adhesive properties of the intact molecule, a 440,000-M, protein (1,2). Work from Wright and Meyer suggested that Fn enhancement of phagocytosis also involved the Arg-Gly-Asp–containing domain (3). We confirmed this finding (4), and then we went on to apply another technique pioneered by the Ruoslahti laboratory, Arg-Gly-Asp–affinity chromatography (5,6), to monocytes and neutrophils. In these studies, we compared receptors isolated on Arg-Gly-Asp–Sepharose to those which bound to an affinity column on which the Fn domain containing Arg-Gly-Asp, rather than the synthetic Arg-Gly-Asp peptide itself, was the affinity ligand. We showed that for PMN, the same receptor adhered to either ligand, whereas for monocytes, two distinct receptors were isolated (7). The receptor which adhered to the Fn cell-binding domain was the "classic" Fn receptor, $\alpha_5\beta_1$, while the receptor which adhered to the Arg-Gly-Asp peptide affinity column resembled that isolated from PMN. Because both PMN and monocytes enhanced phagocytosis in response to Fn, we favored the idea that the Arg-Gly-Asp–binding receptor was responsible for signal transduction in this system. To test this, we examined multiple Arg-Gly-Asp–containing proteins for their ability to stimulate phagocytosis. We found that many could do so, and, ultimately, we found that the synthetic peptide alone could stimulate ingestion, if properly presented (7,8).

Leukocyte Response Integrin

The ability of multiple Arg-Gly-Asp–containing proteins and of multivalent, synthetic Arg-Gly-Asp itself to stimulate phagocytosis, coupled with the rather distinct appearance of the phagocyte receptor which bound to Arg-Gly-Asp–Sepharose, stimulated a more detailed examination of the nature of the receptor. Initial studies showed that the affinity-purified receptor was recognized by polyclonal antibody against the platelet β_3 integrin, gpIIb/IIIa (7). The cross-reactivity appeared to be confined to the β_3 chain. Therefore, we purified the abundant β_3 integrin from placenta, $\alpha_v\beta_3$, in an effort to find antibodies which would cross-react with the phagocytosis-stimulating integrin of monocytes and PMN. We created a series of monoclonal antibodies against α_v, β_3, or the intact $\alpha_v\beta_3$ receptor. To summarize the data with these antibodies, no anti-α_v inhibited

TABLE 8.1. Effect of anti-β_3 on stimulated phagocytosis.

Inhibited by anti-β_3 Synthetic peptides	Unaffected by anti-β_3
Fibronectin Oligomeric GRGDSP	Laminin
Fibrinogen Oligomeric KQAGDV	Entactin
Type IV collagen Oligomeric KGAGDV	Phorbol esters
Von Willibrand's factor	f-Met-Leu-Phe
Vitronectin	C5a
	GM-CSF
	TNF-α

phagocytosis stimulation. Indeed, none of the α_v-specific mAbs even bound to intact PMN or monocytes (8). Several bound to monocytes which had been kept in culture for 5–7 days, consistent with the observation that expression of both $\alpha_v\beta_3$ and $\alpha_v\beta_5$ is induced during differentiation of monocytes to macrophages (9,10). Similar data were obtained with antibodies which recognized the $\alpha_v\beta_3$ complex. However, a mAb, 7G2, which recognized β_3 both bound to PMN and inhibited Arg-Gly-Asp–stimulated phagocytosis; 7G2 also inhibited phagocytosis stimulated by most RGD-containing proteins, without affecting ingestion stimulated by laminin, entactin, phorbol esters, C5a, or f-Met-Leu-Phe (Table 8.1). Similar data were obtained with the polyclonal antibody to $\alpha_v\beta_3$ from placenta. These data suggested that the integrin involved in regulation of phagocytosis by Arg-Gly-Asp was a distinct integrin, closely related to $\alpha_v\beta_3$ and gpIIb/IIIa. Furthermore, the cross-reactivity seemed to be in the β_3 chain. Based on these data, we hypothesized that the phagocyte integrin was unique and previously undescribed; therefore, we named it the leukocyte response integrin (LRI) based on its apparently critical role in signal transduction for phagocyte activation.

Since several families within the integrin superfamily are defined by multiple distinct α chains binding to a single β, the most appealing hypothesis for the structure of LRI is that it is a new member of the β_3 family. While this is still formally possible, there are several pieces of data which make this hypothesis less likely. First, the cross-reactivity between 7G2 and LRI is weaker than the binding of 7G2 to native β_3. While 7G2 will Western blot β_3 after both reduced and unreduced SDS-PAGE, it will Western blot LRI only after unreduced gels have been run. Also, the concentration of mAb required on Western blots to visualize the β chains is about 10-fold different between β_3 and LRI. Second, we have made a polyclonal antibody to LRI, by purifying the receptor from differentiated HL60 cells on Arg-Gly-Asp–Sepharose. This antibody does not cross-react at all with placental $\alpha_v\beta_3$ on ELISA, suggesting that it does not recognize β_3 (11). Thus, we hypothesize that, although immuno-

TABLE 8.2. Reactivity of cells and cell lines with anti-LRI antibody.

Cell type	Fluorescence[a]
Peripheral blood cells	
Monocyte[b]	++
PMN	+++
Lymphocyte[b]	++
Platelet	−
Erythrocyte	−
Cell lines	
HL60	+/−
dHL60[c]	++
THP-1	++
U937	++
Daudi	+/−
Raji	−
K562	+
REH	+
EBV-1[d]	−
EBV-2	−
Namalwa	−
SKW	−
Marbrook	−

a Preimmune serum and anti-LRI were used at a 1:250 dilution, which was determined to be saturating for anti-LRI.
b Elutriated monocytes and lymphocytes.
c HL60 differentiated with 1.2% DMSO for 4 days.
d EBV-transformed PBL.

logically related, LRI β is distinct from β_3. Using this antibody, we have examined the distribution of LRI. It is present on neutrophils, monocytes, and lymphocytes from peripheral blood but is absent from platelets and erythrocytes. It is also absent from many continuous lymphocyte cell lines (Table 8.2).

Peptide Specificity of LRI

To further distinguish LRI from β_3-containing integrins, we have examined the peptide specificity of the receptor, both by functional and direct ligand-binding studies. Initially, we reasoned that the broad Arg-Gly-Asp specificity of the receptor was analagous to platelet gpIIb/IIIa. Since gpIIb/IIIa is known to recognize not only Arg-Gly-Asp but the sequence His-His-Leu-Gly-Gly-Ala-Lys-Gln-Ala-Gly-Asp-Val (HHLG-GAKQAGDV) as well, we tested the latter peptide, derived from the fibrinogen (Fgn) γ chain, as well. This peptide was more potent at inhibition of Fgn- and Fn-stimulated phagocytosis than Arg-Gly-Asp–containing peptides (12). In addition, oligomerized KQAGDV was at least equally potent as oligomerized Arg-Gly-Asp at stimulation of phagocytosis. This experiment demonstrated that LRI recognized the Fgn γ-chain peptide and thus could not be $\alpha_v\beta_3$, which does not bind HHLGGAKQAGDV at all (13). Since gpIIb is expressed only in platelets, the result of this experiment was further evidence against the possibility that LRI was a known member of the β_3 family. Further evidence that LRI was distinct from gpIIb/IIIa came from the study of some systematic substitutions in the γ-chain peptide. Substitution of G for Q completely abrogated function of the peptide in platelet assays, but did not inhibit function in phagocytosis enhancement assays (Table 8.1) (12). Furthermore, substitution of R for A, which restored an Arg-Gly-Asp sequence to the peptide, enhanced its function in platelet assays, but surprisingly significantly inhibited its function in phagocytosis enhancement. The surprising potency of the KGAGDV peptide in assays of phagocytosis enhancement led us to examine its binding to cells directly. Both as free oligomeric peptide and when coupled to an inert surface, KGAGDV shows specific binding to PMN. This binding can be inhibited by 7G2, the cross-reactive anti-β_3 antibody, and by B6H12, an antibody to integrin-associated protein (see below), suggesting that KGAGDV indeed binds to LRI. Thus, LRI has a unique peptide specificity compared to all other known integrins.

Signal Transduction for Enhanced Phagocytosis

PMN phagocytosis has proven to be an excellent model system for the study of signal transduction by integrins. Unstimulated IgG-mediated phagocytosis by PMN is not inhibited by pharmacologic inhibitors of protein kinase C, but Arg-Gly-Asp–stimulated ingestion is inhibited by both H7 and staurosporine (Table 8.3). Arg-Gly-Asp–stimulated, but not unstimulated, ingestion is also inhibited by pertussis toxin, by chelation of intracytoplasmic Ca^{2+} with the cell-permeant EGTA analog MAPTAM, and by complexing phosphoinositides with neomycin. Both unstimulated and stimulated ingestion are inhibited by increasing the concentration of intracytoplasmic cAMP, either with cell-permeant forms of cAMP or by adding adenylate cyclase agonists, such as PGE_1, in the presence of a phosphodiesterase inhibitor. These data suggest the hypothesis that engagement of LRI leads to a trimeric G protein–dependent activation of a phospholipase C, which in turn activates protein kinase C and generates a rise in intracytoplasmic Ca^{2+}. Apparently both increased intracytoplasmic Ca^{2+} and protein kinase C are needed for enhancement of phagocytosis by Arg-Gly-Asp.

One of the most interesting aspects of our studies of signal transduction through LRI was the discovery of the mAb B6H12. This mAb was raised during our initial preparation of mAb antibodies to placental $\alpha_v\beta_3$. However, rather than recognize $\alpha_v\beta_3$, this antibody recognized a 50-kD

TABLE 8.3. Effect of pharmacologic inhibitors on stimulated and unstimulated IGG-mediated ingestion by PMN.

| Inhibitor | None | Stimulator of ingestion | | | |
		RGD[c]	TNF-α	RDBu[b]	fMLP
H7[a]	0	↓	↓	↓	ND
HA1004[a]	0	0	0	0	ND
TFP[b]	0	↓	↓	↓	ND
cAMP	↓	↓	0	0	ND
Staurosporine	0	↓	↓	↓	↓
PT[b]	0	↓	↓	0	↓
CT[b]	0	0	0	0	0
MAPTAM[b]	0	↓	ND	0	↓

aH7 and HA1004 are isoquinalone sulfonamides with increased potency against PKC (H7) and cAMP-dependent protein kinase (HA1004), respectively (25).
bTFP: trifluoroperazine; PT: pertussis toxin; CT: cholera toxin; MAPTAM: bis-(2-amino-5-methylphenoxylethan-N,N,N′,N′-tetraacetic acid tetraacetoxymethyl ester), a cell-permeant EGTA analog; PDBu: phorbol dibutyrate.
cFn, Fg, and multivalent RGD peptides were equivalent in this assay.

protein which was a trace contaminant of our antigen preparation (Fig. 8.1). Interestingly, the antibody inhibited Arg-Gly-Asp–stimulated ingestion completely (Fig. 8.2). Moreover, B6H12 inhibited phagocytosis enhancement by precisely the same ligands as the anti-β_3 mAb 7G2. (See

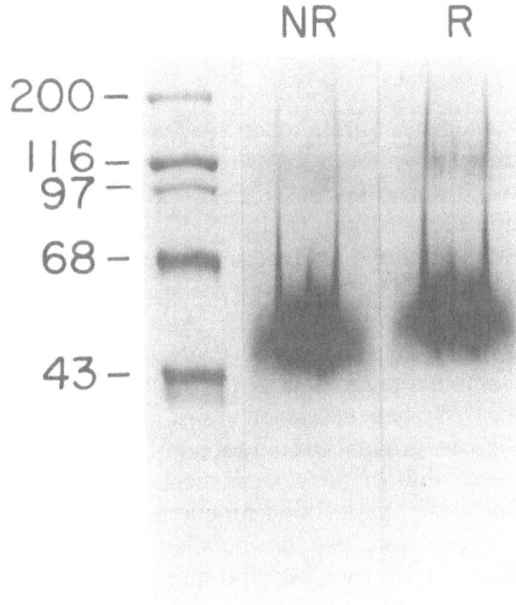

FIGURE 8.1. *SDS-PAGE of purified 50-kD protein.* Protein from placenta was purified by sequential WGA- and mAb-affinity chromatography. The protein shows a broad band on silver stain consistent with its high degree of glycosylation. The apparent M_r increases slightly with reduction. The faint band at 100 kD is a dimer of the 50-kD molecule.

84   Eric J. Brown and Hattie D. Gresham

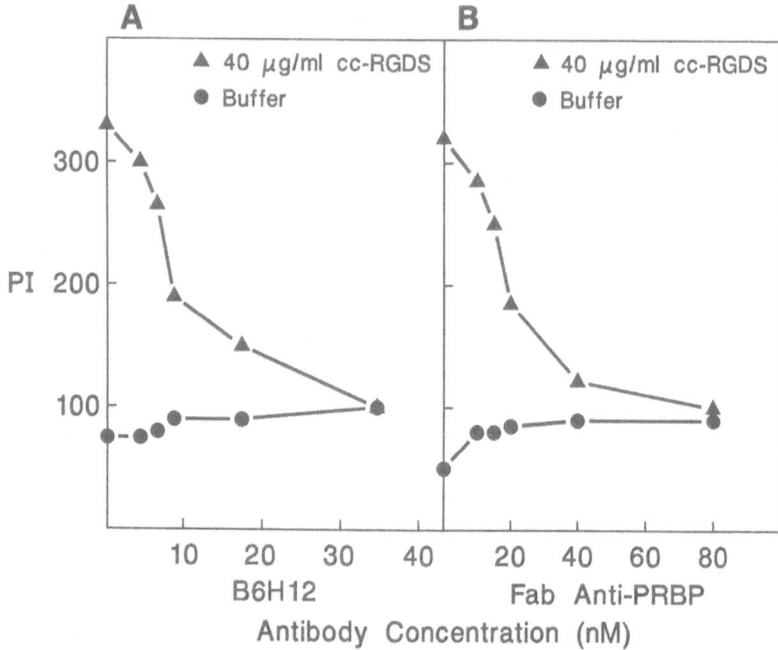

FIGURE 8.2. *Inhibition of Arg-Gly-Asp–stimulated ingestion by mAb B6H12.*
PMNs were incubated with either B6H12 (Panel A) or Fab polyclonal anti-$\alpha_v\beta_3$
(Panel B). IgG-opsonized targets were then added, and the mixture was incu-
bated with (▲) or without (●) 40 μg/ml of oligomeric Arg-Gly-Asp. After 30 min
at 37°C, phagocytosis was assessed as PI, the number of ingested targets per 100
PMNs. Reprinted from (8).

Table 8.1). Careful controls demonstrated that this inhibition was not because of antibody cross-reactivity with LRI. Indeed, mAb made against
two distinct epitopes of the 50-kD protein inhibited Arg-Gly-Asp stimulation of ingestion (Table 8.4). Thus, we concluded that the 50-kD protein
was somehow involved in the function of LRI for enhancement of phagocytosis. Our current hypothesis is that the 50-kD protein is closely physically associated with LRI. Anti–50-kD mAb B6H12 inhibited binding of
Arg-Gly-Asp ligands to PMN (8). Interestingly, we have been able to
coimmunoprecipitate $\alpha_v\beta_3$ with the 50-kD protein from platelets (Fig.
8.3). We believe that the reason we found the 50-kD protein contaminat-
ing our original preparations of $\alpha_v\beta_3$ from placenta is because of a co-
association of the two molecules in that tissue as well (14). Because of
this close physical and functional association of the 50-kD molecule with
two integrins, we have called this protein integrin-associated protein
(IAP). It is obviously intriguing that the two receptors with which IAP

TABLE 8.4. Anti-IAP antibodies.

AB	Epitope[a]	Inhibition[b]
B6H12	1	Y
1F7	1	Y
3E9	1	Y
3G3	1	N
2E11	2	Y
2D3	3	N

a Epitopes defined by cross-inhibition in ELISA.
b Inhibitory in RGD-stimulated phagocytosis assay (8,14).

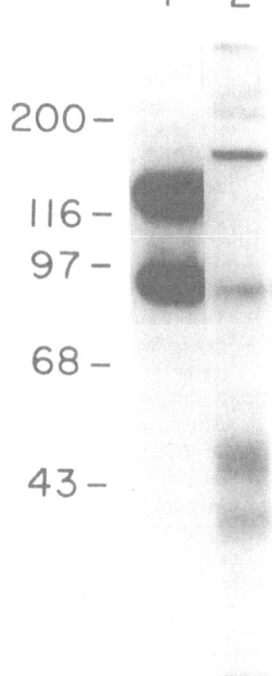

FIGURE 8.3. *Immunoprecipitation of platelets with B6H12*. Platelets were surface-labeled with [125]I (24) and then immunoprecipitated as described (14). In lane 1, anti-gpIIb/IIIa was used; in lane 2, B6H12, the monoclonal anti–50-kD protein, was used. In addition to the 50-kD band, two bands of hiher M_r are seen in lane 2. The ~95-kD band can be immunoblotted with anti-β_3 antibody (14). The ~165-kD band is larger than gpIIb in lane 1, suggesting that the integrin co-immunoprecipitated by B6H12 is $\alpha_v\beta_3$.

has been associated have β_3 or β_3-like β chains. However, we have not found convincing evidence for association of IAP with gpIIb/IIIa on platelets, another β_3 integrin. Obviously, it will be of considerable interest to determine the nature of the association of IAP with specific integrin chains.

TABLE 8.5. Assays in which B6H12 has been used.

Assay	Investi- gator	Result	Putative integrin
Substrate binding			
Endothelial cell (Vn, Fgn substrate)	Swerlick	Inhibits	$\alpha_v\beta_3$
Endothelial cell (Fn substrate)	Swerlick	No effect	$\alpha_5\beta_1$
Lymphocyte (Fn substrate)	Ballard	No effect	$\alpha_4\beta_1$
PMN (laminin substrate)	Bohnsack	No effect	$\alpha_6\beta_1$
PMN (BSA substrate)	Brown	No effect	$\alpha_M\beta_2$
THP-1 (Fn substrate)	Brown	No effect	$\alpha_5\beta_1$
THP-1 (Vn substrate)	Brown	Inhibits	$\alpha_v\beta_5$?
IMR90 (fibroblast) (Vn)	Roman	Inhibits	$\alpha_v\beta_3$
PMN (recombinant entactin)	Senior	Inhibits	LRI
Cell–cell interaction			
Lymphocyte endothelium	Swerlick	Inhibits	$\alpha_v\beta_3$
Macrophage-agglutinating-factor activity	Godfrey	Inhibits	?
Phagocytosis of apoptotic PMN	Savill	No effect	$\alpha_v\beta_3$
Other assays			
ECM-primed, TNF-stimulated respira- tory burst (PMN)	Nathan	No effect	$?\alpha_M\beta_2$
RGD-stimulated phagocytosis (PMN)	Gresham	Inhibits	LRI
Fn-stimulated phagocytosis (mono- cytes)	Brown	Inhibits	LRI
Fn-stimulated phagocytosis of group B streptococci (PMN)	Hill	Inhibits	?
HMW-BCGF–induced proliferation of activated B lymphocytes	Ambrus	Inhibits	?
LMW-BCGF-induced proliferation of activated B lymphocytes	Ambrus	No effect	?
r-entactin–induced PMN chemotaxis	Senior	Inhibits	LRI
Arg-Gly-Asp–induced PMN chemo- taxis	Senior	Inhibits	LRI

Bob Swerlick is at Emory University, Atlanta; Henry Godfrey is at New York Medical College, Valhalla, NY; John Savill is at Imperial Cancer Research Fund, London, England; Carl Nathan is at Cornell Medical College, New York, NY; Laura Ballard is at University of Tennessee, Memphis, TN; Harry Hill and John Bohnsack are at University of Utah, Salt Lake City, UT; Julian Ambrus, Jesse Roman, and Bob Senior are at Washington University, St. Louis, MO.

Function, Structure, and Distribution of IAP

Anti-IAP antibody B6H12 has been tested in a variety of assays, some of which are summarized in Table 8.5. The summary of these functional assays is that anti-IAP inhibits LRI functions including adhesion and chemotaxis, as well as phagocytosis enhancement. It also inhibits some α_v-related functions, as well as some functions, such as HMW-BCGF–dependent B-lymphocyte proliferation, which are not obviously integrin-dependent. IAP apparently has a wide distribution, including platelets, erythrocytes, endothelial cells, epithelial cell lines, and endothelial cells as well as the myeloid cells on which it was first found. We have purified the 50-kD protein and subjected it to some structural analysis. It is highly glycosylated, with a core M_r of ~31 kD. It is also highly hydrophobic and shows a marked tendency to self-aggregation, especially under denaturing conditions. Finally, we have obtained partial primary sequence information. A Kyte-Doolittle hydrophobicity plot of the COOH terminal 265 amino acids is shown in Figure 8.4. This plot predicts that IAP spans the membrane three to five times.

Conclusions

The study of the mechanisms of regulation of phagocytosis has led us to discover a new phagocyte integrin and an associated protein apparently

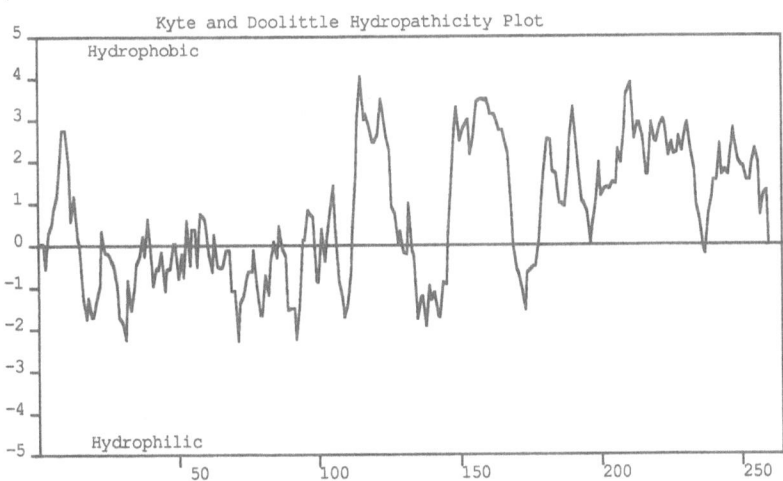

FIGURE 8.4. *Kyte-Doolittle hydropathicity plot of IAP.* The COOH terminal 265 amino acids of a partial IAP cDNA clone were plotted by the Kyte-Doolittle algorithm. Several regions of increased hydrophobicity are noted, consistent with three to five transmembrane domains.

necessary for signal transduction from ligand binding to this integrin. The integrin, LRI, appears to be most closely related to the β_3 family of integrin receptors, but has structural features which suggest that it is a unique, previously undescribed integrin. Indeed, even the nature of the interaction with anti-β_3 antibodies suggests that the LRI β is not truly identical to β_3. Clearly, further work is needed to determine the primary structure of this integrin, its cellular distribution, and its role in binding various Arg-Gly-Asp–containing ligands.

The peptide specificity of LRI is unique. It recognizes the Fgn γ-chain peptide KGAGDV as an alternative to Arg-Gly-Asp, a property shared only by the platelet-specific integrin gpIIb/IIIa. In contrast to gpIIb/IIIa, LRI continues to recognize the peptide KGAGDV and recognizes KQRGDV very poorly. The uniqueness of the peptide specificity of this receptor is important in considering possible therapeutic implications of its inhibition. Since Arg-Gly-Asp is a ligand for many integrins, peptide inhibitors based on this sequence would at least potentially interrupt many different ligand–receptor interactions. However, it is possible that the unique specificity of LRI would allow for the development of inhibitors which are specific for this receptor.

In what circumstances might LRI contribute to pathology? Together with Dr. Robert Senior of Washington University, we have shown that *unactivated* PMNs, adhere to entactin, a ubiquitous basement-membrane protein, via LRI. This raises the possibility that LRI is involved in a very early step in the emigration of PMN from the vasculature into areas of inflammation. If this is true, interference with LRI might be of benefit in many chronic inflammatory diseases characterized by neutrophil accumulation, such as arthritis and inflammatory bowel disease.

A major focus of our work has been on the molecular mechanisms of signal transduction from LRI binding to Arg-Gly-Asp ligands. A hypothesis outlining a model for signal transduction from LRI is shown in Figure 8.5. This hypothesis involves activation of phospholipase(s) and protein kinase C. An interesting sidelight to this hypothesis is the emerging concept that some forms of adhesion and stimulated phagocytosis involve the activation of phospholipase A2 as well as phospholipase C (15,16). Thus, it may be that activation of a phospholipase A2 is an important component of signal transduction through LRI; this is currently being tested in our laboratories. Whether activation of phospholipase A2 and phospholipase C is independent or interrelated is also a question toward which we are currently devoting effort. Another major question concerns the role of IAP in this signal transduction. Is IAP involved in signal transduction strictly, i.e., without influencing the affinity of an individual integrin receptor for its ligand, or is it best thought of as a third component of the ligand receptor? While our data which show that anti-IAP inhibits the binding of polyvalent ligands suggest that IAP may modulate receptor affinity, it is also possible that IAP affects receptor clustering, receptor

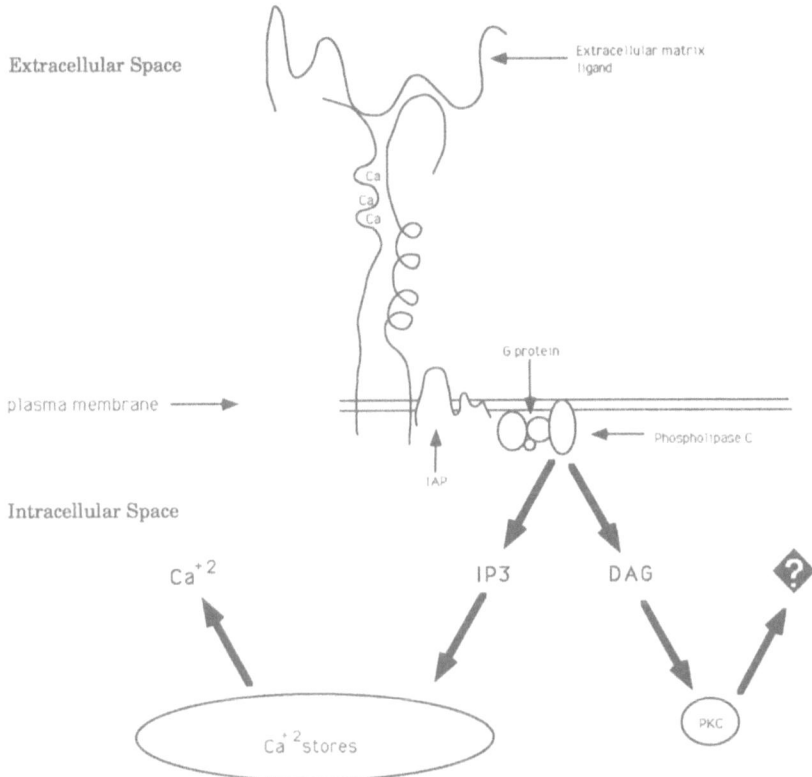

FIGURE 8.5. *Hypothetical model for signal transduction through LRI.* A model is shown for LRI-mediated signal transduction which incorporates IAP and the hypothetical involvement of phospholipase C, protein kinase C, and Ca^{2+}, all of which are thought to be mediators of enhanced phagocytosis on the basis of pharmacologic experiments.

activation, or receptor conformation, and that this function explains the effect of anti-IAP antibodies on ligand binding. The fact that IAP spans the membrane multiple times raises the possibility that it is homologous in function to a membrane channel or to a connexin. The latter possibility is especially intriguing, since connexins are also involved in cell-adhesive phenomena and are apparently functionally closely associated with other adhesion molecules, such as LCAM and NCAM (17,18). Finally, it should be mentioned that a new family of membrane molecules which spans the membrane multiple times is emerging (19–23). This family consists of proteins which apparently have four transmembrane domains, three of which are at the NH_2 terminal of the molecule. Since IAP has three predicted transmembrane domains at its COOH terminus, it is the "mirror image" of this family. Although the functions of these molecules are not known, one is closely associated with the high-affinity IgE receptor on mast cells, and antibody to this protein inhibits signal transduction through the IgE receptor (21). It will be very interesting to determine whether IAP is functionally homologous to this family of membrane proteins. Current efforts in the laboratory are directed toward understanding the structure and function of IAP in more detail.

Acknowledgments. The work in our laboratories is supported by NIH grants AI24674 (E.J.B.), GM38330 (E.J.B.), and AI23790 (H.D.G.), by the Medical Research Service, Department of Veterans Affairs (H.D.G.), and by the Washington University–Monsanto Cooperative Agreement.

References

1. Pierschbacher, M.D., E.G. Hayman, and E. Ruoslahti. 1983. Synthetic peptide with cell attachment activity of fibronectin. *Proc. Natl. Acad. Sci. USA* 80:1224.
2. Pierschbacher, M.D. and E. Ruoslahti. 1984. Cell attachment activity of fibronectin can be duplicated by small synthetic fragments of the molecule. *Nature* 309:30.
3. Wright, S.D. and B.C. Mayer. 1985. Fibronectin receptors of human macrophages recognize the sequence Arg-Gly-Asp-Ser. *J. Exp. Med.* 162:762.
4. Bohnsack, J.F., T. Takahashi, and E.J. Brown, 1986. The cell binding domain of human fibronectin is necessary but inefficient for enhancement of CR1 mediated phagocytosis by human monocyte-derived macrophages. *J. Immunol.* 136:3793.
5. Pytela, R., M.D. Pierschbacher, and E. Ruoslahti. 1985. A 125/115-kDa cell surface receptor specific for vitronectin interacts with the arginine-glycine-aspartic acid adhesion sequence derived from fibronectin. *Proc. Natl. Acad. Sci. USA* 82:5766.
6. Pytela, R., M.D. Pierschbacher, M.H. Ginsberg, E.F. Plow, and E. Ruoslahti. 1986. Platelet membrane glycoprotein IIb/IIIa: member of a family of Arg-Gly-Asp-specific adhesion receptors. *Science* 231:1559.
7. Brown, E.J. and J.L Goodwin. 1988. Fibronectin receptors of phagocytes: characterization of the Arg-Gly-Asp binding proteins of human monocytes and polymorphonuclear leukocytes. *J. Exp. Med.* 167:777.
8. Gresham, H.D., J.L Goodwin, D.C. Anderson, and E.J. Brown, 1989. A novel member of the integrin receptor family mediates Arg-Gly-Asp-stimulated neutrophil phagocytosis. *J. Cell Biol.* 108:1935.
9. Krissansen, G.W., M.J. Elliott, C.M. Lucas, F.C. Stomski, M.C. Berndt, D.A. Cheresh, A.F. Lopez, and G.F. Burns. 1990. Identification of a novel integrin β subunit expressed on cultured monocytes (macrophages): evidence that one α subunit can associate with multiple β subunits. *J. Biol. Chem.* 265:823.
10. Savill, J., I. Dransfield, N. Hogg, and C. Haslett. 1990. Vitronectin receptor-mediated phagocytosis of cells undergoing apoptosis. *Nature* 343:170.

11. Carreno, M.P., H.D. Gresham, and E.J. Brown. 1991. Characterization of a novel inegrin involved in regulation of phagocytosis. *FASEB J.* 5:A549.(Abstract)
12. Gresham, H.D. and E.J. Brown. 1990. Fibrinogen gamma chaim peptide KQAGDV stimulates neutrophil phagocytosis. *Clin. Res.* 38:480a (Abstract).
13. Smith, J.W., Z.M. Ruggeri, T.J. Kunicki, and D.A. Cheresh. 1990. Interaction of integrins αvβ3 and glycoprotein IIb-IIIa with fibrinogen: differential peptide recognition accounts for distinct binding sites. *J. Biol. Chem.* 265:12267.
14. Brown, E.J., L Hooper, T. Ho, and H. Gresham. 1990. Integrin-associated protein: a 50-kD plasma membrane antigen physically and functionally associated with integrins. *J. Cell Biol.* 111:2785.
15. Lefkowith, J.B., M. Rogers, M.R. Lennartz, and E.J. Brown. 1991. Essential fatty acid deficiency impairs macrophage spreading and adherence: role of arachidonate in cell adhesion. *J. Biol. Chem.* 266:1071.
16. Lennartz, M.R. and E.J. Brown. 1991. Arachidonic acid is essential for Fc-receptor-mediated phagocytosis by human monocytes. *J. Immunol.* 147:621.
17. Musil, L.S. and D.A. Goodenough. 1990. Gap junctional intercellular communication and the regulation of connexin expression and function. *Curr. Opin. Cell Biol.* 2:875.
18. Musil, L.S., B.A. Cunningham, G.M. Edelman, and D.A. Goodenough. 1990. Differential phosphorylation of the gap junction protein connexin43 in junctional communication-competent and -deficient cell lines. *J. Cell Biol.* 111:2077.
19. Hotta, H., N. Takahashi, and M. Homma. 1989. Transcriptional enhancement of the human gene encoding for a melanoma-associated antigen (ME491) in association with malignant transformation. *Jpn. J. Cancer Res.* 80:1186.
20. Hotta, H., A.H. Ross, K. Huebner, M. Isobe, S. Wendeborn, M.V. Chao, R.P. Ricciardi, Y. Tsujimoto, C.M. Croce, and H. Koprowski. 1988. Molecular cloning and characterization of an antigen associated with early stages of melanoma tumor progression. *Cancer Res.* 48:2955.
21. Kitani, S., E. Berenstein, S. Mergenhagen, P. Tempst, and R.P. Siraganian. 1991. A cell surface glycoprotein of rat basophilic leukemia cells close to the high affinity IgE receptor (Fc epsilon RI): similarity to human melanoma differentiation antigen ME491. *J. Biol. Chem.* 266:1903.
22. Boucheix, C., B. Benoit, P. Frachet, M. Billard, R.E. Worthington, J. Gagnon, and G. Uzan. 1991. Molecular cloning of the CD9 antigen. A new family of cell surface proteins. *J. Biol. Chem.* 266:117.
23. Classon, B.J., A.F. Williams, A.C. Willis, B. Seed, and I. Stamenkovic. 1989. The primary structure of the human leukocyte antigen CD37, a species homologue of the rat MRC OX-44 antigen. *J. Exp. Med.* 169:1497.
24. Thompson, J.A., A.L. Lau, and D.D. Cunningham. 1987. Selective radiolabeling of cell surface proteins to a high specific activity. *Biochemistry* 26:743.
25. Hidaka, H., M. Inagaki, S. Kawamoto, and Y. Sasaki. 1984. Isoquinolinesulfonamides, novel and potent inhibitors of cyclic nucleotide dependent protein kinase and protein kinase C. *Biochemistry* 23:5036.

9
Interations of the *Plasmodium falciparum*—Infected Erythrocyte with ICAM-1

ANTHONY R. BERENDT, ALISON MCDOWALL, ALISTER G. CRAIG, PAUL A. BATES, CHRISTOPHER I. NEWBOLD, AND NANCY HOGG

Introduction

Erythrocytes infected with the mature forms of the malarial parasite *Plasmodium falciparum* do not circulate in the peripheral blood but instead adhere to postcapillary venular endothelium (1). The phenomenon is analogous in many ways to lymphocyte adhesion to endothelium in that it is mediated by specific receptor–ligand interactions and must take place under shear flow conditions. Unlike lymphocytes, however, the infected erythrocytes do not transmigrate, but remain adherent for a period of approximately 24 h, during which time parasite development continues, until rupture of the erythrocyte and the release of the next generation of blood-stage parasites. Cytoadherence is thought to be an important contributor to the pathogenesis of severe malaria, since high local concentrations of parasitized cells may significantly affect flow, deplete the blood of metabolites, and release toxic products. Several complications of severe disease, such as cerebral malaria, are characterized by dense packing of infected erythrocytes in the vascular beds of affected organs (2), implying that the ability of the infected cells to localize in deep-tissue vasculature has an important bearing on the clinical outcome.

Since *Plasmodium falciparum* malaria kills approximately half-a-million children under 5 each year in Africa alone (3) and is rapidly becoming resistant to established treatments (4), there are urgent medical reasons for elucidating the molecular basis of its virulence with a long-term view toward novel therapies and potential vaccines. For these reasons, the molecular mechanisms underlying malarial adhesion to endothelial cells have come under intense scrutiny. Relatively little is known about the parasite-derived adhesion molecule(s) expressed at the surface of the infected erythrocyte except that it is a target of a strain-specific immune response (5), can undergo clonal antigenic variation (ref. 6 and D.J. Roberts et al., in preparation), and is a high-molecular-weight protein detectable by surface and metabollic labelling (7). In contrast, three endothelial receptors, the glycoproteins thrombospondin (8), CD36 (or platelet glycoprotein IIIb or IV) (see ref. 9), and ICAM-1 (10), have been identified in studies in vitro. All fulfill the necessary criteria for consideration as potential in vivo receptors, but it is not yet clear whether adhesion to any one of these is linked to the ability of the parasite to cause severe disease. There is some evidence that isolates from adults with severe but noncerebral disease preferentially adhere to CD36 compared to ICAM-1 (11), and cerebral malaria in children appears to be linked to an additional cytoadherence phenotype, that of adhesion to uninfected erythrocytes (12). However, systematic studies on the adhesion of field isolates to identified cytoadherence receptors have still not been carried out, and the relative levels of expression of the different receptors on different vascular beds in mild and severe disease are also unknown. Some evidence suggests that cerebral vessels do not express amounts of CD36 comparable to elsewhere in the body (13,14), though it is still possible that there is enough to mediate adhesion. ICAM-1 is widely expressed on vascular endothelium and is capable of dramatic up-regulation in response to inflammatory cytokines, levels of which are raised in acute malaria (15).

With these considerations, we elected to define the binding site on ICAM-1 for the infected erythrocyte (16). The question is of both biological and potential therapeutic interest should ICAM-1–mediated adhesion prove to be of pathological significance. Our strategy has been to map the epitopes of inhibitory and noninhibitory monoclonal antibodies (mAbs) and then to perform mutagenesis in targeted regions, analyzing the effects on the binding of mAbs and infected erythrocytes. Throughout, the interaction of ICAM-1 with the infected cell has been compared with the binding of ICAM-1 to its physiological counter-receptor, the leukocyte integrin LFA-1. In addition we have constructed a molecular model of the first two domains of ICAM-1, based on sequence alignments with the variable region of the immunoglobulin REI and the second domain of CD4. The data localised a critical component of the binding site on ICAM-1 to the first amino-terminal domain, in a region distinct from the LFA-1 binding site, and predict that it may be possible to develop non–LFA-1-blocking, infected erythrocyte-blocking therapies if appropriate. In addition they suggest that as for the CD4/HLA class II, CD2/LFA-3, and V_H/V_L interactions, the LFA-1/ICAM-1 interaction is mediated by one face of the β-sandwich of the first domain of ICAM-1.

Results

In initial studies on the role of ICAM-1 as an endothelial adhesion receptor for infected erythrocytes we noted that certain anti–ICAM-1 mAbs were unable to block adhesion. Since these included the LFA-1–blocking mAb RR1/1 we decided to systematically examine mAbs for their ability to block the interactions of ICAM-1 with both the infected erythrocyte and with LFA-1–expressing cells. LFA-1/ICAM-1 adhesion was studied using four different but comparable experimental systems: homotypic aggregation of PMA-stimulated JY cells, adhesion of activated T cells to fibroblasts, antigen-specific T-cell proliferation, and adhesion of activated T cells to an ICAM-1–Fc chimeric protein. Infected erythrocyte/ICAM-1 adhesion was studied using transiently transfected COS cells. The results, shown in Figure 9.1, demonstrate that many mAbs that block adhesion of LFA-1 do not block adhesion of infected erythrocytes, and vice versa. This implies that the two sites are to a large extent distinct, and interpretation borne out by the results of mAb epitope cross-blocking studies. Those mAbs which block only LFA-1 form a portion of the map which is distinct from those that block only infected erythrocytes. (see Fig. 9.2.) Linking these two portions of the map are a group of serologically indistinguishable antibodies which include the mAb LB2 and are hereafter referred to as the LB2 cluster. In keeping with their apparent location between the two sites (on the epitope map), members of the LB2 cluster block binding of both LFA-1 and infected erythrocytes. (See Fig. 9.1.) We wished to localize the epitopes for the mAbs of interest, and were able to map two out of eight mAbs to specific sequences using a nested set of synthetic octapeptides based on the ICAM-1 sequence, differing by a single amino acid per peptide (i.e., residues 1–8, 2–9, 3–10, etc.) and spanning the whole of the first two domains. The mAb 7F7 localized to the octapeptide L42LLPGNNR and the LFA-1–blocking mAb 8.4A6 reacted with three consecutive octapeptides, implying recognition of a core pentapeptide E162LDLR. Other mAbs of interest failed to react with the peptides, probably because they recognize conformational epitopes. We therefore constructed a series of different domain expression constructs and used the pattern of reactivity of each mAb with the entire set to deduce the domain or combination of domains containing the epitope. Because constructs encoding domain 1 or 2 individually did not express, chimeric exchanges were made between human and murine ICAM-1 such that the first domain was transferred to the remaining four domains of the other species. All the malaria-blocking mAbs were localized to domain 1, in contrast to LFA-1–blocking mAbs which localized to domain 1, but also, in the cases of the mAbs R6.5D6 and 8.4A6, to domain 2. MAbs which blocked neither interaction localized to domains 3–5. (See Fig. 9.3).

We examined adhesion of infected erythrocytes to COS cells transfected with the different domain constructs. (See Fig. 9.3.) Binding is dependent on the presence of domains 1 and 2, though wild-type levels of adhesion require the third domain as well, probably for reasons of receptor length and flexibility. Interestingly, although the epitope mapping locates the binding site entirely in domain 1, the chimeric molecule containing human domain 1 attached to murine domains 2–5 only mediated 10%

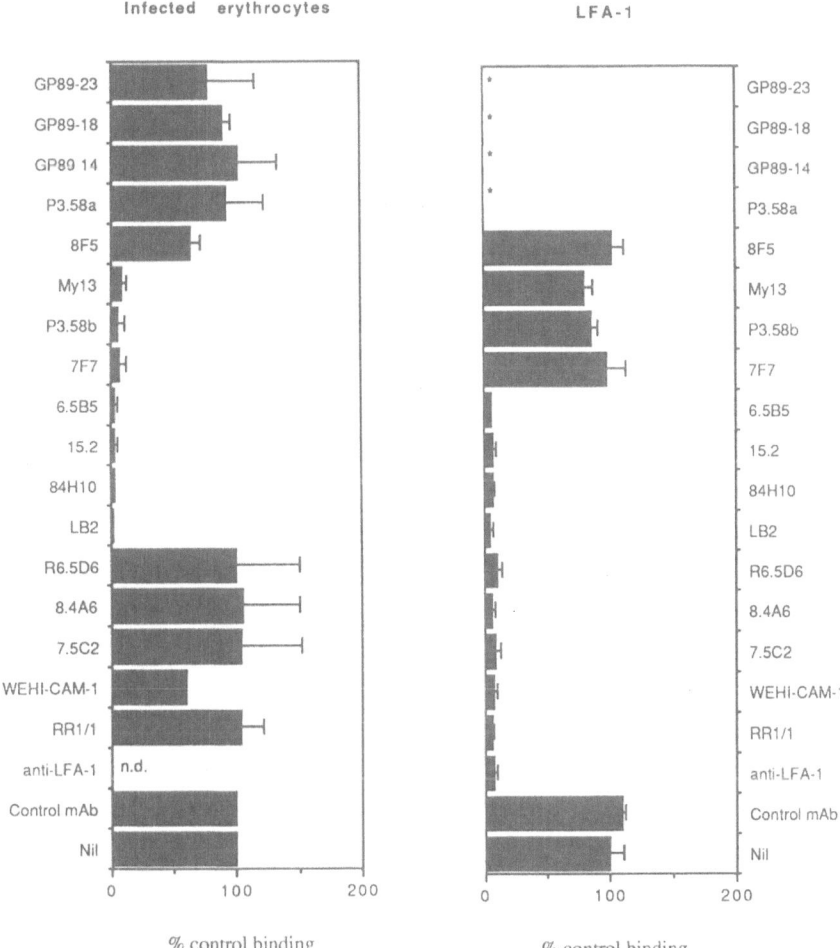

FIGURE 9.1. Monoclonal antibody inhibition of adhesion of infected erythrocytes and LFA-1 to ICAM-1. Adhesion of infected erythrocytes to ICAM-1–transfected COS cells was assessed after preincubation of transfectants with the indicated mAbs (ascites at a dilution of 1:100, culture supernatants undiluted, and purified antibodies at 10 μg/ml). Means of triplicate coverslips were determined and normalized to percentage of control. A minimum of two experiments were performed except for WEHI–CAM-1, where lack of antibody made only a single experiment possible (n.d. = not done). LFA-1/ICAM-1 binding was assessed by four different assays (see text) which gave essentially comparable results. Results shown here are for adhesion of activated T cells to recombinant purified ICAM-1 (ICAM-1-Fc). The symbol * indicates that blocking of adhesion to ICAM-1-Fc was not applicable for these mAbs as they map to domains 4 and 5 of ICAM-1. (ICAM-1-Fc contains domains 1, 2, and 3 fused to the human IgG1 Fc fragment.) The other assays show that P3.58[a] and GP89-14, -18, and -23 do not block LFA-1/ICAM-1 binding (data not shown). Figure reproduced, with permission, from *Cell* (ref. 16).

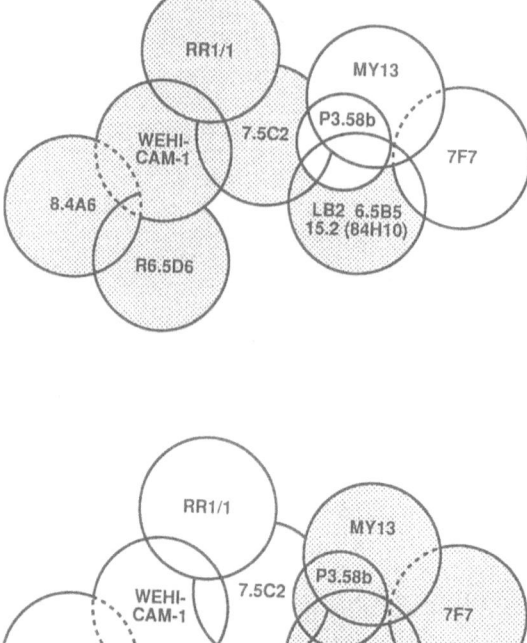

FIGURE 9.2. Epitope map of ICAM-1 mAbs recognizing overlapping epitopes. Schematic representation of the relationship between the epitopes detected by 12 anti–ICAM-1 mAbs. MAbs which block LFA-1 are shaded in (a) and those that block *Plasmodium falciparum*–infected erythrocyte binding are shaded in (b). The extent of cross-blocking is either complete (———), partial (-- --) or the information is not known (stippled). Figure reproduced, with permission, from *Cell* (ref. 16).

of wild-type adhesion. It therefore seems likely that domain 2 plays an important role in the structure of the binding site for the infected erythrocyte, either by directly contributing residues or by its effects on the conformation of the site in domain 1. We further localised relevant mAb epitopes by site-directed mutagenesis. Because the malaria-blocking mAb 7F7 had been mapped to the sequence L42LLPGNNR, we made a number of substitutions in this region, replacing residues with the corresponding murine sequences (homologue-scanning). Mutations were principally made in the loops connecting the strands of the predicted immunoglobulin fold. Some of these mutations affected individual mAb epitopes, as seen in Table 9.1 and Figure 9.4. All the mAbs which block binding of infected erythrocytes were mapped to areas spanned by residues K40-R49 (which corresponds to the CDR2 loop of an Ig fold, see below). Thus the LB2 cluster was sensitive to the mutations K40/D and L43LPGN/ES-GP, 7F7 To L43LPGN/ES-GP (in confirmation of the PEPSCAN result), My 13 to L43LPGN/ES-GP, and P3.58b to K40/D and R49/W. Although some mutations affected several individual mAbs, in all cases where this

FIGURE 9.3. Mapping of mAbs and direct binding to ICAM-1 constructs. a. Schematic diagram showing the domain and chimeric constructs of ICAM-1. The positions of the junctions are indicated. b. A summary of the assignment of mAbs to the ICAM-1 domains, based on the patterns of mAb reactivity with the domain expression constructs. c. A graph of the relative *P. falciparum*–infected erythrocyte binding to the ICAM-i constructs expressed in COS cells. Figure reproduced, with permission, from *Cell* (ref. 16).

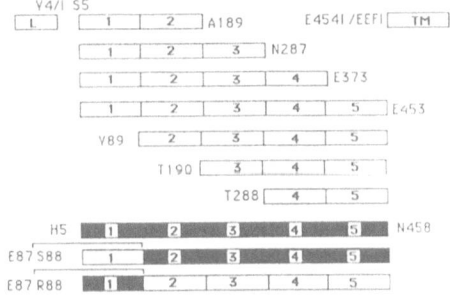

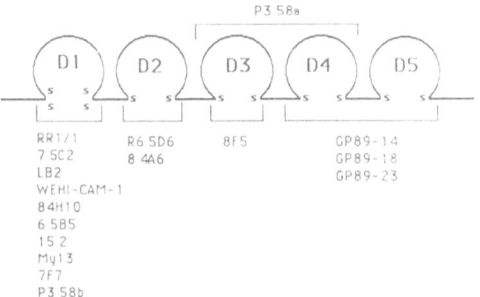

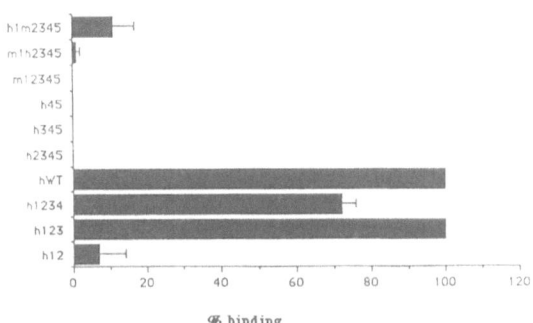

% binding

occurred the mAbs show cross-blocking on the epitope map. We therefore think it is likely that the mutations, rather than disrupting overall conformation, identify critical residues for these overlapping epitopes. It is interesting to note that in a recent homologue-scanning study on the interaction of human rhinovirus with ICAM-1 (17), one of the blocking mAbs was sensitive to the mutations K40/D and L43/E, which is remarkably similar to the behavior of the LB2 cluster with respect to K40/D and L43LPGN/ES-GP.

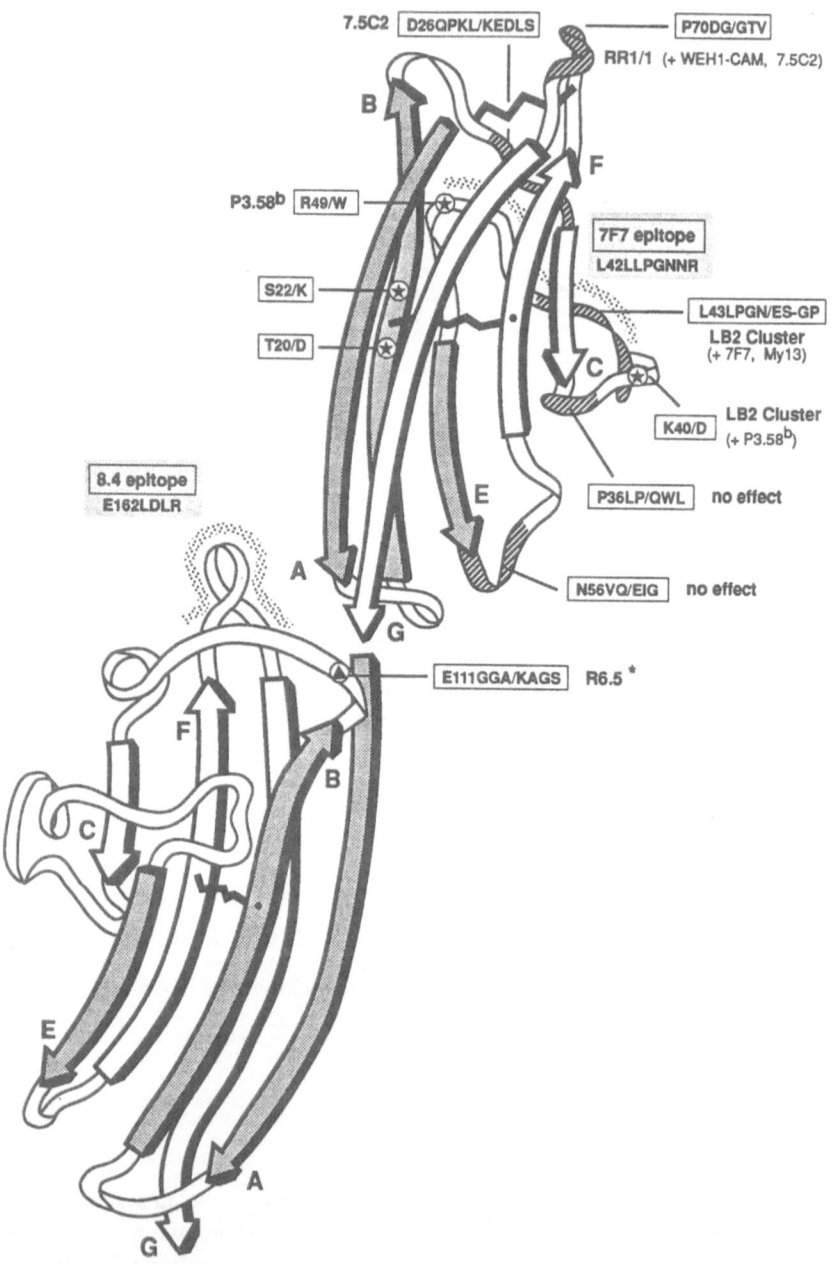

FIGURE 9.4. Model of domains 1 and 2 of ICAM-1. A model of the N-terminal two domains of ICAM-1. The ribbon diagram shows regions subjected to mutagenesis (boxed) and the mAb epitopes affected by each mutant. Single point mutations are shown by the symbol * and deletion/insertion mutations by hatching. The β-strands are labeled in the normal convention for Ig-folds, A to G;

TABLE 9.1. Binding of mutated ICAM-1 expressed in COS-1 cells to ICAM-1 mAbs and *P. falciparum*–infected erythrocytes.

Mutation	Epitope loss	% WT expression	% WT binding	
			Infected erythrocyte binding	
T20CS/DCK	Nil	267 ± 119	**4 ± 3**	(n = 3)
D26QPKL/KEDLS	7.5C2	176 ± 18	104 ± 40	(n = 2)
P36LP/QWL	Nil	160 ± 66	**30 ± 8**	(n = 5)
K40/D	LB2 cluster, P3.58[b]	182 ± 25	**27 ± 13**	(n = 4)
L43LPGN/ESGP	LB2 cluster 7F7, My13	165 ± 24	**1 ± 1**	(n = 3)
R49/W	P3.58[b]	93 ± 19	99 ± 19	(n = 4)
N56VQ/EIG	Nil	188 ± 65	77 ± 18	(n = 5)
P70DG/GTV	RR1/1 (7.5C2, WEHI–CAM-1)	112 ± 23	64 ± 6	(n = 3)

Specific ICAM-1 epitope loss was determined by comparing immunofluorescence intensities of a panel of 12 anti-ICAM-1 mAbs on mutant and wild-type (WT) ICAM-1 COS cell transfectants. For a given mutant, y, the % expression of an epitope for a mAb, x, was calculated, using the domain 3–specific mAb 8F5 or the domain 2–specific mAb 8.4A6 as reporters, by the formula:

$$\frac{\text{MFI mAbx (mutant y)}}{\text{MFI reporter mAb (muant y)}} \times \frac{\text{MFI reporter mAb (WT)}}{\text{MFI mAbx (WT)}} \times 100$$

Complete epitope loss was determined as <10% of WT expression, partial loss (shown in brackets) was 40% (for WEHI-CAM-1) or 37% (for 7.5C2) and no loss 95-120% of WT expression. In experiments to assess binding of infected erythrocytes, level of expression was determined with the domain 3–specific mAb 8F5, and % of WT binding was calculated by comparing mutant binding with the predicted binding to WT ICAM–1 at the same MFI. Mock-transfected COS cells bound at 4.0 ± 3.3% (n = 9) of WT ICAM-1 levels. Binding is represented as % of WT binding, corrected for the level of expression, ±SEM. (n = number of experiments.) Figures in bold highlight significant reductions in binding. Reproduced, with permission, from *Cell* (ref. 16).

Analysis of binding of the mutants pointed to the same region as the malaria-blocking mAb epitopes. Mutations at P36LP/QWL and K40/D both reduced binding by 70%, and the mutation L43LPGN/ES-GP reduced adhesion by over 95%. (See Table 9.1.) An additional mutation, T20CS/DCK, affected no mAb epitopes but also reduced binding by 95%. This mutation was designed to mutate the polar threonine and serine residues which flank one of the predicted intradomain disulphide

◁————————————————————————————————

strands A, B, and E, which form one sheet, are shaded. The "CDR2-like" regions join strands C and E. Disulphide bonds are denoted by dark zigzag lines. The regions incorporating the 7F7 epitope, L42LLPGNNR in domain 1 and the 8.4A6 epitope in domain 2, are shown by overshadowing. The localization of the R6.5D6 epitope to residues 111–114 is from reference 28. Modified and reproduced, with permission, from *Cell* (ref. 16).

bridges and which modeling studies predicted were externally directed. This region is in close proximity with the 7F7, My13, and P3.58[h] epitopes, and thus appears to define a region of the ICAM-1 molecule which does not form an epitope for any of the currently identified mAbs but which plays a critical role in the binding site for infected erythrocytes.

Our data are in broad agreement with those of Ockenhouse, Staunton, and colleagues, who have analyzed a large panel of ICAM-1 mutants (18). In their study, which also identified domain 1 as critical, mutations at S16 and L18 had dramatic effects on binding. The conservative mutation T20CS/ACT did not affect binding implying that disruption of binding in our T20CS/DCK mutant is due to the charged residues introduced. They did not implicate the CDR2-like region in binding, and resolution of this discrepancy will probably require the analysis of further mutations in this region.

An additional feature of this study has been the integration of the data with a novel molecular model of the structure of the first two domains of ICAM-1. (See Fig 9.4.) ICAM-1 is a member of the immunoglobulin (Ig) superfamily and has been assigned to the C2 set because of its similarities to both constant (C) and variable (V) domains of Ig molecules (19). A critical alignment of the first two domains of ICAM-1 with the highly homologous murine ICAM-1 (MALA-2) (ref. 20) and human ICAM-2 (ref. 21) (60% and 34% identity, respectively), both of which bind LFA-1, was used to locate the conserved core of the domain—namely, the β-strands which fold in two sheets to form a β-sandwich. This alignment was further refined and anchored by including the V-region of the immunoglobulin REI and the second domain of CD4, the structures of which have both been resolved at the X-ray crystal level (22–24). A molecular model for the first domain of ICAM-1 has already been produced, using a similar approach, but based on a C-region (25). Differences between the two models would be expected to be small, but an important feature of our model is that the structure that links the C and E strands, which in a C-domain would be designated the D-strand and in a V-domain the C'-, C"-, and D-strands, is not built to run in either of the β-sheets and instead forms a structure almost at right angles to the plane of the sheets. It is this region of the molecule which forms the CDR2-like loop and which our studies implicate as being involved in adhesion of infected erythrocytes as well as forming part of the rather-larger LFA-1 footprint. The other region, from L16 to S22, which Ockenhouse et al and we have identified, lies on the B-strand, close to domain 2.

Our model also makes the prediction that domains 1 and 2 of ICAM-1 form a rigid unit stabilized by interdomain hydrophobic interactions, as has recently been found for CD4 (23,24). The first two domains of CD4 contain hydrophobic residues, unusual to immunoglobulin domains, which orient outward and interact to lock the two domains together. The alignment of ICAM-1, ICAM-2, and murine ICAM-1 with CD4 demonstrates conservation of these residues, and the model predicts that the key hydrophobic side-chains are similarly directed outward. We therefore propose a similar packing arrangement of the first two domains of ICAM-1 to that for CD4. The arrangement is fully compatible with the epitope map, with cross-blocking mAbs spatially close to each other. In the case of the domain 2-specific, LFA-1–blocking mAbs R6.5D6 and 8.4A6, cross-blocking with WEHI–CAM-1 and RR1/1 is explained by the assignment of these epitopes to regions of domain 1 which lie on the same face of the two-domain unit as the R6.5D6 and 8.4A6 epitopes.

The model thus indicates that the epitopes for LFA-1–blocking mAbs all lie on the apex and one face of the ICAM-1 molecule, which also contains residues already identified as being important in the direct interaction with LFA-1. This "LFA-1–binding" face is analogous to that which mediates contact in other Ig-superfamily interactions (CD2/LFA-3 [ref.26], CD4/class II [ref.27], V_H/V_L [ref.28]) and appears to be a primordial binding region of Ig-superfamily members. By contrast, the mutations that localize malaria-blocking mAbs and the mutations with direct effects on binding of infected erythrocytes all lie on the opposite face, at essentially 180° to the LFA-1–binding face. We would therefore predict that it will be possible, should it prove therapeutically desirable, to generate reagents such as mAbs, peptides, or other compounds based on the malarial binding site, which will disrupt binding of infected erythrocytes, but not of LFA-1, and hence preserve immune function. The fact that Ockenhouse and colleagues have demonstrated that soluble ICAM-1 and short peptides based on the V17–T20 sequence can block adhesion of infected erythrocytes (18) appears to validate this prediction, and opens up exciting and realistic possibilities of therapeutic intervention if adhesion to ICAM-1 proves to have a major role in the pathogenesis of severe malaria.

Conclusions

We have demonstrated that a critical component of the binding site for *Plasmodium falciparum*–infected erythrocytes on ICAM-1 lies in domain 1 and that this site is distinct from the LFA-1–binding site. Domain 2 also appears to play an important role, which may be direct or due to its conformational interaction with domain 1. Mutations in domain 1 have identified regions important in the direct binding of infected erythrocytes and in the binding of mAbs which block this interaction. A novel molecular model of the first two domains of ICAM-1 suggests that the region involved in adhesion of infected erythrocytes lies on the opposite face of the first domain to the LFA-1–binding site, so selective anti-adhesion therapy may be possible, if clinically indicated.

Acknowledgments. We thank all our colleagues in Oxford and London for support and helpful discussions in the course of this work and Don Staunton for discussion of unpublished data. This work was supported by the Wellcome Trust, the Imperial Cancer Research Fund, and the Lister Institute for Preventive Medicine. A.R.B. is a Lister Institute Research Fellow.

References

1. Miller, LH. 1969. Distribution of mature trophozoites and of *Plasmodium falciparum* in the organs of *Aotus trivigatus*, the night monkey. *Am. J. Trop. Med. Hyg.* 18:860.
2. MacPherson, G.G., Warrell, M.J., White, N.J., Looareesuran, S. and D.A. Warrell. 1985. Human cerebral malaria. A quantitative ultrastructural analysis of parasitized erythrocyte sequestration. *Am. J. Pathol.* 119:385.
3. Greenwood, B.M. 1990. Populations at risk. *Parasitol. Today* 6:188.
4. Newbold, C.I. 1990. Malaria—the path of drug resistance. *Nature* 345:202.
5. Hommel, M., David, P.D., Oligino. L.D. and J.R. David. 1982. Expression of strain-specific antigens on *Plasmodium falciparum* infected erythrocytes. *Parasite Immunol.* 4:409.
6. Biggs, B.-A., Gooze, L., Wycherley, K., Wollish, W., Southwell, B., Leech, J.H. and G.V. Brown 1991 Antigenic variation in *Plasmodium falciparum*. *Proc. Nat Acad. Sci. USA.* 88:9171.
7. Leech, J.H., Barnwell, J.W., Miller, L.H. and R.J. Howard 1984 Identification of a strain-specific malarial antigen exposed on the surface of *Plasmodium falciparum*–infected erythrocytes. *J. Exp. Med.* 159:1567.
8. Roberts, D.D., Sherwood, J.A., Spitalnik, S.L., Panton, L.J., Howard, R.J., Dixit, V.M., Frazier, W.A., Miller, L.H. and V. Ginsberg 1985 Thrombospondin binds falciparum malaria parasitized erythrocytes and may mediate cytoadherence. *Nature* 318:64.
9. Barnwell, J.W., Ockenhouse, C.F. and D.M. Knowles 1985 Monoclonal antibody OKM5 inhibits the in vitro binding of *Plasmodium falciparum*–infected erythrocytes to monocytes, endothelial and C32 melanoma cells. *J. Immunol.* 135:3494.
10. Berendt, AR, Simmons, DL, Tausey, J, Newbold, CI, and IL. Marsh. 1989. Intercellular adhesion molecule-1 is an endothelial cell adhesion receptor for *Plasmodium falciparum*. *Nature* 341:57.
11. Ockenhouse, C.F., Ho, M., Tandon, N.N., Van Seventer, G., Shaw, S., White, N.J., Jamieson, G.A., Chulay, J.D. and H.K. Webster 1991 Molecular basis of sequestration in severe and uncomplicated malaria: differential adhesion of *Plasmodium falciparum* infected erythrocytes to CD36 and ICAM-1. *J. Infect. Dis.* 164:163
12. Carlson J., Helmby, H., Hill, A.V.S., Brewster, D., Greenwood, B.M. and M. Wahlgren 1990 Human cerebral malaria: association with erythrocyte rosetting and lack of anti-rosetting antibodies. *Lancet* 336:1455.
13. Parravicini, C.L., Soligo, D., Cattoretti, G., Berti, E., Gaiera, G. and P. Biberfeld. Endothelial reactivity of CD31, CD36 and anito-GMP-140 (CD62) mAb in normal tissues an vascular neoplasms. In Knapp, W., Dorken, B., Gilks, W.R., Rieber, E.P., Schmidt, R.E., Stein, H. and von dem Borne, A.E.G.

(eds.) *Leukocyte Typing IV. White Cell Differentiation Antigens.* Oxford, Oxford University Press, p. 985.

14. McCarthy, S.A., Kuzu, I., Gatter, K.C. and R. Bicknell 1991 Heterogeneity of the endothelial and its role in organ preference of tumour metastasis *TiPS* 12:462.

15. Grau, G.E., Taylor, T.E., Molyneux, M.E., Wirima, J.J., Vassalli, P., Hommel, M. and P.H. Lambert 1989 Tumour necrosis factor and disease severity in children with falciparum malaria. *N. Engl. J. Med.* 320:1586.

16. Berendt, A.R., McDowall, A., Craig, A.G., Bates, P., Sternberg, M., Marsh, K., Newbold, C.I.N. and N. Hogg. 1992. The binding site on ICAM-1 for Plasmodium falciparum-infected erythrocytes overlaps, but is distinct from, the LFA-1-binding site. *Cell* 68: *in press.*

17. McClelland, A., deBear, J., Connolly Yost, S., Meyer, A.M., Marlour, C.W. and J.M. Greve 1991 Identification of monoclonal antibody epitopes and critical residues for rhinovirus binding in domain 1 of intercellular adhesion molecule 1. *Proc. Nat. Acad. Sci. USA* 88:7993.

18. Ockenhouse, C.F., Betageri, R., Springer, T.A. and D.E. Staunton. 1992. Plasmodium falciparum-infected erythrocytes bind ICAM-1 at a site distinct from LFA-1, Mac-1, and human rhinovirus. *Cell* 68: *in press.*

19. Williams, A.F., and A.N. Barclay 1988 The immunoglobulin superfamily—domains for cell surface recongition. *Annu. Rev. Immunol.* 6:381.

20. Horley, K.J., Carpenito, C., Baker, B., and F. Takei 1989 Molecular cloning of murine intercellular adhesion molecule (ICAM-1). *EMBO J.* 8:2889.

21. Staunton, D.E., Dustin, M.L. and T.A. Springer 1989 Functional cloning of ICAM-2, a cell adhesion ligand for, LFA-1 homologous to ICAM-1. *Nature* 339:61.

22. Epp, O., Lattman, E.E., Schiffer, M., Huber, R. and W. Palm 1975 The molecular structure of a dimer composed of the variable portions of the Bence-Jones protein REI refined at 2.0-Å resolution. *Biochemistry* 14:4943.

23. Wang, J., Yan, Y., Garrett, T.P.J., Liu, J., Rodgers, D.W., Garlick, R.L., Tarr, G.E., Husain, Y., Reinherz, E.L., and S.C. Harrison 1990 Atomic structure of a fragment of human CD4 containing two immunoglobulin-like domains. *Nature* 348:411.

24. Ryu, S.-E., Kwong, P.D., Truneh, A., Porter, T.G., Arthos, J., Rosenberg, M., Dai, X., Xuong, N.-H., Axel, R., Sweet, R.W., and W.A. Hendrickson 1990 Crystal structure of an HIV-binding recombinant fragment of human CD4. *Nature* 348:419.

25. Giranda, V.L., Chapman, M.S., and M.G. Rossman 1990 Modeling of the human intercellular adhesion molecule-1, the human rhinovirus major group receptor. *Proteins* 7:227.

26. Driscol, P.C., Cyster, J.G., Campbell, I.D. and A.F. Williams 1991 Structure of domain 1 of rat T lymphocyte CD2 antigen. *Nature* 353:762.

27. Fleury, S., Lamarre, D., Meloche, S., Dyu, S.-E., Cantin, C., Hendrickson, W.A. and R.-P. Sekaly 1991 Muatational analysis of the interaction between CD4 and class II MHC: class II antigens contact CD4 on a surface opposite the gp120-binding site. *Cell* 66:1037.

28. Edmundson, A.B., Ely, K.R., Abola, E.E., Schiffer, M. and N. Panagiotopoulos 1975 Rotational allomerism and divergent evolution of domains in immunoglobulin light chains. *Biochemistry* 14:3953.

Part 2
Structure and Function of the Selectins

10
The Selectins

TAKASHI KEI KISHIMOTO

Neutrophils represent a mobile frontline defense against microbial pathogens. By way of the circulatory system, the neutrophil can gain access to virtually any tissue. In addition to the circulating neutrophils, which account for more than half of the white blood cells, new recruits are readily available from the bone marrow. Neutrophils have a short life span of approximately 2–3 days. It is estimated that 80 million neutrophils are generated each minute. Neutrophils which localize to a site of infection become activated to kill or ingest the foreign pathogen in an antigen-nonspecific manner. A misguided neutrophil can cause great damage to healthy tissues. Thus two events in the neutrophil's short life must be tightly regulated: the site of neutrophil localization and the site of neutrophil activation. Adhesion molecules play a crucial role in neutrophil localization. Three major familes of adhesion molecules involved in neutrophil–endothelial cell interactions have been defined: the leukocyte integrins (LFA-1, Mac-1, and p150,95), the intercellular adhesion molecules (ICAM-1,-2,-3), and the selectins (L-selectin, E-selectin, and P-selectin). The leukocyte integrins are crucial in leukocyte localization, as evidenced by patients who are genetically deficient in the expression of all three molecules (reviewed in 1,2). These leukocyte adhesion deficiency (LAD) patients are highly suceptible to severe bacterial infections. Infected lesions are devoid of neutrophils, yet circulating neutrophil levels are elevated, suggesting a defect in the ability of these neutrophils to migrate through the endothelium. ICAM-1 is an inducible endothelial cell ligand for LFA-1 (3–5) and Mac-1 (6,7). ICAM-1 is a member of the immunoglobulin family, most closely related to other adhesion molecules, such as VCAM-1 (8) and NCAM. The leukocyte integrins has been extensively reviewed in the first proceedings of this workshop (9), and more recently by Arnaout (10), Springer (11), and Kishimoto and Anderson (2). This review will focus on the selectin family, which has figured prominently in the past several years.

Three recent developments have fueled the intense research on selectins: (1) The recognition of the selectin gene family. The genes encoding L-selectin (12,13), E-selectin (14), and P-selectin (15) were identified by independent groups yet published within months of each other. It was immediately apparent that these three seemingly unrelated molecules were members of a highly conserved gene family involved in leukocyte interactions with endothelium or platelets. (2) The identification of carbohydrate ligands for the selectins. A role for carbohydrate recognition by L-selectin had long been suspected by Rosen, Stoolman, and colleagues (40). The identification of a C-type lectin domain encoded by the L-selectin cDNA not only confirmed this hypothesis but also suggested that E-selectin and P-selectin would also bind specific carbohydrates. Numerous independent reports demonstrated that a sialylated, fucosylated carbohydrate, similiar or identical to the sialyl Lewis X antigen, was a ligand for E-selectin (16–20) and P-selectin (21–24). (3) Recognition of specific roles of selectins vs. integrins in leukocyte adhesion to endothelium. At the first international symposium on the structure and function of adhesion molecules, Smith (25) presented evidence that the leukocyte integrins were not involved in neutrophil adhesion to endothelium under shear stress—conditions which simulate venous flow but were crucial for transendothelial migration. It was originally proposed that L-selectin is involved in the initial binding of neutrophils to endothelium (adhesion under flow) (26,27). Recent data, from a number of different laboratories, suggest that all three selectins are involved in supporting neutrophil rolling and adhesion under flow. Subsequent stimulation of the neutrophil by chemotactic factors results in a transition in adhesion activity, resulting in CD18-dependent adhesion strengthening and transendothelial migration. This multistep model provides specific mechanisms for regulating neutrophil localization and activation.

Nomenclature

The nomenclature of the selectins was a source of great confusion, with as many as 10 different names to descibe L-selectin. In 1991 an ad hoc group of scientists, including those involved in the discovery, early characterization, and cloning of these molecules, agreed to adopt the "selectins" as a unified nomenclature (28). The term selectin reflects the role of these molecules in "selective" interactions of leukocytes with endothelium or platelets as well as their role as membrane "lectins." L-, E-, and P-selectin denote leukocyte-selectin, endothelial-selectin, and platelet-selectin, respectively, to reflect the first cell type in which these molecules were described. However, P-selectin is expressed on both activated platelets and endothelial cells. Previous names for this gene family include the LEC-CAMs.

L-selectin

L-selectin was first identified in the mouse as a peripheral-lymph-node homing receptor (PLNHR) or lymphocyte homing receptor (LHR or HR). It was defined by the MEL-14 MAb and thus commonly referred to as the MEL-14 antigen, or gp90^{MEL} (29). The Ly-22 antigen defines an allotype of the mouse L-selectin (30). The human homologue of L-selectin was identified by cDNA cloning and by MAb by several independent groups. It was referred to variously as HR, LAM-1, LECAM-1, or LEC-CAM-1. The human lymphocyte differentiation antigens Leu-8 and TQ-1 are identical to L-selectin (31,32).

E-selectin

E-selectin was first identified by MAb against an endothelial cell activation antigen (33). It was subsequently shown to mediate neutrophil adhesion and thus termed "endothelial-leukocyte adhesion molecule-1" (ELAM-1) (34).

P-selectin

P-selectin was described by two groups as a platelet granule protein, and termed "granule matrix protein–140" (GMP-140) (35,36) and "platelet-activation-dependent granule external membrane" (PADGEM) (37,38). P-selectin has a cluster desgnation (CD62) from the International Workshop on Human Leukocyte Differentiation Antigens.

Initial Characterization of the Selectins

L-selectin

L-selectin was first discovered by Gallatin et al. as a tissue-specific lymphocyte homing receptor (29). This topic has been extensively reviewed by others (39,40) and will only briefly be discussed here. Early in vivo studies showed that lymphocytes migrate into lymph nodes in a nonrandom fashion. Perhaps the most striking example is the tissue-specific migration of some lymphoid cell lines to peripheral vs. mucosal lymphoid tissues. Lymphocytes enter lymphoid tissue via specialized high endothelial venules, which are characterized by plump, cuboidal endothelial cells. Determination of the molecular basis for this interaction was made possible by the Stamper-Woodruff assay (41), an ex vivo adhesion assay to measure specific lymphocyte binding to HEV structures on frozen, thin sections of lymphoid tissues. The MEL-14 MAb against L-selectin defined a peripheral-lymph-node–specific lymphocyte homing receptor (29). Anti-L-selectin MAb specifically blocks lymphocyte binding to peripheral but not mucosal HEV in the Stamper-Woodruff assay, and blocks lymphocyte migration to peripheral but not mucosal lymph nodes in vivo.

Rosen, Stoolman, and colleagues made an insightful hypothesis that adhesion that can be measured under the conditions of the Stamper-Woodruff assay might involve carbohydrate recognition (40). Over the last 15 yr, they have provided strong evidence for this model. Lymphocyte adhesion to HEV could be blocked by treating the lymph-node sections with neuraminidase to remove terminal sialic-acid residues (42,43). Intravenous administration of a sialidase disrupts lymphocyte homing in vivo (44). Moreover, lymphocyte–HEV adhesion could be inhibited by monosaccharides, such as mannose-6-phosphate but not mannose-1-phophate, albeit at relatively high concentrations (45,46). Large polysaccharides containing sulfated fucose (fucoidin) or mannose-6-phosphate (PPME, a yeast mannan core polysaccharide) blocks adhesion and binds directly to L-selectin (45,47). Most recently, the use of purified L-selectin or an L-selectin–IgG chimera has allowed direct demonstration of a lectin activity (43,48–50).

The distribution of L-selectin on lymphocytes correlates with the ability of lymphoid cell lines to bind to peripheral-lymph-node HEV. Curiously, however, L-selectin is also expressed on virtually all monocytes, neutrophils, and eosinophils (51). These myeloid cells do not normally traffic to peripheral lymph nodes, raising the obvious questions of what the role of L-selectin is on myeloid cells and why these cells do not migrate to lymph nodes. The first question was addressed by the early study of Lewinsohn et al. (51), which showed that in vivo administration of anti–L-selectin MAb inhibits neutrophil localization to cutaneous sites of inflammation. Jutila et al. further demonstrated that neutrophil migration to the inflamed peritoneum is inhibited to the same extent by in vivo administration of antibodies against Mac-1 or L-selectin (26). Watson et al. found that an L-selectin–IgG chimeric molecule, containing two L-selectin extracellular domains genetically engineered onto the Fc portion of immunoglobulin, directly blocks neutrophil response to acute inflammation (52). Although neutrophils do not normally migrate to peripheral lymph nodes, they do bind specifically to HEV in the Stamper-Woodruff assay (26,51). These results suggest that in vivo a second signal, perhaps a chemotactic factor, is required for neutrophil migration across the endothelium. This factor would be present in an acute inflammatory lesion, but not in normal lymphoid tissues.

E-selectin

The development of techniques to culture endothelial cells in vitro was a crucial step for the major advances in the study of leukocyte–endothelial cell interactions (53). Human umbilical cord provided an abundant, easily accessible source of human vascular endothelial cells. The laboratories of Gimbrone and Pober showed that inflammatory cytokines, such as I1-1, IFN-y, and TNF, induced profound changes in clutered endothelial cells, including induction of class II MHC antigens (in response to IFN-y), procoagulatant activity, and, perhaps most strikingly, a dramatic increase in the ability of endothelial cells to support neutrophil adhesion (33,54–56). These studies indicated that activation of endothelial cells by cytokines induces expression of cell adhesion molecules. Bevilacqua and colleagues raised MAb against stimulated endothelium and defined E-selectin as an inducible cell adhesion molecule (34). MAbs against E-selectin specifically block neutrophil and myeloid cell line (HL-60) adhesion to Il-1– and TNF-stimulated endothelial cells (34,57). E-selectin expression is prominent in acute inflammatory lesions in vivo and correlates with the large influx of neutrophils (58–62). More recent studies have demonstrated that in vivo administration of MAb against E-selectin inhibits neutrophil accumulation in the lung in a primate model of late phase response (63) and in a rat model of immune-complex–mediated lung injury (64). Although E-selectin appears to be primarily associated with acute inflammatory lesions, where it can be induced almost anywhere, recent studies have shown E-selectin expression in some chronic inflammatory lesions, notably the inflamed skin (61,65) and the synovium from patients with arthritis (66). Furthermore, a small subset of memory T lymphocytes, defined by the HECA-452 MAb, bind specifically to E-selectin (65). These studies indicate that in some circumstances, E-selecin can mediate lymphocyte traffic to chronic inflammatory sites.

P-selectin

P-selectin was first defined as a marker for activated platelets (35,37). P-selectin was localized to the α-granules of platelets (36,38) and later to the Weibel-Palade bodies of endothelial cells (67–69). In both cell types, cell activation results in a rapid recruitment of these granules to the cell surface (36,38,69). These observations had suggested an important role for P-selectin in a rapid cellular reponse; however, the precise role of P-selectin was not elucidated until more recently. The identification of the selectin gene family suggested that P-selectin, like L-selectin and E-selectin, would be involved in cell adhesion. McEver, Furie, and their respective colleagues quickly demonstrated that P-selectin is involved in mediating neutrophil–platelet interactions (21,70,71) and in neutrophil–endothelial cell interactions in vitro (72–74). These results indicate that P-selectin may play a central role in bridging hemostasis and acute inflammation very early in the response to vascular injury or insult. Although MAb against P-selectin has been used to image thrombi in vivo, there are no in vivo studies demonstrating inhibition of inflammation or hemostasis (75).

Structural Features of the Selectins

The near-simultaneous cloning of the genes encoding the selectins (12–15,76–78) and the realization that these are closely related adhesion molecules fueled the recent and intense interest in these molecules. All of the selectin genes share common structural features—most prominently, an N-terminal C-type lectin domain (about 120 amino acids), followed by a domain homologous to the epidermal growth factor (about 30 amino acids), a variable number of short consensus repeats (SCR)—which is a motif found in many complement regulatory proteins (each about 62 amino acids), a conventional membrane-spanning domain, and a C-terminal cytoplasmic domain (Fig. 10.1). This is a highly conserved gene family—over 60% amino acid identity in the lectin and EGF domains among the three family members. As discussed below, the lectin domain is a central figure in the carbohydrate-binding properties of all three selectins. The lectin domain motif belongs to the C-type lectin family described by

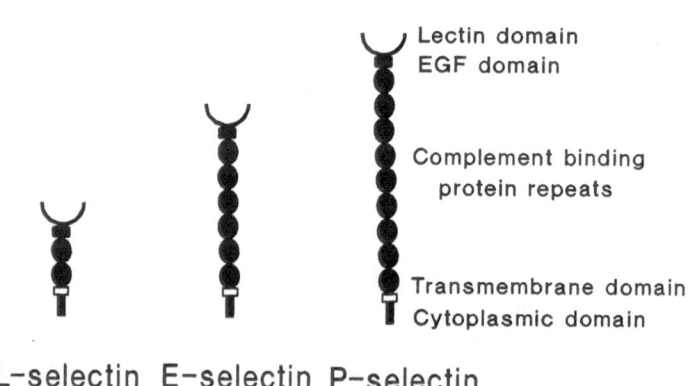

FIGURE 10.1. The selectin family of adhesion molecules.

Drickamer et al. (79). This Ca^{2+}-dependent lectin domain accounts for, at least in part, the requirement of Ca^{2+} in adhesion mediated by all three selectins. The SCRs are less well conserved—about 40% amino acid identity—but they differ from SCRs found in the C3/C4-binding proteins of the complement system in having six cysteine residues per SCR vs. four cysteine residues found in most other SCRs. Some of the selectins have alternatively spliced forms and distinct post translational modifications, which will be discussed below.

L-selectin

The genes encoding both mouse (12,13) and the human L-selectin (32,76–78) have been cloned by several independent groups. L-selectin is the smallest member of the selectin family, with most of the size difference attributed to having only 2 SCRs (vs. 6 for E-selectin and 8–9 for P-selectin). Mouse and human L-selectin are highly conserved, with 77% amino-acid identity (32,76–78). The finding of an N-terminal lectin domain encoded by the gene is a tribute to Rosen and colleagues, who for many years had predicted a direct role for L-selectin in carbohydrate binding (40). Not surprising, structure–function analysis has demonstrated that the carbohydrate-binding activity is asscociated with the lectin domain (80,81). MAbs which define functional epitopes of L-selectin map to the lectin domain or the EGF domain (30,80,81). Siegelman et al. (30) demonstrated that Ly22 MAb recognizes an allotypic epitope of L-selectin. Sequence analysis showed that a single amino-acid substitution in the EGF domain accounts for the Ly22 allotype. The Ly22 MAb specifically blocks lymphocyte adhesion to peripheral-node HEV but does not block binding of PPME. Thus the EGF domain appears to be important either directly for binding ligand, perhaps to a noncarbohydrate epitope, or indirectly for providing structural support for the lectin domain. Molecular chimeric molecules also define the lectin and EGF domains as the minimal structures required for ligand binding (80,81). The role of the SCRs is not known; however, one report suggests that they too enhance ligand-binding affinity by providing a structural scaffolding for the lectin–EGF domains (82). Activated leukocytes rapidly down-regulate L-selectin (27). The cleaved molecule, recovered from the cell supernatant, is slightly smaller than the intact form, suggesting the presence of at least one major protease cleavage site close to the transmembrane domain. No specific functions have been attributed to the transmembrane or cytoplasmic domains, although it is notable that these regions are highly conserved between mouse and human L-selectin.

In addition to the classic transmembrane form of L-selectin, Camerini et al. described a cDNA encoding a PI-linked form of L-selectin (32). They suggest that the PI-linkage might be one mechanism with which to achieve rapid shedding of L-selectin. However, the PI-linked form of L-selectin remains controversial. Patients with PNH, who have a defect in expression of PI-linked proteins, show normal expression of L-selectin and normal shedding upon activation (83). In addition, Tedder and colleagues have analyzed the genomic organization of L-selectin and find no evidence of an alternatively spliced mRNA encoding a PI-linked form (77).

The L-selectin cDNA encodes a protein with a predicted molecular mass of 37,000 daltons. It is apparent that L-selectin undergoes substantial posttraslational modification. Most strikingly, the relative molecular mass of L-selectin from lymphocytes is significantly different from that of neutrophils or monocytes (51). Yet the primary sequence of L-selectin from lymphocytes and neutrophils is the same (77). It has been speculated that this difference in molecular mass might account for some of the difference in the migration characteristics of lymphocytes (which home to peripheral lymph nodes) and neutrophils (which home to acute inflammatory lesions) (51). At least one significant difference is that neutrophil L-selectin displays the SLeX carbohydrate antigen, a ligand for ELAM-1 (see below), while lymphocyte L-selectin does not (84).

There are also several reports suggesting that L-selectin is physically associated with ubiquiton (85,86). Ubiquiton, as the name suggests, is expressed ubiquitously in many cell types. Normally, ubiquiton is associated with intracellular proteins, and this association is thought to direct protein degradation. It is interesting to speculate as to what role ubiquitin may have in L-selectin function or regulation.

E-selectin

The E-selectin gene encodes a protein of 610 amino acids, including six SCR repeats (14,87). The mature E-selectin protein has an Mr of 90,000, suggesting heavy glycosylation. MAbs which inhibit E-selectin function also map to the lectin-EGF domains.

P-selectin

P-selectin is the largest of the selectin family, with nine SCRs as a major product, as well as an alternatively spliced form with eight SCRs (15). The mature protein, of Mr = 140,000 daltons, is heavily glycosylated, like the other selectins. Presumably, the P-selectin transmembrane or cytoplasmic domain encodes a signal for targeting P-selectin to the Weibel-Palade bodies of the endothelium and α-granules of platelets. In addition to the classic transmembrane form, Johnston et al. (15) have identified a cDNA encoding a soluble form of CD62, which lacks the transmembrane and cytoplasmic domains. The functional significance of this soluble form is not known.

Genomic Organization

The genomic structure of all three selectins have been elucidated (87–89) and they are strikingly similar. The lectin domain, EGF domain, and each SCR are encoded by separate exons. Moreover, all three genes are located within 300 k bases of each other on the long arm of chromosome 1, bands q21–24 (90). These data suggest that the selectin mosaic structure arose from gene shuffling of exons from disparate proteins, and that gene amplification resulted in three closely related family members.

Regulation of Expression

Adhesive interactions of immune cells must be transient in nature. Thus the expression and functional activity of adhesion receptors must be tightly regulated. Interestingly, despite the fact that the selectin genes are closely related and physically linked on chromosome 1, the selectins have different tissue distributions and utilize several fundamentally distinct strategies to modulate expression. These observations suggest that distinct regulatory strategies evolved with each of the selectins to allow futher specialization.

L-selectin

L-selectin is constitutively expressed on resting neutrophils in a seemingly functional form. Freshly isolated neutrophils can bind to stimulated endothelium at a reduced temperature (4–7°C) in vitro (91,92). However, within minutes of neutrophil exposure to low levels of chemotactic factors, L-selectin is rapidly down-regulated from the cell surface (27). Near complete down-regulation of L-selectin can be detected within minutes in vitro. A large fragment of L-selectin can be recovered from the supernatant of activated cells, suggesting that L-selectin is proteolytically clipped close to the transmembrane domain (27). A broad range of activating agents including C5a, fMLP, TNF, GM-CSF, and Il-8 are effective at inducing this response (26,27,93,94). Interestingly, the rapid shedding of L-selectin follows the kinetics of Mac-1 up-regulation from intracellular stores.

   Analysis of neutrophils which are recovered from the inflamed peritoneum in vivo (26) and immunohistological analysis of neutrophils in inflamed skin sites (27) suggest that this inverse regulation of adhesion molecules occurs in vivo as well. These observations led to the proposal of a two-step adhesion model (see below).

   Lymphocytes and monocytes can shed L-selectin upon activation, although the kinetics are significantly slower (95–97). Regulation of L-selectin on lymphocytes is more complex. Although activated lymphocytes have down-regulated L-selectin levels, L-selectin can be reexpressed on memory lymphocytes (98). Moreover, expression of L-selectin is thought to contribute to the tissue-specific homing patterns of memory cells (reviewed in 39,40).

A putative protease involved in cleaving L-selectin from the cell surface of activated cells has not been identified. Selective proteolysis of L-selectin can be induced by low levels of exogenous chymotrypsin (99). Chymotrypsin-treated neutrophils fail to home to inflammatory sites in vivo, although it is certainly possible that other chymotrypsin-sensitive proteins are involved. It is not clear that chymotrypsin precisely mimics the cleavage of the endogenous protease, since classic chymotrypsin inhibitors, such as TPCK, PMSF, and aprotinin, do not inhibit L-selectin down-regulation by activated neutrophils.

A second major means by which L-selectin function is regulated involves an increased affinity for ligand. Tedder and colleagues have reported that just prior to L-selectin shedding, there is a transient increased affinity for carbohydrate ligands, such as PPME (100), and adhesion to HEV. This increased affinity is presumably transmitted through a conformational change in the L-selectin extracellular domain. This "inside-out" signaling is reminiscent of that seen with the leukocyte integrins (101). How this type of signaling is transmitted is not known.

E-selectin

E-selectin is normally absent from endothelial cells. However, upon stimulation with inflammatory cytokines, endothelial cells express E-selectin within several hours. E-selectin is synthesized de novo and is blocked by protein synthesis inhibitors (34). This up-regulation of E-selectin is similar to that seen with other endothelial adhesion molecules, such as ICAM-1 and VCAM-1. However, in contrast to these other adhesion molecules which remain highly expressed for over 24 h, E-selectin expression peaks at 3–4 h and then is down-modulated by 8–24 h in vitro (34,102). The mechanism of E-selectin down-modulation is not well characterized. The time course of E-selectin expression is similar to the time course of neutrophil infiltration into acute inflammatory sites in vivo. These results suggest that E-selectin is involved primarily in the acute inflammatory response. E-selectin expression is also rapidly inducible in vivo and coincides with the influx of neutrophils (58–62,103). However, in some chronic inflammatory lesions, notably some inflamed skin and synovial sites, E-selectin expression is quite prominent (58,65,66,103). Recent studies suggest that monocytes, as well as a subpopulation of memory T cells (66,104), defined by the HECA-452 antigen (66), can bind selectively to E-selectin. These studies indicate that there is a distinct but as-yet-undefined mechanism that induces chronic E-selectin expression in some inflammatory responses.

P-selectin

P-selectin, like E-selectin, is expressed by activated endothelium. However, the activation signals and the kinetics of expression are completely different for these two events. P-selectin is stored in intracellular Weibel-Palade bodies (67–69). Stimulation of endothelial cells with histamine or thrombin induces a rapid recruitment of Weibel-Palade bodies to the cell surface, resulting in surface expression of P-selectin within minutes after stimulation. Similarly, P-selectin is also stored in the α-granules of platelets, which are recruited to the cell surface within seconds of platelet activation (36,38). This near-instantaneous up-regulation of P-selectin suggests a critical role in the earliest events of inflammation and hemostasis. Surface expression of P-selectin on endothelium is also extremely transient. Within 30 min, P-selectin is down-modulated, apparently by receptor endocytosis rather than surface proteolysis. Low levels of hydrogen peroxide induce prolonged expression of P-selectin; however, this may be at least in part associated with endothelial cell damage (73). However, Stoolman and colleagues find that chronic levels of P-selectin may be important in some disease states, such as arthritis (chapter 16).

Ligands for the selectins

L-selectin

L-selectin was first described as a lymphocyte homing receptor, so it is not surprising that the most intense search for an L-selectin ligand has been on high endothelial venules of lymph nodes. In a series of elegant studies, Rosen and colleagues demonstrated a clear role for carbohydrates in L-selectin–mediated adhesion, long before the lectin domain of L-selectin was known (40). Lymphocyte adhesion to HEV is sensitive to neuraminidase treatment of the HEV (42,43). Charged sugars, such as mannose-6-phosphate, and polymers of charged sugars such as PPME (mannose-6-phosphate-rich) and fucoidin (fucose-sulfate-rich), bind specifically to L-selectin and block lymphocyte adhesion to HEV (45–47). Rosen, Lasky, and colleagues have recently utilized soluble L-selectin, in the form of a bivalent L-selectin–IgG chimeric molecule, to immunopurify a 50-kd protein from lymph-node tissue. This protein is heavily glycosylated and contains sulfated sugars (50). Recognition of this 50-kd protein by the L-selectin–IgG chimera is sensitive to neuraminidase treatment.

In a parallel line of research, Butcher and colleagues have developed a MAb, MECA-79, which specifically stains peripheral-lymph-node HEV, but not mucosal HEV, and blocks lymphocyte adhesion to peripheral-node HEV (105). The MECA-79 MAb is an IgM and appears to recognize a carbohydrate determinant. Western blot analysis and immunoaffinity purification of reactive antigen shows several distinct bands—a major band of 105 kd, with minor bands of 65, 90, 150, and 200 kd. The purified MECA-79 antigen supports tissue-specific lymphocyte binding, which is blocked by both MECA-79 and by anti–L-selectin MAb. Interestingly, the major 50-kd product isolated by with the L-selectin–IgG chimeric molecule by Rosen et al. also immunoreacts with the MECA-79 MAb.

A ligand for L-selectin on stimulated HUVEC has not been defined. The MECA-79 MAb, which cross-reacts with human-lymph-node HEV, does not appear to stain stimulated HUVEC. Yet L-selectin–dependent adhesion of neutrophils to stimulated HUVEC has been demonstrated by several independent groups. There is general consensus that L-selectin-dependent neutrophil adhesion to HUVEC occurs only with cytokine stimulation of the endothelium (91–93). The adhesion occurs readily at low temperature and appears to be enhanced by mild, rotational shear force (91,92). Spertini et al. (92) have shown that neuraminidase treatment of the stimulated endothelial cells inhibits L-selectin–mediated adhesion. There are conflicting reports as to whether L-selectin–dependent adhesion peaks at 4 h and then is down-modulated by 8–24 h (106) or whether it is sustained for over 24 h (92). Curiously, in vivo intravital microscopy studies in normal animals demonstrate L-selectin–dependent adhesion in the absence of any applied cytokine stimulation (107,108). These results suggest that at least one L-selectin ligand is constitutively expressed or is rapidly inducible (within minutes) simply by the anesthesia or surgical manipulation necessary to prepare the animal for intravital microscopy.

E-selectin

The identification of a lectin domain in the E-selectin structure led to an intense, focused search for a carbohydrate ligand expressed by myeloid cells. Numerous independent groups reported that sialyl Lewis X (SLeX), or related sialylated, fucosylated sugars, serve as specific ligands for E-selectin (Fig 10.2). SLeX is appropriately expressed by neutrophils and monocytes. Several complementary lines of evidence support these claims. Phillips et al. demonstrated that liposomes composed of glycolipids containing the SLeX structure are capable of inhibiting E-selectin-mediated adhesion (16). Furthermore, the LEC11 cell line, a variant of CHO cells, expresses SLeX and binds to HUVEC in an E-selectin-dependent manner. MAbs against the SLeX determinant inhibit E-selectin–mediated adhesion (16,17). Lowe et al. demonstrated that transfection of a 1,3; 1,4 fucosyltransferase into COS or CHO cells enables these cells to synthesize SLeX and to bind E-selectin (18). This particular fucosyltransferase is not expressed in myeloid cells, and thus is not likely

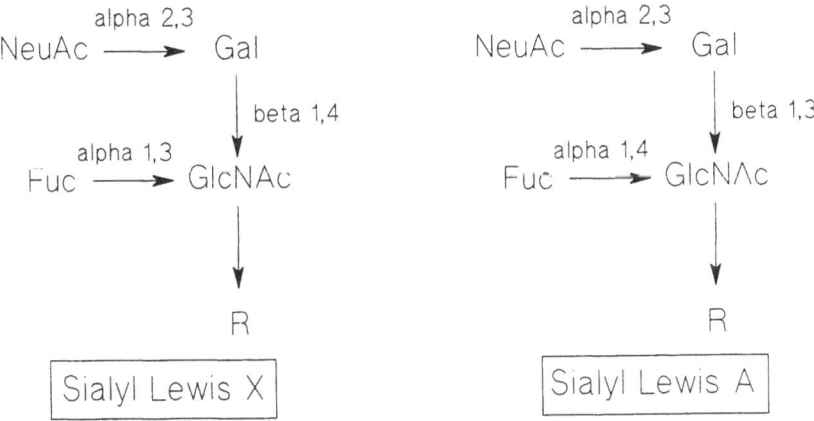

FIGURE 10.2. Sialyl Lewis X and sialyl Lewis A: carbohydrate ligands of E-selectin.

to be the physiologically relevant fucosyltransferase. Lobb et al. screened an expression library, prepared from HL-60 myeloid-cell mRNA, and identified a myeloid-specific 1, 3 fucosyltransferase which allowed transfected COS and CHO cells to bind E-selectin (19). However, there is some confusion as to whether this is the entire story, because other groups have reported that transfection of this gene alone does not enable COS or CHO cells to synthesize SLeX or bind E-selectin (109). Brandley et al. purified glycolipids from myeloid cells and identified fractions which support E-selectin adhesion (110). They identified vim-2, another sialylated, fucosylated structure, as an E-selectin ligand. However, it appears that vim-2 is not a major ligand of E-selectin (17,109). More recently, sialyl Lewis A, an SLeX-related carbohydrate, has been implicated as a ligand for E-selectin (20,111). Computer modeling suggests that the sialic acid and fucose residues are oriented in the same manner in both SLeX and sialyl Lewis A. Both SLeX and sialyl Lewis A are recognized by the HECA-452 MAb which defines a subpopulation of lymphocytes capable of binding to E-selectin (20).

P-selectin

A similar search for carbohydate ligands of P-selectin followed the recognition that P-selectin possessed an N-terminal lectin domain. Initially it was reported that Lewis X antigen (CD15) was a major ligand for P-selectin (21). However, several groups have shown that P-selectin–dependent adhesion is sensitive to neuraminidase treatment of the target cell, suggesting a requirement for sialic acid (22–24). Indeed, P-selectin shows significantly higher affinity for sialyl Lewis X than for Lewis X

(111). Although both E-selectin and P-selectin bind to SLeX, the binding characteristics are distinct (111). Thus E-selectin and P-selectin are not identical in ligand specifity. More recently, Aruffo et al. (112) have reported that soluble P-selectin–IgG chimeric molecules bind specifically to sulfatides (3-sulfated galactosyl ceramides) derived from myeloid cells and some tumor cell lines. Treatment of HL-60 cells with selenate, an inhibitor of sulfation, reduces binding to P-selectin but not to E-selectin. These studies suggest that sulfatides may be a major ligand for P-selectin, but not for E-selectin. Interestingly, neutrophils synthesize and excrete large amounts of sulfatides, a mechanism which may regulate deadhesion (112).

Noncarbohydrate Determinants in Selectin-Mediated Adhesion

Defining the fine carbohydrate structures involved in selectin ligands has been a major breakthrough. While it is clear that SLeX, either on proteins or lipids, can support E-selectin–mediated adhesion in vitro, it is unclear whether all SLeX on the neutrophil cell surface can bind with equal efficiency and affinity to E-selectin. One possibility is that SLeX may be added to a variety of proteins and lipids, but only a subset is presented in a favorable conformation, thus creating a hierarchy in which E-selectin preferentially binds to a subset of the total available SLeX. This hierarchy might reflect the accessibilty of the SLeX or the protein sequences adjacent to the SLeX. Based primarily on MAb blocking data, we have found evidence that L-selectin and E-selectin appear to operate in the same CD18-independent adhesion pathway (106). Anti–L-selectin and anti–E-selectin MAb have nonadditive blocking effects, while both are additive in combination with anti-CD18 and anti–ICAM-1 MAb. Picker et al. (84) extended these studies and showed that L-selectin isolated from neutrophils but not from lymphocytes bears the SLeX carbohydrate and can support E-selectin–dependent adhesion. Moreover, neutrophils treated with chymotrypsin, which cleaves L-selectin from the cell surface but does not significantly affect overall SLeX levels, bind poorly to E-selectin. Similarly activated neutrophils, which have significant SLeX on their surface, bind significantly less than unstimulated neutrophils to E-selectin. Moore et al. (24) have found that protease treatment of neutrophils completely eliminates binding of fluid-phase P-selectin, even though significant SLeX remains on glycolipids. Patel et al. (73) observed that both neutrophils and HL-60 cells, which express SLeX, bind to TNF-stimulated endothelial cells expressing E-selectin; however, only neutrophils, not HL-60 cells, bind to peroxide-stimulated endothelial cells expressing P-selectin (73). Thus there are several clear examples in which the expression of SLeX is not sufficient for optimal binding to E- and P-selectin. Siegelmann et al. have hypothesized that the lectin domain of the selectins provides the carbohydrate specifity while the EGF domain may bind to protein determinates (30). Thus it may be possible to observe adhesion to carbohydrate alone in vitro, but optimal adhesion activity may require both protein and carbohydrate determinants. What specific protein determinants are necessary still remains to be resolved.

A Model for Neutrophil-Endothelial Cell Interactions

In 1989, a two-step adhesion model for neutrophil interaction with endothelium was proposed (Fig. 10.3), based primarily on the observation that Mac-1 and L-selectin are inversely regulated by exposure to chemotactic factors (26,27, reviewed in 113,114). The rapid down-regulation of L-selectin with a concomitant up-regulation of Mac-1 suggested that these adhesion molecules mediate distinct but complementary adhesion events. We had proposed that L-selectin mediates the initial interaction of the resting, circulating neutrophil with the activated endothelium, thus guiding the unstimulated neutrophil to the appropriate site of inflammation (26,27). This intial binding or rolling event would slow the neutrophil down and expose it to low levels of chemotactic factors released at the site of inflammation. The chemoattractants would provide the signal for the neutrophil to enter the inflamed tissue and trigger the transition from L-selectin–mediated adhesion to CD18-mediated adhesion. The CD18 integrins, LFA-1 and Mac-1, are largely responsible for subsequent adhesion strengthening, neutrophil aggregation, and transendothelial migration. The appeal of this model is that it provides a mechanism to closely

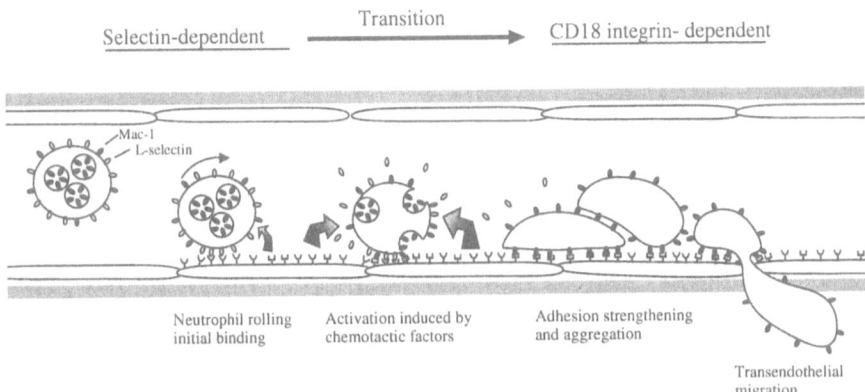

FIGURE 10.3. A dynamic model for neutrophil interaction with endothelial cells at sites of inflammation reprinted with permission from the Journal of NIH Research, Washington D.C. (Kishimoto et al. ref #113).

regulate neutrophil localization and neutrophil activation, thus ensuring minimal damage to surrounding healthy tissue. Data from a number of independent investigators support and extend this model.

Adhesion Under Flow

In venules, neutrophils must resist the shear forces of the flowing blood. Smith, McIntire, and colleagues developed an in vitro flow chamber to measure neutrophil adhesion to cultured endothelial cells under conditions which simulate blood flow (115,116). They made the significant observation that neutrophil adhesion to endothelium, under a defined wall shear stress of 2 dynes/cm^2, is a CD18/ICAM-1–independent event (116). Neutrophils from patients with a genetic deficiency in CD18 expression adhere as well as normal cells under conditions of flow. However, the CD18-deficient neutrophils fail to transmigrate across the endothelial monolayer. These observations suggested that neutrophil adhesion to endothelium has both CD18-dependent and CD18-independent components. Similarly, Arfors et al. utilized intravital microscopy to demonstrate that anti-CD18 MAb inhibited neutrophil entry into a site of inflammation but did affect neutrophil rolling (117).

Smith et al. then showed that anti–L-selectin MAb significantly reduced neutrophil adhesion and rolling under conditions of flow (93). Activation also reduced neutrophil adhesion under flow, consistent with the known down-regulation of L-selectin from activated neutrophils. Significantly, anti–L-selectin MAb did not affect the transmigration of those neutrophils which did bind to the endothelium (93), although Spertini et al. (92) do find a L-selectin component in transmigration. Neonatal neutrophils, which have diminished levels of L-selectin, show reduced adhesion under flow conditions (118). In addition, canine neutrophil adhesion under flow to canine endothelium is specifically reduced by anticanine L-selectin MAb (119).

Two independent groups have also evaluated the role of L-selectin in mediating neutrophil rolling and adhesion in vivo using intravital microscopy. Ley et al. found that a polyclonal rabbit serum which reacts with L-selectin significantly reduces neutrophil rolling (108). Furthermore, soluble L-selectin, in the form of an L-selectin–IgG chimeric molecule, also blocked neutrophil rolling in vivo, while a CD4-IgG chimeric control had no effect. Moreover, the inhibition is reversible if the antagonist is no longer administered. Similarly von Adrian et al. found that the DREG-200 MAb, which cross-reacts with rabbit L-selectin, inhibits the number of neutrophils rolling as well as their velocity of rolling (107).

Since L-selectin is involved in neutrophil rolling and adhesion under flow, it seems reasonable to propose that the other selectins, P-selectin and E-selectin, may also mediate neutrophil rolling. Lawrence et al.

showed that purified P-selectin, incorporated into planar lipid membranes, supports neutrophil rolling in a flow chamber in vitro (120). Similarly, we have shown that mouse L-cell fibroblasts, transfected with the E-selectin gene, support neutrophil rolling and adhesion under flow to levels comparable to that seen with Il-1–stimulated endothelial cells (Abbassi, Kishimoto, McIntire, and Smith, submitted). The interaction of neutrophils with the E-selectin transfectants was completely blocked with anti–E-selectin MAb. Significantly, ICAM-1–transfected L cells did not support neutrophil rolling or adhesion under flow. Further intravital microscopy studies will be required to demonstrate E-selectin and P-selectin involvement in neutrophil rolling in vivo.

Neutrophil Aggregation and Transendothelial Migration

A hallmark of the acute inflammatory reponse is the accumulation of neutrophils and their entry into the inflamed tissue. Some vessels at sites of inflammation are so filled with neutrophils that the vessels become occluded. This aggregation of the neutrophils presumably helps to slow blood flow and allow for further neutrophil accumulation. Neutrophils enter the inflamed tissue by first migrating between endothelial junctions and then through the basement membrane. Both neutrophil aggregation (2,10,121,122) and transendothelial migration (6,25,57,121–123) are CD18 integrin-dependent events. Neutrophils treated with anti-CD18 MAb and neutrophils isolated from CD18-deficient patients fail to aggregate and to transmigrate in vitro. Thus these later stages of neutrophil localization to inflammatory sites require CD18 function. These events require distinct CD18 ligands, since MAbs against ICAM-1 block neutrophil transendothelial migration but do not affect neutrophil aggregation.

Whether CD18–ICAM-1 interactions are sufficient for neutrophil transmigration is controversial. It is widely agreed that anti–CD18 MAb both reduces the number of adherent neutrophils and inhibits all transmigration of the remaining adherent cells (6,92,93,106,121,123,124). Furthermore, both anti–L-selectin and anti–E-selectin MAbs reduce the number of neutrophils which can bind to stimulated endothelial cells. However, Smith and colleagues find that those neutrophils that are adherent in the presence of antiselectin MAb are not affected in their ability to migrate across the endothelial monolayer (93,106). Furie et al. find similar results in neutrophil migration across the endothelium in response to an applied chemotactic gradient (122). In contrast, Luscinskas and colleagues find profound inhibition of transmigration with anti–E-selectin (124) and, to a lesser extent, with anti–L-selectin MAb (92). The reason for this discrepancy is not clear, but may reflect differences in the culture and assay systems used.

The Trigger

The transition from selectin-mediated adhesion to CD18-mediated adhesion occurs rapidly in vitro. A variety of chemotactic factors can cause quantitative down-regulation of L-selectin at the same time that Mac-1 is up-regulated. This transition is reflected in neutrophil adhesion to endothelial cells in vitro. Under static conditions, where no shear stress is applied, resting neutrophil adhesion is partially selectin-dependent and partially CD18/ICAM-1–dependent (57,92,93,106). Upon activation of the neutrophil, the adhesion becomes almost entirely CD18/ICAM-1–dependent (25,125). Similarly, activated neutrophils bind poorly under conditions of flow (25). The physiologically relevant trigger of this transition in vivo is unknown. It is likely that there are multiple mediators, perhaps utilized in different types of inflammatory events. It would be most efficient if the stimulated endothelial cells, themselves, could produce the appropriate chemoattractant. This would ensure appropriate localization of the neutrophil to the inflammatory site. Endothelial cells are capable of producing several neutrophil chemoattractants, including Il-8, GM-CSF, and PAF.

Smith and co-workers demonstrated that co-culture of freshly isolated neutrophils with Il-1–stimulated endothelial cells for 30 min induces a dramatic down-regulation of L-selectin, an upregulation of Mac-1, and an accompanying neutrophil shape-change response (93). Furthermore, the conditioned media from Il-1–stimulated endothelial cells can directly induce this transition in neutrophils (93). Huber et al. (126) recently reported that an anti–Il-8 serum significantly blocked neutrophil transmigration across stimulated endothelial cells. These results suggest that in 3–4-h-stimulated endothelial cultures, Il-8 is the major chemotactic factor involved. Another interesting possibility is that neutrophil adhesion to E-selectin is sufficient to trigger activation of the CD18 integrins (127).

Zimmerman, McIntyre, and colleagues have studied the role of PAF in neutrophil adhesion and activation (128). Both PAF and P-selectin are induced within minutes on thrombin- or histamine-activated endothelium, correlating with increased neutrophil adhesiveness (129,130). These results suggest that in the earliest events in vascular insult, PAF and P-selectin may function cooperatively to mediate neutrophil localization. These investigators propose that P-selectin acts as a tether to stop the circulating neutrophil, allowing PAF to induce the transition to CD18-dependent transmigration (74).

Summary

The selectin field is one of the fastest-growing areas in adhesion research. Three key events have sparked this recent interest: (1) The recognition that the selectins are closely related members of a gene family involved in leukocyte interactions with endothelium and platelets. (2) The identification of tissue-specific carbohydrate ligands of the selectins. (3) A dynamic model for neutrophil interaction with inflamed endothelium, with distinct selectin- and integrin-mediated adhesion events. Not surprisingly, numerous biotechnology and pharmaceutical companies are intensely focused on selectin research. Novel biological antiinflammatory drugs might include carbohydratelike antagonists, receptor–IgG chimeras, and inhibitors of specific glycosyltransferases. Most of the research on selectins to date has focused on in vitro models. While these studies are invaluable for dissecting adhesion mechanisms, ultimately the potential for selectin antagonists must be tested in clinically relevant animal models. In addition, selective knockout mutations of individual selectin genes and transgenic models will provide invaluable insight and almost certainly open new doors for basic research.

References

1. Anderson, D.C., and T.A. Springer. 1987. Leukocyte adhesion deficiency: An inherited defect in the Mac-1, LFA-1, and p150,95 glycoproteins. *Ann. Rev. Med.* 38:175.
2. Kishimoto, T.K., and D.C. Anderson. 1992. The role of integrins in inflammation. In *Inflammation: Basic Principals and Clinical Correlates.* 2nd ed. J.I. Gallin, I.M. Goldstein, and R. Snyderman, eds. Raven Press, New York, p. in press.
3. Rothlein, R., M.L. Dustin, S.D. Marlin, and T.A. Springer. 1986. A human intercellular adhesion molecule (ICAM-1) distinct from LFA-1. *J. Immunol.* 137:1270.
4. Dustin, M.L., R. Rothlein, A.K. Bhan, C.A. Dinarello, and T.A. Springer. 1986. Induction by IL-1 and interferon, tissue distribution, biochemistry, and function of a natural adherence molecule (ICAM-1). *J. Immunol.* 137:245.
5. Marlin, S.D., and T.A. Springer. 1987. Purified intercellular adhesion molecule-1 (ICAM-1) is a ligand for lymphocyte function-associated antigen 1 (LFA-1). *Cell* 51:813.
6. Smith, C.W., S.D. Marlin, R. Rothlein, C. Toman, and D.C. Anderson. 1989. Cooperative interactions of LFA-1 and Mac-1 with intercellular adhesion molecule-1 in facilitating adherence and transendothelial migration of human neutrophils in vitro. *J. Clin. Invest.* 83:2008.
7. Diamond, M.S., D.E. Staunton, S.D. Marlin, and T.A. Springer. 1991. Binding of the integrin Mac-1 (CD11b/CD18) to the third immunoglobulin-like domain of ICAM-1 (CD54) and its regulation by glycosylation. *Cell* 65:961.
8. Osborn, L., C. Hession, R. Tizard, C. Vassallo, S. Luhowskyj, G. Chi-Rosso, and R. Lobb. 1989. Direct expression cloning of vascular cell adhesion molecule 1, a cytokine-induced endothelial protein that binds to lymphocytes. *Cell* 59:1203.
9. Kishimoto, T.K., R.S. Larson, A.L. Corbi, M.L. Dustin, D.E. Staunton, and T.A. Springer. 1989. Leukocyte Integrins. In *Leukocyte Adhesion Mole-*

cules. T.A. Springer, D.C. Anderson, A.S. Rosenthal, and R. Rothlein, eds. Springer-Verlag, New York, p. 7.

10. Arnaout, M.A. 1990. Structure and function of the leukocyte adhesion molecules CD11/CD18. *Blood* 75:1037.

11. Springer, T.A. 1990. Adhesion receptors of the immune system. *Nature* 346:425.

12. Lasky, L.A., M.S. Singer, T.A. Yednock, D. Dowbenko, C. Fennie, H. Rodriguez, T. Nguyen, S. Stachel, and S.D. Rosen. 1989. Cloning of a lymphocyte homing receptor reveals a lectin domain. *Cell* 56:1045.

13. Siegelman, M.H., M. Van de Rijn, and I.L. Weissman. 1989. Mouse lymph node homing receptor cDNA clone encodes a glycoprotein revealing tandem interaction domains. *Science* 243:1165.

14. Bevilacqua, M.P., S. Stengelin, M.A. Gimbrone, and B. Seed. 1989. Endothelial leukocyte adhesion molecule 1: An inducible receptor for neutrophils related to complement regulatory proteins and lectins. *Science* 243:1160.

15. Johnston, G.I., R.G. Cook, and R.P. McEver. 1989. Cloning of GMP-140, a granule membrane protein of platelets and endothelium: Sequence similarity to proteins involved in cell adhesion and inflammation. *Cell* 56:1033.

16. Phillips, M.L., E. Nudelman, F.C.A. Gaeta, M. Perez, A.K. Singhal, S. Hakomori, and J.C. Paulson. 1990. ELAM-1 mediates cell adhesion by recognition of a carbohydrate ligand, sialyl-Le^x. *Science* 250:1130.

17. Walz, G., A. Aruffo, W. Kolanus, M.P. Bevilacqua, and B. Seed. 1990. Recognition by ELAM-1 of the sialyl-Lex determinant on myeloid and tumor cells. *Science* 250:1132.

18. Lowe, J.B., L.M. Stoolman, R.P. Nair, R.D. Larsen, T.L. Berhend, and R.M. Marks. 1990. ELAM-1-dependent cell adhesion to vascular endothelium determined by a transfected human fucosyltransferase cDNA. *Cell* 63:475.

19. Goelz, S.E., C. Hession, D. Goff, B. Griffiths, R. Tizard, B. Newman, G. Chi-Rosso, and R. Lobb. 1990. ELFT: A gene that directs the expression of an ELAM-1 ligand. *Cell* 63:1349.

20. Berg, E.L., M.K. Robinson, O. Mansson, E.C. Butcher, and J.L. Magnani. 1991. A carbohydrate domain common to both sialyl Le^a and sialyl Le^x is recognized by the endothelial cell leukocyte adhesion molecule ELAM-1. *J. Biol. Chem.* 266:14869.

21. Larsen, E., T. Palabrica, S. Sajer, G.E. Gilbert, D.D. Wagner, B.C. Furie, and B. Furie. 1990. PADGEM-dependent adhesion of platelets to monocytes and neutrophils is mediated by a lineage-specific carbohydrate, LNF III (CD15). *Cell* 63:467.

22. Polley, M.J., M.L. Phillips, E. Wayner, E. Nudelman, A.K. Singhal, S. Hakomori, and J.C. Paulson. 1991. CD62 and endothelial cell-leukocyte adhesion molecule 1 (ELAM-1) recognize the same carbohydrate ligand, sialyl-Lewis x. *Proc. Natl. Acad. Sci. USA* 88:6224.

23. Corral, L., M.S. Singer, B.A. Macher, and S.D. Rosen. 1990. Requirement for sialic acid on neutrophils in a GMP-140 (PADGEM) mediated adhesive interaction with activated platelets. *Biochem. Biophys. Res. Commun.* 172:1349.

24. Moore, K.L., A. Varki, and R.P. McEver. 1991. GMP-140 binds to a gly-

coprotein receptor on human neutrophils: evidence for a lectin-like interaction. *J. Cell Biol.* 112:491.

25. Smith, C.W., S.D. Marlin, R. Rothlein, M.B. Lawrence, L.V. McIntire, and D.C. Anderson. 1989. Role of ICAM-1 in the adherence of human neutrophils to human endothelial cells in vitro. In *Leukocyte Adhesion Molecules*. T.A. Springer, D.C. Anderson, A.S. Rosenthal, and R. Rothlein, eds. Springer-Verlag, New York, p. 170.

26. Jutila, M.A., L. Rott, E.L. Berg, and E.C. Butcher. 1989. Function and regulation of the neutrophil MEL-14 antigen in vivo: Comparison with LFA-1 and Mac-1. *J. Immunol.* 143:3318.

27. Kishimoto, T.K., M.A. Jutila, E.L. Berg, and E.C. Butcher. 1989. Neutrophil Mac-1 and MEL-14 adhesion proteins inversely regulated by chemotactic factors. *Science* 245:1238.

28. Bevilacqua, M., E. Butcher, B. Furie, M. Gallatin, M. Gimbrone, J. Harlan, K. Kishimoto, L. Lasky, R. McEver, J. Paulson, S. Rosen, B. Seed, M. Siegelman, T. Springer, L. Stoolman, T. Tedder, A. Varki, D. Wagner, I. Weissman, and G. Zimmerman. 1991. Selectins: A family of adhesion receptors. *Cell* 67:233.

29. Gallatin, W.M., I.L. Weissman, and E.C. Butcher. 1983. A cell-surface molecule involved in organ-specific homing of lymphocytes. *Nature* 304:30.

30. Siegelman, M.H., I.C. Cheng, I.L. Weissman, and E.K. Wakeland. 1990. The mouse lymph node homing receptor is identical with the lymphocyte cell surface marker Ly-22: role of the EGF domain in endothelial binding. *Cell* 61:611.

31. Tedder, T.F., A.C. Penta, H.B. Levine, and A.S. Freedman. 1990. Expression of the human leukocyte adhesion molecule, LAM-1. Identity with the TQ1 and Leu-8 differentiation antigens. *J. Immunol.* 144:532.

32. Camerini, D., S.P. James, I. Stamenkovic, and B. Seed. 1989. Leu-8/TQ-1 is the human equivalent of the Mel-14 lymph node homing receptor. *Nature* 342:78.

33. Pober, J.S., M.P. Bevilacqua, D.L. Mendrick, L.A. Lapierre, W. Fiers, and M.A. Gimbrone Jr. 1986. Two distinct monokines, interleukin 1 and tumor necrosis factor, each independently induce biosynthesis and transient expression of the same antigen on the surface of cultured human vascular endothelial cells. *J. Immunol.* 136:1680.

34. Bevilacqua, M.P., J.S. Pober, D.L. Mendrick, R.S. Cotran, and M.A. Gimbrone. 1987. Identification of an inducible endothelial-leukocyte adhesion molecule, E-LAM 1. *Proc. Natl. Acad. Sci. USA* 84:9238.

35. McEver, R.P., and M.N. Martin. 1984. A monoclonal antibody to a membrane glycoprotein binds only to activated platelets. *J. Biol. Chem.* 259:9799.

36. Stenberg, P.E., R.P. McEver, M.A. Shuman, Y.V. Jacques, and D.F. Bainton. 1985. A platelet alpha granule membrane protein (GMP-140) is expressed on the plasma membrane after activation. *J. Cell Biol.* 101:880.

37. Hsu-Lin, S.-C., C.L. Berman, B.C. Furie, D. August, and B. Furie. 1984. A platelet membrane protein expressed during platelet activation and secretion. Studies using a monoclonal antibody specific for thrombin-activated platelets. *J. Biol. Chem.* 259:9121.

38. Berman, C.L., E.L. Yeo, J.D. Wencel-Drake, B.C. Furie, M.H. Ginsberg,

and B. Furie. 1986. A platelet alpha granule membrane protein that is associated with the plasma membrane after activation. *J. Clin. Invest.* 78:130.

39. Butcher, E.C. 1986. The regulation of lymphocyte traffic. *Curr. Topics Microbiol. Immunol.* 128:85.

40. Yednock, T.A., and S.D. Rosen. 1989. Lymphocyte homing. *Adv. Immunol.* 44:313.

41. Stamper, H.B., Jr., and J.J. Woodruff. 1976. Lymphocyte homing into lymph nodes: in vitro demonstration of the selective affinity of recirculating lymphocytes for high-endothelial venules. *J. Exp. Med.* 144:828.

42. Rosen, S.D., M. Singer, T.A. Yednock, and L.M. Stoolman. 1985. Involvement of sialic acid on endothelial cells in organ-specific lymphocyte recirculation. *Science* 228:1005.

43. True, D.D., M.S. Singer, L.A. Lasky, and S.D. Rosen. 1990. Requirement for sialic acid on the endothelial ligand of a lymphocyte homing receptor. *J. Cell Biol.* 111:2757.

44. Rosen, S.D., S.I. Chi, D.D. True, M.S. Singer, and T.A. Yednock. 1989. Intravenously injected sialidase inactivates attachment sites for lymphocytes on high endothelial venules. *J. Immunol.* 142:1895.

45. Yednock, T.A., L.M. Stoolman, and S.D. Rosen. 1987. Phosphomanosyl-derivatized beads detect a receptor involved in lymphocyte homing. *J. Cell. Biol.* 104:713.

46. Stoolman, L.M., T. Tenforde, and S.D. Rosen. 1984. Phosphomanosyl receptors may participate in the adhesive interaction between lymphocytes and high endothelial venules. *J. Cell Biol.* 99:1535.

47. Yednock, T.A., E.C. Butcher, L.M. Stoolman, and S.D. Rosen. 1987. Receptors involved in lymphocyte homing: Relationship between a carbohydrate-binding receptor and the MEL-14 antigen. *J. Cell. Biol.* 104:725.

48. Geoffroy, J.S., and S.D. Rosen. 1989. Demonstration that a lectin-like receptor (gp90MEL) directly mediates adhesion of lymphocytes to high endothelial venules of lymph nodes. *J. Cell Biol.* 109:2463.

49. Imai, Y., D.D. True, M.S. Singer, and S.D. Rosen. 1990. Direct demonstration of the lectin activity of gp90MEL, a lymphocyte homing receptor. *J. Cell Biol.* 111:1225.

50. Imai, Y., M.S. Singer, C. Fennie, L.A. Lasky, and S.D. Rosen. 1991. Identification of a carbohydrate-based endothelial ligand for a lymphocyte homing receptor. *J. Cell Biol.* 113:1213.

51. Lewinsohn, D.M., R.F. Bargatze, and E.C. Butcher. 1987. Leukocyte-endothelial cell recognition: Evidence of a common molecular mechanism shared by neutrophils, lymphocytes, and other leukocytes. *J. Immunol.* 138:4313.

52. Watson, S.R., C. Fennie, and L.A. Lasky. 1991. Neutrophil influx into an inflammatory site inhibited by a soluble homing receptor-IgG chimaera. *Nature* 349:164.

53. Gimbrone, M.A. Jr. 1976. Culture of vascular endothelium. *Prog. Hemost. Thromb.* 3:1.

54. Bevilacqua, M.P., J.S. Pober, M.E. Wheeler, R.S. Cotran, and M.A. Gimbrone Jr. 1985. Interleukin 1 acts on cultured human vascular endothelium

to increase the adhesion of polymorphonuclear leukocytes, monocytes, and related leukocyte cell lines. *J. Clin. Invest.* 76:2003.

55. Bevilacqua, M.P., J.S. Pober, G.R. Majeau, R.S. Cotran, and M.A. Jr. Gimbrone. 1984. Interleukin 1 (IL-1) induces biosynthesis and cell surface expression of procoagulant activity in human vascular endothelial cells. *J. Exp. Med.* 160:618.

56. Pober, J.S., and M.A. Gimbrone, Jr. 1982. Expression of Ia-like antigens by human vascular endothelial cells is inducible in vitro: demonstration by monoclonal antibody binding and immunoprecipitation. *Proc. Natl. Acad. Sci. USA* 79:6641.

57. Luscinskas, F.W., A.F. Brock, M.A. Arnaout, and M.A. Gimbrone. 1989. Endothelial-leukocyte adhesion molecule-1-dependent and leukocyte (CD11/CD18)-dependent mechanisms contribute to polymorphonuclear leukocyte adhesion to cytokine-activated human vascular endothelium. *J. Immunol.* 142:2257.

58. Cotran, R.S., M.A. Gimbrone, M.P. Bevilacqua, D.L. Mendrick, and J.S. Pober. 1986. Induction and detection of a human endothelial activation antigen in vivo. *J. Exp. Med.* 164:661.

59. Munro, J.M., J.S. Pober, and R.S. Cotran. 1991. Recruitment of neutrophils in the local endotoxin response: Association with *de novo* endothelial expression of endothelial leukocyte adhesion molecule-1. *Lab. Invest.* 64:295.

60. Redl, H., H.P. Dinges, W.A. Buurman, C.J. Van der Linden, J.S. Pober, R.S. Cotran, and G. Schlag. 1991. Expression of endothelial leukocyte adhesion molecule-1 in septic but not traumatic/hypovolemic shock in the baboon. *Am. J. Pathol.* 139:461.

61. Munro, J.M., J.S. Pober, and R.S. Cotran. 1989. Tumor necrosis factor and interferon-gamma induce distinct patterns of endothelial activation and leukocyte accumulation in skin of Papio anubis. *Am. J. Pathol.* 135:121.

62. Leung, D.Y.M., J.S. Pober, and R.S. Cotran. 1991. Expression of endothelial-leukocyte adhesion molecule-1 in elicited late phase allergic reactions. *J. Clin. Invest.* 87:1805.

63. Gundel, R.H., C.D. Wegner, C.A. Torcellini, C.C. Clarke, N. Haynes, R. Rothlein, C. W. Smith, and L.G. Letts. 1991. Endothelial leukocyte adhesion molecule-1 mediates antigen-induced acute airway inflammation and late-phase airway obstruction in monkeys. *J. Clin. Invest.* 88:1407.

64. Mulligan, M.S., J. Varani, M.K. Dame, C.L. Lane, C.W. Smith, D.C. Anderson, and P.A. Ward. 1991. Role of endothelial-leukocyte adhesion molecule 1 (ELAM-1) in neutrophil-mediated lung injury in rats. *J. Clin. Invest.* 88:1396.

65. Koch, A.E., J.C. Burrows, G.K. Haines, T.M. Carlos, J.M. Harlan, and S.J. Leibovich. 1991. Immunolocalization of endothelial and leukocyte adhesion molecules in human rheumatoid and osteoarthritic synovial tissues. *Lab. Invest.* 64:313.

66. Picker, L.J., T.K. Kishimoto, C.W. Smith, R.A. Warnock, and E.C. Butcher. 1991. ELAM-1 is an adhesion molecule for skin-homing T cells. *Nature* 349:796.

67. McEver, R.P., J.H. Beckstead, K.L. Moore, C.L. Marshall, and D.F. Bainton. 1989. GMP-140, a platelet alpha-granule membrane protein, is also

synthesized by vascular endothelial cells and is localized in Weibel-Palade bodies. *J. Clin. Invest.* 84:92.

68. Bonfanti, R., B.C. Furie, B. Furie, and D.D. Wagner. 1989. PADGEM (GMP140) is a component of Weibel-Palade bodies of human endothelial cells. *Blood* 73:1109.

69. Hattori, R., K.K. Hamilton, R.D. Fugate, R.P. McEver, and P.J. Sims. 1989. Stimulated secretion of endothelial von Willebrand factor is accompanied by rapid redistribution to the cell surface of the intracellular granule membrane protein (GMP-140). *J. Biol. Chem.* 264:7768.

70. Hamburger, S.A., and R.P. McEver. 1990. GMP-140 mediates adhesion of stimulated platelets to neutrophils. *Blood* 75:550.

71. Larsen, E., A. Celi, G.E. Gilbert, B.C. Furie, J.K. Erban, R. Bonfanti, D.D. Wagner, and B. Furie. 1989. PADGEM protein: A receptor that mediates the interaction of activated platelets with neutrophils and monocytes. *Cell* 59:305.

72. Geng, J.-G., M.P. Bevilacqua, K.L. Moore, T.M. McIntyre, S.M. Prescott, J.M. Kim, G.A. Bliss, G.A. Zimmerman, and R.P. McEver. 1990. Rapid neutrophil adhesion to activated endothelium mediated by GMP-140. *Nature* 343:757.

73. Patel, K.D., G.A. Zimmerman, S.M. Prescott, R.P. McEver, and T.M. McIntyre. 1991. Oxygen radicals induce human endothelial cells to express GMP-140 and bind neutrophils. *J. Cell Biol.* 112:749.

74. Lorant, D.E., K.D. Patel, T.M. McIntyre, R.P. McEver, S.M. Prescott, and G.A. Zimmerman. 1991. Coexpression of GMP-140 and PAF by endothelium stimulated by histamine or thrombin: A juxtacrine system for adhesion and activation of neutrophils. *J. Cell Biol.* 115:223.

75. Palabrica, T.M., B.C. Furie, M.A. Konstam, M.J. Aronovitz, R. Connolly, B.A. Brockway, K.L. Ramberg, and B. Furie. 1989. Thrombus imaging in a primate model with antibodies specific for an external membrane protein of activated platelets. *Proc. Natl. Acad. Sci. USA* 86:1036.

76. Siegelman, M.H., and I.L. Weissman. 1989. Human homologue of mouse lymph node homing receptor: Evolutionary conservation at tandem cell interaction domains. *Proc. Natl. Acad. Sci. USA* 86:5562.

77. Tedder, T.F., C.M. Isaacs, T.J. Ernst, G.D. Demetri, D.A. Adler, and C.M. Disteche. 1989. Isolation and chromosomal localization of cDNAs encoding a novel human lymphocyte cell surface molecule, LAM-1. Homology with the mouse lymphocyte homing receptor and other human adhesion proteins. *J. Exp. Med.* 170:123.

78. Bowen, B.R., T. Nguyen, and L.A. Lasky. 1989. Characterization of a human homologue of the murine peripheral lymph node homing receptor. *J. Cell Biol.* 109:421.

79. Drickamer, K. 1988. Two distinct classes of carbohydrate-recognition domains in animal lectins. *J. Biol. Chem.* 263:9557.

80. Bowen, B.R., C. Fennie, and L.A. Lasky. 1990. The Mel-14 antibody binds to the lectin domain of the murine peripheral lymph node homing receptor. *J. Cell Biol.* 110:147.

81. Kansas, G.S., O. Spertini, L.M. Stoolman, and T.F. Tedder. 1991. Molecular mapping of functional domains of the leukocyte receptor for endothelium, LAM-1. *J. Cell Biol.* 114:351.

82. Watson, S.R., Y. Imai. C. Fennie, J. Geoffrey, M. Singer, S.D. Rosen, and L.A. Lasky. 1991. The complement binding-like domains of the murine homing receptor facilitate lectin activity. *J. Cell Biol.* 115:235.

83. Berg, M., and S.P. James. 1990. Human neutrophils release the Leu-8 lymph node homing receptor during cell activation. *Blood* 76:2381.

84. Picker, L.J., R.A. Warnock, A.R. Burns, C.M. Doerschuk, E.L. Berg, and E.C. Butcher. 1991. The neutrophil selectin LECAM-1 presents carbohydrate ligands to the vascular selectins ELAM-1 and GMP-140. *Cell* 66:921.

85. St. John, T., W.M. Gallatin, M. Siegelman, H.T. Smith, V.A. Fried, and I.L. Weissman. 1986. Expression cloning of a lymphocyte homing receptor cDNA: Ubiquitin is the reactive species. *Science* 231:845.

86. Siegelman, M., M.W. Bond, W.M. Gallatin, T. St. John, H.T. Smith, V.A. Fried, and I.L. Weissman. 1986. Cell surface molecule associated with lymphocyte homing is a ubiquitinated branched-chain glycoprotein. *Science* 231:823.

87. Johnston, G.I., G.A. Bliss, P.J. Newman, and R.P. McEver. 1990. Structure of the human gene encoding granule membrane protein-140, a member of the selectin family of adhesion receptors for leukocytes. *J. Biol. Chem.* 265:21381.

88. Tedder, T.F., C.M. Isaacs, T.J. Ernst, G.D. Demetri, D.A. Adler, and C.M. Disteche. 1989. Isolation and chromosomal localization of cDNAs encoding a novel human lymphocyte cell surface molecule, LAM-1. *J. Exp. Med.* 170:123.

89. Collins, T., A. Williams, G.I. Johnston, J. Kim, R. Eddy, T. Shows, M.A. Gimbrone, and M.P. Bevilacqua. 1991. Structure and chromosomal location of the gene for endothelial-leukocyte adhesion molecule-1. *J. Biol. Chem.* 266:2466.

90. Watson, M.L., S.F. Kingsmore, G.I. Johnston, M.H. Siegelman, M.M. Le Beau, R.S. Lemons, N.S. Bora, T.A. Howard, I.L. Weissman, R.P. McEver, and M.F. Seldin. 1990. Genomic organization of the selectin family of leukocyte adhesion molecules on human and mouse chromosome 1. *J. Exp. Med.* 172:263.

91. Hallmann, R., M.A. Jutila, C.W. Smith, D.C. Anderson, T.K. Kishimoto, and E.C. Butcher. 1991. The peripheral lymph node homing receptor, LECAM-1, is involved in CD18-independent adhesion of neutrophils to endothelium. *Biochem. Biophys. Res. Commun.* 174:236.

92. Spertini, O., F.W. Luscinskas, G.S. Kansas, J.M. Munro, J.D. Griffin, M.A. Gimbrone, Jr., and T.F. Tedder. 1991. Leukocyte adhesion molecule-1 (LAM-1, L-selectin) interacts with an inducible endothelial cell ligand to support leukocyte adhesion. *J. Immunol.* 147:2565.

93. Smith, C.W., T.K. Kishimoto, O. Abbassi, B.J. Hughes, R. Rothlein, L.V. McIntire, E.C. Butcher, and D.C. Anderson. 1991. Chemotactic factors regulate lectin adhesion molecule-1 (LECAM-1)-dependent neutrophil adhesion to cytokine-stimulated endothelial cells in vitro. *J. Clin. Invest.* 87:609.

94. Griffin, J.D., O. Spertini, T.J. Ernst, M.P. Belvin, H.B. Levine, Y. Kanakura, and T.F. Tedder. 1990. Granulocyte-macrophage colony-stimulating factor and other cytokines regulate surface expression of the leukocyte adhesion molecule-1 on human neutrophils, monocytes, and their precursors. *J. Immunol.* 145:576.

95. Jung, T.M., W.M. Gallatin, I.L. Weissman, and M.O. Dailey. 1988. Down-regulation of homing receptors after T cell activation. *J. Immunol.* 141:4110.
96. Jutila, M.A., T.K. Kishimoto, and E.C. Butcher. 1990. Regulation and lectin activity of the human neutrophil peripheral lymph node homing receptor. *Blood* 76:178.
97. Kishimoto, T.K., M.A. Jutila, and E.C. Butcher. 1990. Identification of a human peripheral lymph node homing receptor: A rapidly down-regulated adhesion molecule. *Proc. Natl. Acad. Sci. USA* 87:2244.
98. Jung, T.M., and M.O. Dailey. 1988. Reversibility of loss of homing receptor expression following activation. *Adv. Exp. Med. Biol.* 237:519.
99. Jutila, M.A., T.K. Kishimoto, and M. Finken. 1991. Low-dose chymotrypsin treatment inhibits neutrophil migration into sites of inflammation *in vivo*: Effects on Mac-1 and MEL-14 adhesion protein expression and function. *Cell. Immunol.* 132:201.
100. Spertini, O., G.S. Kansas, J.M. Munro, J.D. Griffin, and T.F. Tedder. 1991. Regulation of leukocyte migration by activation of the leukocyte adhesion molecule-1 (LAM-1) selectin. *Nature* 349:691.
101. Dustin, M.L., and T.A. Springer. 1989. T cell receptor cross-linking transiently stimulates adhesiveness through LFA-1. *Nature* 341:619.
102. Pober, J.S., M.A. Gimbrone Jr., L.A. Lapierre, D.L. Mendrick, W. Fiers, R. Rothlein, and T.A. Springer. 1986. Overlapping patterns of activation of human endothelial cells by interleukin 1, tumor necrosis factor and immune interferon. *J. Immunol.* 137:1893.
103. Norris, P., R.N. Poston, D.S. Thomas, M. Thornhill, J. Hawk, and D.O. Haskard. 1991. The expression of endothelial leukocyte adhesion molecule-1 (ELAM-1), intercellular adhesion molecule-1 (ICAM-1), and vascular cell adhesion molecule-1 (VCAM-1) in experimental cutaneous inflammation: A comparison of ultraviolet B erythema and delayed hypersensitivity. *J. Invest. Dermatol.* 96:763.
104. Shimizu, Y., S. Shaw, N. Graber, T.V. Gopal, K. J. Horgan, G. A. Van Seventer, and W. Newman. 1991. Activation-independent binding of human memory T cells to adhesion molecule ELAM-1. *Nature* 349:799.
105. Streeter, P.R., B.T. Rouse, and E.C. Butcher. 1988. Immunohistologic and functional characterization of a vascular addressin involved in lymphocyte homing into peripheral lymph nodes. *J. Cell Biol.* 107:1853.
106. Kishimoto, T.K., R.A. Warnock, M.A. Jutila, E.C. Butcher, C. Lane, D.C. Anderson, and C.W. Smith. 1991. Antibodies against human neutrophil LECAM-1 (LAM-1/Leu-8/DREG-56 antigen) and endothelial cell ELAM-1 inhibit a common CD18-independent adhesion pathway in vitro. *Blood* 78:805.
107. Von Andrian, U.H., J.D. Chambers, L.M. McEvoy, R.F. Bargatze, K.-E. Arfors, and E.C. Butcher. 1991. Two-step model of leukocyte-endothelial cell interaction in inflammation: Distinct roles for LECAM-1 and the leukocyte β_2 integrins *in vivo*. *Proc. Natl. Acad. Sci. USA* 88:7538.
108. Ley, K., P. Gaehtgens, C. Fennie, M.S. Singer, L.A. Lasky, and S.D. Rosen. 1991. Lectin-like cell adhesion molecule 1 mediates leukocyte rolling in mesenteric venules in vivo. *Blood* 77:2553.
109. Lowe, J.B., L.M. Stoolman, R.P. Nair, R.D. Larsen, T.L. Behrend, and R.M. Marks. 1991. A transfected human fucosyltransferase cDNA deter-

mines biosynthesis of oligosaccharide ligand(s) for endothelial-leukocyte adhesion molecule I. *Biochem. Soc. Trans.* 19:649.

110. Tiemeyer, M., S.J. Swiedler, M. Ishihara, M. Moreland, H. Schweingruber, P. Hirtzer, and B.K. Brandley. 1991. Carbohydrate ligands for endothelial-leukocyte adhesion molecule-1. *Proc. Natl. Acad. Sci. USA* 88:1138.

111. Tyrrell, D., P. James, N. Rao, C. Foxall, S. Abbas, F. Dasgupta, M. Nashed, A. Hasegawa, M. Kiso, D. Asa, J. Kidd, and B.K. Brandley. 1991. Structural requirements for the carbohydrate ligand of E-selectin. *Proc. Natl. Acad. Sci. USA* 88:10372.

112. Aruffo, A., W. Kolanus, G. Walz, P. Fredman, and B. Seed. 1991. CD62/P-selectin recognition of myeloid and tumor cell sulfatides. *Cell* 67:35.

113. Kishimoto, T.K. 1991. A dynamic model for neutrophil localization to inflammatory sites. *J. NIH Res.* 3:75.

114. Butcher, E.C. 1991. Leukocyte-endothelial cell recognition: Three (or more) steps to specificity and diversity. *Cell* 67:1033.

115. Lawrence, M.B., L.V. McIntire, and S.G. Eskin. 1987. Effect of flow on polymorphonuclear leukocyte/endothelial cell adhesion. *Blood* 70:1284.

116. Lawrence, M.B., C.W. Smith, S.G. Eskin, and L.V. McIntire. 1988. Effect of venous shear stress on CD18-mediated neutrophil adhesion to culture endothelium. *Blood* 75:227.

117. Arfors, K.-E., C. Lundberg, L. Lindbom, K. Lundberg, P.G. Beatty, and J.M. Harlan. 1987. A monoclonal antibody to the membrane glycoprotein complex CD18 inhibits polymorphonuclear accumulation and plasma leakage in vivo. *Blood* 69:338.

118. Anderson, D.C., O. Abbassi, T.K. Kishimoto, J.M. Koenig, L.V. McIntire, and C.W. Smith. 1991. Diminished lectin-, epidermal growth factor-, complement binding domain-cell adhesion molecule-1 on neonatal neutrophils underlies their impaired CD18-independent adhesion to endothelial cells in vitro. *J. Immunol.* 146:3372.

119. Abbassi, O., C.L. Lane, S. Krater, T.K. Kishimoto, D.C. Anderson, L.V. McIntire, and C.W. Smith. 1991. Canine neutrophil margination mediated by lectin adhesion molecule-1 in vitro. *J. Immunol.* 147:2107.

120. Lawrence, M.B., and T.A. Springer. 1991. Leukocytes roll on a selectin at physiologic flow rates: Distinction from and prerequisite for adhesion through integrins. *Cell* 65:1.

121. Furie, M.B., M.C.A. Tancinco, and C.W. Smith. 1991. Monoclonal clonal antibodies to leukocyte integrins CD11a/CD18 and CD11b/CD18 or intercellular adhesion molecule-1 inhibit chemoattractant-stimulated neutrophil transendothelial migration in vitro. *Blood* 78:2089.

122. Furie, M.B., M.J. Burns, M.C.A. Tancinco, C.D. Benjamin, and R.R. Lobb. 1992. Endothelial-leukocyte adhesion molecule-1 is not required for the migration of neutrophils across Il-1–stimulated endothelium in vitro. *J. Immunol.* In press.

123. Smith, C.W., R. Rothlein, B.J. Hughes, M.M. Mariscalco, F.C. Schmalstieg, and D.C. Anderson. 1988. Recognition of an endothelial determinant for CD18-dependent neutrophil adherence and transendothelial migration. *J. Clin. Invest.* 82:1746.

124. Luscinskas, F.W., M.I. Cybulsky, J.-M. Kiely, C.S. Peckins, V.M. Davis, and M.A. Gimbrone, Jr. 1991. Cytokine-activated human endothelial mono-

layers support enhanced neutrophil transmigration via a mechanism involving both endothelial-leukocyte adhesion molecule-1 and intercellular adhesion molecule-1. *J. Immunol.* 146:1617.

125. Dobrina, A., T.M. Carlos, B.R. Schwartz, P.G. Beatty, H.D. Ochs, and J.M. Harlan. 1990. Phorbol ester causes down-regulation of CD11/CD18-independent neutrophil adherence to endothelium. *Immunology* 69:429.

126. Huber, A.R., S.L. Kunkel, R.F. Todd, and S.J. Weiss. 1991. Regulation of transendothelial neutrophil migration by endogenous Il-8. *Science* 254:99.

127. Lo, S.K., S. Lee, R.A. Ramos, R. Lobb, M. Rosa, G. Chi-Rosso, and S.D. Wright. 1991. Endothelial-leukocyte adhesion molecule-1 stimulates the adhesive activity of leukocyte integrin CR3 (CD11b/CD18, Mac-1, alpha M beta 2) on human neutrophils. *J. Exp. Med.* 173:1493.

128. Zimmerman, G.A., T.M. McIntyre, M. Mehra, and S.M. Prescott. 1990. Endothelial cell-associated platelet-activating factor: A novel mechanism for signalling intercellular adhesion. *J. Cell Biol.* 110:529.

129. Zimmerman, G.A., and T.M. McIntyre. 1988. Neutrophil adherence to human endothelium in vitro occurs by CDw18 (Mol, Mac-1/LFA-1/gp150, 95) glycoprotein-dependent and independent mechanisms. *J. Clin. Invest.* 81:531.

130. Zimmerman, G.A., T.M. McIntyre, and S.M. Prescott. 1985. Thrombin stimulates the adherence of neutrophils to human endothelial cells in vitro. *J. Clin. Invest.* 76:2235.

11
Leukocyte Interactions Mediated by P-Selection

RODGER P. MCEVER

Introduction

Endothelial cells, platelets, and leukocytes all play key roles in hemostatic and inflammatory responses to tissue injury. Many of these responses require cellular activation as well as cell–cell contact. This chapter will focus on the properties of P-selectin (CD62, formerly GMP-140 or PADGEM), a receptor that supports binding of leukocytes to activated platelets and endothelium. P-selectin–mediated adhesive interactions operate in conjunction with cell–cell interactions directed by related molecules and are likely to be important in both hemostatic and inflammatory processes.

Results

Tissue and Subcellular Distribution of P-Selectin

P-selectin was originally identified by monoclonal antibodies that reacted only with activated platelets (1,2). Immunocytochemistry in conjunction with electron microscopy indicated that the protein was located in membranes of α granules in unstimulated platelets but was redistributed within seconds to the cell surface following activation-induced fusion of granule membranes with the plasma membrane (3,4). The protein was originally named GMP-140 (granule membrane protein of 140 kd) (3) or PADGEM (platelet-activation-dependent-granule–external membrane protein) (4) and has been given the cluster designation CD62 (5). The protein has recently been renamed P-selectin as part of a consensus nomenclature for a group of proteins of which P-selectin is now known to be a member.

Immunoperoxidase analysis of normal human tissues indicated that P-selectin was present in endothelial cells as well as in platelets and their precursors, megakaryocytes. A striking finding was that the endothelial cell protein was primarily in postcapillary venules, rather than in capil-

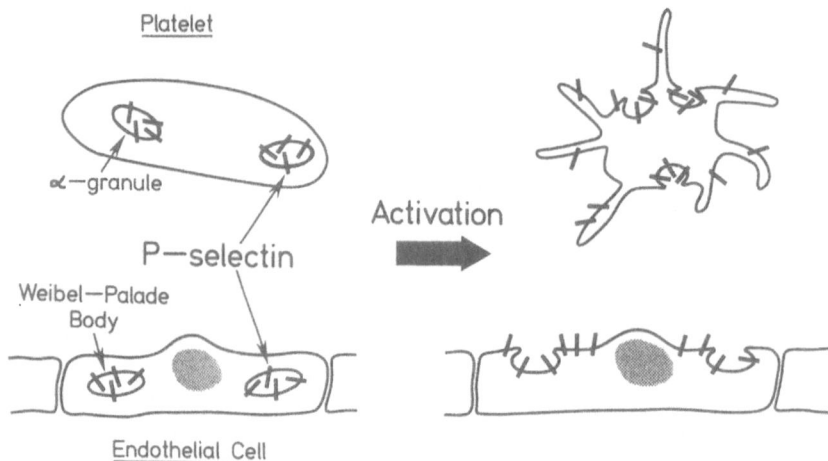

FIGURE 11.1. Redistribution of P-selectin following cellular activation. In unstimulated platelets and endothelial cells, P-selectin is located in membranes of
secretory granules: α granules in platelets and Weibel-Palade bodies in endothelial cells. When these cells are stimulated, the granules rapidly fuse with the plasma membrane, release their contents, and express P-selectin on the cell surface.

laries or in larger veins, arterioles, or arteries. The protein is synthesized
in human umbilical vein endothelial cells, both in vivo and in vitro, which
has facilitated its study (6). Endothelial P-selectin is located in membranes of Weibel-Palade bodies, the secretory granules of endothelium in
which large multimers of von Willebrand factor are stored (6–8). Following cellular activation by agonists such as thrombin and histamine, P-
selectin is rapidly translocated to the cell surface (6,7). Surface appearance is transient, peaking between 3 and 10 min, then declining to basal
levels over 30 min as a result of endocytosis (7). In contrast, P-selectin remains on the surface of platelets stimulated in vitro for at least 1 h following stimulation (9); it is not known whether endocytosis occurs more
rapidly in platelets stimulated in vivo.

In summary, P-selectin is localized in secretory granule membranes of
two cell types in the vascular system, platelets and endothelium (Fig.
11.1). Following activation of these cells by agonists such as thrombin, it
is rapidly redistributed to the cell surface. These properties make monoclonal antibodies to P-selectin useful probes of cellular activation in assays
using radioligand binding or flow cytometry (9,10) and potential markers
of thrombi in vivo with radionuclide imaging procedures (11,12).

Structure of P-Selectin

P-selectin is synthesized by cultured endothelial cells and by HEL cells, a
human cell line with features of megakaryocytes (6,13). Four protein pre-

FIGURE 11.2. Exon–intron boundaries of the gene for P-selectin, shown in relation to the structural domains of the encoded protein. The predicted disulfide bond patterns in each domain are also illustrated. Reproduced from (15) with permission of the publisher.

cursors of slightly different apparent Mr are synthesized by metabolically labeled HEL cells (13). Core high-mannose N-linked oligosaccharides are initially added and then processed into larger complex forms. The mature protein contains 30% carbohydrate by weight (13).

The cDNA-derived amino acid sequence (14) predicts that P-selectin is an elongated molecule with a series of cysteine-rich independently folded domains (Fig. 11.2). Following a cleavable signal peptide, there is an N-terminal domain homologous to Ca^{2+}-dependent lectins such as the asialoglycoprotein receptor, then an epidermal growth factor (EGF)-like motif, nine consensus repeats similar to those in complement-regulatory proteins such as CR1, a transmembrane domain, and a short cytoplasmic tail. Nine nucleotide substitutions have been noted in sequences encoding the consensus repeats (14,15); similar common polymorphisms are present in genes encoding the related repeats of complement-regulatory proteins (16).

The human gene for P-selectin is located on the long arm of chromosome 1 at bands q21–24 (17). It spans over 50 kb and contains 17 exons (15). Most exons encode structurally distinct domains, supporting the concept that P-selectin evolved by exon duplication and rearrangement. Determination of the genomic structure has helped explain two types of variant cDNA clones previously identified, a rare form with a deletion encoding the seventh consensus repeat and a more common form with a deletion encoding the transmembrane domain (14). These deletions are precisely encoded by exons, suggesting that the variants arise by alternative splicing of precursor RNA. The more common variant, which has been found in both platelet and endothelial RNA (15), predicts a soluble form of P-selectin. This potentially important molecule requires further study at the protein level. Very low levels of P-selectin antigen have been measured in both plasma and serum (K.L. Moore and R.P. McEver, unpublished observations). However, it is not known whether the measured antigen represents proteolytic fragments, P-selectin associated with vesicles that are not sedimentable in plasma, or the predicted soluble molecule lacking the transmembrane domain.

P-selectin is structurally related to two vascular cell-surface receptors reviewed elsewhere in this volume, E-selectin, formerly known as endothelial leukocyte adhesion molecule–1 (ELAM-1) (18), and L-selectin, formerly known as the peripheral-lymph-node homing receptor, LECAM-1, murine Mel 14 antigen, human Leu 8 antigen, or LAM-1 (19,20). These three molecules define a new gene family, termed selectins. Each member of the family has been given a prefix letter to indicate the first cell type in which it was identified (platelets for P-selectin, endothelium for E-selectin, and leukocytes for L-selectin). Each selectin contains an N-terminal lectinlike domain, followed by an EGF-like domain, consensus repeats (9 in P-selectin, 6 in E-selectin, and 2 in L-selectin), a transmembrane domain, and a short cytoplasmic tail. The genes have similar intron–exon boundaries, supporting their evolution by gene duplication (15,21,22). In addition, all three genes are tightly clustered in a 300-kb segment of the equivalent regions of mouse and human chromosome 1, suggesting evolutionary pressure to maintain their proximity (17).

Receptor Properties of P-Selectin

The selectins all direct interactions of leukocytes with the blood vessel wall during inflammatory responses. The term selectins was proposed because the molecules mediate selective cell–cell contacts by lectinlike mechanisms. L-selectin is a leukocyte-surface molecule that promotes lymphocyte adhesion to high endothelial venules of peripheral lymph nodes and adherence of neutrophils to cytokine-activated endothelium (19,20,23–26). E-selectin is transiently expressed by cytokine-activated endothelium where it binds neutrophils, monocytes, and memory T cells (18,27–29).

Like E-selectin, P-selectin is a receptor for neutrophils and monocytes (Fig. 11.3). Neutrophils bind to purified, immobilized P-selectin (30), to COS cells transfected with P-selectin cDNA (30), and to activated platelets or endothelial cells expressing P-selectin on their surfaces (30–32). These interactions are inhibited by monoclonal antibodies to P-selectin and by fluid-phase P-selectin. Radiolabeled P-selectin binds reversibly to a saturable number of specific ligands (counterreceptors) on neutrophils, monocytes, and perhaps a subset of lymphocytes. Neutrophil activation does not alter ligand number or the apparent affinity of ligands for P-selectin (33).

Several lines of evidence initially suggested that P-selectin recognizes oligosaccharide structures on target cells. First, the N-terminal location of the lectinlike domain places it in optimal position for binding to another cell component. Second, the epitopes for three monoclonal antibodies to

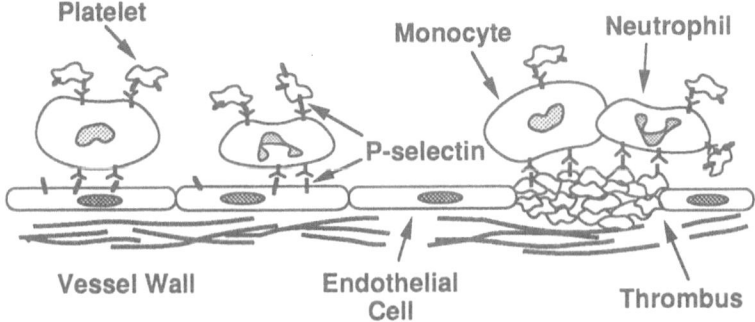

FIGURE 11.3. Adhesion of neutrophils and monocytes to activated platelets and endothelial cells expressing P-selectin. The leukocyte counterreceptors for P-selectin contain sialylated, fucosylated lactosaminoglycans; their structures have not yet been characterized in detail. On the left is an inflammatory site, where leukocytes bind to activated endothelium; the leukocytes, in turn, may recruit activated platelets. On the right is a site of vascular damage and hemorrhage, where adherent, activated platelets may bind neutrophils and monocytes.

P-selectin that block neutrophil recognition (30) have been mapped to the lectin domain (34). Third, binding of P-selectin to leukocytes requires Ca^{2+} (30–33), a finding consistent with the Ca^{2+}-dependent carbohydrate recognition of known "C-type" lectins with similar structures (35). More recently, it has been demonstrated that P-selectin contains two high-affinity binding sites for Ca^{2+} (34). Occupany of these sites alters the conformation of the protein as detected by a decrease in its intrinsic fluorescence emission intensity. This conformational shift includes exposure of an epitope in the lectin domain recognized by a monoclonal antibody capable of blocking leukocyte adhesion to P-selectin. Thus, binding of Ca^{2+} to high-affinity sites on P-selectin modulates the conformation of the lectin domain in a manner that is essential for leukocyte recognition. The actual binding sites for Ca^{2+} have not yet been located, but it is likely that at least one is in the lectin domain.

If P-selectin is a lectin, it could potentially bind to glycoproteins and/or glycolipids on the plasma membranes of leukocytes. However, pretreatment of neutrophils with proteases completely inhibits binding of [125I]P-selectin (33), suggesting that the relevant recognition structures are carried by proteins rather than lipids on these cells. Binding is inhibited by high concentrations of mannose-1-phosphate and certain anionic polysaccharides (33,36), providing supporting but not unequivocal evidence for a lectinlike interaction.

More direct evidence for interaction of P-selectin with carbohydrates comes from the observation that pretreatment of neutrophils with neuraminidases reduces binding of radiolabeled P-selectin (33). This suggests that sialic acid constitutes an essential feature of the ligand for P-selectin. Binding is reduced by neuraminidase from *Vibrio cholerae*, which cleaves sialic acid at $\alpha2,3$, $\alpha2,6$, and $\alpha2,8$ linkages, and from Newcastle disease virus, which cleaves only $\alpha2,3$ and $\alpha2,8$ linkages (33). Since neutrophils do not contain $\alpha2,8$ linkages, at least some of the sialic residues appear to be linked $\alpha2,3$ in the structure recognized by P-selectin. Broad-spectrum neuraminidases also prevent rosetting of activated platelets with neutrophils, a P-selectin–mediated adhesive event. However, neuraminidase from Newcastle disease virus did not prevent rosetting (37). Since the latter enzyme is associated with intact virus which itself binds to cells, it may have directly agglutinated platelets with neutrophils even though it removed a sialic acid linkage required for P-selectin binding.

It has been reported that the oligosaccharide ligand for P-selectin is the abundant myeloid structure Le^x (CD15, lacto-N-fucopentaose III) (38). The core component of this structure, Galβ1,4[Fucα1,3]GlcNAcβ1,3-, is found on glycolipids and several different glycoproteins on neutrophils, generally as part of repeating units in polylactosamine chains (39). The evidence for recognition of Le^x by P-selectin was based on the observation that anti-CD15 antibodies or high concentrations of LNFIII inhibited adhesion of myeloid cells to activated platelets or to P-selectin–transfected cells. Since the neuraminidase studies implicate sialic acid residues in recognition (33,37), it is clear that Le^x per se cannot be the ligand for P-selectin. Furthermore, the protease studies suggest that the critical structure is carried on proteins rather than lipids, whereas Le^x is found on both glycolipids and glycoproteins. Finally, binding of radiolabeled P-selectin to neutrophils is not inhibited by antibodies to CD15 or by a multivalent glycoconjugate consisting of Le^x coupled to albumin by a spacer (33). Therefore, while the Le^x structure might constitute part of the oligosaccharide ligand for P-selectin, it alone cannot provide the required affinity and specificity.

The carbohydrate ligand for the related selectin, E-selectin, has been reported to be sialyl-Le^x, which contains a sialic acid linked $\alpha2,3$ to the Le^x structure (40–43). The hypothesis that sialyl Le^x is also a ligand for P-selectin is supported by the ability of synthetic sialyl Le^x or glycolipids bearing the tetrasaccharide to competitively block P-selectin–mediated interactions of platelets with myeloid cells (44). However, a sialyl-Le^x glycoconjugate does not inhibit binding of radiolabeled P-selectin to neutrophils (33), suggesting that fine modifications or specific presentations of sialyl Le^x or related structures may be required for optimal recognition by P-selectin. This interpretation is supported by studies of Chinese hamster ovary (CHO) cells transfected with a specific fucosyl transferase (45). Unlike parental CHO cells, the transfected cells express sialylated, fucosylated lactosaminglycans, including sialyl Le^x. These cells adhere specifically to immobilized P-selectin and adhesion is inhibited by antibodies to sialyl Le^x. However, the binding of P-selectin to the transfected CHO cells is of much lower affinity than to myeloid cells. Therefore, biologically relevant interactions of P-selectin with opposing cells may require specific modifications, clustering, and/or orientations of sialylated, fucosylated lactosaminoglycan ligands on one or more carrier proteins. It has been recently reported that L-selectin on myeloid cells carries the sialyl Le^x structure and that antibodies to L-selectin partially inhibit neutrophil adhesion to COS cells transfected with P-selectin (46). However, [^{125}I]P-selectin binds to equivalent numbers of sites on resting neutrophils and on stimulated neutrophils under conditions where L-selectin has been quantitatively shed by proteolytic cleavage (33). Thus, further study is required to determine the relevant myeloid cell protein(s) bearing the oligosaccharide ligands for P-selectin as well as the precise carbohydrate structural features that confer optimal recognition by P-selectin.

There is evidence that the EGF domains of L-selectin (47) and E-selectin (48) are required for optimal cell recognition by these molecules, suggesting that the same may be true for P-selectin. Perhaps the affinity and specificity of the interaction of P-selectin with its ligand is due to protein–carbohydrate binding mediated by the lectin domain and protein–protein binding mediated by the EGF domain. Alternatively, the EGF domain may modulate the conformation of the lectin domain to produce a Ca^{2+}-dependent, highly specific interaction with an oligosaccharide ligand. In either of these models, the consensus repeats may mediate other functions not yet characterized. They may also position the lectin domain at a sufficient distance from the platelet or endothelial membrane to allow efficient binding to its ligand on leukocytes.

Physiologic Role of P-Selectin

During inflammation, leukocytes must first adhere to endothelium and then migrate into tissues at sites of infection or injury. This is a temporally regulated process, with migration of neutrophils and monocytes occurring earlier than lymphocytes. Because P-selectin is constitutively synthesized by endothelium and stored in Weibel-Palade bodies, it can be translocated to the cell surface within minutes after endothelial stimulation with agonists such as thrombin and histamine. Therefore it might be the first leukocyte adhesion molecule expressed by activated endothelium during acute inflammation. The surface exposure of P-selectin is also transient because of endocytosis of the receptor within 30 min after its initial appearance. This leads to rapid dampening of the proinflammatory surface on the endothelium. Persistent infection or tissue injury leads to the elaboration of inflammatory cytokines that induce synthesis and surface expression of E-selectin during the next 4 to 6 h (23). P-selectin and E-selectin may therefore work in concert to direct early, regionally specific adherence of neutrophils and monocytes at sites of acute inflammation. The concept that selectins mediate initial attachment of leukocytes to activated endothelium requires that selectins be able to bind to their ligands under flow rates found in postcapillary venules. A recent in vitro model indicates that immobilized P-selectin can support reversible attachment, or rolling, of neutrophils at physiological venular flow rates (49). Other in vitro and in vivo models indicate that E-selectin and L-selectin can also support reversible adhesion (rolling) under conditions of flow (25,50–52). The in vivo studies indicate that L-selectin supports rolling of neutrophils on postcapillary venules within minutes after exteriorization of rat or rabbit mesentery (51,52). The early onset of this rolling response suggests that L-selectin may interact with a rapidly up-regulated counter-receptor on the endothelial surface. This molecule might be P-selectin itself (46) or another molecule not yet characterized.

Tight adhesion of neutrophils to endothelium requires the interaction of the leukocyte $\beta2$ integrins or CD11/CD18 molecules with counter-receptors of the immunoglobulin superfamily on the endothelial surface (53). Integrin-mediated responses are not operative under flow conditions unless the leukocytes have already been transiently arrested by selectin-mediated interactions (49,50,52). Tight adhesion and spreading of leukocytes to endothelium facilitates their subsequent migration into injured tissues. The importance of the $\beta2$ integrins is dramatically illustrated in patients with leukocyte adhesion deficiency, who are congenitally deficient in these molecules. These patients fail to accumulate neutrophils at inflammatory sites and therefore suffer frequent severe infections (54). Although they are constitutively present on the plasma membrane, the CD11/CD18 molecules require neutrophil activation before they can bind to counterreceptors on the endothelial cell surface (55).

The phospholipid-signaling molecule, platelet-activating factor (PAF), is a potent neutrophil agonist. PAF is rapidly synthesized and delivered to the cell surface when endothelium is activated by thrombin or histamine. The expression of PAF is transient due to rapid degradation by a plasma PAF-specific hydrolase (56). Because the kinetics of surface expression of PAF and P-selectin are so similar, it appeared likely that the two molecules could cooperate to promote neutrophil adhesion and migration. This hypothesis is supported by several observations. Neutrophil adhesion to thrombin-activated endothelium is partially inhibited by antibodies to CD18 (57). Although adhesion of neutrophils to P-selectin does not require CD11/CD18 molecules, adherence of PAF-stimulated neutrophils to endothelium is absolutely dependent on these integrins (58). Antibodies to P-selectin and competitive PAF-receptor antagonists each partially block neutrophil adhesion to thrombin- or histamine-stimulated endothelium, whereas combining both inhibitors completely prevents adhesion (58,59). These observations suggest that P-selectin expressed by rapidly activated endothelium promotes attachment of circulating unstimulated neutrophils which are then positioned for activation by PAF. The activated neutrophils develop competent CD11/CD18 molecules that bind to endothelial cell counterreceptors, facilitating spreading and migration of the cells into underlying tissues.

The two-step model of neutrophil adhesion followed by activation assumes that binding of P-selectin does not prevent subsequent neutrophil activation by PAF. It has been proposed that P-selectin acts as a natural antiinflammatory agent, based on observations that the fluid-phase protein prevents adhesion of neutrophils activated by tumor necrosis factor (60). However, other studies suggest that P-selectin, rather than inhibiting neutrophil activation by TNF, actually potentiates the ability of this agent as well as other agonists such as PAF and fMLP to promote neutrophil $\beta2$ integrin function (58).

Unlike the other selectins, P-selectin is also expressed on the surface of activated platelets, where it mediates binding of these cells to neutrophils and monocytes (31,32). The physiologic significance of this process is not established, but there are several possibilities. In vivo, neutrophils support emigration of platelets into acute inflammatory sites (61); this phenomenon might require P-selectin–mediated contacts between neutrophils and activated platelets. The recruitment of platelets could be useful, since these cells release inflammatory mediators such as platelet factor 4 and platelet-derived growth factor (62). At sites of vessel injury, neutrophils and monocytes have been observed adhering to platelet aggregates, an interaction potentially dependent on P-selectin (63). Activated monocytes may promote thrombin generation by developing a surface for prothrombinase components and by expressing tissue factor, the key initiator of the extrinsic pathway of coagulation (64). Close contact between platelets and neutrophils may facilitate transcellular metabolism of leukotrienes and lipoxins that are not produced by either cell type alone (65–67). Finally, expression of P-selectin may provide a mechanism for rapid clearance of activated platelets from the circulation by macrophages in the reticuloendothelial system.

Pathologic Role of P-Selectin

Excessive accumulation of neutrophils has been implicated in a number of inflammatory disorders, including acute respiratory distress syndrome, ischemia-reperfusion injury, Gram-negative septic shock, and rheumatoid arthritis (68). Tissue injury is thought to result from release of oxygen free radicals and proteases from activated neutrophils (69,70). The endothelium is likely to be particularly susceptible to damage because of certain unique biochemical features and its proximity to circulating neutrophils (71). Neutrophils adherent to some surfaces release much larger quantities of free radicals in response to agonists than they do when in suspension (72). In physiologic inflammation, leukocytes adhere only transiently to endothelium before migrating into tissues, thus lessening the risk of vascular damage. However, unregulated expression of adhe-

sion molecules might lead to persistent adherence and activation of neutrophils. If so, interruption of leukocyte adhesion mechanisms may reduce
tissue injury, whatever the proximal cause of disease.

Pathologic inflammatory sites, for example, ischemic regions due to
ruptured atherosclerotic plaque, may produce mediators such as thombin
that can activate platelets and endothelial cells and expose P-selectin.
Nonphysiologic mediators may also be released that can activate these
cells. Insertion of the terminal complement components C5b–9 into cellular membranes activates both platelets and endothelial cells, leading to
granule secretion, expression of P-selectin, and release of procoagulant
microvesicles (73). Low concentrations of oxygen radicals induce prolonged exposure of P-selectin on the endothelial cell surface, leading to
enhanced neutrophil adherence over several hours (74). In vivo, this
unregulated expression of P-selectin could result in sustained neutrophil
recruitment, additional production of oxygen radicals by adherent cells,
and eventual tissue destruction.

Some malignant cells may metastasize by utilizing adhesion mechanisms normally used for leukocyte recruitment. Based on indirect evidence, lectin–carbohydrate interactions have been proposed to promote
spreading of certain experimental tumors (75). More recently, a human
colon carcinoma cell line has been shown to bind specifically to E-selectin
(76). Other malignant cells may potentially express receptors for P-
selectin. Malignant cell-surface display of ligands for leukocyte adhesion
molecules may target spread of tumors to endothelium at inflammatory
sites, where expression of adhesion molecules such as P-selectin and E-
selectin might be induced. Platelets have also been shown to promote experimental tumor metastasis (77). Although the molecules described to
date are integrins, there may be examples where platelet adhesion to
tumor cells is mediated by P-selectin.

Conclusion

P-selectin, like the other selectins, appears to mediate a variety of leukocyte interactions with the blood vessel wall by its ability to bind to cell-
surface oligosaccharide ligands. These cellular interactions may be of fundamental importance for inflammatory as well as certain hemostatic
responses to tissue injury. Pathologic selectin-mediated cell adhesion may
potentiate tissue injury in some inflammatory and thrombotic disorders.
Appropriate in vivo models of inflammation, metastasis, and thrombosis
are required to evaluate the participation of P-selectin and other adhesion
molecules in disease. Should their pathologic roles be confirmed, new
drugs might be designed to interrupt their function. Such pharmaceuticals
might include monoclonal antibodies, peptides, oligosaccharides, or re-

combinant soluble proteins, variously designed to block adhesion molecules on endothelium, leukocytes, platelets, or tumor cells.

References

1. McEver, R.P. and M.N. Martin. 1984. A monoclonal antibody to a membrane glycoprotein binds only to activated platelets. *J. Biol. Chem.* 259:9799.
2. Hsu-Lin, S.-C., C.L. Berman, B.C. Furie, D. August, and B. Furie. 1984. A platelet membrane protein expressed during platelet activation and secretion. Studies using a monoclonal antibody specific for thrombin-activated platelets. *J. Biol. Chem.* 259:9121.
3. Stenberg, P.E., R.P. McEver, M.A. Shuman, Y.V. Jacques, and D.F. Bainton. 1985. A platelet alpha-granule membrane protein (GMP-140) is expressed on the plasma membrane after activation. *J. Cell Biol.* 101:880.
4. Berman, C.L., E.L. Yeo, J.D. Wencel-Drake, B.C. Furie, M.H. Ginsberg, and B. Furie. 1986. A platelet alpha granule membrane protein that is associated with the plasma membrane after activation. *J. Clin. Invest.* 78:130.
5. Knapp, W., B. Dorken, P. Rieber, R.E. Schmidt, H. Stein, and A.E. G. Kr. von dem Borne. 1989. CD Antigens 1989. *Blood* 74:1448.
6. McEver, R.P., J.H. Beckstead, K.L. Moore, L. Marshall-Carlson, and D.F. Bainton. 1989. GMP-140, a platelet alpha-granule membrane protein, is also synthesized by vascular endothelial cells and is localized in Weibel-Palade bodies. *J. Clin. Invest.* 84:92.
7. Hattori, R., K.K. Hamilton, R.D. Fugate, R.P. McEver, and P.J. Sims. 1989. Stimulated secretion of endothelial von Willebrand factor is accompanied by rapid redistribution to the cell surface of the intracellular granule membrane protein GMP-140. *J. Biol. Chem.* 264:7768.
8. Bonfanti, R., B.C. Furie, B. Furie, and D.D. Wagner. 1989. PADGEM (GMP 140) is a component of Weibel-Palade bodies of human endothelial cells. *Blood* 73:1109.
9. George, J.N., E.B. Pickett, S. Saucerman, R.P. McEver, T.J. Kunicki, N. Kieffer, and P.J. Newman. 1986. Platelet surface glycoproteins. Studies on resting and activated platelets and platelet membrane microparticles in normal subjects, and observations in patients during adult respiratory distress syndrome and cardiac surgery. *J. Clin. Invest.* 78:340.
10. Abrams, C.S., N. Ellison, A.Z. Budzynski, and S.J. Shattil. 1990. Direct detection of activated platelets and platelet-derived microparticles in humans. *Blood* 75:128.
11. Miller, D.D., A.J. Boulet, F.O. Tio, O.J. Garcia, R.P. McEver, J.C. Palmaz, K.Y. Pak, D.S. Neblock, H.J. Berger, and P.E. Daddona. 1991. In vivo ^{99m}Technetium S12 antibody imaging of platelet alpha-granules in rabbit endothelial neointimal proliferation after angioplasty. *Circulation* 83:224.
12. Palabrica, T.M., B.C. Furie, M.A. Konstam, M.J. Aronovitz, R. Connolly, B.A. Brockway, K.L. Ramberg, and B. Furie. 1989. Thrombus imaging in a primate model with antibodies specific for an external membrane protein of activated platelets. *Proc. Natl. Acad. Sci. U.S.A.* 86:1036.
13. Johnston, G.I., A. Kurosky, and R.P. McEver. 1989. Structural and biosyn-

theic studies of the granule membrane protein, GMP-140, from human platelets and endothelial cells. *J. Biol. Chem.* 264:1816.

14. Johnston, G.I., R.G. Cook, and R.P. McEver. 1989. Cloning of GMP-140, a granule membrane protein of platelets and endothelium: sequence similarity to proteins involved in cell adhesion and inflammation. *Cell* 56:1033.

15. Johnston, G.I., G.A. Bliss, P.J. Newman, and R.P. McEver. 1990. Structure of the human gene encoding GMP-140, a member of the selectin family of adhesion receptors for leukocytes. *J. Biol. Chem.* 265:21381.

16. Hourcade, D., V.M. Holers, and J.P. Atkinson. 1989. The regulators of complement activation (RCA) gene cluster. In *Advances in Immunology, Vol. 45.* F.J Dixon, ed. Academic Press, New York, p. 381.

17. Watson, M.L., S.F. Kingsmore, G.I. Johnston, M.H. Siegelman, M.M. Le Beau, R.S. Lemons, N.S. Bora, T.A. Howard, I.L. Weissman, R.P. McEver, and M.F. Seldin. 1990. Genomic organization of the selectin family of leukocyte adhesion molecules on human and mouse chromosome 1. *J. Exp. Med.* 172:263.

18. Bevilacqua, M.P., S. Stengelin, M.A. Gimbrone, Jr., and B. Seed. 1989. Endothelial leukocyte adhesion molecule 1: an inducible receptor for neutrophils related to complement regulatory proteins and lectins. *Science* 243:1160.

19. Lasky, L.A., M.S. Singer, T.A. Yednock, D. Dowbenko, C. Fennie, H. Rodriguez, T. Nguyen, S. Stachel, and S.D. Rosen. 1989. Cloning of a lymphocyte homing receptor reveals a lectin domain. *Cell* 56:1045.

20. Siegelman, M.H., M. van de Rijn, and I.L. Weissman. 1989. Mouse lymph node homing receptor cDNA clone encodes a glycoprotein revealing tandem interaction domains. *Science* 243:1165.

21. Ord, D.C., T.J. Ernst, L.-J. Zhou, A. Rambaldi, O. Spertini, J. Griffin, and T.F. Tedder. 1990. Structure of the gene encoding the human leukocyte adhesion molecule-1 (TQ1, Leu-8) of lymphocytes and neutrophils. *J. Biol. Chem.* 265:7760.

22. Collins, T., A. Williams, G.I. Johnston, J. Kim, R. Eddy, T. Shows, M.A. Gimbrone Jr., and M.P. Bevilacqua. 1991. Structure and chromosomal location of the gene for endothelial-leukocyte adhesion molecule 1. *J. Biol. Chem.* 266:2466.

23. Bevilacqua, M.P., J.S. Pober, D.L. Mendrick, R.S. Cotran, and M.A. Gimbrone, Jr. 1987. Identification of an inducible endothelial-leukocyte adhesion molecule. *Proc. Natl. Acad. Sci. U.S.A.* 84:9238.

24. Hallmann, R., M.A. Jutila, C.W. Smith, D.C. Anderon, T.K. Kishimoto, and E.C. Butcher. 1991. The peripheral lymph node homing receptor, LECAM-1, is involved in CD18-independent adhesion of human neutrophils to endothelium. *Biochem. Biophys. Res. Commun.* 174:236.

25. Smith, C.W., T.K. Kishimoto, O. Abbass, B. Hughes, R. Rothlein, L.V. McIntire, E. Butcher, and D.C. Anderson. 1991. Chemotactic factors regulate lectin adhesion molecule 1 (LECAM-1)-dependent neutrophil adhesion to cytokine-stimulated endothelial cells in vitro. *J. Clin. Invest.* 87:609.

26. Kishimoto, T.K., R.A. Warnock, M.A. Jutila, E.C. Butcher, C. Lane, D.C. Anderson, and C.W. Smith. 1991. Antibodies against human neutrophil LECAM-1 (LAM-1/Leu-8/DREG-56 antigen) and endothelial cell ELAM-1 inhibit a common CD18-independent adhesion pathway in vitro. *Blood* 78:805.

27. Hession, C., L. Osborn, D. Goff, G. Chi-Rosso, C. Vassallo, M. Pasek, C. Pittack, R. Tizard, S. Goelz, K. McCarthy, S. Hopple, and R. Lobb. 1990. Endothelial leukocyte adhesion molecule 1: direct expression cloning and functional interactions. *Proc. Natl. Acad. Sci. U.S.A.* 87:1673.

28. Picker, L.J., T.K. Kishimoto, C.W. Smith, R.A. Warnock, and E.C. Butcher. 1991. ELAM-1 is an adhesion molecule for skin-homing T cells. *Nature* 349:796.

29. Shimizu, Y., S. Shaw, N. Graber, T.V. Gopal, K.J. Horgan, G.A. Van Seventer, and W. Newman. 1991. Activation-independent binding of human memory T cells to adhesion molecule ELAM-1. *Nature* 349:799.

30. Geng, J.-G., M.P. Bevilacqua, K.L. Moore, T.M. McIntyre, S.M. Prescott, J.M. Kim, G.A. Bliss, G.A. Zimmerman, and R.P. McEver. 1990. Rapid neutrophil adhesion to activated endothelium mediated by GMP-140. *Nature* 343:757.

31. Larsen, E., A. Celi, G.E. Gilbert, B.C. Furie, J.K. Erban, R. Bonfanti, D.D. Wagner, and B. Furie. 1989. PADGEM protein: a receptor that mediates the interaction of activated platelets with neutrophils and monocytes. *Cell* 59:305.

32. Hamburger, S.A. and R.P. McEver. 1990. GMP-140 mediates adhesion of stimulated platelets to neutrophils. *Blood* 75:550.

33. Moore, K.L., A. Varki, and R.P. McEver. 1991. GMP-140 binds to a glycoprotein receptor on human neutrophils: evidence for a lectin-like interaction. *J. Cell Biol.* 112:491.

34. Geng, J.-G., K.L. Moore, A.E. Johnson, and R.P. McEver. 1991. Neutrophil recognition requires a Ca^{2+}-induced conformational change in the lectin domain of GMP-140. *J. Biol. Chem.* 266:22313.

35. Drickamer, K. 1988. Two distinct classes of carbohydrate-recognition domains in animal lectins. *J. Biol. Chem.* 263:9557.

36. Skinner, M.P., C.M. Lucas, G.F. Burns, C.N. Chesterman, and M.C. Berndt. 1991. GMP-140 binding to neutrophils is inhibited by sulfated glycans. *J. Biol. Chem.* 266:5371.

37. Corral, L., M.S. Singer, B.A. Macher, and S.D. Rosen. 1990. Requirement for sialic acid on neutrophils in a GMP-140 (PADGEM) mediated adhesive interaction with activated platelets. *Biochem. Biophys. Res. Commun.* 172:1349.

38. Larsen, E., T. Palabrica, S. Sajer, G.E. Gilbert, D.D. Wagner, B.C. Furie, and B. Furie. 1990. PADGEM-dependent adhesion of platelets to monocytes and neutrophils is mediated by a lineage-specific carbohydrate, LNF III (CD15). *Cell* 63:467.

39. Fukuda, M. 1985. Cell surface glycoconjugates as onco-differentiation markers in hematopoietic cells. *Biochim. Biophys. Acta* 780:119.

40. Phillips, M.L., E. Nudelman, F.C.A. Gaeta, M. Perez, A.K. Singhal, S. Hakomori, and J.C. Paulson. 1990. ELAM-1 mediates cell adhesion by recognition of a carbohydrate ligand, sialyl-Le^x. *Science* 250:1130.

41. Walz, G., A. Aruffo, W. Kolanus, B. Bevilacqua, and B. Seed. 1990. Recognition by ELAM-1 of the sialyl-Le^x determinant on myeloid and tumor cells. *Science* 250:1132.

42. Lowe, J.B., L.M. Stoolman, R.P. Nair, R.D. Larsen, T.L. Berhend, and R.M. Marks. 1990. ELAM-1-dependent cell adhesion to vascular endothe-

lium determined by a transfected human fucosyltransferase cDNA. *Cell* 63:475.

43. Tiemeyer, M., S.J. Swiedler, M. Ishihara, M. Moreland, H. Schweingruber, P. Hirtzer, and B.K. Brandley. 1991. Carbohydrate ligands for endothelial-leukocyte adhesion molecule 1. *Proc. Natl. Acad. Sci. U.S.A.* 88:1138.

44. Polley, M.J., M.L. Phillips, E. Wayner, E. Nudelman, A.K. Singhal, S. Hakomori, and J.C. Paulson. 1991. CD62 and endothelial cell-leukocyte adhesion molecule 1 (ELAM-1) recognize the same carbohydrate ligand, sialyl-Lewis x. *Proc. Natl. Acad. Sci. U.S.A.* 88:6224.

45. Zhou, Q., K.L. Moore, D.F. Smith, A. Varki, R.P. McEver, and R.D. Cummings. 1991. The selectin GMP-140 binds to sialylated, fucosylated lactosaminoglycans on both myeloid and nonmyeloid cells. *J. Cell Biol.* 115: 557.

46. Picker, L.J., R.A. Warnock, A.R. Burns, C.M. Doerschuk, E.L. Berg, and E.C. Butcher. 1991. The neutrophil selectin LECAM-1 presents carbohydrate ligands to the vascular selectins ELAM-1 and GMP-140. *Cell* 66:921.

47. Siegelman, M.H., I.C. Cheng, I.L. Weissman, and E.K. Wakeland. 1990. The mouse lymph node homing receptor is identical with the lymphocyte cell surface marker Ly-22: Role of the EGF domain in endothelial binding. *Cell* 61:611.

48. Pigott, R., L.A. Needham, R.M. Edwards, C. Walker, and C. Power. 1991. Structural and functional studies of the endothelial activation antigen endothelial leucocyte adhesion molecule-1 using a panel of monoclonal antibodies. *J. Immunol.* 147:130.

49. Lawrence, M.B. and T.A. Springer. 1991. Leukocytes roll on a selectin at physiologic flow rates: Distinction from and prerequisite for adhesion through integrins. *Cell* 65:859.

50. Lawrence, M.B., C.W. Smith, S.G. Eskin, and L.V. McIntire. 1990. Effect of venuous shear stress on CD18-mediated neutrophil adhesion to cultured endothelium. *Blood* 75:227.

51. Ley, K., P. Gaehtgens, C. Fennie, M.S. Singer, L.A. Lasky, and S.D. Rosen. 1991. Lectin-like cell adhesion molecule 1 mediates leukocyte rolling in mesenteric venules in vivo. *Blood* 77:2553.

52. Von Andrian, U.H., J.D. Chambers, L.M. McEvoy, R.F. Bargatze, K.-E. Arfors, and E.C. Butcher. 1991. Two-step model of leukocyte-endothelial cell interaction in inflammation: Distinct roles for LECAM-1 and the leukocyte β_2 integrins in vivo. *Proc. Natl. Acad. Sci. U.S.A.* 88:7538.

53. Kishimoto, T.K., R.S. Larson, A.L. Corbi, M.L. Dustin, D.E. Staunton, and T.A. Springer. 1989. The leukocyte integrins. In *Advances in Immunology*, Vol. 46. F.J. Dixon, ed. Academic Press, New York, p. 149.

54. Anderson, D.C. and T.A. Springer. 1987. Leukocyte adhesion deficiency: an inherited defect in the Mac-1, LFA-1, and p150,95 glycoproteins. *Ann. Rev. Med.* 38:175.

55. Lo, S.K., G.A. Van Seventer, S.M. Levin, and S.D. Wright. 1989. Two leukocyte receptors (CD11a/CD18 and CD11b/CD18) mediate transient adhesion to endothelium by binding to different ligands. *J. Immunol.* 143:3325.

56. Prescott, S.M., G.A. Zimmerman, and T.M. McIntyre. 1990. Platelet-activating factor. *J. Biol. Chem.* 265:17381.

57. Zimmerman, G.A. and T.M. McIntyre. 1988. Neutrophil adherence to hu-

man endothelium in vitro occurs by CDw18 (Mol. MAC-1/LFA-1/GP 150,95) glycoprotein-dependent and -independent mechanisms. *J. Clin. Invest.* 81:531.

58. Lorant, D.E., K.D. Patel, T.M. McIntyre, R.P. McEver, S.M. Prescott, and G.A. Zimmerman. 1991. Coexpression of GMP-140 and PAF by endothelium stimulated by histamine or thrombin: a juxtacrine system for adhesion and activation of neutrophils. *J. Cell Biol.* 115:223.

59. Zimmerman, G.A., T.M. McIntyre, M. Mehra, and S.M. Prescott. 1990. Endothelial cell-associated platelet-activating factor: a novel mechanism for signaling intercellular adhesion. *J. Cell Biol.* 110:529.

60. Gamble, J.R., M.P. Skinner, M.C. Berndt, and M.A. Vadas. 1990. Prevention of activated neutrophil adhesion to endothelium by soluble adhesion protein GMP140. *Science* 249:414.

61. Issekutz, A.C., M. Ripley, and J.R. Jackson. 1983. Role of neutrophils in the deposition of platelets during acute inflammation. *Lab. Invest.* 49:716.

62. Weksler, B.B. 1988. Roles for human platelets in inflammation. In *Platelet Membrane Receptors: Molecular Biology, Biochemistry, and Pathology.* G.A. Jamieson, ed. Alan R. Liss, Inc., New York, p. 611.

63. Henry, R.L. 1961. Leukocytes and thrombosis. *Thromb. Diath. Haemorrh.* 13:35.

64. Edwards, R.L. and F.R. Rickles. 1984. Macrophage procoagulants. In *Progress in Hemostasis and Thrombosis.* 7th ed. T.H. Spaet, ed. Grune and Stratton, New York, p. 183.

65. Maclouf, J.A. and R.C. Murphy. 1988. Transcellular metabolism of neutrophil-derived Leukotriene A$_4$ by human platelets. A potential cellular source of leukotriene C$_4$. *J. Biol. Chem.* 263:174.

66. Serhan, C.N. and K.-A. Sheppard. 1990. Lipoxin formation during human neutrophil-platelet interactions. Evidence for the transformation of leukotriene A$_4$ by platelet 12-lipoxygenase in vitro. *J. Clin. Invest.* 85:772.

67. Marcus, A.J. 1990. Eicosanoid interactions between platelets, endothelial cells, and neutrophils. *Methods Enzymol.* 187:585.

68. Malech, H.L. and J.I. Gallin. 1987. Neutrophils in human diseases. *N. Engl. J. Med.* 317:687.

69. Mullane, K.M., W. Westlin, and R. Kraemer. 1988. Activated neutrophils release mediators that may contribute to myocardial injury and dysfunction associated with ischemia and reperfusion. *Ann. N. Y. Acad. Sci.* 524:103.

70. Henson, P.M. and R.B. Jr. Johnston. 1987. Tissue injury in inflammation: oxidants, proteinases, and cationic proteins. *J. Clin. Invest.* 79:669.

71. Ward, P.A. and J. Varani. 1990. Mechanisms of neutrophil-mediated killing of endothelial cells. *J. Leukocyte Biol.* 48:97.

72. Nathan, C., S. Srimal, C. Farber, E. Sanchez, L. Kabbash, A. Asch, J. Gailit, and S.D. Wright. 1989. Cytokine-induced respiratory burst of human neutrophils: dependence on extracellular matrix proteins and CD11/CD18 integrins. *J. Cell Biol.* 109:1341.

73. Hattori, R., K.K. Hamilton, R.P. McEver, and P.J. Sims. 1989. Complement proteins C5b–9 induce secretion of high molecular weight multimers of endothelial von Willebrand factor and translocation of granule membrane protein GMP-140 to the cell surface. *J. Biol. Chem.* 264:9053.

74. Patel, K.D., G.A. Zimmerman, S.M. Prescott, R.P. McEver, and T.M.

McIntyre. 1991. Oxygen radicals induce human endothelial cells to express GMP-140 and bind neutrophils. *J. Cell Biol.* 112:749.

75. Lotan, R. and A. Raz. 1988. Endogenous lectins as mediators of tumor cell adhesion. *J. Cell. Biochem.* 37:107.

76. Rice, G.E. and M.P. Bevilacqua. 1989. An inducible endothelial cell surface glycoprotein mediates melanoma adhesion. *Science* 246:1303.

77. Boukerche, H., O. Berthier-Vergnes, E. Tabone, J.-F. Dore, L.L. K. Leung, and J.L. McGregor. 1989. Platelet-melanoma cell interaction is mediated by the glycoprotein IIb-IIIa complex. *Blood* 74:658.

12
PMN Extravasation in Acute Inflammation: A Role for Selectin Interaction in Initial PMN– Endothelial Cell Recognition

LOUIS J. PICKER, ULRICH H. VON ANDRIAN, ALAN R. BURNS, CLAIRE M. DOERSCHUK, KARL-E. ARFORS, AND EUGENE C. BUTCHER

Introduction

The selectins are a unique family of cell adhesion molecules characterized functionally by lectin activity and structurally by the juxtaposition of an N-terminal C-type lectin domain, an epidermal growth factor (EGF) domain, and variable numbers of complement-regulatory protein (CRP)-like repeating units (1–4). Known members of this family include a single leukocyte selectin—LECAM-1 (L-selectin; Leu 8 Ag; MEL-14 Ag; LAM-1)—and two vascular selectins—ELAM-1 (E-selectin) and GMP-140 (P-selectin; PADGEM; CD62). Each of these selectins has been shown to participate in polymorphonuclear neutrophilic leukocyte (PMN) interactions with endothelial cells (3,5–13), and is thought to cooperate with β_2-integrins (CD11a,b/CD18) and their endothelial cell (EC) ligands (ICAM-1; ICAM-2) in the mediation of PMN emigration from blood to tissue at sites of acute inflammation. In this chapter, we discuss evidence indicating that the selectins participate in the initial recognition event between PMN and inflamed endothelium, recognition characterized by reversibility and independence from PMN activation, whereas β_2-integrins (CD11a,b/CD18) and their ligands mediate secondary, stable, activation-dependent adhesion and subsequent transmigration. Moreover, we provide evidence that the leukocyte and vascular selectins mutually interact in carrying out their function: both the vascular selectins appear to preferentially recognize oligosaccharide ligands associated with the PMN selectin LECAM-1. This preferential oligosaccharide-presenting function of LECAM-1 may reflect the selective concentration of this molecule on microvillous processes of PMN surface, structures associated with initial PMN–EC contact. Taken together, these observations provide new insight into the role of the selectins in PMN–EC recognition/adhesion and suggest unique mechanisms by which PMN extravasation may be physiologically regulated.

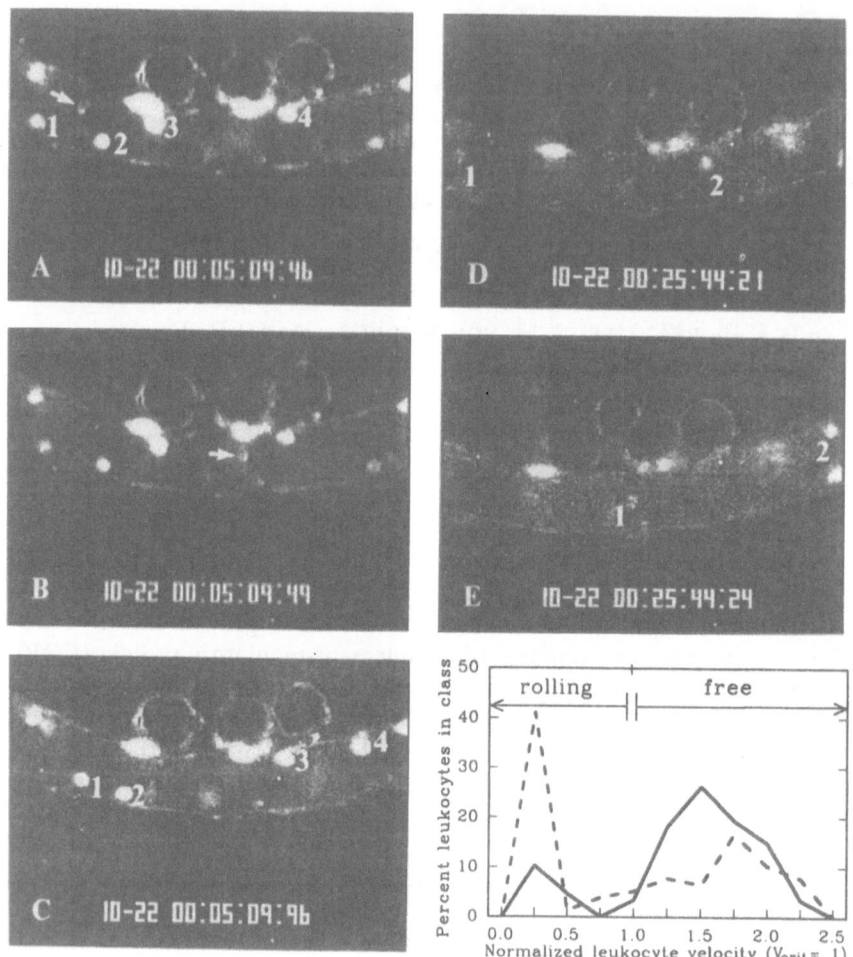

FIGURE 12.1 Leukocyte rolling is a LECAM-1–dependent interaction with endothelial cells. Intravital photomicrographs (using stroboscopic epiillumination) showing fluoresecent leukocytes (labeled in vivo with the systemic injection of acridine red; endothelial cell nuclei are also fluorescent) within a 28-μm-diameter venule (600× magnification) in the rabbit mesentery before (A–C) and 10 min after (D,E) intravenous injection of the anti–LECAM-1 mAb Dreg-200 (16). The blood flow is from left to right. A. Several rolling (#1–#4) and one free (arrow) leukocyte can be seen. B. 33 ms later, the noninteracting leukocyte has traveled approximately 70 μm, whearas the position of the rolling cells is virtually unchanged. C. 500 ms later, the interacting leukocytes, each with a different velocity, have slowly moved downstream. D,E. After antibody injection, the frequency of rolling leukocytes has decreased; most cells passing through the vessel segment do not interact with the vascular wall. (Compare the position of cells #1 and #2 after 30 ms.) The diagram shows a velocity profile of leukocytes passing through a venule during 1 min before (broken line) and after (solid line) anti–LECAM-1 antibody treatment. The velocity of each cell was measured and the critical veloc-

Results and Discussion

Selectins Mediate Initial PMN Adhesion

The earliest-recognizable interaction of PMN with stimulated endothelium in vivo is the rolling of these cells along the endothelial surface of postcapillary and small collecting venules (14–17). The average velocity of rolling PMN is reduced by 80–90% from that of freely flowing PMN in the bloodstream, indicating a significant adhesive interaction. Rolling PMN may be released back into the bloodstream, or may stop and transmigrate into the tissues. β_2-integrins have been implicated in the latter step of the extravasation process. They do not function effectively under shear force, and antibodies to these integrins block firm attachment and diapedesis of PMN, but do not affect rolling (15,16,18,19). β_2-integrin function has been shown to be activation-dependent, requiring PMN stimulation with chemoattractants or other activating factors for optimal adhesion (16,20–22). Indeed, it is thought that the fate of a rolling PMN—either to detach and reenter the bloodstream or to firmly adhere and transmigrate—is determined by the presence of such activating factors associated with inflamed endothelium or diffusing into the vascular lumen from the inflamed tissues.

The participation of the selectins in PMN–EC adhesion has been shown in vitro to be distinct from that of the β_2-integrins (7–9,11,19,21). Among other differences, selectins function well under shear force and do not require PMN activation for effective function. Indeed, PMN activation results in rapid cleavage of LECAM-1 from the PMN surface and loss of LECAM-1–dependent adhesion (6–9,23–25). These observations suggested LECAM-1, as well as the vascular selectins, as logical candidates for the mediation of initial PMN–EC adhesion as manifested by PMN rolling. To test this hypothesis, we utilized intravital videomicroscopy to assess the effect of LECAM-1–specific mAbs on the in vivo interaction of leukocytes with mesenteric venules in the rabbit (16). As illustrated in Figure 12.1, the rolling interaction of rabbit leukocytes with venular endothelium was markedly diminished after treatment with anti-LECAM-1. An anti–CD18 mAb, used as a control, had no effect (not shown). The inhibition mediated by anti–LECAM-1 was manifested by both a decrease in the total number of leukocytes interacting at all with

◁―――――――――――――――――――――――――――――――――――――――

ity (V_{crit}) was determined. V_{crit} is the theoretical minimal velocity which a leukocyte can assume in a parabolic flow profile when it travels next to the vascular wall without interacting with the endothelium. Leukocytes with a velocity less than V_{crit} are rolling; conversely, velocities greater than V_{crit} characterize free cells. Treatment with Dreg 200 resulted in a marked shift of the velocity profile toward a higher percentage of free, noninteracting cells.

endothelium and an increase in the velocity of those cells still capable of rolling.

A similar dependence of leukocyte rolling on LECAM-1 has been recently shown in a rat model; in this report (17), both a soluble, recombinant LECAM-1/IgG chimeric molecule and a polyclonal anti–LECAM-1 antisera specifically reduced leukocyte rolling. In vivo data on the role of the vascular selectins in the PMN rolling phenomena are not yet available, but both ELAM-1 and GMP-140 have been shown to support PMN rolling in in vitro models (19; T.K. Kishimoto and C.W. Smith, personal communication).

The LECAM-1 and Vascular Selectin Adhesion Pathways Overlap

As mentioned above, LECAM-1, ELAM-1, and GMP-140 are all functionally characterized by their lectin activity. Both ELAM-1 and GMP-140 have been shown to recognize sialylated, fucosylated lactosamines, including sialyl Lewis X [sLeX; NeuNAcα2–3Galβ1–4(Fucα1–3)GlcNAc; 26–30]. In the context of its role as lymphocyte homing receptor for peripheral lymph node (PLN), LECAM-1 has been demonstrated to recognize sialylated oligosaccharides associated with the PLN vascular addressin—a tissue-specific determinant expressed by PLN high endothelial venules (31,32). Oligosaccharide ligands for PMN LECAM-1 on acutely inflamed endothelial cells are not well defined, but structural and functional studies suggest that the lectin specificity of PMN LECAM-1 is similar, if not identicial, to its lymphocyte counterpart (5,33).

Thus, given the critical role oligosaccharide recognition plays in the function of these adhesion moelcules, it was anticipated that LECAM-1 and the vascular selectins would function independently in their mediation of the PMN–EC rolling interaction. GMP-140, which is externalized from intracellular stores in EC within minutes of exposure to rapidly produced agents such as thrombin and histamine (12,21), would mediate PMN–EC interaction in the intial phases of the acute inflammatory response, whereas ELAM-1, which requires several hours to up-regulate on EC in response to interleukin 1 or tumor necrosis factor α (10,11), would function in later phases. LECAM-1 might be expected to bridge both early and late phases, recognizing distinct ligands on the activated EC surface. [In the rabbit in vivo model, LECAM-1–dependent leukocyte rolling is found both shortly after initial manipulation of the mesentery and after 5 h of exposure to intraperitoneal interleukin 1 (U. von Andrian and K.-E. Arfors, in preparation).]

However, Kishimoto et al. (9) have presented evidence of possible overlap of the LECAM-1 and vascular selectin-mediated adhesion pathways. This report demonstrated that monoclonal antibodies (mAbs) against LECAM-1 and ELAM-1 show little additivity in their ability to

FIGURE 12.2. PMN binding
to ELAM-1–transfected
COS cells is inhibited by
both LECAM-1–specific
mAbs and by removal of
PMN LECAM-1 by low-
dose chymotrypsin treat-
ment. A. Two-color flow
cytometry of LECAM-1
(PE–Leu 8) vs. sLeX
(FITC–HECA-452) ex-
pression after control treatment (left) or treatment
with low-dose chymotryp-
sin (right; 0.03–0.06 units/
10^6 cells for 5'; 34,35).
Under conditions reducing
LECAM-1 reactivity to
background levels, no
change in sLeX reactivity
is observed. B. The loss of
LECAM-1 (but not total
cell surface, immunoreac-
tive sLeX) seen under
these conditions is correlated with a marked reduction in the ability of treated PMN to bind ELAM-
1 transfectants, an inhibition that is similar in magnitude to that observed
when PMNs pretreated
with the anti–LECAM-1
mAb Dreg-56 are incu-
bated in parallel.

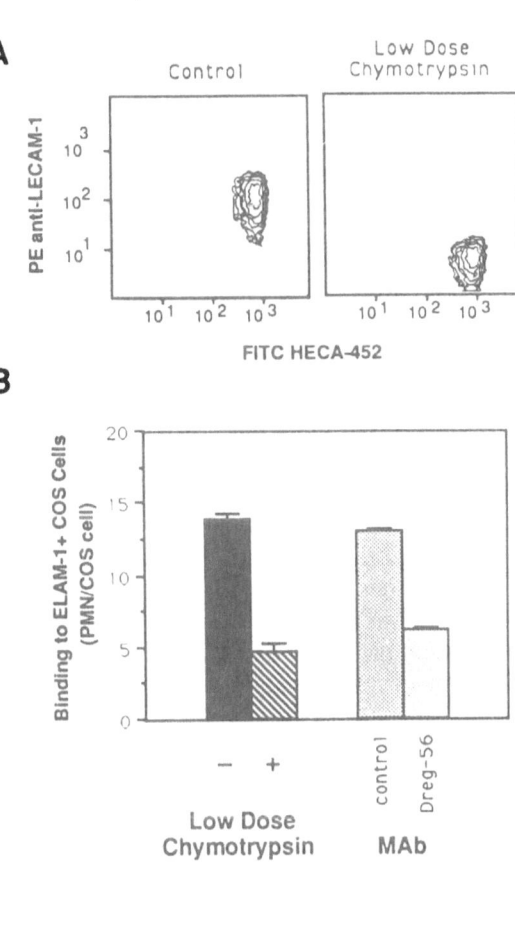

block neutrophil binding to activated EC and that LECAM-1–specific
mAbs reproducibly inhibit PMN binding to ELAM-1–transfected L-cells.
We have independently comfirmed this observation in a COS transfection
system, showing that the anti–LECAM-1 mAb Dreg 56 inhibits up to
60% of the binding to ELAM-1 transfectants as compared to PMN binding control mAbs (34; Fig. 12.2B). Similiar inhibition is seen with
paraformaldehyde-fixed PMN, indicating that mAb-induced down-
regulation of distinct ELAM-1 ligands is unlikely. To further investigate
this observation, we made use of the fact that treatment of PMN with
low-dose chymotryspin (0.03–0.06 units/10^6 cells for 5'; 34,35) completely
removes surface LECAM-1 without alterations in PMN viability, morphology, binding to plastic, random migration, and expression of a varie-

ty of PMN cell-surface molecules, including, most importantly, immunochemically defined cell-surface sLeX (mAb HECA-452; Fig. 12.2A). As illustrated in Figure 12.2B, such treatment inhibited up to 70% of PMN binding to ELAM-1, slightly more than the 60% inhibition seen with the anti–LECAM-1 mAb Dreg 56 in the same experiment. Thus, interference with PMN LECAM-1 function with either mAbs or physical removal reproducibly decreases the ability of PMN to recognize cell-associated ELAM-1, findings suggesting that at least part of PMN binding to ELAM-1 is mediated by cell-surface LECAM-1.

Given the recent observation that the oligosaccharide binding specificity of GMP-140 overlaps with that of ELAM-1—both vascular selectins recognizing sLeX, for example (26–30)—we next asked whether PMN binding to GMP-140 transfectants was also depedent on PMN LECAM-1. In side-by-side comparsions of fresh and fixed PMN binding to ELAM-1 vs. GMP-140–transfected COS cells, the anti–LECAM-1 mAb Dreg 56 was similarly active in inhibiting PMN binding, resulting in 40–60% less PMN bound than with control mAbs (34). These data suggest that a substantial percentage of the ability of cell-associated ELAM-1 and GMP-140 to bind intact PMN is dependent on the surface display of LECAM-1, a finding suggesting that the vascular selectins and LECAM-1 interact, likely as receptor–ligand pairs. This is not true for T-cell interactions with ELAM-1, which are clearly not related to the cell-surface display of the lymphocyte form of LECAM-1 (34,36,37). Thus, PMN LECAM-1 must have structural modifications that mediate these additional functions.

PMN, but not Lymphocyte, LECAM-1 Is Decorated by Oligosaccharide Ligands for ELAM-1 and GMP-140

The direct involvement of LECAM-1 in neutrophil recognition of ELAM-1 and GMP-140 could be explained if these vascular selectins preferentially recognized oligosaccharide ligands present on PMN (but not lymphocyte) LECAM-1 vs. other PMN surface molecules. As mentioned above, ELAM-1 and GMP-140 specifically recognize sLeX or related structures, which are distributed on both glycoprotein and glycolipid components of the PMN cytoplasmic membrane (38,39). The availability of mAbs specific for these oligosaccharide ligands—HECA-452 and CSLEX-1—allowed us to investigate immunochemically the possibility that PMN LECAM-1 was "decorated" with these structures (34,40,41). On western blots of whole PMN lysates, mAb HECA-452 recognizes an array of bands at M_r ranging from 50 to 180 kD (Fig. 12.3A), including a band at about 85–95 kD that comigrates with PMN LECAM-1. Preclearing PMN lysates with mAb Dreg-56, against LECAM-1, specifically removes this HECA-452–reactive band (Fig. 12.3A). Densitometric comparison of HECA-452 reactivity in immunoblots of control and Dreg-56–

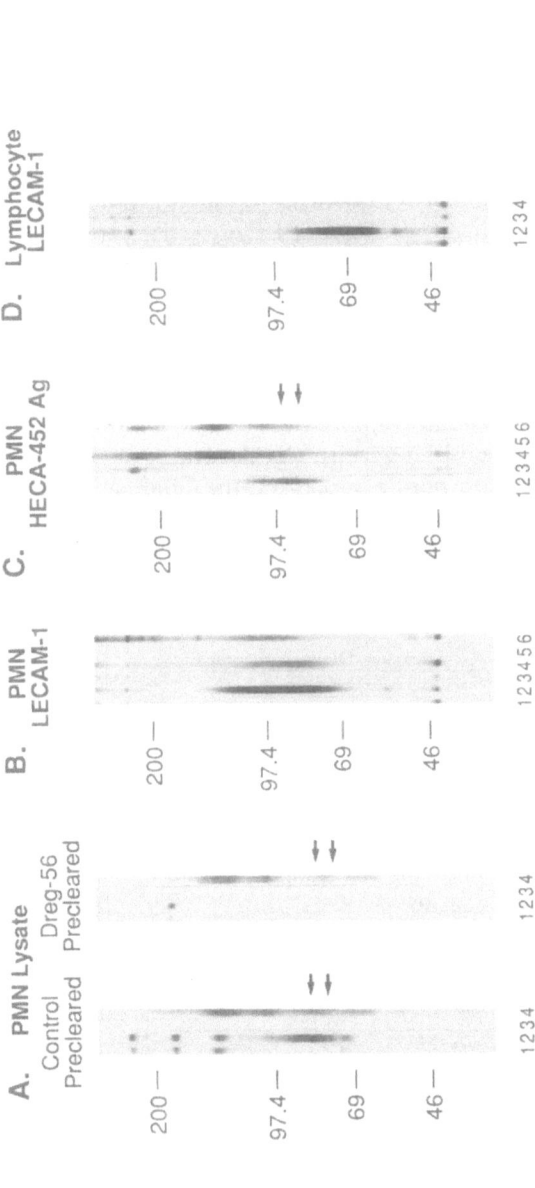

FIGURE 12.3. Western blot analysis of crude PMN lysates (panel A); Dreg-56 (anti-LECAM-1) and HECA-452 (anti-sLeX) immunoprecipitates of PMN lysates (panels B and C, respectively); and Dreg-56 immunoprecipitates of tonsil lymphocyte lysates (panel D). The blots were probed with mAb Dreg-56 vs. a class-matched control (lanes 2 vs. 1, respectively), HECA-452 vs. control (lanes 4 vs. 3), and CSLEX-1 (anti-sLeX) vs. control (lanes 6 vs. 5). On control precleared crude PMN lysates (precleared with mAb L3B12; against CD45; panel A), mAb HECA-452 recognizes multiple glycoproteins of diverse M_r, including a species which comigrates with LECAM-1. Complete preclearing with Dreg-56–Sepharose specifically removes this species (arrows). Analysis of Dreg-56 immunoprecipitates confirms the specific binding of the HECA-452 and CSLEX-1 mAbs to PMN LECAM-1 (panel B), but not the lower M_r lymphocyte LECAM-1 species (panel D). HECA-452 immunoprecipitates show similar HECA-452– and CSLEX-1–reactive species, including a 85–100-kD component reactive with Dreg-56 (arrows). Immunoprecipitations with isotype-matched control mAbs coupled to Sepharose did not show reactivity with any of the detection mAbs used in this analysis (data not shown). All gels were run under nonreducing conditions; the relative positions of the M_r markers are indicated at left ($\times1,000$ kD).

precleared lysates indicates that 5% or less of HECA-452 reactivity in the 50–200-kD M_r range is associated with mAb Dreg-56–defined LECAM-1. Western analyses of LECAM-1 immunoprecipitates revealed that PMN, but not tonsil lymphocyte, LECAM-1 was modified with HECA-452 (and CSLEX-1)–reactive determinants (Fig. 12.3B and D). Reciprocally, HECA-452 immunoprecipitates of PMN (Fig. 12.3C), but not lymphocyte (not shown), lysates contained coprecipitated LECAM-1. These findings were confirmed more quantitatively by ELISA analysis of affinity-purified PMN and tonsil lymphocyte LECAM-1 (34). Thus, although other glycoprotein species also bear high levels of sLeX, LECAM-1 is one major neutrophil glycoprotein decorated by these carbohydrate ligands. This modification is neutrophil-selective, as lymphocyte LECAM-1 is not detectably recognized by sLeX-reactive mAbs.

In order to substantiate the functional significance of the immunochemical demonstration of putative ELAM-1 ligands on LECAM-1, we examined the ability of purified PMN and lymphocyte LECAM-1 to support the binding of ELAM-1–transfected L1-2 cells. As illustrated in Figure 12.4, PMN LECAM-1 adsorbed to plastic supported the binding of L1-2ELAM-1 cells (Fig. 12.4A), but not the control L1-2vector (Fig. 12.4B). In contrast, lymphocyte LECAM-1 preparations, adjusted to identical titers of LECAM-1 immunoreactivity, did not mediate binding of either trans-

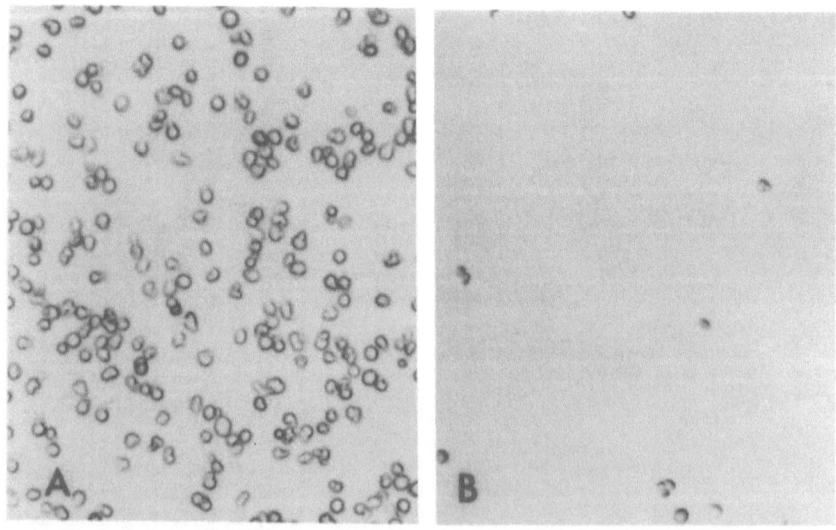

FIGURE 12.4. Isolated PMN LECAM adsorbed to LAB-Tek slides (A and B) supports the binding of ELAM-1 transfectants (L1-2ELAM-1; A), but not control transfectants (L1-2vector; B).

fected or control L1-2 cells (34). PMN LECAM-1 preparations could be diluted 8–10-fold and still show significant ELAM-1 binding over background, indicating that minor differences in LECAM-1 concentration did not account for the inability of lymphocyte LECAM-1 to interact with ELAM-1. PJ-32 Ag, a high-density PMN cell-surface Ag that lacks CSLEX-1 or HECA-452 immunoreactivity, did not support L1-2^{ELAM-1} binding (34).

The specificity of the interaction between isolated PMN LECAM-1 and L1-2^{ELAM-1} was confirmed with mAb-inhibition studies (34). Pretreatment of the L1-2^{ELAM-1} cells with mAbs against ELAM-1 (mAbs CL2 and CL3) completely abrogated the specific interaction between isolated PMN LECAM-1 and the ELAM-1 transfectants. Pretreatment of the adsorbed LECAM-1 with either the HECA-452 or CSLEX-1 mAbs, but not isotype-matched control mAbs, inhibited binding by over 70%. Significantly, pretreatment of the PMN LECAM-1–coated slides with mAbs against LECAM-1 epitopes common to both PMN and lymphocytes (i.e., mAbs not primarily recognizing specific oligosaccharide epitopes; Dreg-56 and Dreg-200) also resulted in nearly 60% inhibition of L1-2^{ELAM-1} binding. These results clearly demonstrate that PMN, but not lymphocyte, LECAM-1 is a vascular selectin counter-receptor, and that anti–LECAM-1 mAbs, against common LECAM-1 determinants, can interfere with this interaction.

Taken together, these data indicate that not only does PMN LECAM-1 display oligosaccharide ligands for the vascular selectins but that these LECAM-1–associated ligands appear to be physiologically superior in mediating cell–cell adhesion (at least in the context of intact PMN binding to cell-associated ELAM-1 or GMP-140 under conditions of shear) than otherwise identical oligosaccharides associated with other molecular structures. In other words, PMN LECAM-1 appears to be a "preferred presenter" of sLeX to cell-associated ELAM-1 or GMP-140. PMN LECAM-1, however, is clearly *not* the only functional vascular selectin ligand. Most variants of the HL-60 myeloid cell line lack LECAM-1 yet still bind to ELAM-1 transfectants (albeit less well than normal PMN; 9); and, as implied by the data described above, at least 30% of the ability of intact PMN to bind ELAM-1 is retained even after essentially complete removal of cell-surface LECAM-1. Moreover, lymphocyte LECAM-1 is not modified by ELAM-1–binding oligosaccharides, and the CLA (HECA-452)$^{+}$ skin-associated memory T-cell subset that avidly binds ELAM-1 does so through distinct HECA-452 epitope-bearing glycoproteins (36,37). Finally, these data do *not* exclude a role for the LECAM-1 lectin domain in the PMN extravasation process. PMN LECAM-1 lectin domains are clearly functional (5,42), and likely recognize specific (at present poorly characterized) oligosaccharides on the surface of inflamed EC.

LECAM-1 Is Preferentially Localized to Microvillous Processes on PMN

It is possible that the selective oligosaccharide "presenting" ability of LECAM-1 is, at least in part, due to specific molecular features of the LECAM-1 molecule, perhaps including participation of the conserved selectin EGF and CRP domains (1,2), whose functions remain largely unknown. While we do not rule out such a possibility, from our own observations and those of others (27,28), it appears likely that, taken out of the context of an interaction between *intact PMN* and *cell-associated* vascular selectins, LECAM-1–associated ELAM-1 ligands may not be recognized by ELAM-1 any more readily than a variety of other molecular structures bearing the appropriate oligosaccharides. We felt it more likely that the special ability of LECAM-1 to display carbohydrate ligands to vascular selectins may be due to a unique topographical distribution of LECAM-1 on the PMN surface that renders LECAM-1–associated sLeX more "bioavailable." To investigate this possibility, fixed neutrophils

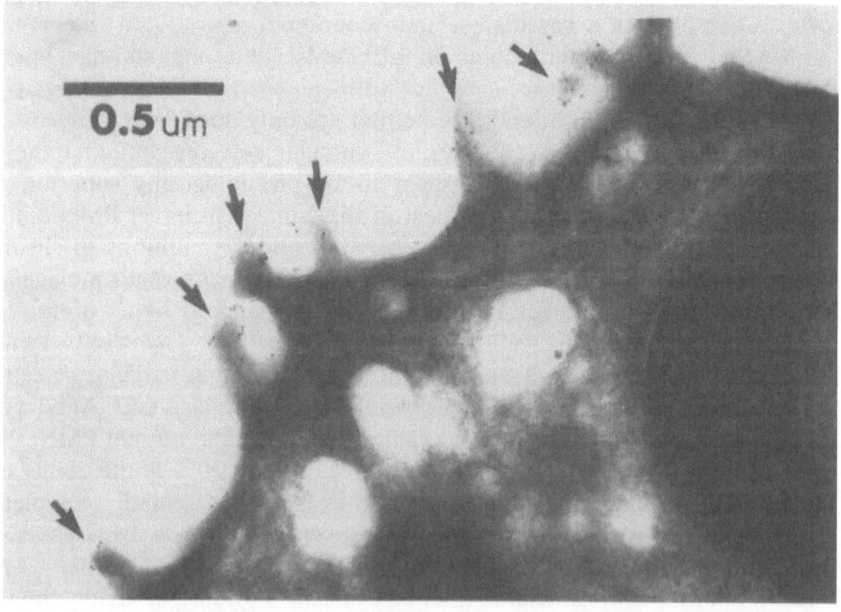

FIGURE 12.5. Immunoelectron microscopic analysis of the distribution of LECAM-1 (mAb Dreg-200) on PMN. LECAM-1 reactivity (delineated by particles of colloidal gold) is concentrated on microvillous processes (arrows) of the PMN surface and is absent from granule membranes. The field shown is representative of the findings for 60 PMN from three donors examined for each mAb. In contrast, CD44, class I MHC antigen, total sLeX, and CD18 reactivity showed no predilection for microvillous processes. (See ref. 34.)

were immunostained (immunogold labeling) with mAbs against LECAM-1, CD18, total sLeX, HCAM (CD44), and major histocompatibility complex (MHC) class I molecules, as well as class-matched controls, and the distribution of these antigens was examined by electron microscopy. Remarkably, LECAM-1 reactivity on the PMN surface was consistently localized to microvillous processes (Fig. 12.5). Only a small fraction of LECAM-1 labeling was identified in "flat" regions of the PMN surface, and no appreciable staining was observed in cytoplasmic granules. This specific distribution of LECAM-1 was apparent in all human PMN examined (60 from three different individuals; 34) and has also been observed on rabbit PMN (A. Burns and C. Doerschuk, unpublished data). In contrast, the cell-surface immunoreactivity of CD18, CD44, class I MHC, and total sLeX was randomly distributed on the PMN plasma membrane (34). CD18 and total sLeX also showed significant immunoreactivity on the membranes of cytoplasmic granules.

The concentration of LECAM-1 on microvillous processes, structures previously demonstrated to mediate the initial contact between leukocyte and EC membranes (43–45), is certainly consistent with the proposed role of this molecule in the initial binding interaction between circulating PMN and EC. Moreover, the localization of LECAM-1–associated selectin ligands to those regions of the PMN membrane most likely to make contact with other cells offers a plausible explanation for the apparent functional predominance of these oligosaccharides over the same structures associated with other molecules on the PMN surface—concentration on microvillous processes might act to preferentially pre-position LECAM-1–associated ligands for effective interaction with vascular selectins on activated EC, particularly under conditions of flow.

Conclusions: A Dynamic Model of PMN–EC Interactions in the Extravasation Process

Data obtained from both in vitro and in vivo studies suggest that neutrophil extravasation at sites of acute inflammation may be considered a two-step process (15–19,22,34). First, PMN within the flowing blood must recognize and bind venular EC activated by a tissue-based inflammatory stimulus. The adhesion strength of this initial interaction may be insufficient to completely overcome intravascular shear force, resulting in the characteristic PMN rolling phenomenon. The second step of this process involves adhesion strengthening—insuring a complete stop of the rolling PMN—followed by migration to an interendothelial cell junction and then diapedesis through the venular wall. A dynamic model of the molecular events mediating this two-step process is presented in Figure 12.6. In this model, the initial interaction of circulating PMN with activated EC—characterized by the rolling phenomenon—is largely selectin-mediated. β_2-integrins (MAC-1 and LFA-1) are present on the PMN

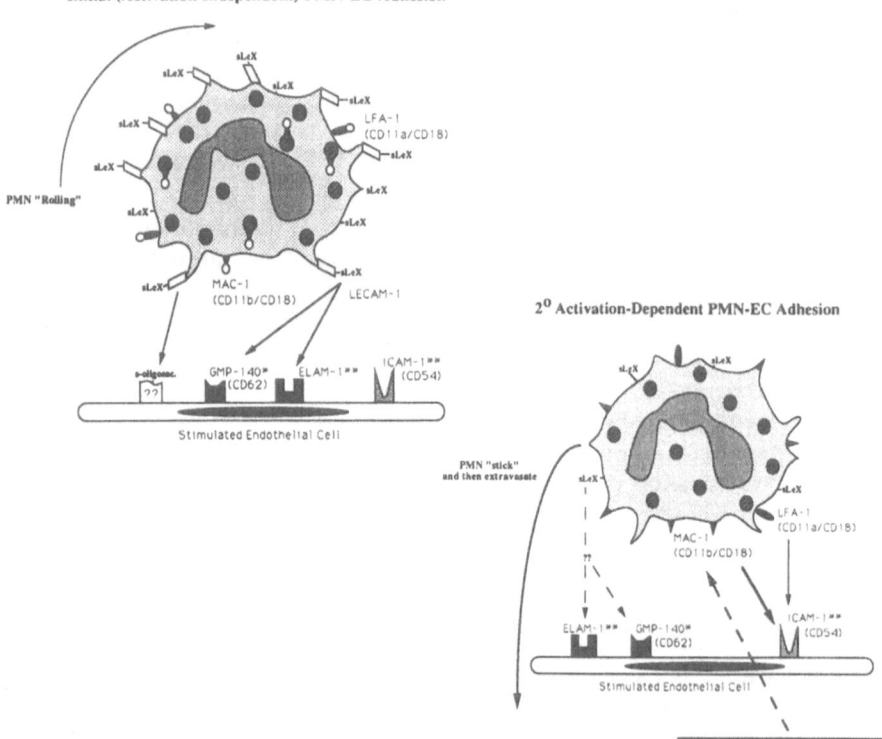

FIGURE 12.6. A dynamic model of PMN–EC interaction. *Early in the course of the inflammatory response, activated ECs display GMP-140 and low levels of the β_2-integrin ligand ICAM-1. **At later time points, ICAM-1 levels are markedly increased, and ELAM-1 replaces GMP-140 as the predominant vascular selectin. Another β_2-integrin ligand, ICAM-2 (not shown), is constitutively present on EC. The initial "rolling" interaction of circulating PMN with inflamed venular endothelium is selectin-mediated, a significant portion of which represents the interaction of the vascular selectins with LECAM-1–associated sLeX. LECAM-1 recognition of distinct sialylated oligosaccharide ligands on the EC likely also contributes to the rolling interaction, as may, to a lesser extent, vascular selectin recognition of sLeX on non–LECAM-1 structures. This initial, unstable PMN–EC interaction allows the PMN to "sample" the local environment, and if appropriate tissue-derived or EC-associated PMN activating factors are present, the next step of the extravasation process is initiated. These activating factors cause an upregulation of β_2-integrin function (see text), associated with LECAM-1 cleavage and down-regulation of selectin-mediated adhesion. (Some residual non–LECAM-1-dependent, ELAM-1– or GMP-140–mediated adhesion may be present; denoted by ?? in lower panel.) This switch from a largely selectin-mediated to a largely β_2-integrin-mediated adhesion results in firm PMN–EC binding and the intravascular arrest of the rolling PMN and is followed by PMN transmigration through inter-EC junctions.

surface, but in a relatively inactive conformation (inactive receptors delineated by open circle in Fig. 12.6). A major component of selectin function would be mediated by LECAM-1—both in its role as a lectin recognizing oligosaccharides on EC, and as a presenter of sLeX to one or the other of the vascular selectins (GMP-140 early in the course of the inflammatory resonse; ELAM-1 at later time points). These initial selectin–mediated interactions may act directly on the PMN to up-regulate β_2-integrin–mediated adhesion (46), or may bring the EC-associated (rolling) PMN within the "concentration boundary layer" (47) of tissue- or EC-derived PMN activating agents that would effect this upregulation. β_2-integrin functional up-regulation has been shown to result from both quantitative increases in cell-surface expression via externalization of intracellular, granule-associated stores, and qualitatively by virtue of molecular or conformational activation (20–22). These activated β_2-integrins may then interact with ICAM-1 and perhaps other ligands on the EC surface with the end result of firm PMN/EC binding and intravascular arrest of the rolling PMN. At the same time, PMN activation would induce the shedding of LECAM-1, down-regulating a large portion of the selectin-mediated adhesion. The same activating agents that up-regulate β_2-integrin frunction are known to induce LECAM-1 shedding (22–25). (This shedding need not be global, but may be restricted to the area of PMN surface in close approximation to the EC.) Down-regulation of selectin-mediated adhesion may be required to facilitate or trigger subsequent diapedesis. The active movement of PMN to EC intercellular junctions and through the venular wall into inflamed tissues would then be largely dependent on β_2-integrin function.

The data presented in this chapter suggest that PMN LECAM-1, in addition to its role as a carbohydrate-binding molecule, serves as a preferred oligosaccharide presenting scaffold for the vascular selectins. From a regulatory perspective, this finding of a linkage between the leukocyte and vascular selectin adhesion pathways offers a mechanism by which the function of the three selectins may be coordinately regulated. Functional down-regulation of all LECAM-1 function and a significant proportion of both ELAM-1 and GMP-140 function can be accomplished by the activation-induced release of LECAM-1. As mentioned above, such down-regulation of selectin function may be important in initiating diapedesis. In addition, coordinate selectin down-regulation may serve as an autoregulatory mechanism that functions to limit PMN extravasation after severe inflammatory stimuli. Permeation of a given vascular bed with neutrophil-activating factors may result in LECAM-1 down-regulation on the freely circulating neutrophils that enter that vascular bed, and thereby reduce their ability to interact with ECs via all three selectin-mediated pathways.

These studies also offer insight into basic molecular mechanisms mediating cell adhesion. Our observation of a selective localization of

LECAM-1 on the PMN surface suggests that a differential distribution of adhesion molecules on the leukocyte surface may regulate the relative participation of some of these structures in various adhesion events. The molecular basis of this phenomenon and its general implications for adhesion molecule function obviously deserve further investigation. Furthermore, these data indicate that the selectin family of mammalian lectins have functional capabilities in addition to oligosaccharide recognition, such as shown here for LECAM-1, a selective ability to present carbohydrate structures to related lectins. Indeed, given the structural and genetic similarities of the three known selectins (1–4), it is tempting to speculate that the modern selectins evolved from a single ancestral molecule that functioned as a homophilic adhesion molecule recognizing self-carbohydrate structures.

References

1. Siegelman, M.H., Van De Rijn, M., Weissman I.L. 1989. Mouse lymph node homing receptor cDNA clone encodes a glycoprotein revealing tandem interaction domains. *Science* 243:1165–1172.
2. Lasky, L.A., Singer, M.S., Yednock T.A., Dowbenko D., Fennie C., Rodriquez H., Nguyen T., Stachel S., Rosen S.D. 1989. Cloning of a lymphocyte homing receptor reveals a lectin domain. *Cell* 56:1045–1055.
3. Bevilacqua, M.P., Stengelin, S., Gimbrone, Jr, M.A., Seed, B. 1989. Endothelial leukocyte adhesion molecule 1: An inducible receptor for neutrophils related to complement regulatory proteins and lectins. *Science* 243:1160–1165.
4. Johnston, G.I., Cook, R.G., McEver, R.P. 1989. Cloning of GMP-140, a granule membrane protein of platelets and endothelium: Sequence similarity to proteins involved in cell adhesion and inflammation. *Cell* 56:1033–1044.
5. Lewinsohn, D.M., Bargatze, R.F., Butcher, E.C. 1987. A common endothelail cell recognition system shared by neutrophils, lymphocytes, and other leukocytes. *J. Immunol.* 138:4313–4321.
6. Jutila M.A., Rott, L., Berg E.L., Butcher, E.C. 1989. Function and regulation of the neutrophil MEL-14 antigen in vivo: Comparison with LFA-1 and MAC-1. *J. Immunol.* 143:3318–3324.
7. Hallman, R., Jutila, M.A., Smith, C.W., Anderson, D.C., Kishimoto, T.K., Butcher, E.C. 1991. The peripheral lymph node homing receptor, LECAM-1, is involved in CD18-independent adhesion of human neutrophils to endothelium. *Biochem. Biophys. Res. Commun.* 174:236–243.
8. Smith, C.W., Kishimoto, T.K., Abbass, O., Hughes, B., Rothlein, R., McIntire, L.V., Butcher, E., Anderson, D.C. 1991. Chemotactic factors regulate lectin adhesion molecule 1 (LECAM-1)-dependent neutrophil adhesion to cytokine-stimulated endothelial cells in vitro. *J. Clin. Invest.* 87:609–618.
9. Kishimoto, T.K., Warnock, R.A., Jutila, M.A., Butcher, E.C., Lane, C., Anderson, D.C., Smith, C.W. 1991. Antibodies against human neutrophil LECAM-1 (LAM-1/Leu-8/DREG-56 antigen) and endothelial cell ELAM-1

inhibit a common CD-18-independent adhesion pathway in vitro. *Blood* 78:805–811.

10. Bevilacqua, M.P., Pober, J.S., Mendrick D.L., Cotran R.S., Gimbrone, Jr. M.A. 1987. Identification of an inducible endothelial-leukocyte adhesion molecule, E-LAM-1. *Proc. Natl. Acad. Sci. USA* 84:9238–9242.

11. Dobrina, A., Schwartz, B.R., Carlos, T.M., Ochs, H.D., Beatty, P.G., Harlan, J.M. 1989. CD11/CD18-independent neutrophil adherence to inducible endothelial-leucocyte adhesion molecules (E-LAM) in vitro. *Immunol.* 67:502–508.

12. Moore, K.L., Varki, A., McEver, R.P. 1991. GMP-140 binds to a glycoprotein receptor on human neutrophils: evidence for a lectin-like interaction. *J. Cell Biol.* 112:491–499.

13. Watson, S.R., Fennie, C., Lasky, L.A. 1991. Neutrophil influx into an inflammatory site inhibited by a soluble homing receptor-IgG chimaera. *Nature.* 349:164–167.

14. Atherton A., Born, G.V.R. 1972. Quantitative investigation of the adhesiveness of circulating polymorphonuclear leukocytes to blood vessel walls. *J. Physiol.* 222:447–474.

15. Arfors, K.E., Lundberg, C., Lindbom, L., Lundberg, K., Beatty, P.G., Harlan, J.M. 1987. A monoclonal antibody to the membrane glycoprotein complex CD18 inhibits polymorphonuclear leukocyte accumulation and plasma leakage in vivo. *Blood* 69:338–340.

16. Von Andrian, U.H., Chambers, J.D., McEvoy, L., Bargatze, R.F., Arfors, K.E., Butcher, E.C. 1991. A two step model of leukocyte-endothelial cell interaction in inflammation: Distinct roles for LECAM-1 and the leukocyte $\beta2$ integrins in vivo. *Proc. Natl. Acad. Sci. USA* 88:7538–7542.

17. Ley, K., Gaehtgens, P., Fennie, C., Singer, M.S., Lasky, L.A., Rosen, S.D. 1991. Lectin-like cell adhesion molecule 1 mediates leukocyte rolling in mesenteric venules in vivo. *Blood* 77:2553–2555.

18. Lawrence, M.B., Smith, C.W., Eskin, S.G., McIntire, L.V. 1990. Effect of venous shear stress on CD18-mediated neutrophil adhestion to cultured endothelium. *Blood* 75:227–237.

19. Lawrence, M.B., Springer, T.A. 1991. Leukocytes roll on a selectin at physiologic flow rates: Distinction from and prerequisite for adhesion through integrins. *Cell* 65:859–873.

20. Kishimoto T.K., Larson, R.S., Corbi, A.L., Dustin, M.L., Staunton D.E., Springer, T.A. 1989a. The leukocyte integrins. *Adv. Immunol.* 46:149–182.

21. Lorant, D.E., Patel, K.D., McIntyre, T.M., McEver, R.P., Prescott, S.M., Zimmerman, G.A. 1991. Coexpression of GMP-140 and PAF by endothelium stimulated by histamine or thrombin: A juxtacrine system for adhesion and activation of neutrophils. *J. Cell Biol.* 115:223–234.

22. Kishimoto, T.K. 1991. A dynamic model for neutrophil localization to inflammatory sites. *J. N.I.H. Res.* 3:75–77.

23. Kishimoto, T.K., Jutila, M.A., Berg, E.L., Butcher, E.C. 1989. Neutrophil Mac-1 and MEL-14 adhesion proteins inversely regulated by chemotactic factors. *Science* 245:1238–1241.

24. Berg, M., James S.P. 1990. Human neutrophils release the Leu-8 lymph node homing receptor during cell activation. *Blood* 76:2381–2388.

25. Griffin, J.D., Spertini, O., Ernst, T.J., Belvin, M.P., Levine, H.B., Kanakura, Y., Tedder, T.F. 1990. Granulocyte-macrophage colony-stimulating factor and other cytokines regulate surface expression of the leukocyte adhesion molecule-1 on human neutrophils, monocytes, and their precursors. *J. Immunol.* 145:576–584.

26. Lowe, J.B., Stoolman, L.M., Nair, R.P., Larsen, R.D., Berhend, T.L., Marks, R.M. 1990. ELAM-1 dependent cell adhesion to vascular endothelium determined by a transfected human fucosyltransferase cDNA. *Cell* 63:475–484.

27. Phillips, M.L., Nudelman, E., Gaeta, F., Perez, M., Singhal, AK., Hakomori, S., Paulson, JC. 1990. ELAM-1 mediates cell adhesion by recognition of a carbohydrate ligand, sialyl-Le^X. *Science* 250:1130–1132.

28. Walz, G., Aruffo, A., Kolanus, W., Bevilacqua, M., Seed, B. 1990. Recognition of ELAM-1 of the sialyl-Le^x determinant on myeloid and tumor cells. *Science* 250:1132–1135.

29. Goelz, S.E., Hession, C., Goff, D., Griffiths, B., Tizard, R., Newman, B., Chi-Rosso, G., Lobb, R. 1990. ELFT: A gene that directs the expression of an ELAM-1 ligand. *Cell* 63:1349–1356.

30. Polley, M.J., Phillips, M.L., Wayner, E., Nudelman, E., Singhal, A.K., Hakamori, S.I., Paulson, J.C. 1991. CD62 and ELAM-1 recognize the same carbohydrate ligand, sialyl-Le^x. *Proc. Natl. Acad. Sci. USA* 88:6224–6228.

31. Berg, E.L., Robinson, M.K., Warnock, R.A., Butcher, E.C. 1991. The human peripheral lymph node addressin is a ligand for LECAM-1, the peripheral lymph node homing receptor. *J. Cell Biol.* 114:343–349.

32. Imai, Y., Singer, M.S., Fennie, C., Lasky, L.A., Rosen, S.D. 1991. Identification of a carbohydrate-based endothelial ligand for a lymphocyte homing receptor. *J. Cell Biol.* 113:1213–1221.

33. Ord, D.C., Ernst, T.J., Zhou, L., Rambaldi, A., Spertini, O., Griffin, J., Tedder, T.F. 1990. Structure of the gene encoding the human leukocyte adhesion molecule-1 (TQ1, Leu-8) of lymphocytes and neutrophils. *J. Biol. Chem.* 265:1–8.

34. Picker, L.J., Warnock, R.A., Burns, A.R., Doerschuk, C.M., Berg, E.L., Butcher, E.C. 1991. The neutrophil selectin LECAM-1 presents carbohydrate ligands to the vascular selectins ELAM-1 and GMP-140. *Cell* 66:921–933.

35. Jutila, M.A., Kishimoto, T.K., Finken, M. 1990. Low-dose chymotrypsin treatment inhibits neutrophil migration into sites of inflammation in vivo: effects on Mac-1 and MEL-14 adhesion protein expression and function. *Cell. Immunol.* 132:201–214.

36. Picker, L.J., Kishimoto, T.K., Smith, C.W., Warnock, R.A., Butcher, E.C. (1991). ELAM-1 is an adhesion molecule for skin-homing T cells. *Nature* 349:796–799.

37. Berg E.L., Yoshino, T., Rott, L.S., Robinson, M.K., Warnock, R.A., Kishimoto, T.K., Picker, L.J., Butcher E.C. 1991. The cutaneous lymphocyte antigen, CLA, is a skin lymphocyte homing receptor for the vascular lectin ELAM-1. *J. Exp. Med.* 174:1461–1466.

38. Fukuda, M., Spooncer, E., Oates, J.E., Dell, A., Klock, J.C. 1984. Structure of sialylated fucosyl lactosaminoglycan isolated from human granulocytes. *J. Biol. Chem.* 259:10925–10935.

39. Symington, F.W., Hedges, D.L., Hakomori, S.I. 1985. Glycolipid antigens of

human polymorphonuclear neutrophils and the inducible HL-60 myeloid cell line. *J. Immunol.* 134:2498–2506.

40. Berg, E.L., Robinson, M.K., Mansson, O., Butcher, E.C., Magnani, J.L. 1991. A carbohydrate domain common to both sialyl Le^A and sialyl Le^X is recognized by the endothelial cell leukocyte adhesion molecule, ELAM-1. *J. Biol. Chem.* 266:14869–14872.

41. Fukushima, K., Hirota, M., Terasaki, P.I., Wakisaka, A., Togashi, H., Chia, D., Suyama, N., Fukushi, Y., Nudelman, E., Hakomori, S.I. 1984. Characterization of sialosylated Lewis^x as a new tumor-associated antigen. *Cancer Res.* 44:5279–5285.

42. Spertini, O., Luscinskas, F.W., Kansas, G.S., Munro, J.M., Griffin, J.D., Gimbrone, Jr., M.A., Tedder, T.F. 1991. Leukocyte adhesion molecule-1 (LAM-1, L-selectin) interacts with an inducible endothelial cell ligand to support leukocyte adhesion. *J. Immunol.* 147:2565–2573.

43. Beesley, J.E., Pearson, J.D., Carleton, J.S., Hutchings, A., Gordon, J.L. 1987. Interaction of leukocytes with vascular cells in culture. *J. Cell Sci.* 33:85–101.

44. Beesley, J.E., Pearson, J.D., Hutchings, A., Carleton, J.S., Gordon, J.L. 1979. Granulocyte migration through endothelium in culture. *J. Cell Sci.* 38:237–248.

45. Ewijk, W.V. 1980. Immunoelectron-microscopic characterization of lymphoid microenvironments in the lymph node and thymus. in *Blood Cells and Vessel Walls: Functional Interactions.* (Ciba Foundation symposium) 71:21–37.

46. Lo, S.K., Lee, S., Ramos, R.A., Lobb, R., Rosa, M., Chi- Rosso, G., Wright, S.D. 1991. Endothelial-Leukocyte Adhesion Molecule 1 stimulates the adhesive activity of Integrin CR3 (CD11b/CD18, Mac-1, $\alpha_m\beta_2$) on human neutrophils. *J. Exp. Med.* 173:1493–1500.

47. Richardson, P.D. 1980. Dynamic theory of leukocyte adhesion. in *Blood cells and Vessel Walls: Functional Interactions.* (Ciba Foundation Symposium) 71:313–335.

13
Interaction of LAM-1 With an Inducible Ligand Supports Leukocyte Adhesion to Endothelium

GEOFFREY S. KANSAS, OLIVIER SPERTINI, FRANCIS W. LUSCINSKAS, MICHAEL GIMBRONE, AND THOMAS F. TEDDER

Introduction

The recruitment of circulating leukocytes into sites of inflammation is of critical importance to host defense. This complex process is initiated by, and regulated largely at the level of, interactions between leukocytes and specially modified regions of the vessel wall. Recently, considerable effort by numerous investigators has begun to reveal the molecular basis of leukocyte–endothelial interactions. It is now clear that adhesion between leukocytes and endothelium is governed principally by members of three gene families: the integrins, the immunoglobulin gene superfamily, and the selectins. (For reviews see 1–4.) While the integrins and Ig gene family each comprise a large gene family, there are three selectins: ELAM-1, which is expressed on cytokine-stimulated endothelium (5–7); PADGEM, also called GMP-140 or CD62, which is expressed on activated platelets as well as activated endothelium (8,9); and the leukocyte adhesion molecule–1 (LAM-1), the human homologue of the MEL-14–defined peripheral lymph node homing receptor (10), which is expressed on all classes of leukocytes (11–14). The selectins are characterized by an interesting and unique mosaic of protein motifs, including an amino-terminal C-type lectin domain, a single EGF-like domain, and a variable number of short concensus repeats (SCR) homologous to those found in complement-binding proteins (15–21). Although the presence of a prominent C-type lectin domain is consistent with a considerable array of functional evidence implicating carbohydrates as ligands for each of the selectins (22–32), the identity of the physiologic ligands remains controversial, and their biochemical characterization remains incomplete. In addition, the precise role each selectin plays in the adhesion of leukocytes to endothelium remains unclear. Here we present evidence that LAM-1, in addition to its well-known role in mediating the binding of lymphocytes to the high endothelial venules (HEV) of peripheral lymph nodes (10,33,34), is also important in the binding of leukocytes to endothelium at sites of inflammation. Specifically, we propose that LAM-1 mediates the initial attachment of all classes of leukocytes to endothelium at sites of inflammation, and that this adhesion is mediated through a novel, inducible glycoprotein.

Results

In order to study the role of LAM-1 (and other adhesion molecules) at sites of inflammation, it was essential to develop an assay which represented a good model of inflammation and in which the action of LAM-1 (and other adhesion receptors) could be detected. In this regard, it seemed clear that the standard HEV frozen-section assay did not meet these requirements. We modified the classical HEV frozen-section assay (35,36) by substituting cytokine-treated human umbilical vein endothelial cell (HUVEC) monolayers for the frozen sections. As described later, this new assay met all of the requirements for an assay in which the function of the known leukocyte and endothelial cell adhesion receptors could be studied. In addition, the use of this assay allowed us to compare different pathways utilized by different types of leukocytes, and to functionally characterize a ligand for LAM-1 which is induced on HUVEC by several proinflammatory cytokines.

LAM-1 Is Involved in the Binding of Leukocytes to Activated Endothelium

Early passage HUVECs were plated on gelatinized glass slides and allowed to grow to confluence in normal HUVEC growth media. The cells were then washed, and new media containing 100 U/ml TNF-α (or other cytokines), or no stimulus, was added, and the cells were incubated for an additional 4–6 h. Lymphocytes or neutrophils were then added while the slides were gently rotated (64 rpm) for 15–30 min. The nonadherent cells were gently tipped off, and the slides were fixed in PBS/1% glutaraldehyde at 4°C, and stained with hematoxylin. Adhesion was quantitated by counting cells bound to the endothelium in multiple high-power fields (0.09 mm^2). When this assay was performed at 4°C, where LFA-1–dependent adhesion is inactive (37,38), the binding of both lymphocytes and neutrophils was inhibited ~50% by mAb LAM1-3 (Fig. 13.1), which we have previously shown inhibits the binding of lymphocytes to HEV and of PPME and fucoidin to LAM-1 (34), and which identifies an epitope within the lectin domain (39). Treatment of lymphocytes with a variety of control mAbs, including those which define functionally silent epitopes of LAM-1 (34), were without effect. Thus, LAM-1 is utilized by both lymphocytes and neutrophils for binding to activated endothelium.

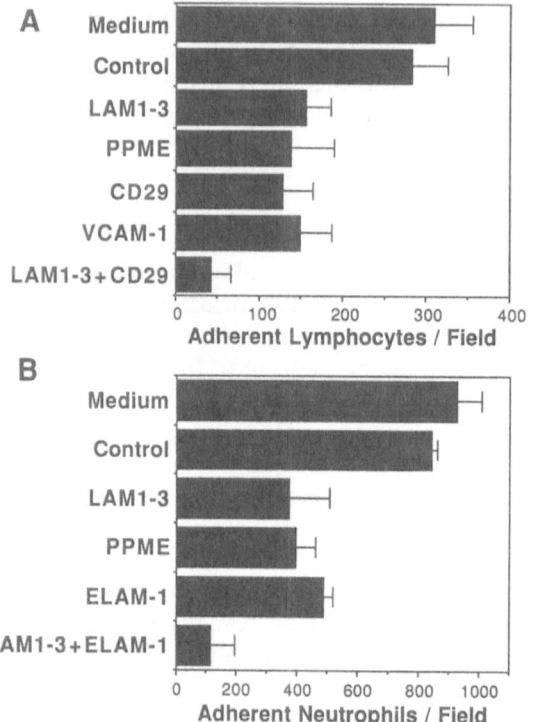

FIGURE 13.1. Inhibition of leukocyte adhesion to activated endothelium by mAb and PPME. Confluent endothelial monolayers were stimulated for 6 h with 100 U/ml rh-TNF-α and cultured with medium, control (anti–MHC class I + anti–LAM1-10 + anti–LAM1-11) mAb, anti–VCAM- 1, or anti–ELAM-1 mAb. Leukocytes and endothelium were treated with the indicated mAb or PPME prior to and during the assay. Both lymphocyte and neutrophil binding is significantly inhibited by mAb to LAM-1 or by PPME.

LAM-1 Mediates the Initial Attachment of Leukocytes to Activated Endothelium

We have been unable to demonstrate a significant role for LAM-1 using traditional static adhesion assays of the sort which have been used by other investigators to characterize other adhesion receptors. We therefore examined the effects of rotation of the slides on LAM-1–dependent adhesion. Inhibition by LAM1-3 mAb was observed only when the slides were rotating (Fig. 13.2). These results suggest that LAM-1 is involved in the initial attachment of leukocytes to activated endothelium. No inhibition of adhesion by LAM1-3 is detected when the assays are performed under static conditions, where no requirement exists for arrest of leukocytes prior to firm adhesion. These results are consistent with recently described in vivo data demonstrating an important role for LAM-1 in neutrophil rolling (40). It has also been shown that PADGEM (GMP-140, CD62) can support neutrophil rolling (41). These observations collectively raise the possibility that the selectins mediate primarily the initial attachment of leukocytes to endothelium, whereas subsequent events, including firm adhesion, are mediated principally by integrins.

FIGURE 13.2. LAM-1–
mediated attachment of
lymphocytes (A) and neut-
rophils (B) to endothelium
was observed only when
the adhesion assays were
carried out with rotation.
Adhesion assays were car-
ried out as described above
except that the leukocytes
and endothelial mono-
layers were incubated
together without (shaded
bars) or with rotation (64
rpm) (filled bars) of the
slides.

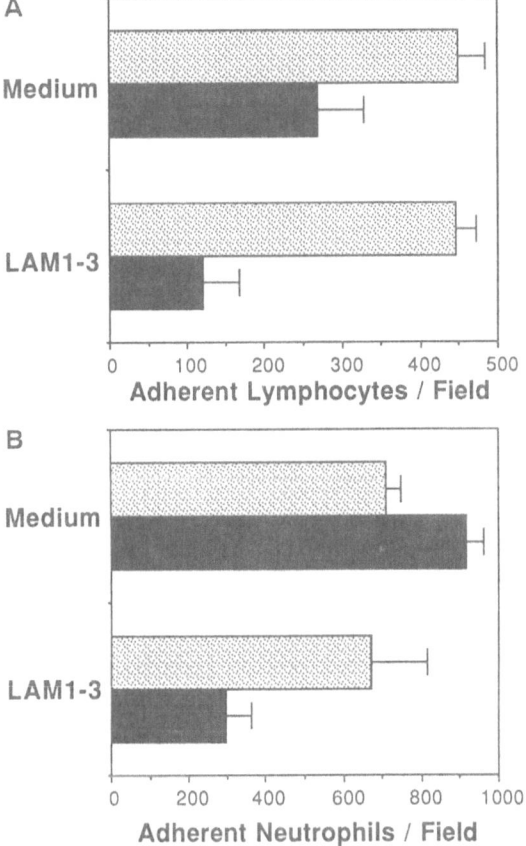

Identical Epitopes of LAM-1 Are Involved in Binding of Lymphocytes to HEV or Activated Endothelium

We have previously described mAbs which identify functionally distinct epitopes of LAM-1 present in different regions of the LAM-1 molecule (34) which are involved in binding of lymphocytes to HEV. To determine if the same epitopes of LAM-1 are involved in lymphocyte binding to en-dothelium, the effect of these mAbs on lymphocyte binding to HEV was compared with their effect on attachment to activated endothelium (Table 13.1). Anti–LAM1-3, -4, and -6, which inhibited >90% of lymphocyte adhesion to HEV, inhibited ~40% of lymphocyte attachment to HUVEC. Similarly, anti-LAM1-1 and LAM1–7, which partially but signi-ficantly inhibited binding to HEV, also significantly inhibited lymphocyte attachment. MAbs which define functionally silent epitopes of LAM-1 in the HEV frozen-section assay were similarly without effect in the HUVEC attachment assay. These data indicate that the same epitopes of

TABLE 13.1. mAb inhibition of lymphocyte binding to activated HUVEC and HEV.

mAb	HUVEC attachment		HEV attachment	
	lymphocytes[a] bound/field	% inhibition[b]	lymphocytes[c] bound/HEV	% inhibition
None	347 ± 21		2.4 ± 1.0	
LAM 1-1	180 ± 78	48**	0.6 ± 0.4	75*
LAM 1-3	188 ± 53	45**	0.3 ± 0.2	87***
LAM 1-4	221 ± 39	36***	0.2 ± 0.1	91*
LAM 1-5	352 ± 40	0	2.1 ± 0.4	12
LAM 1-6	200 ± 49	42***	0.2 ± 0.2	91*
LAM 1-7	196 ± 11	43***	1.7 ± 0.6	29
LAM 1-8	303 ± 10	12	3.0 ± 0.92	0
LAM 1-10	337 ± 36	3	2.0 ± 0.9	16
LAM 1-11	311 ± 18	11	2.4 ± 0.3	0

[a] These data are the means ± SD of results obtained from four fields.
[b] Statistical significance of inhibition: * $p < 0.025$, ** $p < 0.01$, *** $p < 0.005$.
[c] These data represent the mean ± SD obtained in three to eight experiments.

LAM-1 are important in lymphocyte binding to HEV and leukocyte attachment to activated endothelium.

LAM-1 Functions in Concert With Other Adhesion Receptors

We compared the contribution of LAM-1 with that of other adhesion receptors known to be involved in binding to endothelium (Fig. 13.3). At 4°C, binding of lymphocytes to HUVEC was inhibited ~50% by mAbs against either LAM-1 or VLA-4, and the combination of these mAbs inhibited >90% of lymphocyte attachment. Similarly, neutrophil binding was inhibited partially by mAbs to either LAM-1 (~60%) or ELAM-1 (30%), and the combination of these mAbs also inhibited >90% of neutrophil binding. (See also Fig. 13.1.) At 25°C or 37°C, where CD18-dependent adhesion is active, mAbs to CD18, VLA-4, and LAM-1 each partially inhibited lymphocyte adhesion to activated HUVEC, and the combination of these mAbs again inhibited >90% of lymphocyte binding. Similarly, neutrophil binding was partially inhibited by mAbs against LAM-1, ELAM-1, or CD18, and the combination of these mAbs again inhibited adhesion by >90%. Importantly, complete or nearly complete inhibition of adhesion was not achieved unless cells were treated with all of the relevant mAbs. Thus, LAM-1 acts in concert with other adhesion receptors to mediate binding to activated endothelium, and, at least in

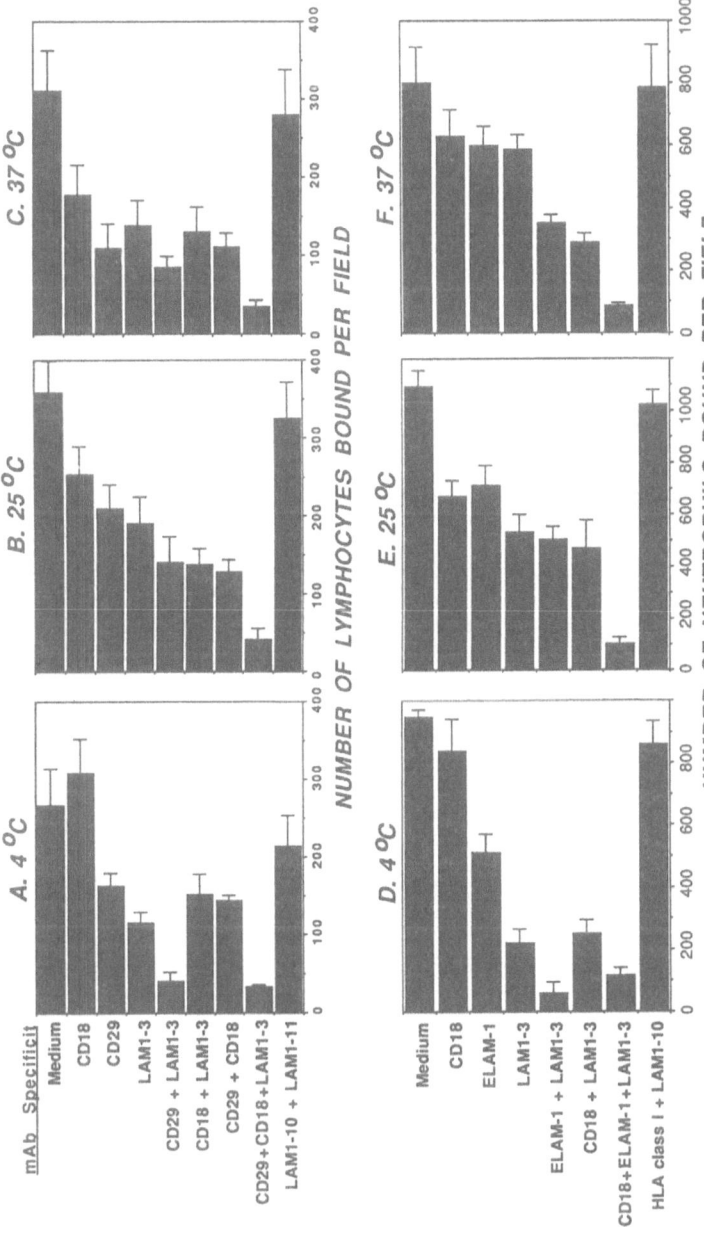

FIGURE 13.3. LAM-1 acts in concert with other adhesion receptors to mediate lymphocyte (A–C) and neutrophil (D–F) adhesion to activated endothelial cells. Assays were carried out as described above at 4°C (A,D), room temperature (B,E), and 37°C (C,F). Less than ten leukocytes were bound/field to unactivated monolayers at 4 and 25°C, while <10 lymphocytes and 150 ± 45 neutrophils were bound/field at 37°C.

A

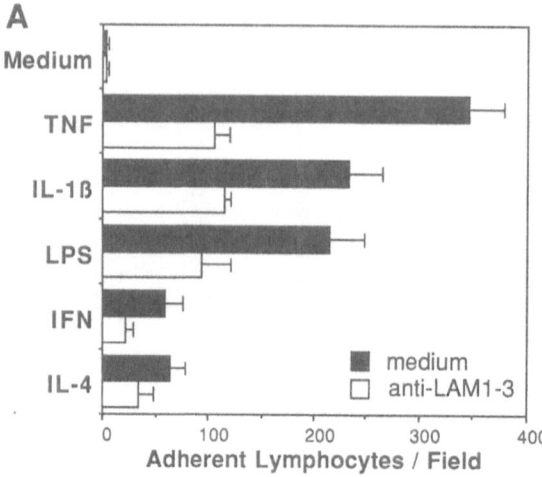

FIGURE 13.4. Induction of a functional LAM-1 ligand by agents that activate endothelial cells. Endothelial monolayers were cultured with medium, TNF-α, IL-1β, LPS, γ-IFN, or IL-4 for 6 h prior to assessment of lymphocyte (A) and neutrophil (B) attachment at 4°C. LAM-1–mediated attachment was assessed using the anti–LAM1-3 mAb to inhibit binding (open bars).

B

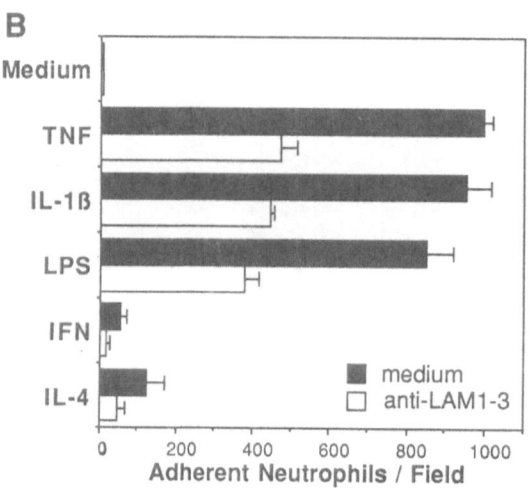

this assay, these adhesion receptors can account for >90% of total leukocyte adhesion.

Induction of a Ligand for LAM-1 by Inflammatory Cytokines

Various cytokines which have previously been shown to induce the expression of other endothelial cell adhesion receptors, including ELAM-1, VCAM-1, and ICAM-1 (6,7,42,43), were tested for their ability to induce LAM-1–dependent adhesion (Fig. 13.4). Very few lymphocytes or neutrophils bound to unstimulated HUVEC. Treatment of HUVEC with TNF-α, IL-1β, or LPS induced a dramatic increase in the level of leuko-

FIGURE 13.5. Kinetics of endothelial cell expression of a LAM-1 ligand. Endothelial monolayers were stimulated with TNF-α for varying periods of time before lymphocyte binding was assessed at 4°C. Separately, endothelium was cultured with cycloheximide (10 mg/ml) for 30 min prior to and during cytokine treatment. Similar to VCAM-1, a LAM-1 ligand is functionally detectable between 2 and 4 h following stimulation of endothelium, and remains for at least 24 h.

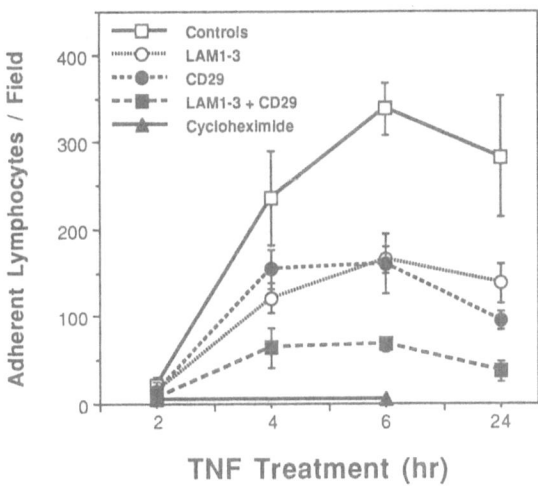

TNF Treatment (hr)

cytes bound, with TNF generally inducing somewhat higher levels of adhesion than either IL-1 or LPS. Treatment of HUVEC with IL-4 or interferon-γ induced much lower, although still significant, levels of binding. In each case, the binding of lymphocytes and neutrophils was inhibited ~50% by LAM1-3 mAb. Thus, treatment of HUVEC with inflammatory cytokines and agents induced expression of a functional LAM-1 ligand, and in this system, none of these stimuli preferentially induced binding by one class of leukocytes.

Kinetics of Expression of LAM-1 Ligand

Endothelium were cultured in the presence of TNF for varying periods of time and the adhesion of lymphocytes was assessed at 4°C in the presence of mAb to LAM-1, VLA-4, or both (Fig. 13.5). Significant adhesion was not detected before 4 h of TNF treatment. Beginning at 4 h and continuing for at least 24 h, high levels of adhesion were observed, and this adhesion was partially inhibited (~50%) by mAbs to LAM-1 or VLA-4, and nearly completely by the combination of these mAbs. Significantly, LAM-1–dependent adhesion was also abrogated by treatment of the endothelium with 10 μg/ml cycloheximide, suggesting that induction of a functional LAM-1 ligand is dependent on de novo protein synthesis. The pattern of expression of the LAM-1 ligand therefore parallels that of VCAM-1, the known cellular counterreceptor for VLA-4 (44), but is clearly distinct from the pattern of expression of ELAM-1, which peaks at 4–6 h and declines to baseline levels by 12–24 h (6,7,45). Unlike VCAM-1, however, the LAM-1 ligand supports neutrophil adhesion.

Transfection With LAM-1 cDNA Confers the Ability to Bind to Activated Endothelium

To confirm the role of LAM-1 in binding of leukocytes to activated endothelium, the NALM-6 pre–B-cell line was transfected with LAM-1 cDNA. The level of expression of LAM-1 on the surface of these transfected cells was comparable to that of normal lymphocytes, and the transfected gene product retains lectin domain activity, as measured by binding of PPME (Fig. 13.6A). The parent NALM-6 cell line expresses VLA-4 but does not express the CD11/CD18 integrins or LAM-1 (data not shown). VLA-4 on the surface of NALM-6 mediates adhesion to TNF-stimulated endothelium via binding to VCAM-1 (Fig. 13.6B). Transfection of NALM-6 with LAM-1 cDNA resulted in an additional large increment of adhesion to TNF-treated endothelium above that which can be accounted for by VLA-4, and this enhanced level of binding is inhibited by LAM1-3 mAb (Fig. 13.6B). These data support our hypothesis that a functional LAM-1 ligand is induced by inflammatory stimuli on endothelium.

Conclusions

These results directly demonstrate that LAM-1 is involved in the adhesion of multiple classes of leukocytes to cytokine-activated endothelium. The observation that rotation of the slides was essential for LAM-1–dependent adhesion suggests that LAM-1 mediates the initial interactions of leukocytes with the endothelium, with other adhesion receptors being engaged subsequently. The activity of these other adhesion receptors, including CD11/CD18 and VLA-4 on lymphocytes, CD11/CD18 on neutrophils, and ELAM-1 and VCAM-1 on cytokine-treated endothelium, was easily detectable in our assays. Importantly, LAM-1, VLA-4/VCAM-1, and LFA-1 appeared to function independently in supporting lymphocyte adhesion to activated endothelium, and LAM-1, ELAM-1, and CD11/CD18 appeared to function independently in supporting neutrophil adhesion. In addition, transfection of LAM-1–lymphoblastoid cell lines with LAM-1 cDNA conferred on these cell lines the ability to bind to cytokine-treated endothelium in a LAM-1–dependent manner. LAM-1 therefore functions as an endothelial receptor on multiple classes of leukocytes, in addition to its previously well-characterized role in binding of lymphocytes to lymph node HEV, and functions in concert with other adhesion receptors to support leukocyte attachment to activated endothelium.

The binding of lymphocytes to HUVEC was inhibited by the same anti–LAM-1 mAbs which we have previously shown (34) to inhibit adhesion to HEV. In addition, the binding of lymphocytes to HEV or

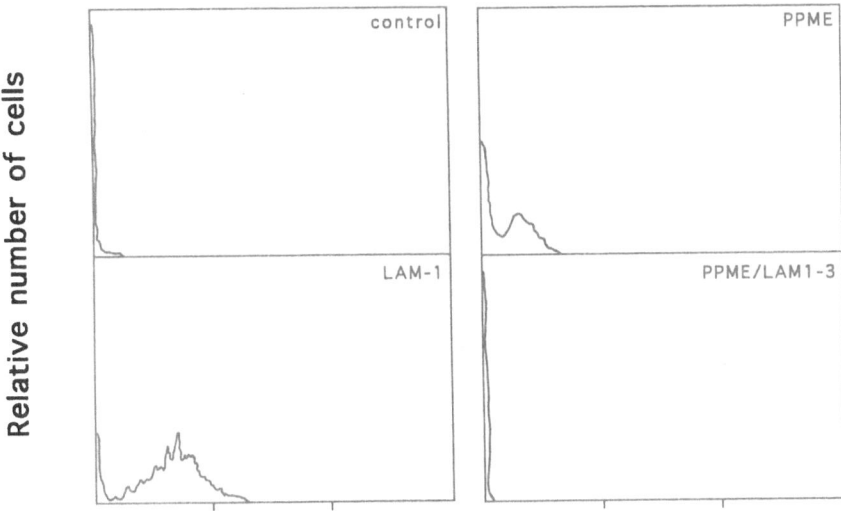

Relative Fluorescence Intensity

FIGURE 13.6. Transfection of LAM-1–lymphoblastoid cell lines with LAM-1 cDAN confers the ability to bind to activated HUVEC. A. Expression and lectin activity of NAML-6/LAM-1. LAM-1 is expressed at levels similar to that of lymphocytes (lower left), and binds PPME (top right); the binding of PPME is blocked by LAM1-3 mAb (lower right).

FIGURE 13.6. B. Binding of NALM-6 or NALM-6/ LAM-1 to HUVEC. Expression of LAM-1 on NALM-6 confers a large increment of binding above that attributable to VLA-4, and this binding is blocked by LAM1-3 mAb.

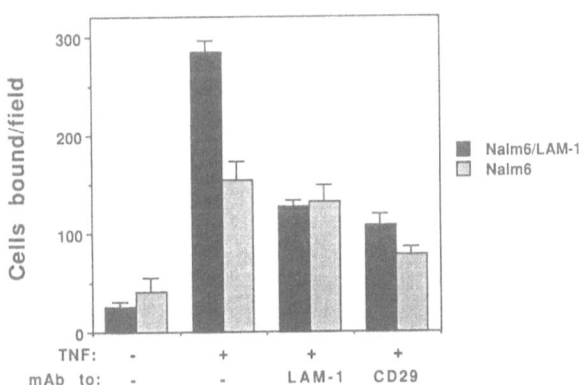

HUVEC, or of neutrophils to HUVEC, is inhibited by PPME or by prior treatment of the HUVEC with neuraminidase (Fig. 13.1, data not shown, and 26,29). These results suggest that the molecular mechanism of LAM-1–mediated adhesion is identical in these three leukocyte-endothelial interactions; i.e., that the lectin domain of LAM-1 present on all classes of leukocytes recognizes a specific carbohydrate ligand on the surface of endothelium.

A functional LAM-1 ligand was induced on HUVEC stimulated with the proinflammatory cytokines TNF-α and IL-1β and by LPS. This ligand was first functionally detectable between 2 and 4 h and remained functionally detectable for at least 24 h. Although the kinetics of appearance of this LAM-1 ligand is similar to that of ELAM-1 and VCAM-1, and although the appearance of each of these inducible endothelial adhesion molecules requires de novo protein synthesis, the LAM-1 ligand can nonetheless be distinguished from ELAM-1 and VCAM-1. The LAM-1 ligand supports neutrophil binding, unlike VCAM-1 (46–48), and is expressed for at least 24 h, unlike ELAM-1 (6,7,43). In addition, mAbs to ELAM-1 or VCAM-1 exhibit inhibition of adhesion of neutrophils and lymphocytes, respectively, which is additive with inhibition due to mAbs to LAM-1. These results indicate that a novel, inducible glycoprotein on the surface of cytokine-treated endothelium functions as a ligand for LAM-1 on all classes of leukocytes to support leukocyte attachment to endothelium. Although the biochemical nature of this ligand is presently unknown, it would not be surprising if this ligand were structurally and functionally homologous to the ligand on HEV defined by Watson et al. and Imai et al. using a MEL-14/IgG chimeric protein (49,50). In summary, LAM-1 and its inducible ligand constitute a new adhesion receptor system which functions to support leukocyte attachment to endothelium at sites of inflammation.

References

1. Springer, T.A. 1990. Adhesion receptors of the immune system. *Nature.* 346:425.
2. Springer, T.A., and L.A. Lasky. 1991. Sticky sugars for selectins. *Nature.* 349:196.
3. Osborn, L. 1990. Leukocyte adhesion to endothelium in inflammation. *Cell.* 62:3.
4. Hemler, M.E. 1988. Adhesive protein receptors on hematopoietic cells. *Immunol. Today.* 9:109.
5. Bevilacqua, M.P., J.S. Pober, D.L. Mendrick, R.S. Cotran, and M.A. Gimbrone Jr. 1987. Identification of an inducible endothelial-leukocyte adhesion molecule. *Proc. Natl. Acad. Sci. USA* 84:9238.
6. Pober, J.S., M.A. Gimbrone, L.A. Lapierre, D.L. Mendrick, W. Fiers, R. Rothlein, and T.A. Springer. 1986. Overlapping patterns of activation of human endothelial cells by interleukin-1, tumor necrosis factor and immune interferon. *J. Immunol.* 137:1893.
7. Pober, J.S., M.P. Bevilacqua, D.L. Mendrick, L.A. Lapierre, W. Fiers, and M.A. Gimbrone. 1986. Two distinct monokines, interleukin 1 and tumor necrosis factor, each independently induce biosynthesis and transient expression

of the same antigen on the surface of cultured human vascular endothelial cells. *J. Immunol.* 136:1680.

8. Hattori, R., K.K. Hamilton, R.D. Fugate, R.P. McEver, and P.J. Sims. 1989. Stimulated secretion of endothelial von Willebrand factor is accompanied by rapid redistribution to the cell surface of the intracellular granule membrane protein GMP-140. *J. Biol. Chem.* 264:7768.

9. Hsu-Lin, S.-C., C. Berman, B. Furie, D. August, and B. Furie. 1984. A platelet membrane protein expressed during platelet activation and secretion. *J. Biol. Chem.* 259:9121.

10. Gallatin, W.M., I.L. Weissman, and E.C. Butcher. 1983. A cell-surface molecule involved in organ-specific homing of lymphocytes. *Nature.* 304:30.

11. Kansas, G.S., G.S. Wood, D.M. Fishwild, and E.G. Engleman. 1985. Functional characterization of human T lymphocyte subsets distinguished by monoclonal anti-Leu-8. *J. Immunol.* 134:2995.

12. Lewinsohn, D.M., R.F. Bargatze, and E.C. Butcher. 1987. Leukocyte-endothelial cell recognition: evidence of a common molecular mechanism shared by neutrophils, lymphocytes, and other leukocytes. *J. Immunol.* 138:4313.

13. Tedder, T.F., A.C. Penta, H.B. Levine, and A.S. Freedman. 1990. Expression of the human leukocyte adhesion molecule, LAM1. Identity with the TQ1 and Leu-8 differentiation antigens. *J. Immunol.* 144:532.

14. Griffin, J.D., O. Spertini, T.J. Ernst, M.P. Belvin, H.B. Levine, Y. Kanakura, and T.F. Tedder. 1990. GM-CSF and other cytokines regulate surface expression of the leukocyte adhesion molecule-1 on human neutrophils, monocytes, and their precursors. *J. Immunol.* 145:576.

15. Siegelman, M.H., and I.L. Weissman. 1989. Human homologue of mouse lymph node homing receptor: evolutionary conservation at tandem cell interaction domains. *Proc. Natl. Acad. Sci. USA* 86:5562.

16. Tedder, T.F., T.J. Ernst, G.D. Demetri, C.M. Isaacs, D.A. Adler, and C.M. Disteche. 1989. Isolation and chromosomal localization of cDNAs encoding a novel human lymphocyte cell-surface molecule, LAM1: Homology with the mouse lymphocyte homing receptor and other human adhesion proteins. *J. Exp. Med.* 170:123.

17. Siegelman, M.H., M. van de Rijn, and I.L. Weissman. 1989. Mouse lymph node homing receptor cDNA clone encodes a glycoprotein revealing tandem interaction domains. *Science* 243:1165.

18. Lasky, L.A., M.S. Singer, T.A. Yednock, D. Dowbenko, C. Fennie, H. Rodriguez, T. Nguyen, S. Stachel, and S.D. Rosen. 1989. Cloning of a lymphocyte homing receptor reveals a lectin domain. *Cell* 56:1045.

19. Bowen, B., T. Nguyen, and L.A. Lasky. 1989. Characterization of a human homologue of the murine peripheral lymph node homing receptor. *J. Cell. Biol.* 109:421.

20. Johnston, G.I., R.G. Cook, and R.P. McEver. 1989. Cloning of GMP-140, a granule membrane protein of platelets and endothelium. *Cell* 56:1033.

21. Bevilacqua, M.P., S. Stengelin, M.A. Gimbrone Jr., and B. Seed. 1989. Endothelial leukocyte adhesion molecule 1: an inducible receptor for neutrophils related to complement regulatory proteins and lectins. *Science* 243:1160.

22. Goelz, S.E., C. Hession, D. Goff, B. Griffiths, R. Tizard, B. Newman, G.

Chi-Rosso, and R. Lobb. 1990. ELFT: A gene that directs the expression of an ELAM-1 ligand. *Cell* 63:1349.

23. Imai, Y., D.D. True, M.S. Singer, and S.D. Rosen. 1990. Direct demonstration of the lectin activity of gp90mel, a lymphocyte homing receptor. *J. Cell Biol.* 111:1225.

24. Larsen, E., T. Palabrica, S. Sajer, G.E. Gilbert, D.D. Wagner, B.C. Furie, and B. Furie. 1990. PADGEM-dependent adhesion of platelets to monocytes and neutrophils is mediated by a lineage-specific carbohydrate, LNF III (CD15). *Cell* 63:467.

25. Lowe, B., L.M. Stoolman, P.N. Rajan, R.D. Larsen, T.L. Berhend, and R.M. Marks. 1990. ELAM-1-dependent cell adhesion to vascular endothelium determined by transfected human fucosyltransferase cDNA. *Cell* 63:475.

26. Rosen, S.D., M.S. Singer, Y.A. Yednock, and L.M. Stoolman. 1985. Involvement of sialic acid on endothelial cells in organ-specific lymphocyte recirculation. *Science* 228:1005.

27. Phillips, M.L., E. Nudelman, F.C.A. Gaeta, M. Perez, A.K. Singhal, S.-I. Hakomori, and J.C. Paulson. 1990. ELAM-1 mediates cell adhesion by recognition of a carbohydrate ligand, Sialyl-Le^x. *Science* 250:1130.

28. Stoolman, L.M., T.A. Yednock, and S.D. Rosen. 1987. Homing receptors on human and rodent lymphocytes—evidence for a conserved carbohydrate-binding specificity. *Blood* 70:1842.

29. True, D.D., M.S. Singer, L.A. Lasky, and S.D. Rosen. 1990. Requirement for sialic acid on the endothelial ligand of a lymphocyte homing receptor. *J. Cell. Biol.* 111:2757.

30. Walz, G., A. Aruffo, W. Kolanus, M. Bevilacqua, and B. Seed. 1990. Recognition by ELAM-1 of the Sialyl-Le^x determinant on myeloid and tumor cells. *Science* 250:1132.

31. Corral, L., M.S. Singer, B.A. Macher, and S.D. Rosen. 1990. Requirement for sialic acid on neutrophils in a GMP-140/PADGEM mediated adhesive interaction with activated platelets. *Bchem. Bphys. Res. Comm.* 172:1349.

32. Polley, M.J., M.L. Phillips, E. Wayner, E. Nudelman, A.K. Singhal, S.I. Hakomori, and J.C. Paulson. 1991. CD62 and endothelial cell leukocyte adhesion molecule-1 (ELAM-1) recognize the same carbohydrate ligand, sialyl-Lewis x. *Proc. Natl. Acad. Sci. USA.* 88:6224.

33. Kishimoto, T.K., M.A. Jutila, and E.C. Butcher. 1990. Identification of a human peripheral lymph node homing receptor: a rapidly down-regulated adhesion molecule. *Proc. Natl. Acad. Sci. USA.* 87:2244.

34. Spertini, O., G.S. Kansas, K.A. Reimann, C.R. Mackay, and T.F. Tedder. 1991. Function and evolutionary conservation of distinct epitopes on the leukocyte adhesion molecule-1 (LAM-1) that regulate leukocyte migration. *J. Immunol.* 147:942.

35. Stamper, H.B., Jr, and J.J. Woodruff. 1976. Lymphocyte homing into lymph nodes: in vitro demonstration of the selective affinity of recirculating lymphocytes for high-endothelial venules. *J. Exp. Med.* 144:828.

36. Jalkanen, S., and E. Butcher. 1985. In vitro analysis of the homing properties of human lymphocytes: developmental regulation of functional receptors for high endothelial venules. *Blood* 66:577.

37. Dustin, M.L., and T.A. Springer. 1988. Lymphocyte function-associated antigen-1 (LFA-1) interaction with intercellular adhesion molecule-1 (ICAM-

1) is one of at least three mechanisms for lymphocyte adhesion to cultured endothelial cells. *J. Cell Biol.* 107:321.

38. Dustin, M.L., and T.A. Springer. 1989. T-cell receptor cross-linking transiently stimulates adhesiveness through LFA-1. *Nature* 341:619.

39. Kansas, G.S., O. Spertini, L.M. Stoolman, and T.F. Tedder. 1991. Molecular mapping of functional domains of the leukocyte receptor for endothelium, LAM-1. *J. Cell Biol.* 114:351.

40. Ley, K., P. Gaehtgens, C. Fennie, M.S. Singer, L.A. Lasky, and S.D. Rosen. 1991. Lectin-like cell adhesion molecule 1 mediates leukocyte rolling in mesenteric venules in vivo. *Blood* 77:2553.

41. Lawrence, M.B., and T.A. Springer. 1991. Leukocytes roll on a selectin at physiologic flow rates: distinction from and prerequisite for adhesion through integrins. *Cell* 65:859.

42. Pohlman, T.H., K.A. Stanness, G.G. Beatty, H.D. Ochs, and J.M. Harlan. 1986. An endothelial cell surface factor(s) induced in vitro by lipopolysaccharide, interleukin 1, and tumor necrosis factor-alpha increases neutrophil adherence by a CDw18-dependent mechanism. *J. Immunol.* 136:4548.

43. Wellicome, S., M. Thornhill, C. Pitzalis, D. Thomas, J. Lanchbury, G. Panayi, and D. Haskard. 1990. A monoclonal antibody that detects a novel antigen on endothelial cells that is induced by tumor necrosis factor, IL-1, or lipopolysaccharide. *J. Immunol.* 144:2558.

44. Elices, M.J., L. Osborn, Y. Takada, C. Crouse, S. Luhowskyj, M.E. Hemler, and R.R. Lobb. 1990. VCAM-1 on activated endothelium interacts with the leukocyte integrin VLA-4 at a site distinct from the VLA-4/fibronectin binding site. *Cell* 60:577.

45. Luscinskas, F.W., A.F. Brock, M.A. Arnaout, and M.A. Gimbrone Jr. 1989. Endothelial-leukocyte adhesion molecule-1 dependent and leukocyte (CD11/CD18)-dependent mechanisms contribute to polymorphonuclear leukocyte adhesion to cytokine-activated human vascular endothelium. *J. Immunol.* 142:2257.

46. Rice, G.E., J.M. Munro, C. Corless, and M.P. Bevilacqua. 1991. Vascular and nonvascular expression of INCAM-110: A target for mononuclear leukocyte adhesion in normal and inflamed human tissues. *Am. J. Pathol.* 138:385.

47. Osborn, L., C. Hession, R. Tizard, C. Vassallo, S. Luhowskyj, G. Chi-Rosso, and R. Lobb. 1989. Direct expression cloning of vascular cell adhesion molecule-1, a cytokine-induced endothelial protein that binds to lymphocytes. *Cell* 59:1203.

48. Carlos, T.M., B.R. Schwartz, N.L. Kovach, E. Yee, M. Rosa, L. Osborn, G. Chi-Rosso, B. Newman, R. Lobb, and J.M. Harlan. 1990. Vascular cell adhesion molecule-1 mediates lymphocyte adherence to cytokine-activated cultured human endothelial cells. *J. Immunol.* 76:965.

49. Watson, S.R., Y. Imai, C. Fennie, J.S. Geoffroy, S.D. Rosen, and L.A. Lasky. 1990. A homing receptor-IgG chimera as a probe for adhesive ligands of lymph node high endothelial venules. *J. Cell Biol.* 110:2221.

50. Imai, Y., M.S. Singer, C. Fennie, L.A. Lasky, and S.D. Rosen. 1991. Identification of a carbohydrate-based endothelial ligand for a lymphocyte homing receptor. *J. Cell Biol.* 113:1213.

14
The Contribution of L-Selectin to Leukocyte Trafficking *In Vivo*

Bruce Walcheck and Mark A. Jutila

Introduction

Leukocyte entry into tissues is the hallmark of all types of acute and chronic inflammatory events. Both inflammation and leukocyte specificity are exhibited in this process. In most situations, cells such as monocytes, neutrophils, and eosinophils do not constitutively traffic through tissues in significant numbers. However, their numbers in tissues can be dramatically increased after certain inflammatory insults. Leukocyte specificity is illustrated by the predominance of neutrophils in sites of early acute inflammation, mononuclear cells in sites of chronic inflammation, and eosinophils in sites of inflammation associated with allergies or parasitic infection.

The first event in the process of leukocyte extravasation is adhesion of the circulating cell to the vascular endothelium. Attention has been given to characterizing the adhesion proteins that regulate this event in hopes of gaining insight to the molecular basis for the specific interactions described above. Two types of adhesion proteins expressed by the circulating leukocyte appear to be important. One type is the leukocyte integrins [LFA-1 (CD11a) and Mac-1 (CD11b)], which regulate adhesion to endothelial cells by binding the inducible endothelial adhesion molecule ICAM-1. The second is leukocyte L-selectin (previously called LECAM-1, LAM-1, gp90MEL-14), which is a member of the selectins, a new family of adhesion proteins characterized by a N-terminal domain homologous to mammalian type-c lectins. This family also includes the vascular selectins, E-selectin (formerly called ELAM-1) and P-selectin (formerly called GMP-140 or PADGEM). The carbohydrate binding activity of selectins appears to be important in their adhesive functions.

Much of what is known concerning the molecular nature of leukocyte integrins and selectins is reviewed in other sections of this book and will not be repeated here. Instead, we will briefly review our functional and regulatory studies in the context of the neutrophil, which have provided, in part, the basis for the current models on the function of these two types of adhesion molecules. Particular emphasis will be given to our work on leukocyte L-selectin in the mouse and human. We will also summarize the molecular characterization of L-selectin in the sheep and cow.

Regulation of Neutrophil L-Selectin Expression

The surface expression of neutrophil L-selectin is uniquely regulated by activating factors, which provides important clues to the function of the molecule in vivo. Neutrophils, isolated from an inflammatory site, express little L-selectin, whereas cells in circulation express high levels of the antigen (1). In response to chemoattractants in vitro neutrophils rapidly shed L-selectin from their cell surface (1,2). In contrast, the expression and functional activity of Mac-1 increases upon activation (1,2). The transition of a L-selectin–high/Mac-1–low neutrophil to a L-selectin–low/ Mac-1–high phenotype occurs rapidly (within minutes) in vitro and in vivo, and before the cell enters into the inflamed tissue (2). Based on these results, it is likely that L-selectin's role in extravasation occurs during the very early interactions between the neutrophil and the vascular endothelium. In contrast, integrin activity, which increases after activation, may be primarily involved in events subsequent to the interactions mediated by L-selectin. These two points will be addressed below.

We (2) and others (3) have proposed that the release of L-selectin from the cell surface after activation is due to proteolysis; although, to date, this has not been conclusively shown. The identification of the putative protease has not been established, nor has the definitive site on the molecule which is cleaved been characterized. However, we have shown that very low doses of chymotrypsin can cause the release of L-selectin (4), and a predicted cleavage site in L-selectin, based upon our earlier studies, contains the chymotrypsin-sensitive amino acid tyrosine (2). Further analysis is obviously needed in order to completely define the role of proteases in the unique regulation of L-selectin expression.

L-Selectin Is Involved in Leukocyte–Endothelial Cell Interactions and Leukocyte Migration to Sites of Inflammation

Dave Lewinsohn, in Eugene Butcher's laboratory, provided the original demonstration of a potential functional role for L-selectin in neutrophil interactions with endothelial cells (5). Using the Stamper and Woodruff ex vivo HEV adhesion assay (6), Lewinsohn showed that, like lymphocytes, bone marrow neutrophils actively bind to high endothelial venules

(HEV) in frozen sections of mouse lymph nodes and other lymphoid tissues (5). The binding to peripheral lymph node HEV was completely blocked by the MEL-14 anti-mouse L-selectin mAb (6). In follow-up studies comparing and contrasting the effects of MEL-14 with mAb against leukocyte integrins (LFA-1 and Mac-1), only MEL-14 significantly blocked binding under conditions of the assay (binding done at 4°C, under constant agitation) (1).

Though the ex vivo HEV binding assay provided the first demonstration that neutrophil L-selectin could specifically interact with endothelial cells, binding was not inflammation specific: lymph nodes were taken from healthy mice and the neutrophils were not activated with inflammatory mediators. Therefore, it was of particular interest to test the effect of anti–L-selectin mAb in leukocyte–endothelial cell adhesion systems that more accurately reflected inflammation. This required the use of endothelial culture systems. At the time, most of the well-characterized systems were established using human umbilical cord endothelial cells, but blocking reagents specific for human L-selectin were not available. Monoclonal antibodies to human L-selectin were first developed (7), and R Hallman et al. (8) showed that they blocked neutrophil adhesion to IL-1–activated cultured endothelial cells under assay conditions which mimic the HEV binding assay (binding done at 4°C and under constant agitation). This work was extended by W. Smith, T.K. Kishimoto, D. Anderson, and colleagues in different models of leukocyte adhesion in which they have quite convincingly shown that neutrophil L-selectin controls a major component of neutrophil binding to cytokine-stimulated endothelial cells under conditions of flow (9,10). Further, Kishimoto has provided evidence that L-selectin participates in the same adhesion pathway as E-selectin expressed by the vascular endothelium (11), and L.J. Picker et al. (12) has recently demonstrated that L-selectin may preferentially present SLeX carbohydrate ligands to the vascular selectins. Each of these latter points are covered in detail in other sections of this book.

Importantly, anti–L-selectin mAbs have been shown to block leukocyte migration to sites of inflammation in vivo. In Lewinsohn's earlier paper he showed that injection of MEL-14 into mice reduced thioglycollate-induced inflammation in the peritoneal cavity without affecting the circulating levels of neutrophils (5). Further, the blocking by MEL-14 given in vivo was equal to that of mAbs to the leukocyte integrins (1). In contrast, when unactivated neutrophils were removed from the animal, precoated with mAbs, and then injected into a second animal with an ongoing inflammatory response, MEL-14, but not anti-integrin mAbs, blocked the accumulation of the transferred neutrophils into the inflamed tissues (1). This latter result suggests that blocking the available integrins (Mac-1) on the surface of unactivated neutrophils is not sufficient to block neutrophil homing to sites of inflammation, which provides in vivo evidence that the increase in cell-surface Mac-1 that occurs after leukocyte activation does participate in extravasation.

The in vivo blocking results discussed have been repeated in different laboratories and in different animal models. S. Watson et al. (13) have shown that a soluble L-selectin–Ig chimeric construct blocks neutrophil entry into the thioglycollate-inflamed peritoneum equally as well as MEL-14. Don Anderson has recently demonstrated the same level of inhibition with antibodies to L-selectin and Mac-1 in a rabbit model of peritonitis (reported in this volume). Finally, two separate groups have shown by intravital microscopy that anti–L-selectin antibodies specifically inhibit the initial interaction of the neutrophil with the inflamed vascular endothelium in vivo (14,15).

The importance of L-selectin in homing to sites of inflammation in vivo has also been shown by means other than mAb blocking studies. Based on our regulation studies summarized above, we found that L-selectin could be removed from the neutrophil cell surface by activation with chemotactic factors (1). L-selectin–negative neutrophils generated by activation no longer exhibit the capacity to home to sites of inflammation when reinjected back into the animal (1). Low doses of chymotrypsin also specifically removed L-selectin from the neutrophil cell surface, but did not cause the secondary effects induced by activation, such as increased integrin activity. Furthermore, this treatment had no demonstrable effect on the expression or function of the leukocyte integrins. However, treatment of neutrophils with low doses of chymotrypsin inhibited their ability to bind endothelial cells in the ex vivo assay and migrate to sites of inflammation in vivo (4).

Model

The antibody-blocking and regulation studies briefly summarized above provided us with the basis for the hypothesis presented in Kishimoto et al. (2) and Jutila et al. (1)—that L-selectin is predominantly involved in the initial interaction of the neutrophil with inflamed vascular endothelium. This event may allow the neutrophil to interact with inflammatory mediators released by the inflamed tissues, which then cause a rapid increase in integrin activity. Integrins would then mediate a strengthening of the binding event, cause leukocyte aggregation, and regulate migration of the neutrophil into the inflamed tissues. The rapid shedding of L-selectin may be an important signal that allows the cell to release from the endothelium and migrate into the underlying tissue. Support for this model has recently come from a number of different studies, some of which appear in this book.

Future Directions

It is clear that new models of acute and chronic inflammatory disease will be needed to fully characterize how the adhesion molecules defined to date are orchestrated in vivo to regulate the exquisite specificity of leukocyte–endothelial cell interactions and whether or not these events can be manipulated to the ultimate benefit of the animal. We have begun to characterize the selectin family in the sheep, goat, and cow, since unique leukocyte-trafficking models are available in these systems. For example: 1) the leukocyte adhesion deficiency syndrome (LAD) has been recently defined in cattle (16) and, thus, cattle offer the first model in which we can examine the function of L-selectin in vivo in the absence of leukocyte integrins, 2) ruminants have very high numbers of circulating γ/δ T cells (17), therefore the characterization of the unique homing phenotype of this cell population can be more readily done in these animals, and 3) models of chronic inflammation, such as arthritis, are available (18).

We have initially identified and characterized mAbs which recognize bovine, ovine, and caprine L-selectin. In a separate report we showed that the DREG 56 and Leu-8 antihuman L-selectin mAbs stain bovine leukocytes (19). The bovine DREG antigen was regulated like human L-selectin, and DREG 56 specifically blocked bovine lymphocyte adhesion to peripheral lymph node HEV (19). Bovine leukocyte binding of the mannose-6-PO_4–rich phosphomannan, PPME, was also blocked by DREG 56. Finally, we have recently identified new mAbs that recognize L-selectin in goats and sheep, and we have shown that these molecules exhibit the same characteristics as the bovine and human protein (data not shown).

Sequence comparisons of bovine and sheep L-selectin lectin domain cDNAs (cloned by PCR), with the published sequences of mouse and human L-selectin lectin domains, demonstrated a high level of identity at the nucleotide level (Fig. 14.1). The predicted amino acid sequences were also highly conserved. The major difference between the ruminant and human sequences was a potential additional N-linked glycosylation site (19). Interestingly, even though there is >97% identity between the sheep and cow L-selectin lectin domains, four mAbs, whose epitopes have been mapped to the human L-selectin lectin domain and recognize bovine L-selectin, did not recognize the sheep molecule.

We used the reagents characterized above to study the expression of L-selectin on different leukocyte populations in newborn calves (<1 month) and mature animals (>1 yr). As shown in Table 14.1 and reported in (19) all neutrophils and monocytes in newborns express L-selectin. The same expression was seen on neutrophils and monocytes from mature animals (data not shown). Essentially, all circulating lymphocytes in newborns expressed L-selectin, but in mature animals a characterisitc bimodal dis-

```
                                                      lectin domain
                                                    ───────────▶
BOVINE   5'  T  TTGCTCATCGTGGCA  CCGATTGC  TGGACT  30
OVINE        *  ****G**********  ********  ******
MOUSE    C  *GATA**C*A***A*  *TC*C**T  ******
HUMAN    C  *G**A****A***A*  **T*C***  ******


TACCATTATTCTAAA  AGACCCATGCCCTGG  GAAAAGGCTAGAGCG  75
*************** ***************  ***G***********
***********G**  *AG******AA****  *****T******AA*
***********G**  *A*******AA****  C***G*******AGA


TTCTGCAGGGAAAAT  TACACAGATTTAGTT  GCCATACAAAACAAA  120
********AG****  ***************  **************G
*******A*C****  **************C  **************G
*****C*A**C***  ***************  **************G


GGAGAGATCGAATAC  CTGAATAAGACACTT  CCCTTCAGCCGTACT  165
*************** ***************  ********T******
A****A**T**G*GT  T*AG*G**T***T*G  ***AAA****C*TA*
*CG**A**T**G**T  ***G*G*****T**G  ********T***T**


TACTACTGGATTGGA  ATCCGGAAAGTAGAA  GGGGTGTGGACTTGG  210
*************** ***************  ***************
***********A***  ***A*****A*T*GG  AAAA*******A***
***********A***  *******GA***G*  **AA*A****G***


GTGGGAACCAACAAA  TCTCTCACTGAAGAA  GCAAAGAACTGGGGT  255
*************** ***************  ***************
*************** A********A*****  ***G***********
*************** ***************  ***G*********A


GCAGGGGAGCCCAAC  AACAGGAAGAGTAAG  GAGGACTGTGTGGAA  300
*A************* ***************  ***************
**T***C*******  ****A****TCC***  **************G
*AT**T********  ****A*****AC***  ********C****G


ATCTATATCAAGAGG  AACAAAGACTCGGGG  AAATGGAATGATGAT  345
*************** *T************  ********C******
*************** GAACG******T***  ***CC**C****C
************A  ********TG*A**C  ********C****C


GCCTGCCACAAAGCA  AAGAAA  XXXXXX  3'                    366
*************** ******
*****T******CG*  ***GC*
***********CT*  ***GC*
```

FIGURE 14.1. Comparison of the nucleotide sequences of bovine, sheep, mouse, and human L-selectin lectin domains. The lectin domains of bovine and sheep LECAM-1 were cloned from single-strand lymphocyte cDNA by PCR using primers based on 5' and 3' flanking regions of the lectin domain taken from the published human L-selectin cDNA sequence (7). The details of these procedures are contained in a separate report (19). The PCR products were sequenced and compared to the published sequences of mouse and human L-selectin. Bovine and sheep lectin domains both showed >80% identity to both human and mouse.

TABLE 14.1. Expression of L-selectin on bovine leukocytes.

Leukocyte[a]	Antigen	Percent positive[b]	Mode flourescence[c]
PMN	L-selectin	98 ± 2.4[e]	233 ± 63
Monocyte	L-selectin	94 ± 5.3	289 ± 44
Lymphocyte (<1 mo)[d]	L-selectin	88 ± 4.1	214 ± 21
(>1 yr)	L-selectin	42 ± 19.7	n.d.

[a] Leukocytes were harvested from the peripheral blood via Histopaque separation, washed, and stained with anti-L-selectin (DREG56) and control mouse mAbs for flow cytometric analysis (FACScan). Leukocyte subsets were identified by their distinctive forward and light side scatter profiles.
[b] Percent positive is the percentage of cells which exhibited fluorescence above background given with the negative control mAb.
[c] Mode fluorescence of the positive cell population.
[d] Age of animals that were used as sources of leukocytes.
[e] Values are means ± SEM from six different animals.

tribution was seen. The percent of L-selectin–positive lymphocytes in mature animals in these experiments ranged from 17 to 60% (Table 14.1). The expression of L-selectin on small resting lymphocytes isolated from various lymphoid tissues was also examined. In the peripheral lymph node >80% of the cells were L-selectin–positive, whereas in ileal Peyer's patch, the percentage was <20%. Other than the exquisite tissue-specific expression, the distribution of L-selectin in cattle is very similar to that seen in the human and mouse.

Overall, L-selectins in sheep and cattle are remarkably similar to the human and mouse homologues. This suggests that information derived from studies in these animals will be directly applicable to human medicine. An example of the usefulness of the large animal in studies of unique leukocyte homing properties is demonstrated in our study of the γ/δ T cell. Gamma/δ T cells, which can comprise up to 70% of the circulating pool of T cells in the cow (17), do not home efficiently to secondary lymphoid organs, such as peripheral lymph nodes. However, we have shown that they express three to five times the level of L-selectin as other lymphocytes and the molecule appears to be functional. Gamma/δ T cells bind venules that support extravasation into secondary lymphoid tissues in vivo, but they simply do not migrate across the endothelium. Currently, we are examining the molecular basis for the unique homing of γ/δ T cells which may give insights into the function of this poorly understood population of lymphocytes.

Summary

The study of leukocyte–endothelial cell interactions has entered a new era. Many of the molecules involved in these events have been identified, cloned, and studied at the molecular level. The next step is to determine how the expression and function of these molecules are regulated in vivo to control leukocyte extravasation into diverse sites of inflammation. The development of new and effective therapeutics will then likely follow. Finally, it is still possible (and perhaps likely) that the molecules defined to date are not sufficient to account for all of the unique specificities of leukocyte extravasation exhibited in vivo and that new ones await our discovery.

Acknowledgments. The authors wish to acknowledge the excellent technical assistance of Sandy Kurk, Gayle Watts, and Kathryn Jutila, and the stimulating discussions with Drs. T.K. Kishimoto and L.J. Picker. The contributions of the following former co-workers and collaborators, who were instrumental in much of the work summarized here, are also greatly appreciated: Rupert Hallman, Ellen Berg, Robert Bargatze, Franz Kroese, Lusijah Rott, and Eugene C. Butcher. Part of these studies were supported by grants from the U.S.D.A. (CRGO-90-01666), Pardee Research Foundation and the Montana Agricultural Experiment Station to M.A.J.

References

1. Jutila. M.A., L. Rott, E.L. Berg, and E.C. Butcher. 1989. Function and regulation of the neutrophil Mel-14 antigen in vivo: Comparison with LFA-1 and Mac-1. *J. Immunol.* 143:3318.
2. Kishimoto, T.K., M.A. Jutila, E.L. Berg, and E.C. Butcher. 1989. Neutrophil Mac-1 and Mel-14 adhesion proteins inversely regulated by chemotactic factors. *Science* (Wash. D.C.) 245:1238.
3. Jung. T.M. and M.O. Dailey. 1990. Rapid modulation of homing receptors (gp90MEL-14) induced by activators of protein kinase C: Receptor signalling due to accelerated proteolytic cleavage at the cell surface. *J. Immunol.* 144:3130.
4. Jutila, M.A., T.K. Kishimoto, and M. Finken. 1990. Low dose Chymotrypsin treatment inhibits neutrophil migration into sites of inflammation in vivo: Effects on Mac-1 and Mel-14 adhesion protein expression and function. *Cell. Immunol.* 132:201.
5. Lewinsohn, D.M., R.F. Bargatze, and E.C. Butcher. 1987. Leukocyte-endothelial cell recognition: evidence of a common molecular mechanism shared by neutrophils, lymphocytes and other leukocytes. *J. Immunol.* 138:4313.
6. Stamper, H.B. and J.J. Woodruff. 1976. Lymphocyte homing into lymph nodes: in vitro demonstration of the selective affinity of recirculating lymphocytes for high endothelial venules. *J. Exp. Med.* 144:828.
7. Kishimoto, T.K., M.A. Jutila, and E.C. Butcher. 1990. Identification of a human peripheral lymph node homing receptor: a rapidly down-regulated adhesion molecule. *Proc. Natl. Acad. Sci. USA* 87:2244.
8. Hallman, R., M.A. Jutila, C.W. Smith, D.C. Anderson, T.K. Kishimoto,

and E.C. Butcher. 1990. CD18-independent adhesion of human neutrophils to endothelium involves a cell adhesion molecule related to the lymph node homing receptors. *Biochem. Biophy. Res. Commun.* 174:236.

9. Abbassi, O., T.K. Kishimoto, C.L. Lane, L.V. McIntire, and C.W. Smith. 1991. Endothelial-leukocyte adhesion molecule-1 supports neutrophil rolling in vitro under conditions of flow. Submitted.

10. Smith, C.W., T.K. Kishimoto, O. Abbass, B. Hughes, R. Rothlein, L.V. McIntire, E.C. Butcher, and D.C. Anderson. 1991. Chemotactic factors regulate lectin adhesion molecule 1 (LECAM-1)-dependent neutrophil adhesion to cytokine-stimulated endothelial cells in vitro. *J. Clin. Invest.* 87:609.

11. Kishimoto, T.K., R.A. Warnock, M.A. Jutila, E.C. Butcher, D.C. Anderson, and C.W. Smith. 1990. Antibodies against human neutrophil LECAM-1 (DREG56/LAM-1/Leu-8 antigen) and endothelial cell ELAM-1 inhibit a common CD18-independent adhesion pathway in vitro. *Blood* 78:805.

12. Picker, L.J., R.A. Warnock, A.R. Burns, C.M. Doerschuk, E.L. Berg and E.C. Butcher. 1991. The neutrophil selectin LECAM-1 presents carbohydrate ligands to the vascular selectin ELAM-1 and GMP-140. *Cell* 66:921.

13. Watson, S.R., C. Fennie, and L.A. Lasky. 1991. Neutrophil influx into an inflammatory site inhibited by a soluble homing receptor-IgG chimaera. *Nature* (London) 349:164.

14. Ley, K., P. Gaehtgens, C. Fennie, M.S. Singer, L.A. Lasky, and S.D. Rosen. 1991. Lectin-like adhesion molecule-1 mediates leucocyte rolling in mesenteric venules in vivo. *Blood* 77:2553.

15. von Andrian, U.H., J.D. Chambers, L. McEvoy, R.F. Bargatze, K.E. Arfors, and E.C. Butcher. 1991. Two step model of leukocyte-endothelial cell interaction in inflammation: Distinct roles for LECAM-1 and leukocyte beta-2 integrins in vivo. *Proc. Natl. Acad. Sci. USA* 88:7538.

16. Kehrli, M.E., F.C. Schmalstieg, D.C. Anderson, M.J. Van Der Maaten, B.J. Hughes, M.R. Ackermann, C.L. Wilhelmsen, G.B. Brown, M.G. Stevens, and C.A. Whetstone. 1990. Molecular definition of the bovine granulocyto-pathy syndrome: Identification of deficiency of the Mac-1 (CD11b/CD18) gly-coprotein. *Am.J. Vet. Res.* 51:1826.

17. Hein, W.R. and C.R. Mackay. 1991. Prominence of gamma/delta T cells in the ruminant immune system. *Immunol. T.* 12:30.

18. Jutila, M.A. and K.L. Banks. 1988. Increased synovial fluid macrophage division in Caprine Arthritis-Encephalitis virus infection. *J. Infect. Disease* 157:1193.

19. Walcheck, B., M. White, S. Kurk, T.K. Kishimoto, and M.A. Jutila. 1992. Characterization of the bovine peripheral lymph node homing receptor: a lectin cell adhesion molecule (LECAM). *Eur. J. Immunol.* 22:469.

15
L-Selectin as a Mediator of Leukocyte Binding to Myelinated Tracts of the Central Nervous System: Possible Pathophysiologic Significance

Steven D. Rosen and Kun Huang

Introduction

The study of lymphocyte homing receptors (1–4) has been greatly expedited by the in vitro adherence assay developed by Stamper and Woodruff (5). In this assay, lymphocytes that are overlain on cryostat-cut sections of secondary lymphoid organs are observed to adhere selectively to profiles of high endothelial venules (HEV) exposed within the sections. This assay led to the identification of L-selectin, the peripheral lymph node homing receptor (6–9). L-selectin, a member of the selectin family of cell–cell adhesion proteins, is a calcium-dependent, lectinlike molecule which functions by recognizing carbohydrate-based ligands on lymph node HEV (10–16). Three years after the original description of the in vitro adherence assay, Kuttner and Woodruff (17) reported that rat lymphocytes are able to bind to myelinated regions of the central nervous system (CNS) which are exposed in crystat-cut sections. It was suggested that this interaction could potentially promote the accumulation of damage-inducing lymphocytes in the vicinity of myelinated axons and thereby might be a contributing factor in demyelinating diseases.

Results

With the advent of antibodies and other specific probes to a variety of leukocyte adhesion proteins, it has been possible to reexamine the original findings of Kuttner and Woodruff. We were able to extend the original observations by showing that mouse lymphocyte and a human Jurkat T cell are also able to bind to cerebellar white matter regions (Fig. 15.1). Cerebellum represents a convenient target tissue because it possesses a clearly defined white matter tract with abundant myelinated axons.

The original studies (17) reported that thymocytes bind much more

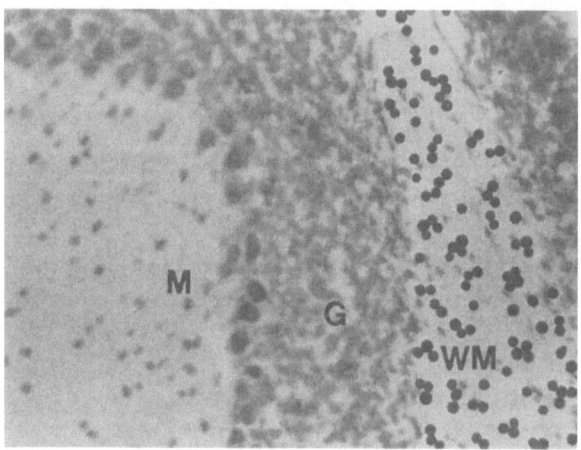

FIGURE 15.1. Human Jur-kat T-cell binding to cyostat section of mouse cerebellum. The binding assay was carried out as described in Huang et al. (18). The exogenous cells bind selectively to the white matter region (WM), which contains abundant myelinated axons. The granular (G) and molecular layers (M) do not support lymphocyte attachment.

weakly than peripheral lymphocytes to myelinated brain regions, which was compatible with the participation of a lymphocyte homing receptor (4). Lymphocyte homing receptors are known to be calcium-dependent (4). In the case of L-selectin, the calcium dependency is thought to be due to the presence of a calcium-type lectin domain in the amino terminus of the protein. As shown in Figure 15.2, chelation by EGTA chelation substantially reduces mouse lymphocyte attachment to cerebellar white matter (18). Potential ligand sites for L-selectin were shown to be present within white matter regions by the use of a L-selectin/IgG chimera as a histochemical staining reagent. This reagent has previously been used to stain HEV in lymph nodes (19). Myelinated tracts of cerebellum, cerebrum, and spinal cord are selectively stained by the homing receptor chimera and the staining is calcium-dependent (18). Strikingly, the chimera reacts selectively with white matter regions of the CNS (18). Little or no staining is observed on myelinated tracts of the peripheral nervous system (PNS).

To investigate directly whether L-selectin on lymphocytes mediates lymphocyte attachment to white matter regions, we employed a monoclonal antibody and a polyclonal antibody (Mel-14 and polyMEL) that react with mouse L-selectin (18). These antibodies are active, but surpri-

--→

FIGURE 15.3. Human peripheral blood lymphocyte attachment to cerebellar white matter. Human PBL binding to mouse cerebellar white matter was measured in the presence of saturating concentrations of TQ-1 mAb (anti–L-selectin), anti–VLA-4 (α_4 subunit–specific), anti–β_2 integrin, or 10 mM EGTA. Binding is presented as lymphcytes bound per unit area of white matter. Control indicates binding level in the absence of additives. The error bars indicate SEM ranges. The data are taken from Huang et al. (18).

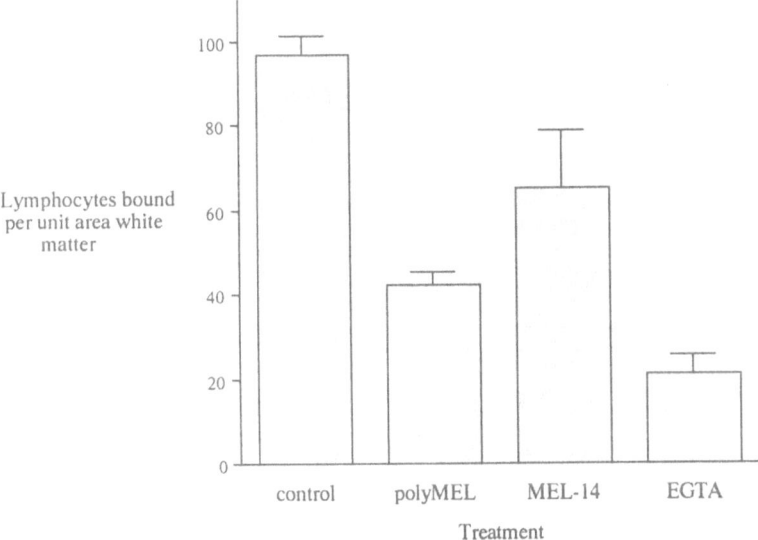

FIGURE 15.2. Mouse peripheral lymphocyte binding to cerebellar white matter. Lymphocyte attachment to mouse cerebellar white matter was measured in the presence of MEL-14 mAb, polyMEL (a polyclonal antibody developed against mouse L-selectin), or 10 mM EGTA. Binding was determined as a percentage of control binding in the absence of additives. The error bars indicate SEM ranges. The data are taken from Huang et al. (18).

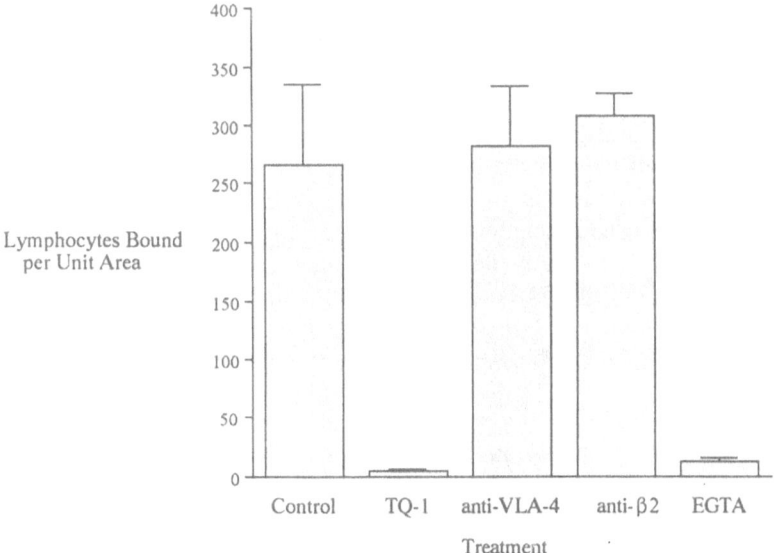

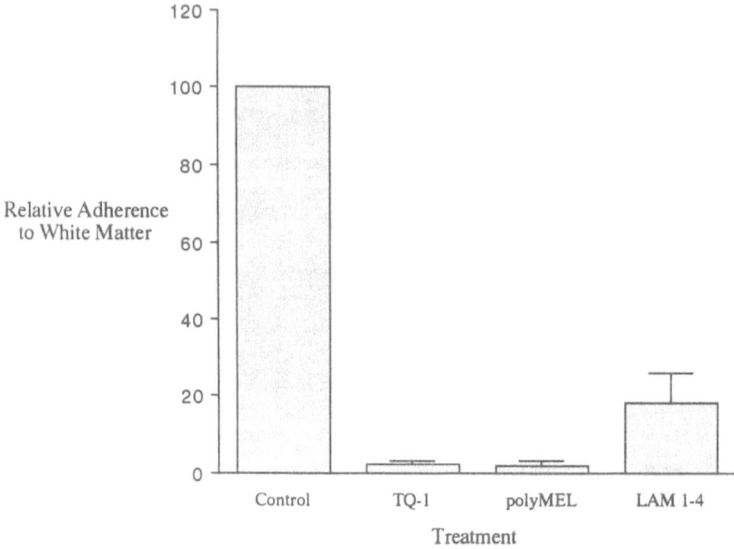

FIGURE 15.4. Human Jurkat T-cell attachment to cerebellar white matter. Jurkat cell binding to mouse cerebellar white matter was measured in the presence of three anti-L-selectin antibodies: TQ-1 mAb, polyMEL, or LAM 1.4 mAb (33). Binding was determined as a percentage of control binding in the absence of additives. The error bars indicate SEM ranges. The data are taken from Huang et al. (18).

singly, they block lymphocyte binding only partially (30–50% inhibition), even at concentrations that are sufficient to completely prevent lymphocyte binding to lymph node HEV (Fig. 15.2). This partial reduction is consistently less than the effect produced by the removal of calcium ($\approx$80%), suggesting the participation of other calcium-dependent adhesion receptors in the interaction. In contrast, when human lymphocyte or a human T cell line (Jurkat) are employed in the adherence assay (18), antibodies against L-selectin are almost completely effective (>90% reduction), as is calcium chelation (Fig. 15.3). Strong inhibition is observed with three independently derived antibodies, including polyMEL (Fig. 15.4). Equivalent inhibition is seen whether the target brain sections are of mouse or human origin. Antibodies against other leukocyte cell adhesion proteins such as the β_1 and β_2 integrins do not affect adherence to white matter.

Further evidence for the participation of L-selectin in the adherence of human lymphocytes to white matter regions is based on the use of a phorbol ester (18). It is well established that phorbol ester treatment of lymphocytes causes a rapid down-regulation of L-selectin through a shedding process (20,21). When Jurkat cells are treated with PMA, they lose their

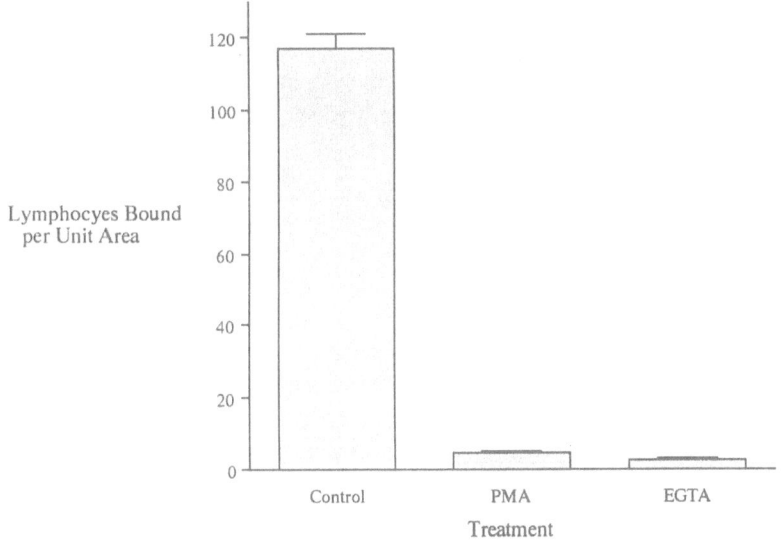

FIGURE 15.5. Binding of Jurkat cells to cerebellar white matter after phorbol ester treatment. Jurkat cells were treated with PMA and then tested for binding to mouse cerebellar white matter. The effect of 10 mM EGTA was also determined. Binding was determined as a percentage of control binding in the absence of additives. The error bars indicate SEM ranges. The data are taken from Huang et al. (18).

ability to bind to cerebellar white matter, in parallel with the loss of L-selectin from the cell surface (Fig. 15.5).

Finally, we examined the ability of specific carbohydrates to inhibit lymphocyte binding to CNS myelinated regions. Mannose-6-phosphate (M6P) and the structurally related fructose-1-phosphate (F1P) exhibit ligand activity for L-selectin and block its adhesive function (14,22). As shown in Figure 15.6, these sugars also prevent Jurkat cell binding to cerebellar white matter, whereas the control sugar, galactose-1-phosphate (G1P), has no effect (Fig. 15.6).

Conclusion

A major question in neuroimmunology concerns the mechanisms by which inflammatory cells that invade the CNS effect demyelination. Recent studies of experimental allergic encephalomyelitis (EAE) by Cross, Raine, and colleagues have examined the role of brain antigen-specific lymphocytes in the onset of the disease (23,24). These cells appear to be among the first to arrive in the CNS but are largely confined to perivascu-

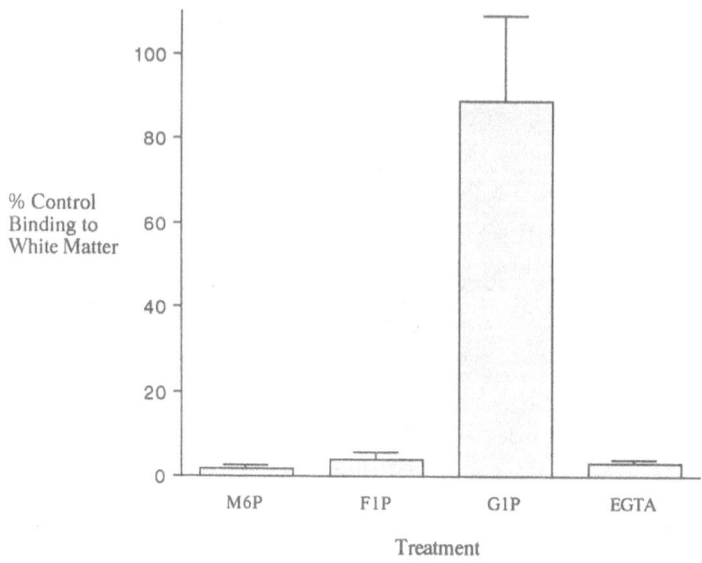

FIGURE 15.6. Binding of Jurkat cells to cerebellar white matter in the presence of phosphorylated monosaccharides. Jurkat cell binding to mouse cerebellar white matter was measured in the presence of 10 mM mannose-6-phosphate (M6P), fructose-1-phosphate (F1P), or galactose-1-phosphate (G1P). Binding was determined as a percentage of control binding in the absence of added sugar. The error bars indicate SEM ranges based on four replicates.

lar sites and represent a minority of the leukocytes that eventually gain access into the CNS. The majority of cells to be recruited are not known to be brain-antigen specific and are sometimes referred to as bystander cells (23,25). It is these bystander cells that predominate at the sites of demyelination lesions and whose arrival correlates with the onset of clinical signs (24). Cross et al. (24) speculate that the early arriving antigen-specific cells, through recognition of specific antigen presented by class II+ perivascular cells, induce the up-regulation of general adhesion molecules on the endothelium of CNS venules, precipitating the influx of a large number of nonspecific leukocytes. Thus, the antigen-specific lymphocytes essentially orchestrate the recruitment of the bystander cells.

Since the bystander cells are implicated in lesion formation by their localization and predominance, the key question is how these cells might participate in the pathological process. The finding reviewed herein, that L-selectin can mediate lymphocyte attachment to myelinated tracts, offers a new perspective on addressing this question (18). We propose that L-selectin may be responsible for targeting lymphocytes, regardless of their antigen specificity to myelinated axons. Since other leukocytes also express L-selectin (26–28), targeting of these cells may also occur. The basis for the adhesive recognition would be the fortuitous expression on myelin membranes of a carbohydrate determinant that is recognized by the lectin domain of L-selectin. This postulated carbohydrate would be structurally related, but need not be identical, to the carbohydrates of the lymph node HEV-ligand. A second aspect of the hypothesis is that once proximity is established via L-selectin, myelin destruction could result from a direct cytolytic process or through the release of cytokines from the leukocyte. With respect to the latter possibility, it is noteworthy that TNF and lymphotoxin, which are products of activated macrophages and lymphocytes, are known to be cytotoxic for oligodendrocytes, the glial cells responsible for CNS myelin formation (29,30). It may be envisioned that the leukocyte is activated prior to the adhesive interaction with the myelinated cell, and that contact via L-selectin triggers its effector mechanisms (cytotoxic granule exocytosis or cytokine secretion). Another possibility is that the leukocyte may attain effector cell status as a consequence of an L-selectin-mediated adhesive interaction with myelinated axons. Both mechanisms would require that L-selectin play a role in signaling as well as adhesion, a possibility which gains support from the recent demonstration that L-selectin on cytotoxic lymphocytes can participate in redirected cell lysis (31), that L-selectin is associated with the TCR/CD3 complex on the T-cell surface (34) and from the effects of L-selectin antibodies on lymphocyte functions (32).

A myelin-targeting mechanism involving L-selectin might be relevant to a variety of conditions (infection, hemorrhage, autoimmune reactivity, etc.) in which leukocytes gain entry to the normally privileged regions of the CNS. Perhaps the fact that demyelination is a frequent sequela of leukocyte trafficking to the brain is due, at least in part, to the fortuitous interaction between a myelin-associated ligand and the widely distributed leukocyte receptor, L-selectin, although other factors undoubtedly contribute to the initiation of the process. The restriction of ligand sites for L-selectin to CNS myelinated tracts, as compared to PNS myelin (18), may provide an explanation of why some demyelinating diseases, such as multiple sclerosis, selectively affect the CNS. Given the availability of a variety of reagents that react with L-selectin and its ligands, the validity of the highly speculative proposals presented herein should be testable.

Acknowledgments. The research was supported by grants from the NIH (GM23547 and Multipurpose Arthritis Center Grant P60 AR20684) and from Genentech, Inc., to S.D.R.

References

1. Stoolman, L.M. 1989. Adhesion molecules controlling lymphocyte migration. *Cell* 56:907.

2. Yednock, T.A., and S.D. Rosen. 1989. Lymphocyte homing. *Adv. Immunol.* 44:313.
3. Butcher, E. 1986. The regulation of lymphocyte traffic. *Curr. Top. Microbiol. Immunol.* 128:85.
4. Woodruff, J.J., L.M. Clarke, and Y.H. Chin. 1987. Specific cell-adhesion mechanisms determining migration pathways of recirculating lymphocytes. *Ann. Rev. Immunol.* 5:201.
5. Stamper, H., and J. Woodruff. 1976. Lymphocyte homing into lymph nodes: In vitro demonstration of the selective affinity of recirculating lymphocytes for high-endothelial venules. *J. Exp. Med.* 144:828.
6. Gallatin, W.M., I.L. Weissman, and E.C. Butcher. 1983. A cell surface molecule involved in organ-specific homing of lymphocytes. *Nature* 303:30.
7. Lasky, L.A., M.S. Singer, T.A. Yednock, D. Dowbenko, C. Fennie, H. Rodriguez, T. Nguyen, S. Stachel, and S.D. Rosen. 1989. Cloning of a lymphocyte homing receptor reveals a lectin domain. *Cell* 56:1045.
8. Siegelman, M.H., M. Van De Rijn, and I.L. Weissman. 1989. Mouse lymph node homing receptor cDNA clone encodes a glycoprotein revealing tandem interaction domains. *Science* 243:1165.
9. Tedder, T.F., C.M. Isaacs, T.J. Ernst, G.D. Demetri, D.A. Adler, and C.M. Disteche. 1989. Isolation and chromosomal localization of cDNAs encoding a novel human lymphocyte cell surface molecule, LAM-1. Homology with the mouse lymphocyte homing receptor and other human adhesion proteins. *J. Exp. Med.* 170:123.
10. Stoolman, L.M., and S.D. Rosen. 1983. Possible role for cell surface carbohydrate-binding molecules in lymphocyte recirculation. *J. Cell Biol.* 96:722.
11. Rosen, S.D., M.S. Singer, T.A. Yednock, and L.M. Stoolman. 1985. Involvement of sialic acid on endothelial cells in organ-specific lymphocyte recirculation. *Science* 228:1005.
12. Yednock, T.A., L.M. Stoolman, and S.D. Rosen. 1987. Phosphomannosyl-derivatized beads detect a receptor involved in lymphocyte homing. *J. Cell Biol.* 104:713.
13. Yednock, T.A., E.C. Butcher, L.M. Stoolman, and S.D. Rosen. 1987. Receptors involved in lymphocyte homing: relationship between a carbohydrate-binding receptor and the MEL-14 antigen. *J. Cell Biol.* 104:725.
14. Imai, Y., D.D. True, M.S. Singer, and S.D. Rosen. 1990. Direct demonstration of the lectin activity of gp90[MEL], a lymphocyte homing receptor. *J. Cell Biol.* 111:1225.
15. Imai, Y., M.S. Singer, C. Fennie, L.A. Lasky, and S.D. Rosen. 1991. Identification of a carbohydrate-based endothelial ligand for a lymphocyte homing receptor. *J. Cell Biol.* 113:1213.
16. Berg, E.L., M.K. Robinson, R.A. Warnock, and E.C. Butcher. 1991. The human peripheral lymph node vascular addressin is a ligand for LECAM-1, the peripheral lymph node homing receptor. *J. Cell Biol.* 114:343.
17. Kuttner, B.J., and J.J. Woodruff. 1979. Selective adherence of lymphocytes to myelinated areas of rat brain. *J. Immunol.* 122:1666.
18. Huang, K., J.S. Geoffroy, M.S. Singer, and S.D. Rosen. 1991. A lymphocyte homing receptor (L-selectin) mediates the in vitro attachment of lymphocytes to myelinated tracts of the central nervous system. *J. Clin. Invest.* 88:1178.
19. Watson, S.R., Y. Imai, C. Fennie, J.S. Geoffroy, S.D. Rosen, and L.A. Las-

ky. 1990. A homing receptor-IgG chimera as a probe for adhesive ligands of lymph node high endothelial venules. *J. Cell Biol.* 110:2221.

20. Jung, T.M., and M.O. Dailey. 1990. Rapid modulation of homing receptors (gp90[MEL-14]) induced by activators of protein kinase C: Receptor shedding due to accelerated proteolytic cleavage at the cell surface. *J. Immunol.* 144:3130.

21. Hung, K., M. Beigi, and R.A. Daynes. 1990. Peripheral lymph node-specific and Peyer's patch-specific homing receptors are differentially regulated following lymphocyte activation. *Reg. Immunol.* 3:103.

22. Stoolman, L.M., T.S. Tenforde, and S.D. Rosen. 1984. Phosphomannosyl receptors may participate in the adhesive interactions between lymphocytes and high endothelial venules. *J. Cell Biol.* 99:1535.

23. Cross, A.H., B. Cannella, C.F. Brosnan, and C.S. Raine. 1990. Homing to central nervous system vasculature by antigen-specific lymphocytes. I. Localization of ^{14}C-labeled cells during acute, chronic, and relapsing experimental allergic encephalomyelitis. *Lab. Invest.* 63:162.

24. Cross, A.H., B. Cannella, C.F. Brosnan, and C.S. Raine. 1991. Hypothesis: Antigen-specific T cells prime central nervous system endothelium for recruitment of nonspecific inflammatory cells to effect autoimmune demyelination. *J. Neuroimmunol.* 33:237.

25. Swanborg, R.H. 1990. Horror autotoxicus and homing: implications for autoimmunity. *Lab. Invest.* 63:141.

26. Lewinsohn, D.M., R.F. Bargatze, and E.C. Butcher. 1987. Leukocyte-endothelial cell recognition: evidence of a common molecular mechanism shared by neutrophils, lymphocytes, and other leukocytes. *J. Immunol.* 138:4313.

27. Ord, D.C., T.J. Ernst, L.J. Zhou, A. Rambaldi, O. Spertini, J. Griffin, and T.F. Tedder. 1990. Structure of the gene encoding the human leukocyte adhesion molecule-1 (TQ1, Leu-8) of lymphocytes and neutrophils. *J. Biol. Chem.* 265:7760.

28. Jutila, M.A., L. Rott, E.L. Berg, and E.C. Butcher. 1989. Function and comparison of the neutrophil MEL-14 antigen in vivo: comparison with LFA-1 and MAC-1. *J. Immunol.* 3318.

29. Selmaj, K.W., and C.S. Raine. 1988. Tumor necrosis factor mediates myelin and oligodendrocyte damage in vitro. *Ann. Neurol.* 23:339.

30. Selmaj, K., C.S. Raine, B. Cannella, and C.F. Brosnan. 1991. Identification of lymphotoxin and tumor necrosis factor in multiple sclerosis. *J. Clin. Invest.* 87:949.

31. Seth, A., L. Gote, M. Nagarkatti, and P.S. Nagarkatti. 1991. T-cell-receptor-independent activation of cytolytic activity of cytotoxic T lymphocytes mediated through CD44 and gp90[MEL-14]. *Proc. Natl. Acad. Sci. USA.* 88:7877.

32. James, S.P., Y. Murakawa, M.E. Kanof, and M. Berg. 1991. Multiple roles of Leu-8/MEL-14 in leukocyte adhesion and function. *Immunol Res.* 10:282.

33. Kansas, G.S., O. Spertini, L.M. Stoolman, and T.F. Tedder. 1991. Molecular mapping of functional domains of the leukocyte receptor for endothelium, LAM-1. *J. Cell Biol.* 114:351.

34. Murakawa, Y., Y. Minarni, W. Strobor, and S.P. James. 1992. Association of human lymph node homing receptor (Leu 8) with the TCR/CO3 complex. *J. Immunol* 148:1771.

16
Selectin-Dependent Monocyte Adhesion in Frozen Sections of Rheumatoid Synovitis

L.M. Stoolman, J. Grober, B. Bowen, P. Reddy, J. Shih, C. Thompson, D.A. Fox, and H. Ebling

Introduction

Rheumatoid synovitis is characterized by marked mononuclear infiltration and microvascular proliferation. Identifiable histologic patterns include heavily infiltrated lymphocyte-rich areas, transitional zones populated by macrophages, plasma cells, and lymphocytes (1), and noninfiltrated collagenous interstitial areas containing macrophages and fibroblasts (2). A superimposed infiltrate of polymorphonuclear leukocytes, particularly in synovial fluid, occurs during acute exacerbation, further compromising tissues. The relapsing acute and progressive chronic inflammation leads to development of an erosive pannus which eventually destroys the joint.

All leukocyte classes contribute to joint pathology. Neutrophils in the synovial fluid elaborate free radicals and release lysosomal enzymes capable of eroding cartilage and bone (3). In model systems of inflammatory arthritis, the transfer of synovial T-lymphocytes to naive animals results in development of joint pathology (4,5). In humans, T-cell ablative therapy partially ameliorates inflammation. The development of autoantibodies and the prominent plasma cell infiltrates observed in chronic synovitis reflect B-cell activation. The heightened susceptibility of individuals with the HLA-DR4 haplotype to rheumatoid arthritis further underscores the contribution of the immune system and raises the possibility that aberrant antigen presentation fosters chronic synovial inflammation. Regardless of the etiology, monocytes clearly play a pivotal role contributing to both the immunologic and inflammatory components of the disease. Synovial fluid monocytes and their tissue counterparts display an activated phenotype including high-density expression of MHC class II (6,7). In tissues, macrophages and dendritic cells capable of antigen presentation are intimately associated with T cells in acellular, transitional, and immunologically active areas of synovium (8). Monocyte products such as IL-1β, IL-6, and TNF-α contribute to a variety of immunopathologic features including T-cell activation, microvascular proliferation (9), and induction of degradative enzyme secretion by chondrocytes and synovial fibroblasts (10,11). Finally, recent studies confirm that OKM-1 (CD11b)-positive mononuclear cells (i.e., monocytes, monocyte-derived macrophages, and dendritic cells) account for the bulk of IL-1β, TNF-α, and IL-8 gene expression in RA synovium (12). Thus blood monocytes are major contributors to synovial inflammation.

The entry of circulating leukocytes into acute and chronic inflammatory lesions is commonly referred to as recruitment. The process encompasses several stages (13). Attachment to the lumenal surface of postcapillary venules in target organs initiates the process. The immobilized leukocytes then detach from the surface, migrate through the endothelial layer, and penetrate the basement membrane. Each stage involves adhesive interactions between the leukocyte and the vessel wall. The current paradigm holds that the selectins (LEC-CAMs) and their carbohydrate-based counter-receptors initiate contacts between circulating leukocytes and the vessel wall (14–17). The integrins and their counter-receptors in the immunoglobulin gene family augment adhesion and support migration through the vessel wall (18).

In general, recruitment involves broad-spectrum adhesion molecules on both the leukocytes and endothelial cells. This contrasts with the related process of lymphocyte recirculation in which organ-selective attachments between lymphocytes and the high endothelial venules of lymphoid organs result in regional trafficking patterns (19,20). The pertinent leukocyte adhesion receptors implicated in recruitment include the $\beta2$ integrins (LFA-1, Mo-1/Mac-1, and p150,95), the $\beta1$ integrin VLA-4 (21), and L-selectin. The pertinent endothelial receptors include the ICAMs (22), VCAM-1 (23,24), and two members of the selectin (LEC-CAM) family, E-selectin (ELAM-1) and P-selectin (GMP-140/PADGEM) (13).

Intravital microscopy, in vitro adhesion assays, and recent studies in rodent models provide insight into the mechanism of neutrophil recruitment. The process begins when leukocytes decelerate and attach to the lumenal surface of postcapillary venules. In vivo, this initial interaction manifests as "rolling." All three members of the selectin (LEC-CAM) family, P-selectin, E-selectin and L-selectin, may participate in the initial attachment of neutrophils to endothelium. A subsequent interaction between activated $\beta2$ integrins and endothelial ICAM-1 appears necesary to arrest rolling neutrophils and promote migration through endothelial monolayers (14,15,18,25,26). In rodent models, Mabs specific for E-selectin and the $\beta2$ integrins partially inhibit neutrophil-mediated lung injury (27,28). However, the effects are partial and dependent on the inciting agent, suggesting that the contributions of various adhesion molecules vary with the disease process.

Monocyte-endothelial interactions during recruitment are the most complex and least well understood. Monocytes bind to both P-selectin and E-selectin; however, the physiologic significance of these interactions remains to be determined. P-selectin mediates attachment of thrombin-activated platelets to both monocytes and neutrophils (29,30). In addition, it is contained in the Weibel-Palade bodies of postcapillary venules in most tissues (31) and rapidly mobilized to the surface of human umbilical vein endothelium (HUVE) in response to a variety acute-inflammatory stimuli including thrombin, histamine, products of complement-activation, and oxidants (32,33). This translocation results in the rapid and transient adherence of neutrophils to treated HUVE, giving rise to speculation that P-selectin functions exclusively during the earliest phase of leukocyte recruitment. In contrast to P-selectin, E-selectin is transcriptionally induced by a variety of cytokines in HUVE (34) and at sites of acute and chronic inflammation in vivo (35–38). Monocytes bind to E-selectin on cytokine-treated HUVE in vitro (34) and accumulate around E-selectin–positive venules in vivo. Both P-selectin and E-selectin recognize sialylated oligosaccharides expressed on a variety of cellular glycoproteins and glycolipids (13,39–41).

L-selectin on leukocytes interacts with inflamed microvasculature in vitro; however, the significance of this interaction to monocyte recruitment has not been established (42–45). Both the carbohydrate-recognition domain (CRD) and the oligosaccharide side chains of this receptor are capable of binding to endothelial counter-receptors. Blockade of the CRD of L-selectin inhibits monocyte attachment to inflamed lymph node endothelium in the mouse (45). In humans, blockade of the CRD on lymphocytes with the Lam series of Mabs inhibits attachment to cytokine-treated HUVE, implying that oligosaccharide-bearing counter-receptors for L-selectin are induced on HUVE in response to cytokines (46). Finally, recent studies show that L-selectin on neutrophils carries oligosaccharides capable of binding to the CRDs of both E-selectin and P-selectin (47). This interaction may account for 60% of E-selectin–mediated neutrophil attachment to cytokine-treated HUVE at physiologic levels of shear-stress (48). Thus L-selectin may act as both a receptor (for endothelial carbohydrate) and a carbohydrate-based counter-receptor (for the endothelial selectins).

Finally, both the $\beta 1$ and $\beta 2$ integrins contribute to monocyte-endothelial attachment in vitro (49–55). The $\beta 2$ family is essential to recruitment of neutrophils but may not be essential to recruitment of mononuclear leukocytes in vivo. Individuals with the LAD syndrome lack functional receptors yet are still capable of mounting delayed-type hypersensitivity reactions (56). Thus recruitment of circulating monocytes and lymphocytes must still take place in these individuals. Neutrophil recruitment, on the other hand, is severely compromised, leading to recurrent pyogenic infections. A member of the $\beta 1$ family of integrins, VLA-4, may contribute to these differences. Monocytes and lymphocytes, but not neutrophils, bind to the inducible endothelial molecule VCAM-1 through this receptor (24,34,56). VCAM-1 is strongly expressed in cutaneous hypersensitivity lesion in humans (57) and a VLA-4–specific Mab partially inhibits the entry of circulating lymphocytes into such lesions in the rat

(58,59). In summary, studies in model systems implicate a variety of adhesion molecules on monocytes and endothelium in the recruitment process. Multiple recruitment pathways and unsuspected synergism amongst receptor families may exist in vivo, highlighting the need to evaluate receptor function in actual disease processes.

The principal tool for studying recruitment in diseased human tissues has been immunoperoxidase staining for expression of inducible endothelial adhesion molecules. The detection of ICAM-1, ELAM-1, and VCAM-1 in the microvasculature of cutaneous delayed hypersensitivity lesions, rheumatoid synovitis, and other inflammatory conditions is frequently cited as support for a role in recruitment (36–38,57). This approach establishes the potential for involvement but cannot confirm the functional activity of adhesion receptors. For example, traditional immunoperoxidase techniques generate a soluble pigment which diffuses away from the point of generation. Therefore, one cannot reliably distinguish cell surface from cytoplasmic antigen. In addition, both functional receptors at the lumenal surface and inactive cytoplasmic storage pools may stain with Mabs, leading to overestimates of functional activity. This problem is greatest for receptors packaged in cytoplasmic granules such as P-selectin; however, transient pools of inducible receptors (e.g., E-selectin, ICAM, or VCAM) generated during synthesis or recycling may also be detected. Finally, staining does not necessarily reflect function. The distribution (e.g., clustering or co-patching with other adhesion receptors) and activation state may alter receptor avidity without impacting on staining intensity. These considerations point out the need for assays which measure the binding activity of endothelial adhesion receptors in situ.

The current study shows that the Stamper-Woodruff assay, initially used to detect organ-specific adhesion involved in lymphocyte recirculation, detects selectin- and integrin-mediated monocyte-microvascular adhesion in rheumatoid synovitis. The most striking finding was the preeminent role of P-selectin in mediating monocyte attachment. The studies revealed lesser but significant contributions from E-selectin, L-selectin, and the leukocytic integrins. In addition, they suggest that P-selectin stabilizes bonds formed by E-selectin, L-selectin, and the $\beta 2$ integrins. Thus the functional activity and interactions of endothelial adhesion molecules can be investigated directly in frozen sections of human inflammatory disease.

Results

Initial experiments with the frozen-section assay determined optimal conditions for attachment of monocytes to the microvasculature. Specifically, we defined conditions which maintained binding to synovial venules (SV) while minimizing attachment to stromal and synovial lining cells. Attachment to these nonendothelial tissues should not be construed as "nonspecific" and in all probability reflects receptor-mediated interactions of potential functional significance. However, the current study required conditions which promoted selective interactions with the microvasculature.

From this standpoint, optimal adhesion was achieved using (1) elutriated monocytes, (2) air-dried, unfixed tissue sections on BSA-coated glass slides, and (3) 7–10°C incubations with rotary agitation. Monocytes isolated by Percoll-based techniques showed selective, high-density microvascular binding. However, staining for the platelet antigen gp IIb/IIIa with Mab AP-3 revealed substantial contamination of the monocyte surface by platelets and platelet fragments. Use of centrifugal elutriation to isolate monocytes eliminated contamination by intact platelets and fragments (determined by the absence of staining with AP-3), facilitating interpretation of the adhesion experiments. Under optimal conditions, specific and reproducible binding of monocyte-enriched fractions (>90% purity) to SV was evident at concentrations as low as 5×10^5 cells/ml (5×10^4 cells/section). In contrast, the lymphocyte-enriched fraction (>90% purity) showed significantly lower affinity for SV at concentrations as high as 1×10^7 cells/ml (1×10^6 cells/section).

Previous studies documented expression of at least three endothelial adhesion molecules in SV—ICAM-1, E-selectin, and VCAM-1 (37). We found a similar pattern of expression in our specimens with strong staining for ICAM-1, much weaker staining for VCAM-1, and patchy positivity for E-selectin. In addition, P-selectin was strongly expressed in virtually all synovial venules. In contrast to Koch et al. (37), fewer than half of our specimens revealed endothelial E-selectin, and the intensity of staining was significantly below that for either ICAM-1 or P-selectin. The staining pattern in acetone-fixed tissues suggested that P-selectin was present in both cytoplasmic granules and along the lumenal endothelial surface. The focal deposition of the platelet epitope gpIb/IX along the vessel walls indicated that platelets contributed some of the P-selectin. However, most P-selectin–positive vessels were negative for gpIb/IX; thus the principle source for P-selectin was the endothelium. High-resolution microscopy will be needed to confirm the distribution and source for P-selectin associated with the SV. Nonetheless, immunocytochemical data identified P-selectin, E-selectin, ICAM-1, and VCAM-1 as potential contributors to monocyte–SV adhesion in the frozen-section assay (Table 16.1).

Antibody-blocking experiments showed complete inhibition by the P-selectin–specific MAb G1 (30) and partial inhibition by two Mabs specific for E-selectin, CL-2 (26,27) and BB11 (60,61). Fab'2 fragments of the P-selectin–specific Mab G1 completely inhibited binding in all RA synovial specimens tested (n = 11). In contrast, Fab'2 fragments specific for a nonfunctional epitope on P-selectin, S12 (62), had no effect (n = 2). Fab'2

TABLE 16.1. Immunocytochemical staining of rheumatoid synovium from patient with chronic disease (specimens obtained at the time of therapeutic synovectomy or joint replacement).

Antigen	Positive	Distribution
P-selectin	10/10	Diffuse, strong on endothelium + platelet plugs
E-selectin	6/20	Patchy, weak to moderate on endothelium
ICAM-1	3/3	Diffuse on endothelium
VCAM-1	8/18	Weak on endothelium (synovial lining strongly positive)
gpIb/IX	2/2	Focal on endothelium + platelet plugs

Mabs specific for the E-selectin epitopes CL2 and BB11 reduced monocyte–SV adhesion by 0–50% compared with Fab'2 fragments specific for the nonfunctional epitope CL37 (n = 10). The degree of inhibition varied between patients, between blocks from a single patient and, in some cases, between sections cut from a single block. This behavior may reflect the patchy, relatively weak expression of E-selectin in our series. Thus inhibition studies implicate both P- and E-selectin in monocyte attachment with the former predominating.

Blockade of potential adhesion sites on the monocyte revealed significant, partial blockade by Mabs specific for L-selectin and the $\beta2$ integrins. A Mab specific for the $\beta1$ chain, AIIB2, inhibited slightly. The L-selectin–specific Mab DREG 200 (63) blocked attachment by ~40% (n = 2). In the same tissue, blockade of E-selectin inhibited to an equivalent degree while blockade of P-selectin inhibited >90%. On neutrophils, DREG 200 blocks both the carbohydrate recognition domain of L-selectin as well as the oligosaccharide side chains mediating interactions with E- and, possibly, P-selectin (26,47). Therefore, the relative contributions of the CRD and oligosaccharide side chains of monocytic L-selectin remain to be established. However, the marked discrepancy between inhibition by DREG 200 and G1 indicates that L-selectin is unlikely to be the principle counter-receptor for endogenous P-selectin on monocytes. This observation illustrates the importance of assessing the functional activity of endogenous adhesion receptors.

Mabs specific for LFA-1 (CD11a) and Mol/Macl (CD11b) chains consistently inhibited attachment 25–35% while MAb to the common $\beta2$ (CD18) subunit inhibited ~35–47% relative to a class-matched control directed at vWF. In contrast, MAbs directed at CD14, CD45R, and HLA-DR, surface antigens expressed at equal or greater densities than either L-selectin or the integrins, failed to inhibit significantly. The same degree of inhibition was observed with both Percoll-fractionated and elutriated monocytes for the CD18-specific Mabs. MAbs specific for a functional epitope on the common $\beta1$ chain of the VLA subfamily, AIIB2 (64), inhibited the attachment of Percoll-fractionated monocytes by 30–

50% (n = 3) while inhibiting attachment of elutriated cells by only 10–15% (n = 2). Thus elimination of monocyte-associated platelets via elutriation does not completely eliminate β1-dependent adhesion. Whether VLA-4–mediated attachment to VCAM-1 on SV accounts for this component of adhesion remains to be determined. The marked reduction of β1-dependent adhesion observed with elutriated cells underscores the contribution that the fractionation protocol makes to monocyte behavior. Thus monocyte attachment to SV in frozen sections involves P-selectin primarily with lesser contributions from E-selectin, L-selectin, the β2 integrins and, possibly, the β1 integrins.

P-selectin contributes to monocyte attachment to the microvasculature in frozen sections of other inflamed and noninflamed tissues. However, P-selectin–positive vessels in histologically inflamed tissues show a significantly higher density of monocyte attachment than those in relatively noninflamed tissues. In a limited series, binding to the microvasculature was greatest in chronic RA, followed by inflamed tonsil, normal newborn foreskins, and placenta. The minimally inflamed foreskins showed binding densities 3–30-fold lower than the inflamed tissues and monocytes did not attach to the P-selectin–rich microvessels of normal-term placenta. The RA synovium and tonsils showed clear histologic evidence of inflammation with accumulations of CD-14–positive cells, lymphocytes and, in the tonsil, neutrophils. The surgically removed foreskins showed a variable infiltrate of CD14-positive mononuclear while the placental tissue did not show any histologic evidence of inflammation.

Conclusions

The studies show that selectin- and integrin-dependent monocyte adhesion to postcapillary venules can be detected in frozen sections of human inflammatory disease. In rheumatoid synovitis, Mabs to P-selectin inhibited binding completely in all cases. Mabs to β2 integrins consistently inhibited attachment by 30–40% on their own. Mabs to E-selectin showed less-consistent inhibition but blocked up to 40% of attachment in some specimens. Preliminary studies with Mabs to L-selectin and a functional β1 epitope partially blocked monocyte attachment. Thus a monocyte interaction with P-selectin appears necessary for bond formation through other selectins and the integrins in the frozen-section assay.

One explanation for this synergism is that P-selectin enhances the interactions of other receptor–counter-receptor pairs. It may, for example, approximate the monocyte and endothelial membranes facilitating the formation of bonds between adhesion molecules with shorter working distances. Alternatively, P-selectin may interact directly with oligosaccharide side chains expressed on several different monocyte receptors. L-selectin on neutrophils reportedly interacts with both E- and P-selectin through side chains carrying the sialylated Lewis[x] tetrasaccharide (47). The $\beta2$ integrin family carries the structurally related Lewis[x] trisaccharide (65,66). This structure binds to P-selectin with low affinity (29) and is frequently coexpressed with the sialylated form on leukocytic glycoproteins (67–69). Further studies are needed to clarify the binding sites for P-selectin on monocytes. Nonetheless, P-selectin is clearly the predominant monocyte adhesion receptor detected in tissues using the Stamper-Woodruff assay.

The synergism observed in the Stamper-Woodruff assay implies that the inflammatory milieu of chronic rheumatoid arthritis induces coexpression of P-selectin and other adhesion receptors at the endothelial surface. However, no single inflammatory stimulus has yet been identified which leads to such coexpression on cultured microvasculature. One potential explanation is that synovial venules reflect the concerted action of multiple acute and chronic inflammatory mediators. For example, TNF-α and IL-1β are readily identified in RA synovitis (12). The presence of activated platelets and "platelet plugs" within synovial postcapillary venules in RA further suggests that intravascular thrombin is generated locally. In addition, the phagocyte-rich rheumatoid pannus is a plausible source for oxidants and a considerable body of indirect evidence supports a role for oxygen-derived free radicals in the pathogenesis of RA (3,70). Each of these mediators induces different adhesion molecules in vitro. TNF-α and IL-1β up-regulate E-selectin (71), ICAM-1 (71), VCAM-1 (23,72) and ligands for L-selectin (46) on cultured microvasculature. In contrast, both thrombin (73,74) and oxidants (75) selectively induce P-selectin in HUVE. Thus the combined action of inflammatory mediators could, in theory, result in coexpression of P-selectin with other adhesion receptors in vivo. The synergism detected in tissue sections suggests that monocyte recruitment may be enhanced as a result.

P-selectin differs from other inducible endothelial adhesion receptors in having preformed cytoplasmic stores. EM studies show P-selectin in Weibel-Palade bodies (31). In theory, exogenous monocytes could interact with either cytoplasmic or membrane-associated P-selectin. However, only expression at the endothelial surface is relevant to recruitment in vivo. Monocyte interaction with cytoplasmic P-selectin in the Stamper-Woodruff assay is unlikely for several reasons. First, both the plasma and granule membranes separate this pool from exogenous monocytes in the binding assay. Sectioning would have to breach both of these barriers to expose sequestered receptor. Second, cytoplasmic constituents are notoriously difficult to detect in frozen sections by immunocytochemistry unless membranes are first permeabilized with fixatives such as acetone or methanol. (The binding assay uses unfixed, air-dried sections.) One would expect surface receptors on intact monocytes to have as little access to sequestered molecules as antibodies in solution. Third, if monocytes bound primarily to cytoplasmic P-selectin then one would expect adhesion to all P-selectin–containing venules in tissue sections. On the contrary, the density of binding to venules varies markedly within sections, between specimens within a tissue group (e.g., rheumatoid synovitis) and amongst tissue types. For example, P-selectin–dependent monocyte adhesion is greater on venules in rheumatoid synovium and tonsil than on venules in normal newborn foreskin and placenta. Yet venules in all four tissues are strongly positive for P-selectin by immunocytochemistry. While differences in the density of other endothelial binding sites may contribute, they cannot account for the 5–30-fold-lower binding densities observed in newborn foreskin or for the total absence of attachment to the P-selectin-containing venules of placenta. Finally, preliminary experiments with the confocal laser microscope indicate that P-selectin in nonpermeabilized frozen sections of RA is associated primarily with endothelial membranes (Stoolman and Grober, unpublished). Only after permeabilization of lipid membranes with acetone can one visualize cytoplasmic stores with immunofluorescence. We propose, therefore, that the Stamper-Woodruff assay detects P-selectin associated with the lumenal endothelial surface and that inflammation enhances its mobilization from sequestered cytoplasmic stores. In addition, the assay detects a variety of inducible endothelial adhesion molecules and thus enables one to assess functional activity directly in human inflammatory disease.

References

1. Ishikawa, H., and M. Ziff. 1976. Electron microscopic observations of immunoreactive cells in the rheumatoid synovial membrane. *Arthritis Rheum.* 19(1):1–14.
2. Iguchi, T., M. Kurosaka, and M. Ziff. 1986. Electron microscopic study of HLA-DR and monocyte/macrophage staining cells in the rheumatoid synovial membrane. *Arthritis Rheum.* 29(5):600–613.
3. Greenwald, R.A. 1991. Oxygen radicals, inflammation, and arthritis: pathophysiological considerations and implications for treatment. *Semin. Arthritis Rheum.* 20:219–240.
4. Taurog, J.D., G.P. Sandberg, and M.L. Mahowald. 1983. The cellular basis of adjuvant arthritis. II. Characterization of the cells mediating passive transfer. *Cell Immunol.* 80:198–204.
5. Cohen, I.R., J. Holoshitz, Eden.W. van, and A. Frenkel. 1985. T lymphocyte clones illuminate pathogenesis and affect therapy of experimental arthritis. *Arthritis Rheum.* 28:841–845.
6. Firestein, G.S., and N.J. Zvaifler. 1987. Peripheral blood and synovial fluid monocyte activation in inflammatory arthritis. I. A cytofluorographic study of monocyte differentiation antigens and class II antigens and their regulation by gamma-interferon. *Arthritis Rheum.* 30(8):857–863.
7. Barkley, D., S. Allard, M. Feldmann, and R.N. Maini. 1989. Increased expression of HLA-DQ antigens by interstitial cells and endothelium in the synovial membrane of rheumatoid arthritis patients compared with reactive arthritis patients. *Arthritis Rheum.* 32(8):955–963.
8. Kurosaka, M., and M. Ziff. 1983. Immunoelectron microscopic study of the

distribution of T cell subsets in rheumatoid synovium. *J. Exp. Med.* 158:1191–1210.

9. Koch, A.E., P.J. Polverini, and S.J. Leibovich. 1986. Stimulation of neovascularization by human rheumatoid synovial tissue macrophages. *Arthritis Rheum.* 29(4):471–479.

10. Poubelle, P., M. Damon, F. Blotman, and J.-M. Dayer. 1985. Production of mononuclear cell factor by mononuclear phagocytes from rheumatoid synovial fluid. *J. Rheumatol.* 12:412–417.

11. Dayer, J.-M., B. de Rochemontieix, B. Burrus, S. Demczuk, and C.A. Dinarello. 1986. Human recombinant interleukin 1 stimulates collagenase and prostaglandin E2 production by human synovial cells. *J. Clin. Invest.* 77:645–648.

12. Firestein, G.S., J.M. Alvaro-Gracia, R. Maki, and J.M. Alvaro-Garcia. 1990. Quantitative analysis of cytokine gene expression in rheumatoid arthritis [published erratum appears in J Immunol 1990 Aug 1;145(3):1037]. *J. Immunol.* 144:3347–3353.

13. Stoolman, L.M. 1992. Selectins (LEC-CAMs): Lectin-like receptors involved in lymphocyte recirculation and leukocyte recruitment. In: Cell Surface Carbohydrates and Cell Development, Chapter 3. M. Fukuda, ed. CRC Press, Inc, Boca Raton, FL: 72–98.

14. Lawrence, M.B., and T.A. Springer. 1991. Leukocytes roll on a selectin at physiologic flow rates: distinction from and prerequisite for adhesion through integrins. *Cell* 65:1–20.

15. von Andrian, U.H., J.D. Chambers, L.M. McEvoy, R.F. Bargatze, K.E. Arfors, and E.C. Butcher. 1991. Two-step model of leukocyte-endothelial cell interaction in inflammation: distinct roles for LECAM-1 and the leukocyte beta 2 integrins in vivo. *Proc. Natl. Acad. Sci. U.S.A.* 88:7538–7542.

16. Ley, K., P. Gaehtgens, C. Fennie, M.S. Singer, L.A. Lasky, and S.D. Rosen. 1991. Lectin-like cell adhesion molecule 1 mediates leukocyte rolling in mesenteric venules in vivo. *Blood* 77:2553–2555.

17. Abbassi, O., C.L. Lane, S. Krater, T.K. Kishimoto, D.C. Anderson, L.V. McIntire, and C.W. Smith. 1991. Canine neutrophil margination mediated by lectin adhesion molecule-1 in vitro. *J. Immunol.* 147:2107–2115.

18. Smith, C.W., T.K. Kishimoto, O. Abbass, B. Hughes, R. Rothlein, L.V. McIntire, E. Butcher, and D.C. Anderson. 1991. Chemotactic factors regulate lectin adhesion molecule 1 (LECAM-1)-dependent neutrophil adhesion to cytokine-stimulated endothelial cells in vitro. *J. Clin. Invest.* 87:609–618.

19. Stoolman, L.M. 1989. Adhesion molecules controlling lymphocyte migration. *Cell* 56:907–910.

20. Yednock, T.A., and S.D. Rosen. 1989. Lymphocyte homing. *Adv. Immumol.* 44:313–378.

21. Hemler, M.E., M.J. Elices, C. Parker, and Y. Takada. 1990. Structure of the integrin VLA-4 and its cell-cell and cell-matrix adhesion functions. *Immunol. Rev.* 114:45–65.

22. Smith, C.W., S.D. Marlin, R. Rothlein, C. Toman, and D.C. Anderson. 1989. Cooperative interactions of LFA-1 and Mac-1 with intercellular adhesion molecule-1 in facilitating adherence and transendothelial migration of human neutrophils in vitro. *J. Clin. Invest.* 83:2008–2017.

23. Osborn, L., C. Hession, R. Tizard, C. Vassallo, S. Luhowskyj, G. Chi-

Rosso, and R. Lobb. 1989. Direct cloning of vascular cell adhesion molecule 1 (VCAM1), a cytokine-induced endothelial protein that binds to lymphocytes. *Cell* 59:1203–1211.

24. Elices, M.J., L. Osborn, Y. Takada, C. Crouse, S. Luhowskyj, M.E. Hemler, and R.R. Lobb. 1990. VCAM-1 on activated endothelium interacts with the leukocyte integrin VLA-4 at a site distinct from the VLA-4/fibronectin binding site. *Cell* 60:577–584.

25. Lawrence, M.B., C.W. Smith, S.G. Eskin, and L.V. McIntire. 1990. Effect of venous shear stress on CD18-mediated neutrophil adhesion to cultured endothelium. *Blood* 75:227–237.

26. Kishimoto, T.K., R.A. Warnock, M.A. Jutila, E.C. Butcher, C. Lane, D.C. Anderson, and C.W. Smith. 1991. Antibodies against human neutrophil LECAM-1 (LAM-1/Leu-8/DREG 56 antigen) and endothelial cell ELAM-1 inhibit a common CD18-independent adhesion pathway in vitro. *Blood* 78:805–811.

27. Mulligan, M.S., J. Varani, M.K. Dame, C.L. Lane, C.W. Smith, D.C. Anderson, and P.A. Ward. 1991. Role of ELAM-1 in neutrophil-mediated lung injury in rats. *J. Clin. Invest.* 88:1396–1406.

28. Doerschuk, C.M., R.K. Winn, H.O. Coxson, and J.M. Harlan. 1990. CD18-dependent and -independent mechanisms of neutrophil emigration in the pulmonary and systemic microcirculation of rabbits. *J. Immunol.* 144:2327–2333.

29. Larsen, E., T. Palabrica, S. Sajer, G.E. Gilbert, D.D. Wagner, B.C. Furie, and B. Furie. 1990. PADGEM-dependent adhesion of platelets to monocytes and neutrophils is mediated by a lineage-specific carbohydrate, LNF III (CD15). *Cell* 63:467–474.

30. Hamburger, S.A., and R.P. McEver. 1990. GMP-140 mediates adhesion of stimulated platelets to neutrophils. *Blood* 75:550–554.

31. McEver, R.P., J.H. Beckstead, K.L. Moore, L. Marshall-Carlson, and D.F. Bainton. 1989. GMP-140, a platelet alpha-granule membrane protein, is also synthesized by vascular endothelial cells and is localized in Weibel-Palade bodies. *J. Clin. Invest.* 84:92–99.

32. Geng, J.G., M.P. Bevilacqua, K.L. Moore, T.M. McIntyre, S.M. Prescott, J.M. Kim, G.A. Bliss, G.A. Zimmerman, and R.P. McEver. 1990. Rapid neutrophil adhesion to activated endothelium mediated by GMP-140. *Nature* 343:757–760.

33. Patel, K.D., G.A. Zimmerman, S.M. Prescott, R.P. McEver, and T.M. McIntyre. 1991. Oxygen radicals induce human endothelial cells to express GMP-140 and bind neutrophils. *J. Cell Biol.* 112:749–759.

34. Carlos, T., N. Kovach, B. Schwartz, M. Rosa, B. Newman, E. Wayner, C. Benjamin, L. Osborn, R. Lobb, and J. Harlan. 1991. Human monocytes bind to two cytokine-induced adhesive ligands on cultured human endothelial cells: endothelial-leukocyte adhesion molecule-1 and vascular cell adhesion molecule-1. *Blood* 77:2266–2271.

35. Cotran, R.S., Gimbrone, Jr. MA, M.P. Bevilacqua, D.L. Mendrick, and J.S. Pober. 1986. Induction and detection of a human endothelial activation antigen in vivo. *J. Exp. Med.* 164:661–666.

36. Groves, R.W., M.H. Allen, J.N. Barker, D.O. Haskard, and D.M. MacDonald. 1991. Endothelial leucocyte adhesion molecule-1 (ELAM-1) expression in cutaneous inflammation. *Br. J. Dermatol.* 124:117–123.

37. Koch, A.E., J.C. Burrows, G.K. Haines, T.M. Carlos, J.M. Harlan, and S.J. Leibovich. 1991. Immunolocalization of endothelial and leukocyte adhesion molecules in human rheumatoid and osteoarthritic synovial tissues. *Lab. Invest.* 64:313–320.

38. Picker, L.J., T.K. Kishimoto, C.W. Smith, R.A. Warnock, and E.C. Butcher. 1991. ELAM-1 is an adhesion molecule for skin-homing T cells. *Nature* 349(6312):796–799.

39. Phillips, M.L., E. Nudelman, F.C. Gaeta, M. Perez, A.K. Singhal, S. Hakomori, and J.C. Paulson. 1990. ELAM-1 mediates cell adhesion by recognition of a carbohydrate ligand, sialyl-Lex. *Science* 250:1130–1132.

40. Polley, M.J., M.L. Phillips, E. Wayner, E. Nudelman, A.K. Singhal, S. Hakomori, and J.C. Paulson. 1991. CD62 and endothelial cell-leukocyte adhesion molecule 1 (ELAM-1) recognize the same carbohydrate ligand, sialyl-Lewis x. *Proc. Natl. Acad. Sci. U.S.A.* 88:6224–6228.

41. Lowe, J.B., L.M. Stoolman, R.P. Nair, R.D. Larsen, T.L. Berhend, and R.M. Marks. 1990. ELAM-1-dependent cell adhesion to vascular endothelium determined by a transfected human fucosyltransferase cDNA. *Cell* 63:475–484.

42. Butcher, E.C., D. Lewinsohn, A. Duijvestijn, R. Bargatze, N. Wu, and S. Jalkanen. 1986. Interactions between endothelial cells and leukocytes. *J. Cell Biochem.* 30:121–131.

43. Jutila, M.A., L. Rott, E.L. Berg, and E.C. Butcher. 1989. Function and regulation of the neutrophil MEL-14 antigen in vivo: comparison with LFA-1 and MAC-1. *J. Immunol.* 143(10):3318–3324.

44. Lewinsohn, D.M., R.F. Bargatze, and E.C. Butcher. 1987. Leukocyte- endothelial cell recognition: evidence of a common molecular mechanism shared by neutrophils, lymphocytes and other leukocytes. *J. Immunol.* 138:4313–4321.

45. Jutila, M.A., E.L. Berg, T.K. Kishimoto, L.J. Picker, R.F. Bargatze, D.K. Bishop, C.G. Orosz, N.W. Wu, and E.C. Butcher. 1989. Inflammation-induced endothelial cell adhesion to lymphocytes, neutrophils and monocytes: role of homing receptors and other adhesion molecules. *Transplantation.* 48:727–731.

46. Spertini, O., F.W. Luscinskas, G.S. Kansas, J.M. Munro, J.D. Griffin, Gimbrone, Jr. MA, and T.F. Tedder. 1991. Leukocyte adhesion molecule-1 (LAM-1, L-selectin) interacts with an inducible endothelial cell ligand to support leukocyte adhesion. *J. Immunol.* 147:2565–2573.

47. Picker, L.J., R.A. Warnock, A.R. Burns, C.M. Doerschuk, E.L. Berg, and E.C. Butcher. 1991. The neutrophil selectin LECAM-1 presents carbohydrate ligands to the vascular selectins ELAM-1 and GMP-140. *Cell* 66:921–933.

48. Kishimoto, T.K., R.A. Warnock, M.A. Jutila, E.C. Butcher, C. Lane, D.C. Anderson, and C.W. Smith. 1991. Antibodies against human neutrophil LECAM-1 (LAM-1/Leu-8/DREG-56 antigen) and endothelial cell ELAM-1 inhibit a common CD18-independent adhesion pathway in vitro. *Blood* 78:805–811.

49. Wallis, W.J., P.G. Beatty, H.D. Ochs, and J.M. Harlan. 1985. Human monocyte adherence to cultured vascular endothelium: monoclonal antibody-defined mechanisms. *J. Immunol.* 135:2323–2330.

50. Prieto, J., P.G. Beatty, E.A. Clark, and M. Patarroyo. 1988. Molecules

mediating adhesion of T and B cells, monocytes and granulocytes to vascular endothelial cells. *Immunology* 63:631–637.

51. Harlan, J.M. 1987. Consequences of leukocyte-vessel wall interactions in inflammatory and immune reactions. *Semin. Throm. Hemo.* 13:434–444.

52. Luscinskas, F.W., A.F. Brock, M.A. Arnaout, and M.A. Gimbrone. 1989. Endothelial-leukocyte adhesion molecule-1-dependent and leukocyte (CD11/CD18)-dependent mechanisms contribute to polymorphonuclear leukocyte adhesion to cytokine-activated human vascular endothelium. *J. Immunol.* 142(7):2257–2263.

53. Arnaout, M.A., L.L. Lanier, and D.V. Faller. 1988. Relative contribution of the leukocyte molecules Mol, LFA-1, and p150.95 (LeuM5) in adhesion of granulocytes and monocytes to vascular endothelium is tissue- and stimulus-specific. *J. Cell. Physiol.* 137:305–309.

54. Mentzer, S.J., M.A.V. Crimmins, S.J. Burakoff, and D.V. Faller. 1987. Alpha and beta subunits of the LFA-1 membrane molecule are involved in human monocyte-endothelial cell adhesion. *J. Cell. Physiol.* 130:410–415.

55. Keizer, G.D., A.A. Te Velde, R. Schwartig, C.G. Figdor, and J.E. De Vries. 1987. Role of p150.95 in adhesion, migration, chemotaxis and phagocytosis of human monocytes. *Eur. J. Immunol.* 17:1317–1322.

56. Carlos, T., and J.M. Harlan. 1990. Membrane proteins involved in phagocyte adherence to endothelium. *Immunol. Rev.* 114:1–24.

57. Nickoloff, B.J., and E.M. Griffiths. 1990. Abnormal cutaneous topobiology: the molecular basis for dermatopathologic mononuclear cell patterns in inflammatory skin disease. *J. Invest. Dermatol.* 95:128S–131S.

58. Issekutz, T.B., and A.C. Issekutz. 1991. T lymphocyte migration to arthritic joints and dermal inflammation in the rat: differing migration patterns and the involvement of VLA-4. *Clin. Immunol. Immunopathol.* 61:436–447.

59. Issekutz, T.B. 1991. Inhibition of in vivo lymphocyte migration to inflammation and homing to lymphoid tissues by the TA-2 monoclonal antibody. A likely role for VLA-4 in vivo. *J. Immunol.* 147:4178–4184.

60. Hession, C., L. Osborn, D. Goff, G. Chi-Rosso, C. Vasallo, M. Pasek, C. Pittack, R. Tizard, S. Goelz, K. McCarthy, S. Hopple, and R. Lobb. 1990. Endothelial leukocyte adhesion molecule 1: Direct expression cloning and functional interactions. *Proc. Natl. Acad. Sci. USA* 87:1673–1677.

61. Goelz, S.E., C. Hession, D. Goff, B. Griffiths, R. Tizard, B. Newman, G. Chi-Rosso, and R. Lobb. 1990. ELFT: A gene that directs the expression of an ELAM-1 ligand. *Cell* 63:1349–1356.

62. McEver, R.P., and M.N. Martin. 1984. A monoclonal antibody to a membrane glycoprotein binds only to activated platelets. *J. Biol. Chem.* 259:9799–9804.

63. Kishimoto, T.K., M.A. Jutila, and E.C. Butcher. 1990. Identification of a human peripheral lymph node homing receptor: a rapidly downregulated adhesion molecule. *Proc. Natl. Acad. Sci. USA* 87:2244–2248.

64. Hall, D.E., L.F. Reichardt, E. Crowley, B. Holley, H. Moezzi, A. Sonnenberg, and C.H. Damsky. 1990. The alpha1-beta1 and alpha6-beta1 integrin heterodimers mediate cell attachment to distinct sites on laminin. *J. Cell Biol.* 110:2175–2184.

65. Skubitz, K.M., and 2nd R.W. Snook 1987. Monoclonal antibodies that recognize lacto-N-fucopentaose III (CD15) react with the adhesion-promoting gly-

coprotein family (LFA-1/HMac-1/gp 150,95) and CR1 on human neutrophils. *J. Immunol.* 139:1631–1639.

66. Stocks, S.C., M. Albrechtsen, and M.A. Kerr. 1990. Expression of the CD15 differentiation antigen (3-fucosyl-N-acetyl-lactosamine, LeX) on putative neutrophil adhesion molecules CR3 and NCA-160. *Biochem. J.* 268:275–280.

67. Fukuda, M., E. Spooncer, J.E. Oates, A. Dell, and J.C. Klock. 1984. Structure of sialylated fucosyl lactosaminoglycan isolated from human granulocytes. *J. Biol. Chem.* 259:10925–10935.

68. Carlsson, S.R. H. Sasaki, and M. Fukuda. 1986. Structural variations of O-linked oligosaccharides present in leukosialin isolated from erythroid, myeloid, and T-lymphoid cell lines. *J. Biol. Chem.* 261:12787–12795.

69. Carlsson, S.R., J. Roth, F. Piller, and M. Fukuda. 1988. Isolation and characterization of human lysosomal membrane glycoproteins, h-lamp-1 and h-lamp-2. Major sialoglycoproteins carrying polylactosaminoglycan. *J. Biol. Chem.* 263:18911–18919.

70. Skaleric, U., J.B. Allen, P.D. Smith, S.E. Mergenhagen, and S.M. Wahl. 1991. Inhibitors of reactive oxygen intermediates suppress bacterial cell wall-induced arthritis. *J. Immunol.* 147:2559–2564.

71. Pober, J.S., L.A. Lapierre, A.H. Stolpen, T.A. Brock, T.A. Springer, W. Fiers, M.P. Bevilacqua, D.L. Mendrick, and Gimbrone Jr. MA. 1987. Activation of cultured human endothelial cells by recombinant lymphotoxin: comparison with tumor necrosis factor and interleukin 1 species. *J. Immunol.* 138:3319–3324.

72. Carlos, T.M., B.R. Schwartz, N.L. Kovach, E. Yee, M. Rosa, L. Osborn, G. Chi-Rosso, B. Newman, R. Lobb, M. Rosso, and a.l. et. 1990. Vascular cell adhesion molecule-1 mediates lymphocyte adherence to cytokine-activated cultured human endothelial cells [published erratum appears in *Blood* 1990 Dec 1;76(11):2420]. *Blood* 76:965–970.

73. Hattori, R., K.K. Hamilton, R.D. Fugate, R.P. McEver, and P.J. Sims. 1989. Stimulated secretion of endothelial von Willebrand factor is accompanied by rapid redistribution to the cell surface of the intracellular granule membrane protein GMP-140. *J. Biol. Chem.* 264:7768–7771.

74. Bonfanti, R., B.C. Furie, B. Furie, and D.D. Wagner. 1989. PADGEM (GMP140) is a component of Weibel-Palade bodies of human endothelial cells. *Blood* 73:1109–1112.

75. Patel, K.D., G.A. Zimmerman, S.M. Prescott, R.P. McEver, and T.M. McIntyre. 1991. Oxygen radicals induce human endothelial cells to express GMP-140 and bind neutrophils. *J. Cell Biol.* 112:749–759.

Part 3
Cell Biology of Adhesion Molecules

17
Fc Receptor Function and Cytoplasmic Domain Heterogeneity

IRA MELLMAN

Introduction

It is becoming increasingly apparent that many adhesion proteins play far more complex functions than the simple attachment of one cell to another or to a fixed substrate. Ligand binding to both integrin-type and noninteg-rin adhesion proteins has been associated with a variety cellular events including alterations in morphology, motility, transmembrane signaling, stimulation of effector functions, and possibly endocytosis and virus penetration. Since there is as yet little information about how these activities are transduced, it is useful to consider the features of better-understood leukocyte receptors as models to gain some insight into how other related molecules carry out similar sets of functions.

Structural Features of Fc Receptors for IgG

Among the best-characterized receptors on leukocytes and lymphocytes are receptors for the Fc domain of immunoglobulin G (IgG). Fc receptors (FcR) actually comprise a multigene family of closely related, immuno-globulin (Ig)-like molecules that can be divided into three distinct groups in both human and murine cells (Table 17.1) (1). FcRI is the only recep-tor class that has a high affinity (10^{-8}–10^{-9} M) for monomeric IgG and whose expression is limited almost entirely to macrophages. These recep-tors have an extracellular region that contains three Ig-like domains, a single membrane anchor, and a cytoplasmic tail of ~30 amino acids. In contrast, FcRIII has a very low affinity for IgG and, in fact, will only bind immune complexes or IgG aggregates. Its extracellular portion contains two Ig-like domains, homologous to the first two Ig domains of FcRI. There are two distinct FcRIII genes that differ only by the presence or absence of a single charged amino acid residue within the presumptive membrane-spanning region. In humans, the FcRIII gene expressed in neutrophils leads to the production of a GPI-anchored glycoprotein, while in natural killer cells, FcRIII is expressed a type I membrane pro-

TABLE 17.1. Three major classes of Fc receptors for IgG.

FcR-I	FcR-II	FcR-III
Monomeric IgG high affinity	Multimeric IgG intermediate affinity	Multimeric IgG low affinity
Macrophages	Macrophages, lymphocytes, NK cells, granulocytes	Granulocytes, NK cells
~50-kD transmembrane glycoprotein	40–60-kD transmembrane glycoprotein	40–80-kD transmembrane or PI-linked protein ± γ-chain
Human, mouse	Human, mouse	Human, mouse

tein with an authentic membrane anchor and cytoplasmic tail. Interestingly, the natural killer cell FcRIII is also associated with an accessory subunit, designated γ-chain, that is virtually identical to that expressed by the basophil FcR for IgE and to the ζ-chain that is a component of the antigen receptor complex in T-cells (2). There is now considerable evidence that γ-chain association is required for FcRIII to transmit a transmembrane signal in response to ligand binding. In this regard, γ-chain in FcRIII plays a role homologous to that played by the ζ-chain in the T-cell receptor (3).

The third major class of receptor, FcRII, is also the most abundant and widely distributed, FcRII being the one receptor expressed by almost every FcR-positive cell type included monocytes, macrophages, granulocytes, as well as lymphocytes (B-cells and to a lesser extent T-cells). Certain IgG-transporting epithelia, such as placental syncytiotrophoblasts, also express FcRII (4). Since FcRII when transfected into polarized MDCK cells mediates apical to basolateral transcytosis of IgG (5), it is likely (but not yet proved) that this receptor is responsible for IgG transport across the placenta.

cDNA and genomic cloning of FcRII from both mouse and human cells has indicated that even within this single class, there is considerable structural heterogeneity due to cell-type–specific alternative mRNA splicing. This situation is particularly well characterized in the case of murine FcRII, in which two major splice products are generated in macrophages or in B-cells (1). As shown in Figure 17.1, macrophages express an isoform designated FcRII-B2 consisting of an extracellular region containing two Ig-like domains, a hydrophobic membrane-spanning segment, and a cytoplasmic tail of 47 amino acids. The FcRII-B1 isoform characteristic of B-cells is identical to B2 except for the presence of an in-frame insertion of an additional 47 amino acids at position 6 in the receptor's cytoplasmic domain. In human cells, a similar albeit shorter insertion is expressed corresponding to the amino-terminal half of the murine insertion.

FIGURE 17.1. Structure of the major isoforms of mouse FcRII.

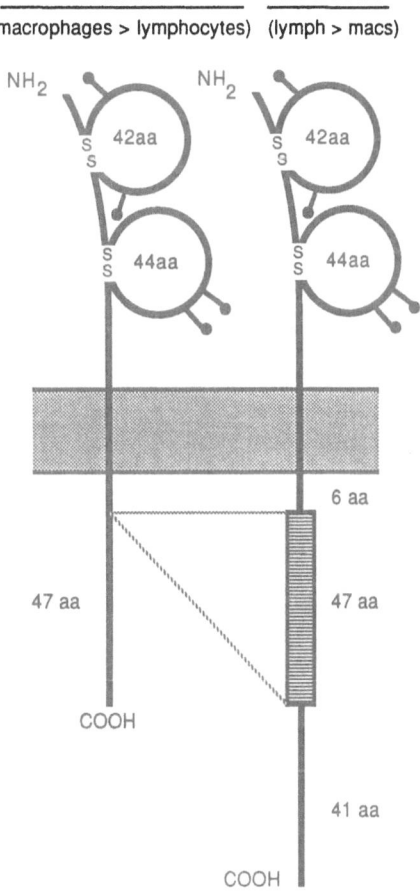

Functions of FcRII

Given the cell-type–restricted pattern of alternative splicing, it seems likely that the two major FcRII isoforms would also be associated with distinct functions. In macrophages, FcRs are well known to be responsible for a variety activities including the endocytosis of soluble IgG-containing antibody–antigen complexes via clathrin-coated pits, the phagocytosis of large IgG-coated particles, triggering the localized polymerization of actin-containing microfilaments beneath the site of particle attachment, and signaling the exocytosis or release of a variety of protein and nonprotein mediators of inflammatory and cytotoxic function (plasminogen activator, elastase, leukotrienes, prostaglandins, H_2O_2, and other active forms of oxygen). All of these activities can thus be associated with FcRII-B2, although FcRI and FcRIII are also likely to be capable of mediating at least the signaling functions of FcRII-B2. Much less is known about FcR functions in B-cells, although it is thought that these receptors (i.e., FcRII-B1) play a role in regulating antigen-triggered activation of B-lymphocytes via surface Ig.

Given the wide array of activities associated with FcRII and the exquisite cell-type specificity in the expression of the two major receptor isoforms, we began a systematic evaluation of the functional activities associated with FcRII-B1 and -B2 as well as an attempt to correlate function with structural differences in the cytoplasmic domains of the two receptors. Our general approach involved the expression of the two receptor isoforms, as well as a series of mutants derived by site-directed mutagenesis, in FcR-negative fibroblasts and B-cell lymphomas.

Results

Only FcRII-B2 Is Capable of Accumulating at Clathrin-Coated Pits

We first sought to determine whether the two receptor isoforms differed in their abilities to mediate endocytosis via coated pits and coated vesicles. Thus, cDNAs encoding FcRII-B1, -B2, or various cytoplasmic tail mutants were transfected into fibroblasts. COS cells were used for transient expression while CHO cells were used to generate permanently expressing stable cell lines. Expression levels were enhanced using an amplifiable expression vector such that the numbers of receptors/cell approximated that found in macrophages and B-cells ($2-6 \times 10^5$/cell). Using immunofluorescence, electron microscopic (EM), and biochemical assays for endocytosis of soluble IgG–immune complexes, we found that only cells transfected with FcRII-B2 were capable of localizing at clathrin-coated pits and mediating rapid internalization of bound ligand (Fig. 17.2) (6). Neither FcRII-B1 nor a deletion mutant lacking the entire cytoplasmic domain (except for the first lysine residue) exhibited appreciable coated-pit localization. Interestingly, however, the FcRII-B1 phenotype was even more severely affected than that of the tail-minus mutant.

To map the region or regions of the FcRII-B2 cytoplasmic domain that were required for coated-pit localization and endocytosis, a series of partial deletion mutants was constructed and analyzed biochemically and by EM. These results suggested that the region between amino acids 18 and 31 contained sequences that were important for localization at coated pits (7). This region also corresponds to the only area of significant homology between the cytoplasmic tails of murine and human FcRII-B2. It also contains a single conserved tyrosine residue; a second tyrosine residue is found at position 43, outside the region of homology. Since tyrosines are an important feature of coated-pit-localization domains in a number of

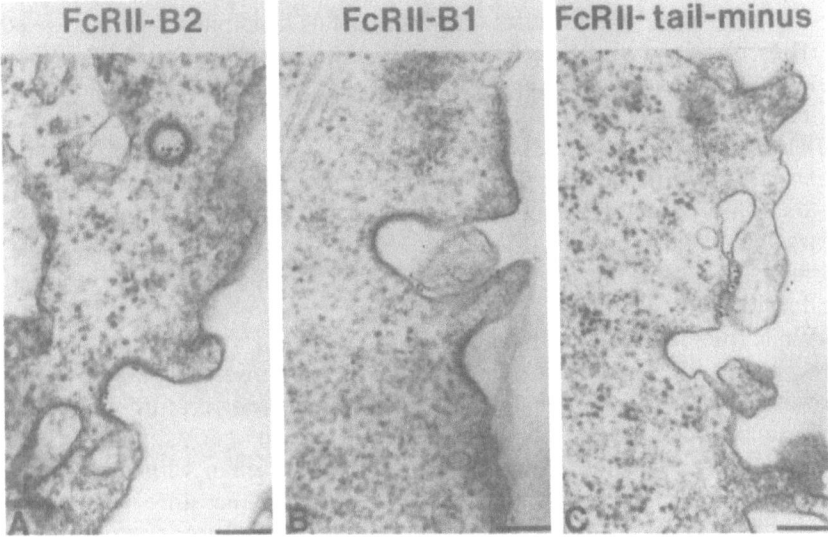

FIGURE 17.2. Localization of colloidal-gold-labeled immune complexes bound to CHO cells expressing FcRII-B1, FcRII-B2, or a tail-minus deletion mutant. Taken from Miettinen et al. (6).

other endocytic receptors, we next expressed FcRII-B2 mutants lacking either or both tyrosines (changed for alanines). Surprisingly, these mutations only partially decreased the ability of FcRII-B2 to mediate coated-pit localization and endocytosis (7). Thus, it is apparent that reasonably efficient endocytosis can occur via coated-pit localization domains that do not rely disproportionately on tyrosine (or other aromatic residues) for activity.

The FcRII-B1 Insert Actively Prevents Coated-Pit Localization and Facilitates Cytoskeleton Attachment

How does the cytoplasmic domain insert in FcRII-B1 act to prevent coated-pit localization? Since it occurred at a site in the cytoplasmic tail which was apparently outside the presumptive coated-pit localization domain, it did not seem likely that the insertion acted by physically disrupting a determinant required for accumulation of the receptor at coated pits. A potential hint, however, came from the quantitative EM immunocytochemistry. As mentioned above, FcRII-B1 was found to be even less effective at coated-pit localization than even the tail-minus deletion mutant (6). Since the tail-minus mutant would be expected to be distributed randomly in coated and uncoated regions of the plasma membrane, this observation suggested that the insert played an active role in preventing coated-pit localization. To test this possibility, we next used PCR to place the insert at the very COOH-terminus of FcRII-B2. When analyzed by EM and biochemistry, this B2/B1 insert mutant was found to exhibit an endocytosis-negative phenotype very similar to wild-type FcRII-B1 (7). Thus, the B1 insertion appeared to exhibit its negative effect on coated-pit localization irrespective of its position in the FcRII cytoplasmic tail, suggesting that it played an active role in preventing entry into coated pits. Further experiments showed that only the NH_3-terminal half of the insert is actually needed to produce the inhibition of endocytosis (K. Matter et al., unpublished).

While the mechanism of action of the FcRII-B1 insert remains unknown, it is interesting that receptors bearing the insert—as well as some downstream sequences—exhibit a ligand- and temperature-dependent association with a detergent-insoluble fraction (7). Since by immunofluorescence these receptors also appear to align with cytoplasmic actin filaments (stained with labeled phalloidin), and since the Triton-insoluble fraction is greatly increased by performing detergent extractions in the presence of phalloidin, it seems likely that detergent insolubility reflects association of the receptor with the cytoskeleton. Certainly, such an association would help prevent entry of FcRII-B1 into clathrin-coated regions; but it is not possible to conclude that cytoskeletal attachment is the mechanism of endocytosis inhibition. Conceivably, the insertion specifies interaction with another protein that blocks coated-pit localization and in turn attaches the complex to a detergent-insoluble fraction. Nevertheless, FcRII represents the first documented example of alternative mRNA splicing, leading to the formation of receptors with different efficiencies of coated-pit localization and presumptive cytoskeleton attachment.

FcRII Endocytosis in B-Cells

Given that the endocytosis-negative FcRII-B1 isoform is the only FcRII species expressed in abundance in B-lymphocytes, one would expect that these cells would also be defective in FcR-mediated endocytosis. In recent experiments, we examined this possibility in detail using the A20 B-cell lymphoma (which expresses endogenous FcRII-B1). In addition, we used an FcRII-negative derivative of A20 cells—IIA.1.6 cells—which bear a deletion of the 5' end of the FcRII gene (8). Importantly, neither A20 cells nor IIA.1.6 cells transfected with the FcRII-B1 cDNA were capable of localizing their surface FcR at coated pits or of mediating rapid internalization of bound ligand. This endocytosis "defect" could be "corrected," however, by transfecting the cells with FcRII-B2 cDNA, under which conditions the B-cells were as efficient at FcR-mediated endocytosis and coated-pit localization as macrophages or FcRII-B2-transfected CHO cells (9). Thus, the only reason B-lymphocytes are unable to internalize their FcR is that they express an internalization-incompetent isoform of the receptor, namely, FcRII-B1. This result establishes the biological relevance of the functional differences observed between the two receptor isoforms when transfected into fibroblasts.

If B-cells are unable to mediate rapid endocytosis of FcRII-B1, one might also expect that they would be unable to process and present exogenous antigen via this receptor. In contrast, B-cells transfected with internalization-competent FcR such as FcRII-B2 should be capable of efficient presentation of antigen delivered as immune complexes. To test this prediction, we determined the ability of IIA.1.6 cells to present a well-characterized antigen, cI λ-repressor, to an antigen-specific T-cell hybridoma. λ-repressor was delivered to the cells in one of two ways: either as soluble antigen or as an immune complex using two monoclonal anti–λ-repressor antibodies directed against distinct epitopes. In a series of experiments just recently completed (9), we found that the efficiency of presentation by both nontransfected IIA.1.6 cells and cells expressing FcRII-B1 was identical whether the cells were incubated with soluble or immune-complexed antigen. In contrast, presentation of the λ-repressor was enhanced almost 10^3-fold when incubated as immune complexes with FcRII-B2–expressing IIA.1.6 cells.

Thus, the ability of B-cells to mediate efficient antigen presentation via their FcRs correlated with the endocytosis competence of the receptor they expressed. This observation has two interesting implications. *First,* primary B-cells which normally express FcRII-B1 would not be expected to present antigen present as immune complexes despite the fact that B-cells are otherwise extremely potent antigen-presenting cells following processing of antigens bound to sIg. This feature would help ensure that presenting B-cells are only stimulated in an antigen-specific fashion: i.e., that they can only present antigens that reflect the specificity of their own sIg molecules. Efficient FcR-mediated presentation would subvert this process by allowing any antigen present in an immune complex to be presented irrespective of the actual specificity of the presenting cell. Release of B-cells from such a restriction has been implicated as a possible cause of certain forms of autoimmunity. *Second,* since FcR-mediated presentation by FcRII-B2–expressing cells occurs at concentrations of IgG and antigen that are both physiologically relevant and experimentally detectable, this approach provides an exciting new strategy with which to approach the analysis of the cell biology of antigen processing and presentation.

Signal Transduction and Capping in B-Cells

Apart from acting to prevent FcR-mediated endocytosis, it is possible that the cytoplasmic domain insert in FcRII-B1 also serves another function-specific FcR in B-cells. Although relatively little is known about

B-cell FcR functions, it is clear that they play a potentially important role in regulating antigen-induced B-cell activation (10,11). When sIg becomes cross-linked by antigen or F(ab)$'_2$ fragments of anti-sIg, most B-cells respond by exhibiting an activation response. In A20 B-cells, this is characterized by a rapid increase in cytosolic free Ca^{2+} and leads to increases in serine and tyrosine phosphorylation as well as in cytokine (IL-2) release. If, however, sIg is cross-linked by an intact IgG anti-sIg—under conditions which permit co–cross-linking of surface FcRII-B1—this activation response is blocked or aborted. We used the transfected IIA.1.6 cells to determine whether this activity required the B-cell–specific cytoplasmic domain insert. Our results, measuring both Ca^{2+} and IL-2 release, indicate that both FcRII-B1 and FcRII-B2 are equally capable of regulating sIg-triggered activation (9). Thus, although the insert contains one or more serine phosphorylation sites that are used in conjunction with the activation process (12), neither these sites nor the insert in general appear to be required for this activity. Instead, it seems that a region of the FcR cytoplasmic tail corresponding to the presumptive coated-pit-localization domain (sequences between residues 18 and 31) also contains information required to mediate activation regulation.

What is the insert actually used for? One possibilty is enhancing the efficiency of B-cell "capping." When FcR (among other surface molecules such as sIg) are cross-linked on B-cells, they first aggregate in small "patches" which then coalesce into a large "cap" at one pole of the cell. By analyzing the abilities of the different receptor isoforms and mutants to cap in the transfected IIA.1.6 cells, we have found that most efficient capping is exhibited by those receptors that retain the FcRII-B1–specific cytoplasmic domain insert. Given that the insert may help immobilize the receptor to the cytoskeleton, this is an interesting possibility since capping has long been thought (but never actually shown) to involve interactions between plasma membrane proteins and the underlying cytoskeletal network.

Phagocytosis in Transfected Fibroblasts

The final FcR-associated function we have analyzed thus far is phagocytosis, i.e., the internalization of large (>0.5 μm) particles coated with IgG. Phagocytosis is a form of endocytosis must closely associated with leukocytes such as macrophages, but we have found that even CHO cells transfected with FcRII cDNAs can be induced to mediate this form of uptake. Interestingly, however, CHO cells will only avidly phagocytose certain particles. One particularly good ligand is *Toxoplasma gondii*, a protozoan parasite that normally gains access to nucleated cells by a modified form of phagocytosis (in the absence of FcR) resulting in the sequestration of the parasite in an intracellular vacuole that is not acidic and is incompetent for fusion with lysosomes or other intracellular organelles (13). If

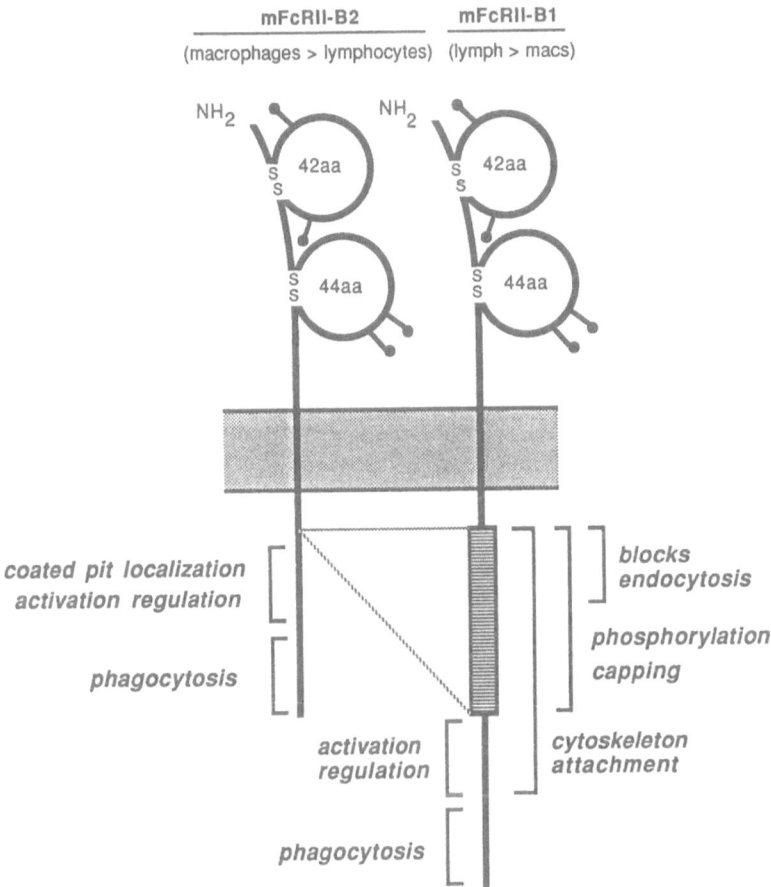

FIGURE 17.3. Hypothetical domain map for functions associated with the cytoplasmic tails of FcRII-B1 and FcRII-B2.

either viable or killed *T. gondii* are coated with IgG and incubated with FcR-expressing CHO cells, however, the opsonized parasites are bound and internalized via FcR. This reflects a more classical phagocytic event since the parasite-containing vacuole so formed is both acidic and fusion-competent (13). By analyzing the abilities of cells transfected with different FcR isoforms and mutants, we have been able to determine that again both FcRII-B1 and FcRII-B2 are equally efficient at mediating this form of uptake. From an analysis of cytoplasmic-domain deletion mutants, however, it appears that phagocytosis requires a region of the cytoplasmic tail which is at or near the very COOH-terminus. Even addition of sequence to the COOH-terminus will greatly reduce this form of phagocytic activity. While it is not clear whether *T. gondii* phagocytosis measured in FcR-transfected CHO cells is truly equivalent to phagocytosis in macrophages, our observations have led us to tentatively conclude that there may be a domain that is required for phagocytic uptake that is entirely distinct from that involved in coated-pit–mediated endocytosis.

Conclusions

Based on our observations thus far, we have prepared a preliminary domain map that attempts to summarize our observations concerning the localization of the multiple functional domains that appear to be present in the FcRII cytoplasmic tail (Fig. 17.3). While all of these assignments must be viewed as tentative and, of course, are subject to change, together they make a more important general point: namely, that even a short cytoplasmic tail that bears no obvious sequence homology to the cytoplasmic tails of other receptors may harbor a wealth of different activities. Thus, when considering the molecular basis for the activities associated with leukocyte proteins typically involved in various cell adhesion events, it will be important to keep in mind the potential for such heterogeneity of structure and function even when faced with cytoplasmic domains that are otherwise undistinguishable.

References

1. Mellman, I. (1988) Relationships between structure and function in the Fc receptor family. *Current Opinion in Immunology* 1:16–25.
2. Ra, C., Jouvin, M.-H.E., Blank, U. and Kinet, J.-P. (1989) A macrophage Fc receptor and the mast cell receptor for immunoglobulin E share an identical subunit. *Nature* 341:752–754.
3. Orloff, D.G., Ra, C., Frank, S.J., Klausner, R.D. and Kinet, J.-P. (1990) Family of disulphide-linked dimers containing the ζ and ϵ chains of the T-cell receptor and the γ chain of Fc receptors. *Nature* 347:189–191.
4. Stuart, S., Simister, N.E., Clarkson, S.B., Kacinski, B.M., Shapiro, M. and Mellman, I. (1989) Human IgG Fc receptor (hFcRII; CD32) exists as multiple isoforms in macrophages, lymphocytes and IgG-transporting placental epithelium. *EMBO J.* 8:3657–3666.
5. Hunziker, W. and Mellman, I. (1989) Expression of macrophage-lymphocyte Fc receptors in MDCK cells: polarity and transcytosis differ for isoforms with or without coated pit localization domains. *J. Cell Biol.* 109:3291–3302.
6. Miettinen, H.M., Rose, J.K. and Mellman, I. (1989) Fc receptor isoforms exhibit distinct abilities for coated pit localization as a result of cytoplasmic domain heterogeneity. *Cell* 58:317–327.
7. Miettinen, H.M., Matter, K., Hunziker, W., Rose, J.K. and Mellman, I. (1991) Fc receptors contain a cytoplasmic domain determinant that actively regulates coated pit localization. *J. Cell Biol.* 118:875–888.
8. Lewis, V.A., Koch, T., Plutner, H. and Mellman, I. (1986) A com-

plementary DNA clone for a macrophage-lymphocyte Fc receptor. *Nature* 324:372–375.

9. Amigorena, S., Bonnerot, C., Drake, J.R., Choquet, D., Hunziker, W., Guillet, J.-G., Webster, P., Sautes, C., Mellman, I. and Fridman, W.H. (1991) Cytoplasmic domain heterogeneity and functions of IgG Fc receptors in B-lymphocytes. *Science* in press.

10. Phillips, N.E. and Parker, D.C. (1984) Cross-linking of B lymphocyte Fcγ receptors and membrane immunoglobulin inhibits anti-immunoglobulin induced blastogenesis. *J. Immunol.* 132:627–632.

11. Bijsterbosch, M.K. and Klaus, G.G.B. (1985) Crosslinking of surface immunoglobulin and Fc receptors on B lymphocytes inhibits stimulation of inositol phospholipid breakdown via the antigen receptors. *J. Exp. Med.* 162:1825–1836.

12. Hunziker, W., Koch, T., Whitney, J.A. and Mellman, I. (1990) Fc receptor phosphorylation during receptor-mediated control of B-cell activation. *Nature* 345:628–632.

13. Joiner, K.A., Fuhrman, S.A., Miettinen, H., Kaspar, L.L., Howe, C.L. and Mellman, I. (1990) Toxoplasma gondii: Fusion competence of parasitophorous vacuoles in Fc receptor transfected CHO cells. *Science* 249:641–646.

18
Regulation of Antigen-Independent Adhesion of CD4 T Cells

Fabienne Mazerolles, F. Amblard, O. Lecomte,
S. Meloche, C. Barbat, P. Hauss, C. Hivroz, R. Sekaly,
and A. Fischer

T-lymphocyte adhesion to antigen-presenting cells and B cells is mediated by contact between adhesion receptors and their cellular ligands (e.g., LFA-1/ICAM-1, -2, CD2/LFA-3, CD28/B7). In addition VLA molecules may be involved through binding to the extracellular matrix (1–5).

Such antigen-independent adhesion can be transiently up-regulated by T-cell-receptor cross-linking, which leads to increased LFA-1–dependent adhesion and also increased VLA-dependent adhesion (5–8). This event is likely of importance in promoting cell-to-cell interaction, i.e., delivering transducing signals for appropriate polarized cytokine secretion (3). It is not clear, however whether weak antigen-independent adhesion precedes antigen recognition (9) and whether the extracellular matrix promotes cell-to-cell adhesion (10).

By studying conjugate formation between resting CD4$^+$ T-cell and B-cell lines, we have previously found that a low percentage (around 10%) of T cells rapidly bind B cells independent of antigen recognition. The adhesion was abrogated by T-cell incubation with a pair of anti–LFA-1 and anti-CD2 antibodies (11,12). In order to better analyze and quantify the weak interaction of resting T cells with B cells, a technique based on the assessment of hydrodynamic elongation required for conjugate disruption was established. It consists of counting conjugates formed using a flow cytometer. Fluidic and electronic modifications were made so that conjugate flow rates could vary over three orders of magnitude, leading to proportional elongation stress. A corresponding force could be determined for which half of the conjugates were disrupted (13,14).

This methodology allows us to observe the transient and temperature-dependent conjugate formation of resting CD4$^+$ T cells with Raji cells. It could be determined by incubation with specific antibody that LFA-1 and CD2 contribute equivalently to binding of resting CD4 T cells. In contrast, PMA-activated CD4 T cells had an increased rate of conjugate formation and a much higher percentage of cells bound to B cells, and such adhesion was found stable. Most of the adhesion was shown to be dependent upon the LFA-1 pathway by antibody-blocking experiments. It was estimated that PMA-activated T-cell adhesion was strengthened by three orders of magnitude as compared to adhesion of resting T cells. This method makes the evaluation of the adhesion strength of T (or other) cells following different types of stimulation feasible.

Adhesion of resting CD4+ T cells to B cells is a transient phenomenon (12). Such transicncy was also found when studying the adhesion of helper T-cell clones or of anti-CD3 antibody-activated T cells (12,15). In contrast, T cells passively activated by the NKIL16 antibody, an antibody able to induce LFA-1−mediated aggregation, were found to adhere strongly but stably (15). Finally, it was found that CD45RA(+) naive CD4+ T cells which weakly adhere to B cells exhibit a stable pattern of adhesion (16).

The reason adhesion of resting memory CD4+ T cells, of T-cell clones, and of anti-CD3 antibody-activated CD4+ T cells is transient is unclear. Dustin and Springer have proposed that following LFA-1−mediated up-regulation of adhesion induced by TCR/CD3 crosslinking, the de-adhesion process occurs passively by "deactivation" of the protein−kinase C−dependent phenomenon (6,10). Alternatively, competiveness of adhesion to other surrounding cells or to the extracellular matrix in vivo may enable cells to detach and to move away (10). These mechanisms are not incompatible with the existence of more specific processes of down-regulation of T-cell adhesion.

In the study of the role of membrane receptors in determining T-cell adhesion, we found evidence for a regulatory role of CD4. Indeed, it was first observed that binding of CD4 to its ligand, MHC class II molecules at the B-cell surface, played no detectable role in determining cell adhe-sion. Resting or activated CD4+ T cells equally bind MHC class II(+) or MHC class II(−) B cells (12). Conversely, MHC class II(+) B cells non-selectively bind the CD4(−) mutant of the CEM T-cell line A201 or of the HUT78 T-cell line and their respective counterparts transfected with the CD4 cDNA (17). That CD4-MHC class II interaction between T and B cells is not strongly involved in mediating T-cell adhesion is not a sur-prising result since MHC class II(+) cell binding to CD4 could only be observed provided that CD4 expression is very high (20–50 times the physiological level) (18).

It has been recently reported that crosslinking of MHC class II mole-cules or superantigen binding to MHC class II induced an LFA-1−mediated up-regulation of adhesion (19). It is interesting to note that the weak CD4−MHC class II interaction cannot mimick this effect.

CD4−MHC class II interaction appears, however, to alter the kinetics of CD4+ T-cell adhesion. As previously mentioned, adhesion of CD4+ T cells to B cells is a transient phenomenon that peaks about 20 min after cell contacts have been initiated. This is followed by a decay in cell conju-gates over the next 60 min. The latter phenomenon requires a 37°C temperature and is not observed at 4°C. In contrast, adhesion of CD4 T cells to MHC class II(−) B cells from different origins (irradiation-induced mutants and naturally occurring mutants) is stable over at least 60 min. The same observation has been made in studying the adhesion of CD4(−) T-cell mutants, and reexpression of CD4 following CD4 cDNA transfection results in a return to transiency of adhesion (17).

It was finally found that CD4 ligands such as certain anti-CD4 antibodies (Leu3a, OKT4a, 13B82), and the gp160 HIV env protein can strongly inhibit the adhesion of CD4$^+$ T cells (12,17,20). This inhibition could be exerted on the adhesion of resting memory CD4$^+$ T cells, anti-CD3 Ab–activated T cells, and CD4$^+$ T-cell lines but not of NKIL16 Ab-activated T cells (12,15,17). Similar results were obtained by studying the effect of a 12-mer peptide mimicking the 35–46 position of the DRβ1 domain which contains the highly preserved RFDS sequence (12,17). The latter effect was neutralized by the addition of soluble CD4 molecules (12).

Together, these results indicate that CD4–MHC class II interaction can down-regulate antigen-independent adhesion of resting and activated T cells. These effects are not related to CD4-TCR/CD3 interaction because they were observed in the study of the adhesion of the A201 CD4$^+$ T cells that do not express the TCR/CD3 complex. These findings lead to the hypothesis of a putative negative signal delivered to T cells through CD4–MHC class II or CD4-gp120 interaction, which could induce down-regulation of T-cell adhesion.

Further data supporting this hypothesis have been collected by studying the role of the p56^{lck} tyrosine kinase in the regulation of adhesion. P56^{lck}, a member of the src family, can associate with CD4 (and CD8). Its tyrosine kinase activity is enhanced by CD4 crosslinking and may result in phosphorylation of different substrates including the CD3 zeta subunit (22,23). CD4(−) T-cell lines in which a mutated form of CD4 cDNA lacking an intracytoplasmic domain has been transfected (CD4) resulted in membrane CD4 expression without interaction with p56^{lck} as shown by coprecipitation studies (24). Such A201 and HUT T-cell lines behave in adhesion assays similarly to CD4(−) T-cell counterparts since adhesion to B cells was stable and since it could not be inhibited by CD4 ligands (17).

Finally, pretreatment of CD4$^+$ T cells with Herbimycin A, a tyrosine kinase inhibitor, resulted in loss of sensitivity to CD4-mediated down-regulation of cell addition, whereas conjugates formation was not altered. These results demonstrate that CD4-mediated down-regulation of T-cell adhesion is dependent upon p56^{lck} function. Substrates of p56^{lck} involved in this regulation remain to be determined. These results lead us to propose the following scheme of T-cell adhesion regulation.

In vivo, multiple contacts of T cells with surrounding B cells or APCs occur in lymphoid organs. These contacts are weak, although mediated by specific adhesion receptors. Such adhesion is transient provided that there is no antigen recognition. De-adhesion occurs because of competiveness with other cells and by an active component, i.e., CD4 p56^{lck}–

mediated signaling to the T cell. Antigen-specific recognition (involving CD4-TCR/CD3 association), in contrast, leads to rapid up-regulation of adhesion, enabling optimal cell-to-cell interaction. CD4 molecules associated to TCR/CD3 complexes are then thought to deliver positive signaling, perhaps because the $p56^{lck}$ is now in contact with distinct substrates (25). Such adhesion is also transient because of deactivation and also because at a certain point "free" CD4 molecules on the surface might prevail upon CD4-TCR/CD3 complexes which are progressively internalized, and might eventually deliver negative signaling.

References

1. Shaw S. et al. 1986. *Nature* 323:262–264.
2. Springer T.A. 1990. *Ann. Rev. Cell. Biol.* 6:359–402.
3. Springer T.A. 1990. *Nature* 343:425–434.
4. Van de Velde H et al. 1991. *Nature* 351:662–665.
5. Shimizu Y. et al. 1990. *Immunol. Rev.* 114:109–143.
6. Dustin M.L., and Springer T.A. 1989. *Nature* 341:619–624.
7. Van Kooyk Y. et al. 1989. *Nature* 342:811–813.
8. Figdor C.G. et al. 1990. *Immunol. Today* 11:277–280.
9. Spits H. et al. 1986. *Science* 232:403–405.
10. Dustin M.L., and Springer T.A. 1991. *Ann. Rev. Immunol.* 9:27–66.
11. Mazerolles F. et al. 1988. *Eur. J. Immunol.* 18:1229–1234.
12. Mazerolles F. et al. 1990. *Eur. J. Immunol.* 20:637–644.
13. Amblard F. et al. *Cytometry*, 1992, 13, 15–22.
14. Amblard F. et al. Submitted.
15. Mazerolles F. et al. 1991. *Eur. J. Immunol.* 21:887–894.
16. Lecomte O., and Fischer A. 1992. *Int. Immunol.* 4:191–196.
17. Mazerolles F. et al. submitted.
18. Doyle C., and Strominger J.L. 1987. *Nature* 330:256–258.
19. Mourad W. et al. 1990. *J. Exp. Med.* 172:1513–1516.
20. Corado J. et al. 1991. *J. Immunol.* 147:475–482.
21. Rudd C.E. et al. 1988. *Proc. Natl. Acad. Sci. (USA)* 85:5190–5194.
22. Veillette A. et al. 1988. *Cell* 55:301–308.
23. Veillette A. et al. 1989. *Nature* 338:257–259.
24. Turner J.M. et al. 1990. *Cell* 60:755–765.
25. Mazerolles F. et al. 1991. *Human. Immunol.* 31:40–46.

19
The Adhesion and Transendothelial Migration of Human T Lymphocytes

NANCY OPPENHEIMER-MARKS, PETER PIETSCHMANN,
LAURIE S. DAVIS, JOHN J. CUSH, ARTHUR F. KAVANAUGH, AND
PETER E. LIPSKY

Introduction

A hallmark of chronic inflammatory reactions is the accumulation of mononuclear cells within the inflamed tissue. One principal mechanism leading to tissue accumulation of cells is an increase in entry of cells into the tissue. Cells enter the tissue by undergoing receptor-mediated interactions with the endothelial cell (EC) lining of small blood vessels. At inflammatory sites there is a marked enhancement of this receptor-mediated transendothelial migration.

One cell that enters inflammatory sites is the T lymphocyte. A number of adhesion receptors have been identified that mediate T-cell adhesion to EC. Thus, lymphocytes, including T cells, express members of the $\beta1$ and $\beta2$ families of integrin adhesion receptors, including leukocyte function–associated antigen-1, LFA-1 (CD11a/CD18), and very late antigen-4, VLA-4 (CD49d/CD29) (1,2). Also expressed by lymphocytes are the gp90 receptors, CD44 and L-selectin (LECAM-1, LAM-1, Leu-8), which mediate lymphocyte adhesion during normal recirculation through perivascular tissue (3–8). EC express the counter-receptors for these molecules, including intercellular adhesion molecule–1, ICAM-1 (CD54) and -2, vascular cell adhesion molecule–1, VCAM-1 (also called INCAM-110), the vascular addressins and sialylated oligosaccharides, respectively (9–17). The mechanism by which adhesion receptors facilitate the entry of lymphocytes into inflammatory sites is the focus of intense investigation.

After T cells bind to EC by utilizing various receptors, a fraction of them migrate between the intercellular junctions of the EC. Transendothelial migration of T cells is also a receptor-mediated process that depends not only on the intrinsic motility of the T cells but also on their capacity to utilize specific adhesion receptors (18–20). Even though adhesion to EC is a necessary prerequisite for transendothelial migration, not all T cells that bind to EC migrate through the EC layer and enter perivascular tissue. This suggests that T-cell subsets manifest different intrinsic motility or that they differ in the capacity to utilize adhesion receptors involved in migration. Previous studies had indicated that different adhesion receptors may be used for adhesion to EC and transendothelial migration. Thus, for example, the adhesion receptors LFA-1 and CD44 are particularly important in mediating both T-cell binding and transendothelial migration, whereas VLA-4 may be uniquely involved in mediating the binding of T cells to EC at inflammatory sites. Moreover, the use of these receptors can be effected by the activation and differentiation status of the T cell and the cytokine exposure of the EC (19). Thus T cells that exhibit a migratory capacity may be at a different stage of activation or differentiation such that they express functional receptors not only for binding but also for transendothelial migration.

Results

The Effect of Cellular Activation on T-Cell Interactions With EC

To examine the role of various receptor counter-receptor pairs during T-cell interactions with EC, experiments were carried out to determine whether monoclonal antibodies (mAb) to specific adhesion molecules inhibited binding or subsequent migration. Previous work had indicated that resting T cells cannot utilize LFA-1 as a binding receptor, whereas several activation stimuli up-regulate its capacity to bind ICAM-1 (21). Experiments utilizing mAb confirmed that activation of LFA-1 is also necessary for its utilization during T-cell binding to EC (19) (Fig. 19.1). The activation of LFA-1 expressed by T cells can result from a variety of specific and nonspecfic stimuli including ligation of the T-cell receptor/ CD3, activators of protein kinase C, or merely culture in fetal-calf-serum–containing medium. After the initial activation of LFA-1, additional activation further increases the capacity of T cells to bind to EC as well as their ability to migrate after initial adherence. Thus prolonged stimulation with activators of protein kinase C (phorbol dibutyrate, PDB) in the presence or absence of the calcium ionophore ionomycin markedly increases the binding of T cells to EC (19). Activation of T cells by PDB, but not the combination of ionomycin and PDB, also increases their transendothelial migratory capacity. This appears to be related to both the stimulation of the intrinsic motility of T cells as well as enhanced utilization of adhesion receptors (19). The findings that the adhesion and transendothelial migratory capacities of T cells are mutable and affected by their activation or differentiation status led to detailed investigation of the various adhesion receptors empolyed.

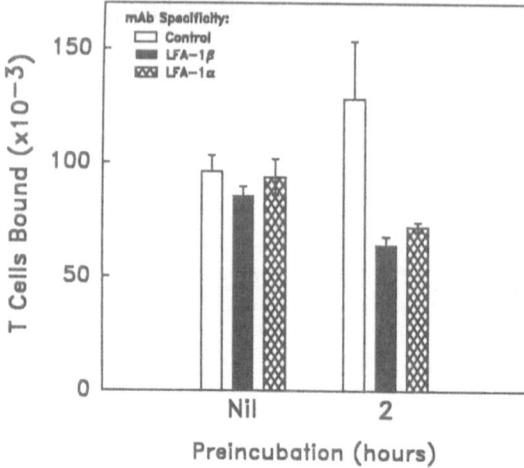

FIGURE 19.1. LFA-1–mediated adhesion of freshly isolated T cells can be induced by culture. Peripheral T cells were isolated from the circulation by standard means and then immediately incubated at 4°C with control mAb or saturating amounts of mAb directed against the α (TS1/22) or β (TS1/18) chains of LFA-1. Subsequently, the T cells were assayed for binding to EC. Alternatively, the freshly isolated T cells were incubated for 2 h at 37°C in 10% fetal-calf-serum-containing medium prior to their assay.

The Role of LFA-1/ICAM-1 and VLA-4/VCAM-1 During the Adhesion and Transendothelial Migration of T Cells

The binding of T cells to unstimulated EC is dependent on the activities of LFA-1 and ICAM-1, as evidenced by the finding that it is inhibited by mAb directed to ICAM-1 or the α or β chain of LFA-1, but not by mAb against VCAM-1 or the α chain of VLA-4 (Fig. 19.2) (19,20). In contrast, when the EC are activated by IL-1, under conditions that induce the expression of VCAM-1 and increase the expression of ICAM-1, the binding of T cells becomes partially mediated by VLA-4/VCAM-1 (Fig. 19.2). Even though IL-1–activated EC express increased amounts of ICAM-1, neither the antibody against ICAM-1 nor anti–LFA-1 antibody blocks binding, indicating that T-cell adhesion to IL-1–activated EC is not mediated by this receptor counter-receptor pair (Fig. 19.2). Thus, IL-1 stimulation increases the capacity of EC to bind T cells and enhances the expression of ICAM-1, but the binding of T cells by means of an interaction between ICAM-1 and LFA-1 is no longer apparent. The explanation for this result is not entirely clear but may reflect an increase in the activity of a variety of other adhesion receptors including VCAM-1/VLA-4.

These findings are in contrast to the role of these molecules during the transendothelial migration of T cells (20). As shown in Figure 19.3, when

FIGURE 19.2. T-cell–EC
binding is dependent on
the activities of LFA-1/
ICAM-1 and VLA-4/
VCAM-1. Peripheral T
cells were isolated from the
circulation by standard
means and maintained at
37°C overnight in 10%
fetal-calf-serum–containing
medium to activate LFA-1.
Subsequently, the T cells
were incubated with con-
trol mAb or with saturat-
ing amounts of mAb
directed against the α
chains of LFA-1 (TS1/22)
or VLA-4 (HP2/1), and
then assayed for binding to
unstimulated or IL-1 activated EC. To measure the
activities of ICAM-1 and
VCAM-1 during T-cell–
EC binding, unstimulated
or IL-1–activated EC were
incubated with control
mAb or with saturating
amounts of mAb directed
against ICAM-1 (RR1/1)
or VCAM-1 (4B9), prior
to the addition of T cells.

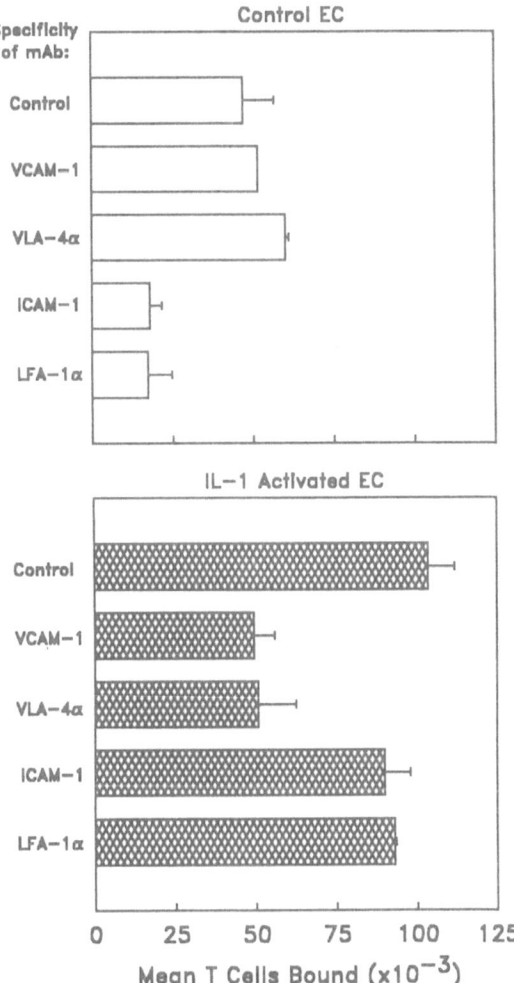

T cells are bound to EC and then exposed to the anti-adhesion receptor
mAb, the transendothelial migration of the bound T cells is partially inhibited by anti–LFA-1 or anti–ICAM-1 mAb. The inhibition of the transendothelial migration of T cells by mAb to LFA-1 or ICAM-1 occurs regardless of whether the EC are resting or activated by IL-1, and thus, is
independent of whether LFA-1 or ICAM-1 mediates the initial cell–cell
binding. These results, therefore, indicate that in addition to its role
in mediating T-cell binding to unstimulated EC, the receptor counter-
receptor pair, LFA-1/ICAM-1, also plays an important role during the
transendothelial migration of T cells regardless of the molecules that
mediate the initial adhesion. In contrast, the adhesion receptor pair,
VLA-4/VCAM-1, may be important in mediating the adhesion of lym-

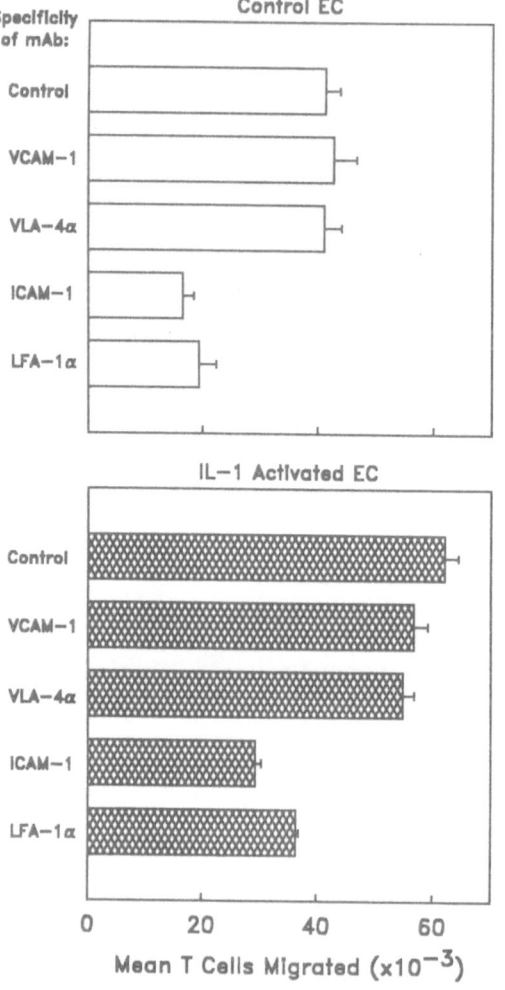

FIGURE 19.3. The transendothelial migration of T cells is dependent on the activities of LFA-1/ICAM-1. T cells were prebound to unstimulated or IL-1–activated EC. Subsequently, control mAb or saturating amounts of mAb directed against ICAM-1 (RR1/1), VCAM-1 (4B9), or the α chains of LFA-1 (TS1/22) or VLA-4 (HP2/1) were added to the cells, and the bound T cells were allowed to migrate in the presence of the mAb.

phocytes to sites of activated endothelium. These results also indicate that specific adhesion receptors operate during different steps involved in the entry of T cells into extravascular sites. Adhesion receptors, such as VLA-4 and VCAM-1, primarily facilitate the binding of T cells to cytokine-activated EC, whereas others, such as LFA-1 and ICAM-1, mediate binding in certain circumstances, but also mediate the transendothelial migration of the bound T cells into the perivascular tissue.

The data indicate that adhesion-receptor-mediated transendothelial migration is promoted by the formation of bonds between molecules on EC and their counter-receptors on T cells. It is likely that transendothelial migration of T cells is mediated by progressive interactions between adhesion receptors on the leading edge of migrating T cells and counter-receptors on EC. Bonds between these receptor counter-receptor pairs, for example, LFA-1 on T cells and ICAM-1 on EC, would be continually formed and then broken as transendothelial migration proceeds and the T cell enters the perivascular space. The mechanism of transendothelial migration would thus be analagous to the "zipper" mechanism that mediates receptor-mediated phagocytosis with the added complexity of requiring a mechanism of receptor de-adhesion of the trailing edge of the advancing T cell (22).

This model predicts that adhesion receptors would be localized on EC surfaces that are in contact with the leading edge of migrating T cells. This has been confirmed by immunoelectron microscopy studies (20). ICAM-1 is prominently expressed at sites of contact between the plasma membrane of IL-1–activated EC and the leading edge of migrating T cells. In contrast, VCAM-1 is absent from these sites. However, VCAM-1 is expressed at sites of contact between EC and adherent lymphocytes, although not uniformly, supporting the conclusion that this adhesion receptor plays a role in mediating the initial binding of T cells to activated EC.

That LFA-1 and ICAM-1 are important during the transendothelial migration of T cells is supported by observations on the behavior of T-cell clones that were derived from a patient with leukocyte adhesion deficiency (LAD), and thus fail to express members of the $\beta2$ family of integrin adhesion molecules, including LFA-1 (23). In comparison to normal T-cell clones, the LAD T-cell clones exhibit a normal capacity to bind to EC, but they exhibit a diminished transendothelial migratory capacity. This is not related to defective intrinsic motility by these cells, but rather appears to be related to the absence of LFA-1.

Identification of Circulating T Cells That Exhibit a Transendothelial Migratory Capacity

It was apparent from a variety of observations that not all circulating T cells are competent to migrate. However, the characteristics of T cells that exhibit the migratory behavior have not been completely delineated. Several reports have indicated that there are differences in the capacities of "naive" and "memory" T cells to bind to EC (24–26). Phenotypic analyses of cells entering sites of inflammation suggest that T cells preferentially accumulating in these sites are "memory" cells (27–30). Thus, it seems likely that circulating CD4+ T cells that either inherently express a migratory phenotype or that are susceptible to its induction by interactions with EC are "memory" cells. This conclusion is supported by in vivo studies that examined the phenotype of T cells that entered into inflammatory foci during the early stages of tuberculin-induced DTH lesions (30).

To identify the phenotype of T cells with a transendothelial migratory

TABLE 19.1. Phenotype of migrating CD4(+) T cells.

CD29^{bright}
CD45RA^{dim}
CD45RO^{bright}
CD45RB^{bright}
CD7^{dim}
L-selectin^{low}

CD4(+) T cells were purified by negative selection and incubated with EC monolayers that had been formed on hydrated collagen gels. Subsequently, the T cells that had migrated into the collagen were recovered following collagenase digestion of the gel and were processed for analysis by flow cytometry. mAbs that were utilized include 4B4 (anti-CD29), 2H4 (anti-CD45RA), UCHL 1 (anti-CD45RO), PD6/26 (anti-CD45RB), T3.3A1 (anti-CD7), and Leu8 (anti–L-selectin).

capacity, an experimental method was developed to recover lymphocytes following their transendothelial migration (31). In this assay, EC are cultured to form a confluent monolayer on hydrated collagen gels and then incubated with peripheral blood T cells. After a 4-h incubation of T cells with EC, approximately two-thirds of the added T cells are recovered as nonadherent and nonmigratory, whereas the remaining T cells are recovered in a 2:1 ratio as those that bind to EC but do not migrate and those that migrate through the EC layer into the collagen gel, respectively. ECs appear to play an important role during transendothelial migration of T cells, since there is minimal T-cell migration into plain collagen gels without an overlying EC monolayer. Moreover, the migrated T cells that are recovered from the collagen gel maintain this functional capacity. Thus, when the migratory capacity of the nonadherent and migrated populations was reassessed in a second assay, the initially migrated T cells were found to migrate much more efficiently than the nonadherent T cells, indicating that the migrated T cells retain this functional capacity. These results also suggest that there is a subset of circulating T cells that stably expresses a transendothelial migratory capacity. The phenotype of these migratory T cells has been extensively characterized.

Analyses of the phenotype of the T cells in the nonadherent, bound, and migrated populations have shown that the majority of the migrated CD4(+) T cells are CD45RA(−) and CD29^{bright}, CD45RO(+), whereas the nonadherent T-cell population contains both CD4(+)/CD45RA(−) and CD45RA(+) T cells as well as CD4(+)/CD29^{bright} and CD29^{dim} and CD45RO(+) and CD45RO(−) T cells. Thus, the migrated population appears to be enriched for T cells that express the "memory" phenotype. As shown in Table 19.1, additional analysis indicates that the migrating population is CD45RB^{bright}, CD7^{dim}, and L-selectin^{low}. Thus, CD4(+) T cells with the capacity to migrate through EC appear to be a unique population of memory cells.

Conclusions

The adhesion of T cells to EC is mediated by a variety of adhesion receptors, including LFA-1/ICAM-1 and VLA-4/VCAM-1. The transendothelial migration of T cells is also a receptor-mediated process and is promoted in part by the receptor counter-receptor pair LFA-1/ICAM-1. It is also apparent that other as-yet-unidentified adhesion receptors play a role in transendothelial migration, since none of the mAb tested, used alone or in combination, completely inhibits this process.

Circulating CD4+ "memory" T cells express an inherent capacity to extravasate out of blood. This migratory behavior accounts, in part, for their accumulation at sites of chronic inflammation. It has been suggested that "memory" T cells may access extravascular sites by binding to the adhesion receptor endothelial-leukocyte adhesion receptor-1 (ELAM-1) that is expressed by cytokine-activated EC (26). However, the role of ELAM-1 during the transendothelial migration of "memory" T cells has not been completely delineated. In contrast, it can be concluded that ICAM-1 and VCAM-1 facilitate the entry of T cells into sites of inflamation by playing distinct roles. VCAM-1 promotes the binding of T cells to activated EC whereas ICAM-1 mediates the binding of T cells to resting EC and also the transendothelial migration of the bound cells into the extravascular sites regardless of the cytokine exposure of the EC or the adhesion molecules used for initial binding to EC. Thus, the repertoire of adhesion receptors expressed by T cells and EC appears to serve distinct functions during the extravasation of T cells.

References

1. Mentzer, S.J., S.J. Burakoff, and D.V. Faller. 1986. Adhesion of T lymphocytes to human endothelial cells is regulated by the LFA-1 membrane molecule. *J. Cell. Physiol.* 126:285.
2. Wayner, E.A., A. Garcia-Pardo, M.J. Humphries, J.A. McDonald, and W.G. Carter. 1989. Identification and characterization of the T lymphocyte adhesion receptor for an alternative cell attachment domain (CS-1) in plasma fibronectin. *J. Cell Biol.* 109:1321.
3. Gallatin, W.M., I.L. Weissman, and E.C. Butcher. 1983. A cell surface molecule involved in organ-specific homing of lymphocytes. *Nature* 304:30.
4. Camerini, D., S.P. James, I. Stamenkovic, and B. Seed. 1989. Leu-8/TQ1 is the human equivalent of the Mel-14 lymph node homing receptor. *Nature* 342:78.
5. Bowen B.R., T. Nguyen, and L.A. Lasky. 1989. Characterization of a human homologue of the murine peripheral lymph node homing receptor. *J. Cell Biol.* 109:421.
6. Tedder, T.F., A.C. Penta, H.B. Levine, and A.S. Freedman. 1990. Expression of the human leukocyte adhesion molecules, LAM1. Identity with the TQ1 and Leu-8 differentiation antigens. *J. Immunol.* 144:532.

7. Jalkanen, S.T., R.F. Bargatze, L.R. Herron, and E.C. Butcher. 1986. A lymphoid cell surface glycoprotein involved in endothelial cell recognition and lymphocyte homing in man. *Eur. J. Immunol.* 16:1195.

8. Picker, L.J., J. De Los Toyos, M.J. Telen, B.F. Haynes, and E.C. Butcher. 1989. Monoclonal antibodies against the CD44 [*In(Lu)*-related p80], and Pgp-1 antigens in man recognize the Hermes class of lymphocyte homing receptors. *J. Immunol.* 142:2046.

9. Dustin, M.L., R. Rothlein, A.K. Bhan, C.A. Dinarello, and T.A. Springer. 1986. Induction by IL-1 and interferon-γ: tissue distribution, biochemistry, and function of a natural adherence molecule (ICAM-1). *J. Immunol* 137:245.

10. Marlin, S.D. and T.A. Springer. 1987. Purified intercellular adheison molecule-1 (ICAM-1) is a ligand for lymphocyte function-associated antigen 1 (LFA-1). *Cell* 51:813.

11. Staunton, D.E., M.L. Dustin, and T.A. Springer. 1989. Functional cloning of ICAM-2, a cell adhesion ligand for LFA-1 homologous to ICAM-1. *Nature* 339:61.

12. Nortamo, P., R. Salcedo, T. Timonen, M. Patarroyo, and C.G. Gahmberg. 1991. A monoclonal antibody to the human leukocyte adhesion molecule intercellular adhesion molecule-2. Cellular distribution and molecular characterization. *J. Immunol.* 146:2530.

13. de Fougerolles, A.R., S.A. Stacker, R. Schwarting, and T.A. Springer. 1991. Characterization of ICAM-2 and evidence for a third counter-receptor for LFA-1. *J. Exp. Med.* 174:253.

14. Carlos, T.M., B.R. Schwartz, N.L. Kovach, E. Yee, M. Rosso, L. Osborn, G. Chi-Rosso, B. Newman, R. Lobb, and J.M. Harlan. 1990. Vascular cell adhesion molecule-1 mediates lymphocyte adherence to cytokine-activated cultured human endothelial cells. *Blood* 76:965.

15. Osborn, L., C. Hesslon, R. Tizard, C. Vassallo, S. Luhowskyj, G. Chi-Rosso, and R. Lobb. 1989. Direct expression cloning of a vascular cell adhesion molecule 1, a cytokine-induced endothelial protein that binds to lymphocytes. *Cell* 59:1203.

16. Rice, G.E., J.M. Munro, and M.P. Bevilacqua. 1990. Inducible cell adhesion molecule 110 (INCAM-110) is an endothelial receptor for lymphocytes. A CD11/CD18 independent adhesion mechanism. *J. Exp. Med.* 171:1369.

17. Elices, M.J., L. Osborn, Y. Takada, C. Crouse, S. Luhowskyj, M.E. Hemler, and R.R. Lobb. 1990. VCAM-1 on activated endothelium interacts with the leukocyte integrin VLA-4 at a site distinct from the VLA-4/fibronectin binding site. *Cell* 60:577.

18. Oppenheimer-Marks, N., and M. Ziff. 1988. Migration of lymphocytes through endothelial cell monolayers: augmentation by interferon-gamma. *Cell. Immunol.* 114:307.

19. Oppenheimer-Marks, N., L.S. Davis, and P.E. Lipsky. 1990. Human T lymphocyte adhesion to endothelial cells and transendothelial migration. Alteration of receptor use relates to the activation status of both the T cell and the endothelial cell. *J. Immunol.* 145:140.

20. Oppenheimer-Marks, N., L.S. Davis, D. Tompkins Bogue, J. Ramberg, and P.E. Lipsky. 1991. Differential utilization of ICAM-1 and VCAM-1 during

the adhesion and transendothelial migration of human T lymphocytes. *J. Immunol.* 147:2913.

21. Dustin, M.L., and T.A. Springer. 1989. T-cell receptor cross-linking transiently stimulates adhesiveness through LFA-1. *Nature* 341:619.

22. Griffin, Jr., F.M., J.A. Griffin, J.E. Leider, and S.C. Silverstein. 1975. Studies on the mechanism of phagocytosis. I. Requirements for circumferential attachment of particle-bound ligands to specific receptors on the macrophage plasma membrane. *J. Exp. Med.* 142:1263.

23. Kavanaugh, A.F., E. Lightfoot, P.E. Lipsky, and N. Oppenheimer-Marks. 1991. The role of CD11/CD18 in adhesion and transendothelial migration of T cells: analysis untilizing CD18 deficient T cell clones. *J. Immunol.* 146:4149.

24. Pitzalis, C., G. Kingsley, D. Haskard, and G. Panayi. 1988. The preferential accumulation of helper-inducer T lymphocytes in inflammatory lesions: evidence for regulation by selective endothelial and homotypic adhesion. *Eur. J. Immunol.* 18:1397.

25. Damle, N.K., and L.V. Doyle. 1990. Ability of human T lymphocytes to adhere to vascular endothelial cells and to augment endothelial permeability to macromolecules is linked to their state of post-thymic maturation. *J. Immunol.* 144:1233.

26. Shimizu, Y., S. Shaw, N. Graber, T.V. Gopal, K.J. Horgan, G.A. Van Seventer, and W. Newman. 1991. Activation-independent binding of human memory T cells to adhesion molecule ELAM-1. *Nature* 349:799.

27. Cush, J.J., and P.E. Lipsky. 1991. Cellular basis for rheumatoid inflammation. *Clin. Orthopaed.* 265:9.

28. Cush, J.J. and P.E. Lipsky. 1988. Phenotypic analysis of synovial tissue and peripheral blood lymphocytes isolated from patients with rheumatoid arthritis. *Arth Rheum.* 31:1230.

29. Cush, J.J., and P.E. Lipsky. 1990. Dual immunofluorescence analysis of lymphocyte subsets eluted from rheumatoid synovium. *Faseb J.* 41:1855.

30. Pitzalis, C., G.H. Kingsley, M. Covelli, R. Meliconi, A. Markey, and G.S. Panayi. 1991. Selective migration of the human helper-inducer memory T cell subset: confirmation by in vivo cellular kinetic studies. *Eur. J. Immunol.* 21:369.

31. Pietschmann, P., J.J. Cush, P.E. Lipsky, and N. Oppenheimer-Marks. 1992. Identification of human T cells capable of enhanced transendothelial migration. *J. Immuol.* 149:1170.

20
Changes in Topography of Cell Adhesion Molecules During Lymphocyte Migration Across Endothelium

STEPHEN J. ROSENMAN, PATRICIA A. HOFFMAN, AND
W. MICHAEL GALLATIN

Introduction

The widely expressed CD44 single-chain transmembrane glycoprotein has been the object of recent studies in the fields of cell adhesion and lymphocyte activation. This molecule commonly occurs as a 37-kDa polypeptide extensively glycosylated with N- and O-linked oligosaccharides and glycosaminoglycans which migrates in SDS-PAGE as an 80–95-kDa major component of the membranes of most cell types examined (1–3). Higher-molecular-weight forms bearing chondroitin sulfate have been described in lymphoid cells. In other cell types alternately spliced isoforms possessing an additional extracellular domain exon of 132 or 162 amino acids have been detected (4,5).

The extracellular domain of the common, 80–95-kDa form of CD44 includes a region of conserved sequence with high homology to cartilage link proteins. Receptor activity for hyaluronic acid appears to reside in this structure (6), as may binding activity for types I and VI collagen (7).

Because of similarities in molecular weights, cell-surface expression by lymphoid cells, immunological cross-reactivity, and apparent involvement in lymphocyte attachment to lymph node postcapillary high endothelial venules (HEV), CD44 was initially proposed to be the primate homologue of the murine lymphocyte peripheral lymph node homing receptor (PLN-HR) first identified with mAb MEL-14 (8,9). Subsequent molecular cloning of the genes encoding CD44 (10–12) and the PLN-HR (13–17) established distinct and independent structures and identities for CD44 and PLN-HR (LECCAM-1, LAM-1), the latter now being designated by the name L-selectin. Two other structurally related selectins (also referred to as lectin cell adhesion molecules or LECAMs) have been identified and are also involved in vascular cell adhesive interactions (reviewed in 3,18).

Here we present data demonstrating that distinct adhesive properties are conferred upon host cells transfected with either CD44 or L-selectin genes. Since both of these cell-surface molecules have been implicated in lymphocyte–endothelial cell interactions, we also studied their localization on the surfaces of normal human peripheral blood T cells prior to and during interaction with cytokine-activated human umbilical vein endothelial cells (HUVEC).

Results

Adhesion-Competent L-Selectin Transfectants but not CD44 Transfectants Bind to Lymph Node HEV

The L-selectin–negative human B-cell lymphoblastoid line Nalm/6 was transfected with a pLxsne⁻ retroviral construct containing the complete coding sequence for macaque (*Macaca nemestrina*) L-selectin (Fig. 20.1). This construct also contained the neomycin resistance gene, permitting selection by growth in G418-containing medium of stable transfectants expressing macaque L-selectin. The flow-cytometric profile of one such transfectant, Nalm/6-VVC6, is shown in Figure 20.2. Expression of L-selectin by VVC6 cells but not by the parent cell line Nalm/6 was demonstrated using the L-selectin–specific mAb Leu8.

When Nalm/6 or Nalm/6-VVC6 cells were evaluated for their relative

FIGURE 20.1. Deduced amino acid sequence of L-selectin from the pigtailed macaque (*M. nemestrina*) compared with human and murine L-selectin sequences (13–17). Only non-conserved residues are shown for macaque and murine sequences. Underlined regions indicate leader sequence (top line) and putative transmembrane domain. Vertical lines separate N-terminal C-type lectin domain, EGF-like domain, and two tandem complement-binding proteinlike consensus repeat units (CRP).

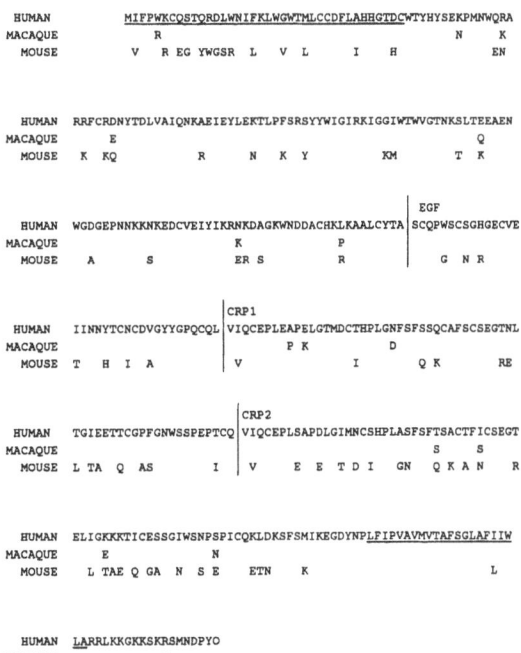

ability to bind to HEV in lymph node sections, only the transfectants displayed significant adhesion to HEV. Further specificity of this adhesion became evident when binding to HEV of PLN and Peyer's patch (PP) sections was assessed: VVC6 cells selectively bound to HEV in PLN sections, consistent with the functional properties of L-selectin predicted by mAb inhibition studies (8).

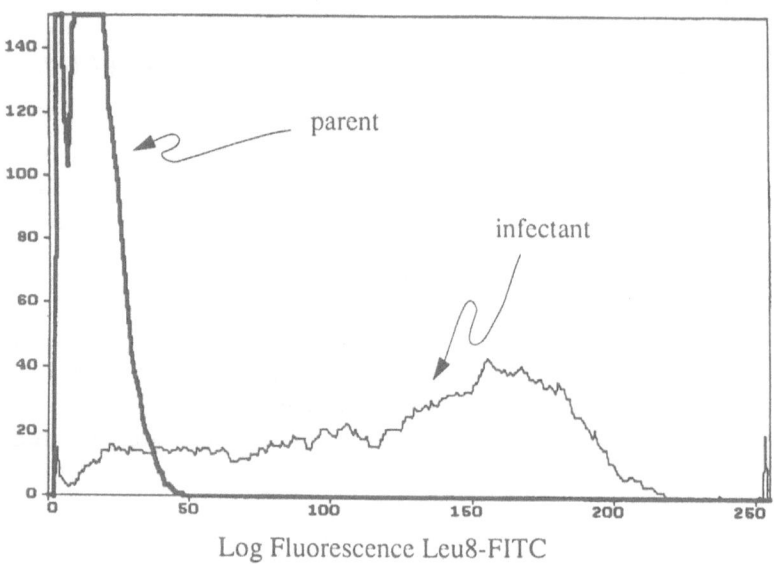

Log Fluorescence Leu8-FITC

FIGURE 20.2. Immunocytofluorimetric profiles of nontransfected (Nalm/6 parent) and L-selectin-transfected (VVC6 transfectant) cells stained with the L-selectin–specific mAb Leu8.

TABLE 20.1. *Syn*-capping of T-lymphocyte cell adhesion molecules.

First CAM (indirect cap)	Second CAM (direct stain)	Co-capping
CD44	CD2	+++
CD2	CD44	—
CD44	LFA-1	—
LFA-1	CD44	—
CD44	L-selectin	+
L-selectin	CD44	—
CD2	LFA-1	—
LFA-1	CD2	+++
CD2	L-selectin	—
L-selectin	CD2	—
LFA-1	L-selectin	—
L-selectin	LFA-1	—

CD44 transfectants were also evaluated for their HEV binding ability. Murine L cells were transfected with baboon CD44 cDNA and used in adhesion assays (19). Expression of CD44 conferred upon L cells no ability to bind to HEV from sections of either PLN or PP. However, the CD44 transfectants acquired a novel ability to form large homotypic aggregates by a CD44-mediated mechanism, since anti–CD44 mAbs were inhibitory. At the molecular level, this aggregation event could be shown to be heterotypic in nature because nontransfected cells, which by themselves did not form aggregates, were able to bind to transfectants and in so doing created aggregates (19). This result suggests the presence on nontransfected cells of one or more CD44 adhesion ligands.

Localization and Redistribution of T Lymphocyte Adhesion Molecules

Human peripheral blood T lymphocytes express multiple classes of cell-surface adhesion molecules (CAMs) which mediate adhesive interactions with an array of adhesion ligands expressed by endothelial cells lining the vasculature. We focused on the T-cell CAMs CD2, CD44, LFA-1, and L-selectin, which display binding affinity for endothelial cell-surface LFA-3, hyaluronic acid and possibly other ligands, the ICAMs, and the vascular addressin Sgp^{50}, respectively (6,20–24). In order to assess associations amongst these CAMs, or their potential redistribution during interaction with HUVEC, we first examined localization of each CAM as well as the ability of each to be aggregated into patches and caps by antibodies. Each of these CAMs displayed uniform T-cell-surface distribution in direct immunofluorescence using FITC-mAb conjugates and the laser-scanning confocal microscope. Each of these CAMs could also be aggregated into patch/caps by indirect immunofluorescence, provided that a multivalent second-stage reagent was used and that metabolic energy was available to the cells (25). These results suggest that cell-surface distribution of each CAM studied is uniform and regulated by the cytoskeleton (26).

We next sought to address the question, again by confocal immunofluorescence microscopy, of whether any physical associations amongst different CAMs take place. The general experimental strategy for these experiments involved capping the first class of CAM by indirect means using a biotinylated mAb and Texas Red–avidin at physiological temperature, after which cells were chilled and maintained in buffer containing 10 mM NaN_3 to prevent any further membrane molecule reorganization. The second CAM was then localized by direct immunofluorescence with FITC-mAb (25). A summary of the results from these experiments is presented in Table 20.1, which shows that certain paired CAMs exhibit unidirectional co-capping, also known as *syn*-capping (27).

The physiological relevance of T-cell CAM *syn*-capping was investi-

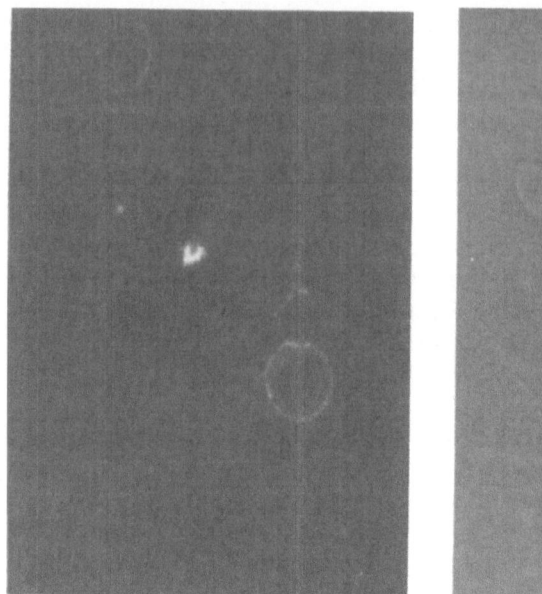 

FIGURE 20.3. Accumulation of L-selectin in the pseudopod of a human peripheral blood T lymphocyte adhering to an endothelial cell. T cells pretreated with FITC-mAb LAM1-3 (kindly provided by Dr. T.F. Tedder, Dana-Farber Cancer Institute, Boston, MA) were incubated with cytokine-activated HUVEC for 90 min, fixed, and examined by laser-scanning confocal microscopy as described (25). Fluorescence (left) and transmitted light images of the same field are shown. Similar sequestration of CD2 and CD44, but not LFA-1, was observed when mAbs to those CAMs were used.

gated in experiments designed to detect alterations in CAM distribution when T cells were co-cultured with cytokine-activated HUVEC. Previous studies demonstrated that preincubation of HUVEC with IL-4 and TNF-α increased their expression of adhesion molecules involved in lymphocyte binding (28–30). During T-cell–HUVEC co-culture we have also observed dramatic lymphocyte shape changes which appear to be cytoskeletally regulated (25). T cells prelabeled with CAM-specific FITC-mAb which had first been shown not to block lymphocyte-EC adhesion were added to HUVEC monolayers and cultured 0.5–3 h, after which nonadherent cells were washed away. When fixed cultures were examined by confocal microscopy, CD2, CD44, and L-selectin had all accumulated in the pseudopodia of deformed lymphocytes but remained uniformly distributed in cells which had retained round morphology (Fig. 20.3 and Table 20.2). By contrast, T cells stained with LFA-1–specific mAb displayed uniform surface staining with FITC regardless of morphology.

TABLE 20.2. Redistribution of T-cell adhesion molecules during co-culture with endothelial cells.

CAM	Accumulation in T-cell pseudopodia
CD2	+++
CD44	+++
L-selectin	+++
LFA-1	—

These lymphocytes therefore appear to selectively regulate CAM distribution during interaction with EC.

Discussion

Isolated genes for two distinct CAMs implicated in lymphocyte–endothelial cell interactions were transfected into heterologous cells which expressed functional cell–surface gene products. L-selectin retained organ-specific HEV adhesive potential and conferred this phenotype upon the lymphoid cell into which it was introduced. A requirement for CD44 in HEV binding, on the other hand, has not previously been demonstrated, nor was CD44 able to confer this ability upon transfectants. CD44 transfectants did, however, spontaneously aggregate, suggesting that there exist contexts in which this molecule is important to intercellular adhesion.

One such adhesive interaction in which these and other CAMs may play discrete and critical roles takes place during lymphocyte extravasation, an important component of lymphocyte recirculation and inflammation. Multiple molecular pathways of lymphocyte–EC adhesion have been discerned (e.g. 31–33) but the precise role of each component thus far identified has not been determined and the discovery of additional CAMs is likely. CAM roles in lymphocyte activation have also been described which, in concert with the *syn*-capping data presented here, implicate physical association of different CAMs as a potential mechanistic basis for their functional cooperation (24,34–37).

The differential redistribution of CD2, CD44, and L-selectin to T-cell pseudopodia during co-culture with HUVEC is reminiscent of the behavior of these CAMs in co-capping experiments; the unidirectional nature of co-capping may signal the relative involvement of each CAM during lymphocyte–endothelial interaction in a temporal sense. Co-localization of CD44 with pseudopod-associated L-selectin was not observed in thymocytes migrating on fibronectin-coated substrates (38). Regulated positioning of CAMs may therefore vary as lymphocytes mature.

248   Stephen J. Rosenman et al.

References

1. Rosenman, S., and T. St. John. 1992. CD44. In Guidebook to the Cytos-keletal, Extracellular Matrix, and Adhesion Proteins. T. Kreis and R. Vale, eds., Sambrook and Tooze, Heidelberg, in press.
2. Gallatin, W.M., Rosenman, S.J., Ganji, A., Rees, G., and T. St. John. 1991. Structure function relationships of the CD44 class of glycoproteins. In Cellular and Molecular Mechanisms of Inflammation, Vol. 2: Vascular Adhesion Molecules. C.G. Cochrance and M.A. Gimbrone, Jr., eds., Academic Press, New York, pp.131–150.
3. Rosenman, S.J. and W.M. Gallatin. 1991. Cell surface glycoconjugates in intercellular and cell-substratum interactions. *Sem. Canc. Biol.* 2:357.
4. Dougherty, G.J., Landsdorp, P.M., Cooper, D.L., and R.K. Humphries. 1991. Molecular cloning of CD44R1 and CD44R2, two novel isoforms of the human CD44 lymphocyte "homing" receptor expressed by hemopoietic cells. *J. Exp. Med.* 174:1.
5. Gunthert, U., Hofmann, M., Rudy, W., Reber, S., Zoller, M., Haussmann, I., Matzku, S., Wenzel, A., Ponta, H., and P. Herrlich. 1991. A new variant of glycoprotein CD44 confers metastatic potential to rat carcinoma cells. *Cell* 65:13.
6. Aruffo, A., Stamenkovic, I., Melnick, M., Underhill, C.B., and B. Seed. 1990. CD44 is the principal cell surface receptor for hyaluronate. *Cell* 61:1303.
7. Gallatin, W.M., Wayner, E.A., Hoffman, P.A., St. John, T., Butcher, E.C., and W.G. Carter. 1989. Structural homology between lymphocyte receptors for high endothelium and class III extracellular matrix receptor. *Proc. Nat. Acad. Sci. USA* 86:4654.
8. Gallatin, W.M., T.P. St. John, M. Siegelman, R. Reichert, E.C. Butcher, and I.L. 1986. Lymphocyte homing receptors. *Cell* 44:673.
9. Jalkanen, S., Bargatze, R.F., de los Toyos, J., and E.C. Butcher. 1987. Lymphocyte recognition of high endothelium: Antibodies to distinct epitopes of an 85–95 kD glycoprotein antigen differentially inhibit lymphocyte binding to lymph node, muscosal, or synovial endothelial cells. *J. Cell Biol.* 105:983.
10. Goldstein L.A., Zhou D.F.H., Picker L.J., Minty C.N., Bargatze R.F., Ding J.F., and E.C. Butcher. 1989. A human lymphocyte homing receptor, the Hermes antigen, is related to cartilage proteoglycan core and link proteins. *Cell* 56:1063.
11. Idzerda R.L., Carter W.G., Nottenburg C, Wayner E.A., Gallatin W.M., and T. St. John. 1989. Isolation and DNA sequence of a cDNA clone encoding a lymphocyte adhesion receptor for high endothelium. *Proc. Nat. Acad. Sci. USA* 86:4659.
12. Stamenkovic I., Amiot M., Pesando J.M., and B. Seed. 1989. A lymphocyte molecule implicated in lymph node homing is a member of the cartilage link protein family. *Cell* 56:1057.
13. Lasky, L.A., Singer, M.S., Yednock, T.A., Dowbenko, D., Fennie, C., Rodriguez, H., Nguyen, T., Stachel, S., and S.D. Rosen. 1989. Cloning of a lymphocyte homing receptor reveals a lectin domain. *Cell* 56:1045.
14. Siegelman, M., van de Rijn M., and I. Weissman. 1989. Mouse lymph node homing receptor cDNA clone encodes a glycoprotein revealing tandem in-

teraction domains. *Science (Wash. DC)* 243:1165.

15. Tedder, T.F., Isaac, C.M., Ernst, T.J., Demetri, G.D., Adler, D.A., and C.M. Disteche. 1989. Isolation and chromosomal localization of cDNAs encoding a novel human lymphocyte cell surface molecule, LAM-1. Homology with the mouse lymphocyte homing receptor and other human adhesion proteins. *J. Exp. Med.* 170:123.

16. Bowen, B., Nguyen T., and L. Lasky. 1989. Characterization of a human homologue of the murine peripheral lymph node homing receptor. *J. Cell Biol.* 109:421.

17. Camerini D., James S.P., Stamenkovic I., and B. Seed. 1989. Leu-8/TQ1 is the human equivalent of the Mel-14 lymph node homing receptor. *Nature* 342:78.

18. Brandley B.K., Swiedler S.J., and P.W. Robbins. 1990. Carbohydrate ligands of the LEC cell adhesion molecules. *Cell* 63:861.

19. St. John T., Meyer J., Idzerda R., and W.M. Gallatin. 1990. Expression of CD44 confers a new adhesive phenotype on transfected cells. *Cell* 60:45.

20. Albelda, S.M., and C.A. Buck. 1990. Integrins and other cell adhesion molecules. *FASEB J.* 4:2868.

21. Hemler, M.E. 1990. VLA proteins in the integrin family: Structures, functions, and their role on leukocytes. *Ann. Rev. Immunol.* 8:365.

22. Imai, Y., Singer, M.S., Fennie, C., Lasky, L.A., and S.D. Rosen. 1991. Identification of a carbohydrate-based endothelial ligand for a lymphocyte homing receptor. *J. Cell Biol.* 113:1213.

23. Selvaraj, P.M., Plunkett, M.L., Dustin, M., Sanders, M.E., Shaw, S., and T.A. Springer. 1987. The T lymphocyte glycoprotein CD2 (LFA-2/T11/E-rosette receptor) binds the cell surface ligand LFA-3. *Nature* 326:400.

24. Springer, T.A. 1990. Adhesion receptors of the immune system. *Nature* 346:425.

25. Rosenman S.J., Ganji A.A., Tedder T.F., and W.M. Gallatin. 1992. Syncapping of human T lymphocyte adhesion/activation molecules and their redistribution during interaction with endothelial cells. *J. Leuk. Biol.* in press.

26. Bourguignon, L.Y.W., and G.J. Bourguignon. 1984. Capping and the cytoskeleton. *Int. Rev. Cytol.* 87:195.

27. Kupfer, A. and S.J. Singer. 1989. Cell biology of cytotoxic and helper T cell functions: Immunofluorescence microscopic studies of single cells and cell couples. *Ann. Rev. Immunol.* 7:309.

28. Masinovsky, B., Urdal, D., and W.M. Gallatin. 1990. IL-4 acts synergistically with IL-1b to promote lymphocyte adhesion to microvascular endothelium by induction of vascular cell adhesion molecule-1. *J. Immunol.* 145:2886.

29. Pober, J.S. and R.S. Cotran. 1990. The role of endothelial cells in inflammation. *Transplantation* 50:537.

30. Spertini, O., Luscinskas, F.W., Kansas, G.S., Munro, J.M., Griffin, J.D., Gimbrone, M.A., Jr., and T.F. Tedder. 1991. Leukocyte adhesion molecule-1 (LAM-1, L-selectin) interacts with an inducible endothelial cell ligand to support leukocyte adhesion. *J. Immunol.* 147:2565.

31. Kavanaugh, A.F., Lightfoot, E., Lipsky, P.E., and N. Oppenheimer-Marks. 1991. Role of CD11/CD18 in adhesion and transendothelial migration of T cells: Analysis utilizing CD18-deficient T cell clones. *J. Immunol.* 146:4149.

32. Oppenheimer-Marks, N., Davis, L.S., and Lipsky, P.E. 1990. Human T lym-

phocyte adhesion to endothelial cells and transendothelial migration: Alteration of receptor use relates to the activation status of both the T cell and the endothelial cell. *J. Immunol.* 145:140.

33. Shimizu, Y., Newman, W., Gopal, T.V., Horgan, K.J., Graber, N., Beall, L.D., van Seventer, G.A., and S. Shaw. 1991. Four molecular pathways of T cell adhesion to endothelial cells: Roles of LFA-1, VCAM-1, and ELAM-1 and changes in pathway hierarchy under different activation conditions. *J. Cell Biol.* 113:1203.

34. Denning, S.M., Le, P.T., Singer, K.H., and B.F. Haynes. 1990. Antibodies against the CD44 p80, lymphocyte homing receptor molecular augment human peripheral blood T cell activation. *J. Immunol.* 144:7.

35. Huet, S., Groux, H., Caillou, B., Valentin, H., Prieur, A., and A. Bernard. 1989. CD44 contributes to T cell activation. *J. Immunol.* 143:798.

36. Shimizu, Y., Van Seventer, G.A., Siraganian, R., Wahl, L., and S. Shaw. 1989. Dual role of the CD44 molecule in T cell adhesion and activation. *J. Immunol.* 143:2457.

37. Bierer, B.E., and S.J. Burakoff. 1991. T cell receptors: Adhesion and signaling. *Adv. Cancer Res.* 56:49.

38. Pilarski, L.M., Turley, E.A., Shaw, A.R.E., Gallatin, W.M., Laderoute, M.P., Gillitzer, R., Beckman, I.G.R., and H. Zola. 1991. FMC46, a cell protrusion-associated leukocyte adhesion molecule-1 epitope on human lymphocytes and thymocytes. *J. Immunol.* 147:136.

21
Cytokines and the Control of Endothelial Cell Adhesiveness for Leukocytes in Inflammation

DORIAN O. HASKARD AND MARTIN H. THORNHILL

Introduction

Over the last few years there has been considerable attention paid to understanding the mechanisms by which circulating leukocytes gain access to the tissues during inflammatory responses (1–3). Direct inspection of the microvasculature in inflammatory tissues has shown that leukocytes "roll" along the vessel wall before becoming flattened on the endothelial surface. Adherent leukocytes then put out pseudopodia and migrate between endothelial cells (EC) into the tissues (4–6). These events appear to be very similar at the cellular level for all types of leukocyte and depend both upon chemoattractive factors that directly activate leukocyte function and upon factors which alter the adhesive state of the endothelium. Important differences, however, may exist in the nature and regulation of endothelial ligands for leukocytes of different lineage.

Endothelial cells can undergo sequential phases of activation, each associated with the induction or up-regulation of surface adhesion ligands for leukocytes. A rapid and transient increase in the adhesiveness of cultured EC for neutrophils can be achieved by stimulation with acute agonists such as histamine and thrombin. Increased EC adhesiveness due to these agents does not require de novo protein synthesis and involves the translocation to the cell surface of the P-selectin, which is stored within intracellular Weibel-Palade bodies (7). In contrast, the cytokines interleukin-1 (IL-1) and tumor necrosis factor (TNF) activate EC through the induction of a number of primary response genes and de novo synthesis of adhesion molecules, resulting in up-regulation of ICAM-1 expression and the appearance on the cell surface of E-selectin and VCAM-1 (8,9).

Although either IL-1 or TNF stimulate EC to become more adhesive for both neutrophils and T lymphocytes, the kinetics of enhanced adhesiveness for these different leukocyte populations are not identical (10). Thus maximal adhesion of neutrophils and T cells occurs after 4–6 h and 6–10 h, respectively. Furthermore, the duration of enhanced EC adhe-

251

siveness for neutrophils is less long-lasting than that for T cells. These observations can be explained at least in part by differences in the kinetics of expression of E-selectin, ICAM-1, and VCAM-1. E-selectin and VCAM-1 have different capacities to support neutrophil and T-cell adhesion, with E-selectin binding neutrophils but only a minority of peripheral blood T cells, while VCAM-1 binds lymphocytes but not neutrophils (11–14).

Activation of endothelium by interleukin-1 and TNF in vivo probably accounts for a subacute stage in the generation of an inflammatory response when the balanced expression of ICAM-1, VCAM-1, and E-selectin can support the adhesion and migration of most leukocyte types. As part of our interest in the pathophysiology of chronic inflammation we have therefore recently attempted to define forms of endothelial activation that may be more specifically associated with the generation of mononuclear-cell–predominant infiltrates in immune-mediated inflammatory responses.

Results

Whereas IL-1 and TNF stimulate EC adhesiveness for most leukocyte types, the lymphokines interferon gamma (IFNγ) and interleukin-4 (IL-4) were found to have a more selective effect, resulting in increased EC adhesiveness for T cells but not neutrophils (10,15,16). Analysis of the adhesion molecules involved in IL-4–stimulated T-EC adhesion showed that IL-4 results in the selective induction of VCAM-1 without induction of E-selectin. Moreover, increased T-cell adhesion to IL-4–stimulated EC could be completely abrogated by anti–VCAM-1 mAb, suggesting that VCAM-1 is the predominant EC adhesion molecule involved (17,18).

Analysis of changes in EC surface molecules in response to stimulation with IFNγ showed a selective up-regulation of ICAM-1 expression without induction of E-selectin or VCAM-1. However, ICAM-1 does not appear to be the only EC molecule involved in the IFNγ effect, as anti–LFA-1 or anti–ICAM-1 mAb only partially inhibited the increased T-EC adhesion. Although IFNγ induces the expression of HLA class II, this occurs at a later time-point and seems unlikely to be responsible for the enhanced T-cell adhesion (Fig. 21.1).

The capacity of IL-4 or IFNγ to stimulate EC adhesiveness is relatively modest when compared with the effects of IL-1 and TNF. We argued that neither IL-4 nor IFNγ would likely be present in tissues in an immune-mediated inflammatory response without the simultaneous presence of other cytokines and that cytokines in combination might have important further effects. EC were therefore stimulated with pairs of cytokines and tested for the ability to support T-cell adhesion. Of all the

FIGURE 21.1. Kinetics of surface changes in human umbilical vein EC cultures stimulated for up to 72 h with human recombinant IFNγ (250 U/ml): (a) increased EC adhesiveness for human peripheral blood T cells and (b) expression of ICAM-1 and HLA class II molecules, as determined by ELISA. Values represent mean ± SD of triplicates.

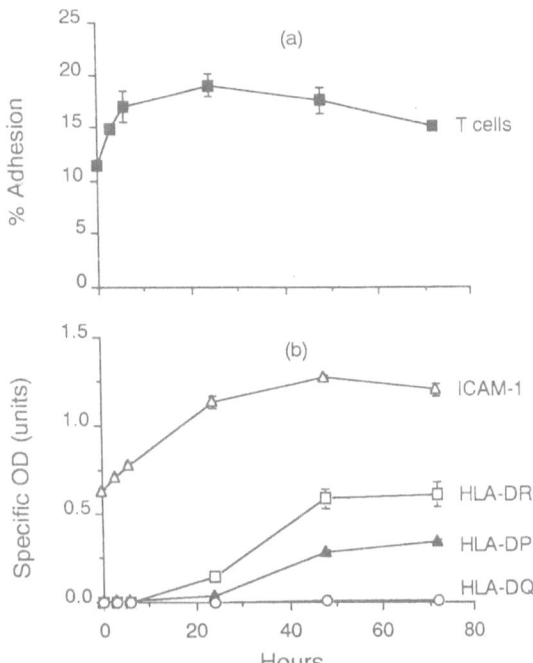

possible pairs involving IL-1, TNF, IL-4, and IFNγ, only TNF + IL-4 and TNF + IFNγ to T-cell adhesion that was greater than that of the more effective cytokine of each pair acting alone (18).

In contrast to their additive effects on EC adhesiveness for T cells, the combinations of TNF + IL-4 and TNF + IFNγ were no better than for TNF alone in stimulating EC adhesiveness for neutrophils. As a consequence, whilst approximately equal proportions of neutrophils and T cells adhered to adjacent cultures of TNF-stimulated EC, the ratio of T cells to neutrophils increased toward 2:1 in adjacent wells stimulated by TNF + IL-4 or, to a lesser extent, TNF + IFNγ.

Parallel experiments examining the changes in cell-surface adhesion molecule expression in response to these cytokine combinations showed that IL-4 increased the capacity of TNF to induce VCAM-1 expression but partially inhibited TNF-induced E-selectin expression and ICAM-1 up-regulation (17). T-cell adhesion to TNF + IL-4–stimulated EC was reduced to the level of adhesion to EC cultures stimulated by TNF alone when the assays were performed in the presence of anti–VCAM-1 mAb, supporting the conclusion that the additive effect of TNF + IL-4 on EC adhesiveness for T cells is mainly mediated through VCAM-1. As with the IFNγ alone, the molecules involved in the capacity of IFNγ to increase TNF-stimulated EC adhesiveness for T cells are not yet clear.

Conclusion

The various patterns of leukocyte migration in different forms and stages of inflammation are probably controlled by the concerted action of chemoattractants and adhesion molecules with selective actions on leukocytes of different lineage. Our observations suggest that in immune-mediated inflammation, the lymphokines IL-4 and IFNγ may have important regulating effects on endothelial activation in response to TNF and that this may critically alter the endothelial surface to one favoring T-cell adhesion and the formation of lymphocytic infiltrates.

References

1. Carlos, T.M. and J.M. Harlan. 1990. Membrane proteins involved in phagocyte adherence to endothelium. *Immunol. Rev.* 114:5.
2. Osborn, L. 1990. Leukocyte adhesion to endothelium in inflammation. *Cell* 62:3
3. Thornhill, M.H. and D.O. Haskard. 1991. Lymphocyte adhesion in inflammation. In Vascular Endothelium: Interactions With Circulating Cells. J.L. Gordon, ed. Elsevier, Amsterdam, pp. 93–112.
4. Cohnheim, J. 1989. Lectures on General Pathology: A Handbook for Practitioners and Students. The New Sydenham Society, London.
5. Florey, H.W. and L.H. Grant. 1961. Leucocyte migration from small blood vessels stimulated with ultra-violet light: an electron microscope study. *J. Path. Bact.* 82:13.
6. Atherton, A. and G.V.R. Born. 1972. Quantitative investigations of the adhesiveness of circulating polymorphonuclear leukocytes to blood vessel walls. *J. Physiol.* 222:447.
7. Geng, J.-G., M.P. Bevilacqua, K.L. Moore, T.M. McIntyre, S.M. Prescott, J.M. Kim, G.A. Bliss, G.A. Zimmerman, and R.P. McEver. 1990. Rapid neutrophil adhesion to activated endothelium mediated by GMP-140. *Nature* 343:757.
8. Pober, J.S. and R.S. Cotran. 1990. Cytokines and endothelial cell biology. *Physiol. Rev.* 70:427.
9. Dixit, V.M., S. Green, V. Sarma, L.B. Holzman, F.W. Wolf, K. O'Rourke, P.A. Ward, E.V. Prochownik, and R.M. Marks. 1990. Tumor necrosis factor-alpha induction of novel gene products in human endothelial cells including a macrophage-specific chemotaxin. *J. Biol. Chem.* 265:2973.
10. Thornhill, M., U. Kyan-Aung, T.L. Lee, and D.O. Haskard. 1990. T cells and neutrophils exhibit differential adhesion to cytokine stimulated endothelial cells. *Immunology* 69:287.
11. Bevilacqua, M.P., S. Stengelin, M.A. Gimbrone, and B. Seed. 1989. Endothelial leukocyte adhesion molecule 1: an inducible receptor for neutrophils related to complement regulatory proteins and lectins. *Science* 243:1160.
12. Picker, L.J., T.K. Kishimoto, C.W. Smith, R.A. Warnock, and E.C. Butcher. 1991. ELAM-1 is an adhesion molecule for skin-homing T cells. *Nature* 349:796.

13. Shimizu, Y., S. Shaw, N. Graber, T.V. Gopal, K.J. Horgan, G.A. Van Seventer, and W. Newman. 1991. Activation-independent binding of human memory T cells to adhesion molecule ELAM-1. *Nature* 349:799.
14. Osborn, L., C. Hession, R. Tizard, C. Vassallo, S. Luhowskyj, G. Chi-Rosso, and R. Lobb. 1989. Direct expression cloning of vascular cell adhesion molecule 1 (VCAM1), a cytokine-induced endothelial protein that binds to lymphocytes. *Cell* 59:1203.
15. Yu, C.-L., D.O. Haskard, D.E. Cavender, A.R. Johnson, and M. Ziff. 1985. Human gamma inteferon increases the binding of T lymphocytes to endothelial cells. *Clin. Exper. Immunol.* 62:554.
16. Thornhill, M., U. Kyan-Aung, and D.O. Haskard. 1990. Interleukin-4 increases human endothelial cell adhesiveness for T cells but not for neutrophils. *J. Immunol.* 144:3060.
17. Thornhill, M.H. and D.O. Haskard. 1990. IL-4 regulates endothelial activation by IL-1, tumor necrosis factor or IFNy. *J. Immunol.* 145:865.
18. Thornhill, M., S.M. Wellicome, D.L. Mahiouz, J.S.S. Lanchbury, U. Kyan-Aung, and D.O. Haskard. 1991. Tumor necrosis factor combines with IL-4 or IFN-y to selectively enhance endothelial cell adhesiveness for T cells: the contribution of VCAM-1 dependent and independent binding mechanisms. *J. Immunol.* 146:592.

22
Co-stimulation of T-Lymphocyte Activation by Adhesion Molecules

Laurie S. Davis and Peter E. Lipsky

Introduction

Members of the integrin family of adhesion molecules modulate T-lymphocyte contact with other cells and with extracellular matrix molecules (1–5). In addition to mediating cell–cell and cell–matrix contact, the integrin molecules also appear to be able to transmit important co-stimulatory signals that may be crucial during T-cell responses to antigen (1,2). Integrin molecules expressed by resting T cells are unable to bind their ligands, requiring stimulation of the T cell to acquire functional competence. Thus, activation of the T cell induces high-affinity binding of integrin molecules to their ligands which in turn can co-stimulate T-cell responsiveness. This complex interactive regulation of receptor binding and signaling competency appears to play a critical role in the induction and modulation of T-cell responses.

Fibronectin Receptors Co-Stimulate T-Cell Activation

Several of the integrin receptors belong to a subfamily known as the VLA (very late antigen) family of adhesion molecules. The name VLA was used because the first two members of the family to be discovered, VLA-1 and VLA-2, were expressed on T cells 2–4 wk after stimulation (6,7). Members of the VLA family are composed of a common $\beta1$ chain (CD29) and have unique α chains (6,8). The VLA molecules can interact with the extracellular matrix proteins fibronectin, collagen, and/or laminin (6). Members of the VLA family found on T cells include VLA-1, VLA-2, VLA-4, VLA-5, and VLA-6. Both VLA-1 and VLA-2 molecules bind collagen and laminin. VLA-4 binds fibronectin and, using a distinct site, binds a molecule expressed on activated endothelial cells and other cell types termed vascular cell adhesion molecule-1 or VCAM-1 (9,10). VLA-4 is expressed on a majority of resting T cells and is unique in that relatively high levels of expression are restricted to lymphoid and myeloid cells (6). VLA-5 has been described as the classical fibronectin receptor and has been found on a variety of cells including many hemopoietic cells, endothelial cells, and fibroblasts (9,11–13). VLA-6 primarily serves as a laminin receptor. It is not known whether VLA-5 or VLA-6 can also bind to specific cell-surface molecules in a manner similar to VLA-4. VLA-4, VLA-5, and VLA-6 are expressed on memory T cells at a level three to four times higher than on naive T cells (4).

The ability of VLA-4 and VLA-5 molecules to deliver co-stimulatory signals during T-cell activation was examined using freshly isolated highly purified human peripheral blood T cells (14). The weakly stimulatory anti-CD3 mAb, OKT3, was used as a suboptimal activation stimulus (15). OKT3 immobilized to the bottom of microtiter wells generated minimal T-cell proliferation in the absence of co-stimulatory signals. By contrast, immobilized OKT3 induced maximal responses in the presence of accessory cells. T-cell proliferation in response to OKT3 required the presence of serum (5–10% serum was optimal) or a serum substitute containing bovine serum albumin (BSA), high-density lipoprotein, oleic acid, and transferrin to induce proliferation (14). The addition of fibronectin alone in soluble or immobilized form could not substitute for the serum requirement. When fibronectin was immobilized along with OKT3 to the bottom of the microtiter well, it co-stimulated T-cell–proliferative responses in a dose-dependent fashion (Fig. 22.1). Immobilized fibronectin could not induce T-cell activation in the absence of co-stimulation with immobilized anti-CD3, regardless of the serum content of the culture medium.

Although normal human serum contains large amounts of soluble fibronectin, the addition of human serum did not inhibit the ability of immobilized fibronectin to co-stimulate responses, indicating that the fibronectin receptors on T cells preferentially bind immobilized but not soluble fibronectin once activated by T-cell antigen receptor–CD3 complex ligation. Immobilized fibronectin alone was not mitogenic for T cells since fibronectin did not stimulate proliferation in cultures containing IL2 and/or phorbol ester (14). It should be noted that activation with phorbol ester did induce binding of T cells to fibronectin-coated surfaces, although proliferation was not observed. These results indicate that fibronectin alone was unable to deliver a mitogenic signal even when receptor competency was induced by phorbol ester co-stimulation.

In the presence of immobilized OKT3, immobilized fibronectin had a particularly potent effect when compared with other substrates. Although gelatin and laminin could also co-stimulate T-cell–proliferative responses, they were not as potent as immobilized fibronectin. When the serum substitute was used to support cultures rather than serum, gelatin was unable to enhance T-cell responses. One likely explanation for this result was that serum fibronectin bound to the gelatin inducing co-stimulation and that gelatin alone was unable to deliver a co-stimulatory signal. Collagen and BSA had no potentiating effect on anti-CD3 responses. These studies

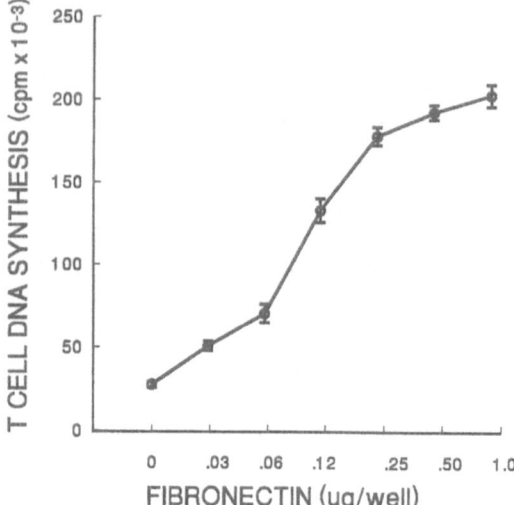

FIGURE 22.1. Immobilized fibronectin augments suboptimal mitogen stimulation. The anti-CD3 mAb, OKT3, was immobilized to microtiter wells at a concentration of 200 ng/well. The wells were rinsed and fibronectin was immobilized to the wells at the concentrations indicated in an additional brief incubation at room temperature. The wells were rinsed and T cells (1×10^5/well) were added in the presence of 5% normal human serum. Data are presented as ^{3}H-thymidine incorporation (cpm $\times 10^{-3} \pm$ SEM) after a 96-h incubation.

indicated that not all matrix proteins had the capacity to modulate T-cell–proliferative responses and suggested that specific receptors for fibronectin were able to provide potent co-stimulatory signals to T cells also activated through the T-cell receptor–CD3 complex. Although anti-CD3 stimulation induced binding of T cells to fibronectin-coated surfaces, this appeared to be insufficient to elicit T-cell proliferation as similar up-regulation of fibronectin receptor competency, but no proliferation was induced by phorbol esters. This suggests that specific signals provided by CD3 ligation beyond these needed to induce fibonectin receptor-binding competency are necessary to permit fibronectin to co-stimulate T-cell activation.

Matsuyama et al. had indicated that only the memory T-cell subset expressed VLA-5 and that this was the primary receptor mediating co-stimulation of T cells by immobilized fibronectin and OKT3 (16). Since VLA-4 was expressed on a majority of T cells, it was of interest to examine the ability of immobilized fibronectin to co-stimulate T cell subsets and the role of VLA-4 and VLA-5 in these responses. Fibronectin was consistently able to co-stimulate both CD4$^+$ and CD8$^+$ subsets of T cells (Fig. 22.2). Moreover, when CD4$^+$ T cells, isolated by negative selection panning, were further separated into naive and memory T cells based on the expression of CD45RA or LFA-3, both subsets were co-stimulated by fibronectin (14). However, the memory subset of T cells always manifested a higher level of proliferation in the presence of fibronectin than

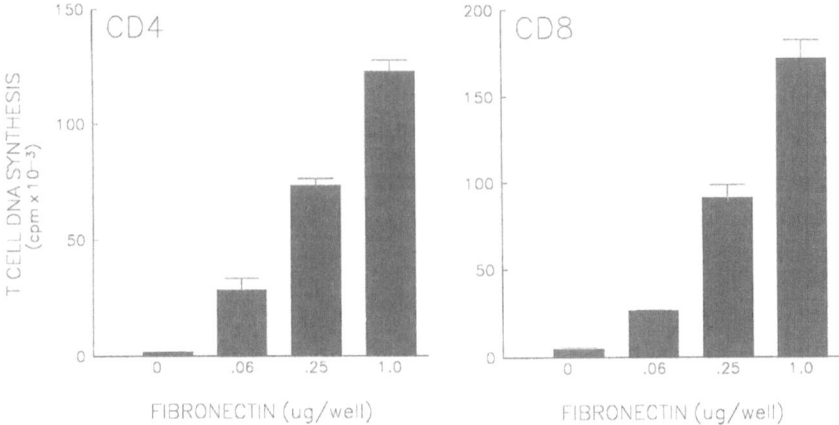

FIGURE 22.2. The ability of fibronectin to co-stimulate CD4+ and CD8+ T-cell subsets. T-cell subsets were isolated by negative selection panning and were cultured in wells containing immobilized OKT3 (200 ng/well) and immobilized fibronectin at the concentrations indicated as described in Figure 22.1. DNA synthesis was assessed after a 96-h incubation.

the naive cells. The augmentation of anti-CD3–induced proliferation of both naive and memory T cells was completely blocked by the addition of the anti-CD29 (β chain) mAb 4B4 (Fig. 22.3). A number of control antibodies including anti-CD18 (60.3), anti–class I (W6/32), and anti-CD45RA (2H4) did not block the fibronectin-mediated co-stimulation of anti-CD3 responses. These results indicate that both memory and naive cells can be co-stimulated by fibronectin and confirm the role of β1 integrins in this response.

The enhancement of anti-CD3–stimulated T-cell proliferation by fibronectin could be mediated by two different receptors. The classical fibronectin receptor, VLA-5, binds the RGDS sequence in one binding domain of the fibronectin molecule. By comparison, VLA-4 binds fibronectin in the IIICS region at a site known as the second cell binding domain or CS-I. In order to determine the ability of these receptors to co-stimulate anti-CD3 responses, monoclonal antibodies (mAb) to either VLA-4 or VLA-5 were employed. The anti–VLA-5 antibody (mAb 16; to the α chain of VLA-5) inhibited the fibronectin-mediated augmentation of anti-CD3–stimulated T-cell responses by at least 50%, whereas the anti–VLA-4 mAb (L25; to the α chain of VLA-4) had much less of an inhibitory effect (14). However, the combination of anti–VLA-4 and anti–VLA-5 mAb was required to inhibit responses to the level caused by anti-CD29 (Fig. 22.4). Similar results have also been reported by Shimizu et al. using mAbs to VLA-4 and VLA-5 in combination with synthetic

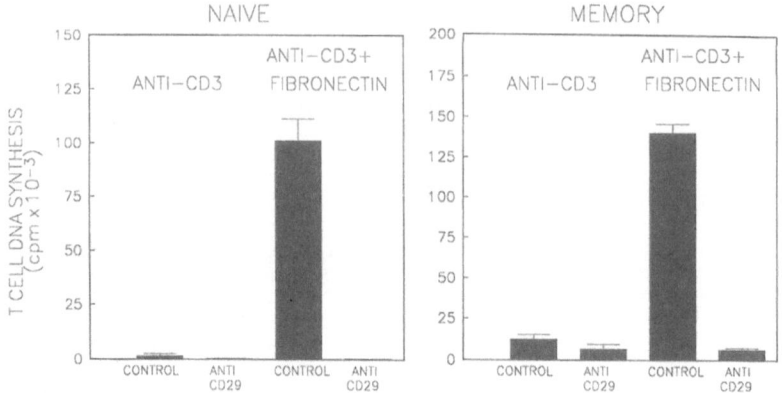

FIGURE 22.3. Effect of fibronectin on anti-CD3 responses induced proliferation of naive and memory CD4+ T cells. Purified CD4+ T cells isolated by negative selection panning were separated into CD45RA dim or CD45RA bright populations and stimulated with immobilized anti-CD3(OKT3, 200 ng/well) in the absence or presence of immobilized fibronectin (1 μg/well). Soluble anti-CD29 mAb, 4B4 (0.25 μg/ml), was added where indicated. T-cell DNA synthesis was assessed after a 4-day incubation.

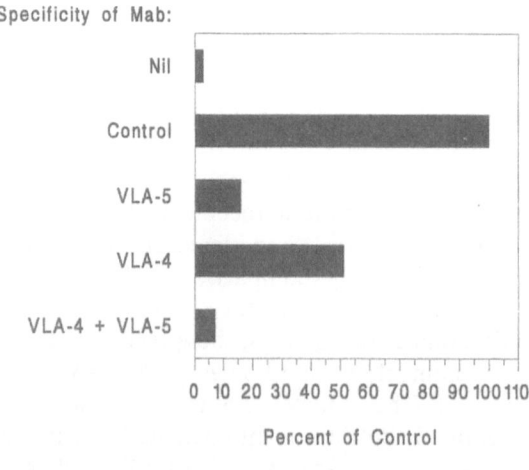

FIGURE 22.4. Effect of anti–VLA-4 and anti–VLA-5 on fibronectin-mediated augmentation of anti-CD3–stimulated cultures. T cells (1 × 10⁵/well) suspended in medium containing 5% normal human serum were cultured with immobilized OKT3(200 ng/well) alone (Nil) or with immobilized OKT3 and immobilized fibronectin (1 μg/well) in the presence of Control mAb (0.5 μg/ml), or anti–VLA-5 α-chain mAb (0.5 μg/ml), or anti–VLA-4 α-chain mAb (0.5 μg/ml) or both of the latter two. Data are presented as percent of control ³H-thymidine incorporation (cpm × 10⁻³ ± SEM) after a 96-h incubation.

TABLE 22.1. Human T-cell proliferation induced by anti-CD3 and anti–VLA-4 mAb.

| | ^{3}H-thymidine incorporation (cpm $\times$ 10^{-3}) | | | |
| | Nil | | GAMIg | |
Specificity of Mab	Nil	PDB	Nil	PDB
P117	0.1 ± 0.0	0.2 ± 0.0	0.1 ± 0.0	0.2 ± 0.0
CD3	0.1 ± 0.0	35.7 ± 7.3	0.1 ± 0.0	94.3 ± 17.3
VLA-4	0.1 ± 0.0	0.1 ± 0.0	0.1 ± 0.0	0.6 ± 0.3
VLA-4 + CD3	0.1 ± 0.0	128.3 ± 10.8	69.8 ± 7.4	123.9 ± 8.4
CD29	0.2 ± 0.0	0.1 ± 0.0	0.1 ± 0.0	0.1 ± 0.0
CD29 + CD3	0.2 ± 0.0	116.5 ± 26.0	2.4 ± 0.7	124.2 ± 3.9

T cells were pulsed with saturating concentrations of the control mAB is P117 or anti-CD3 mAb alone or in combination with anti–VLA-4 mAb or anti-CD29. The T cells (1×10^5/ well) were cultured in microtiter wells alone or in wells coated with goat antimouse Ig (GAMIg) in the presence or absence of PDB (0.5 ng/ml). The T cells were harvested after a 6-day incubation.

peptides of the RGDS and CS-I sequences (17). Thus, these studies indicate that T cells employed both VLA-4 and VLA-5 to bind fibronectin and that each participated in the generation of a co-stimulatory signal.

In additional experiments, the effects of the anti–VLA-4 mAb, L25, were examined in greater detail (14,18,19). L25 enhanced T-cell responses to suboptimal concentrations of immobilized OKT3 or immobilized OKT3 and low concentrations of fibronectin, whereas it consistently inhibited responses to OKT3 and high concentrations of fibronectin. In other studies, L25 was found to co-stimulate T-cell responses when it was immobilized to the microtiter well along with the anti-CD3 mAb (Table 22.1). An additional anti–VLA-4 mAb (HP1/3) was tested and found only to inhibit fibronectin-mediated augmentation of T-cell responses induced by anti-CD3. A previous report by Groux et al. found that some anti-CD29 mAb enhanced anti-CD3–induced T-cell proliferation, whereas other anti-CD29 mAb inhibited responses (20). The current studies demonstrated that the anti-CD29 mAb (4B4) inhibited fibronectin-mediated co-stimulation of T-cell responses when added to the cultures in soluble form (Fig. 22.3), whereas this same antibody augmented T-cell responses when crosslinked or co-immobilized with OKT3 in the presence of phorbol esters (Table 22.1). Taken together these studies suggest that while most anti-VLA mAbs have distinct enhancing or inhibiting activities, one anti–VLA-4 mAb, L25, has weak agonist activity in soluble and immobilized forms. These results provide further evidence that VLA-4 can function to provide co-stimulatory signals to T cells. We have found no evidence that the antibody to VLA-5 exerts a co-stimulatory effect. However, the blocking studies clearly indicate that VLA-5 engagement of fibronectin can also provide a powerful co-stimulation to T cells activated by ligation of the T-cell receptor–CD3 complex.

One possible explanation for the co-stimulatory effects was that fibronectin could have increased binding of the T cell to the coated surface and thereby facilitated the delivery of activation signals through the CD3 complex. Alternatively, fibronectin might have transduced a co-stimulatory signal through VLA-4 and/or VLA-5. In order to distinguish between these two possibilities, experiments were undertaken to determine whether anti-CD3 and fibronectin would co-stimulate T-cell responses when immobilized onto different surfaces (14). Surface-bound fibronectin was employed because it was clear from the studies mentioned above that soluble fibronectin was unable to co-stimulate. Anti-CD3 was coated onto the surface of polystyrene beads and fibronectin was immobilized to the bottom of microtiter wells. If co-stimulation occurred, it would support the conclusion that fibronectin provided an activation signal and was not merely facilitating binding to the stimulatory surface. Under these conditions, fibronectin enhanced responses twofold. These results indicated that fibronectin co-immobilized to the same surface as the anti-CD3 was far more efficient than when presented on a separate surface, but also documented that receptors for fibronectin had the ability to transmit co-stimulatory signals to the T cell. The greater degree of responsiveness noted when fibronectin and anti-CD3 were co-immobilized to the same surface could have a number of explanations. Thus, fibronectin co-immobilized to the same surface may play a role in stabilizing the T-cell receptor–CD3 complex interactions with the anti-CD3 antibody. Alternatively, co-aggregation of stimulated fibronectin receptors with the T-cell antigen receptor–CD3 complex may be required for the most effective signal transduction and induction of proliferation.

Regulatory Role of LFA-1 During T-Cell Activation

One of the principal molecules mediating cell–cell contact is another integrin molecule, LFA-1 (6). LFA-1, like members of the VLA subfamily, shares a common β chain (CD18) with other members of the LeuCAM family, but has a unique α chain (CD11a). LFA-1 is restricted in expression to leukocytes and binds to ICAM-1 (CD54) and ICAM-2 molecules expressed on leukocytes and endothelial cells (3,6,21,22). ICAM-1 expression has also been shown to be up-regulated by cytokines and has been found on a number of activated or transformed cell types (3,6). Inhibiting LFA-1/ICAM-1 adhesion disrupts T-cell–accessory-cell contact. Studies using antibodies to LFA-1 have shown that LFA-1–mediated interactions are critical for cytotoxic T-cell binding to target cells and in T-cell–B-cell collaboration (23,24). T-cell responses to alloantigens are also dependent on LFA-1/ICAM-1 interactions as indicated by the capacity of

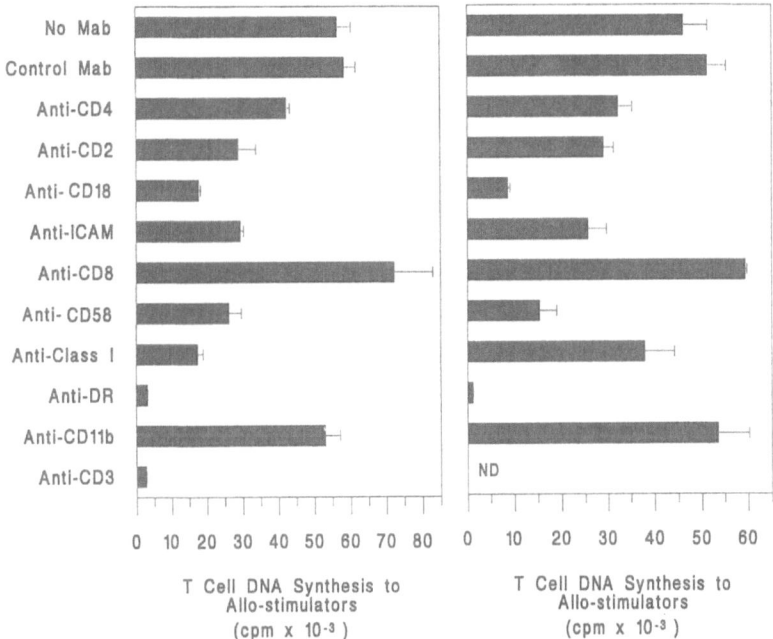

FIGURE 22.5. Effect of anti-CD18 on T-cell responses to alloantigens. Purified CD4+ T cells (5×10^4/well) were cultured with allogeneic irradiated monocytes and B cells (5×10^4/well) in the presence or absence of monoclonal antibodies. The left and right panels show the results of two separate experiments. The purified antibodies P117 (control); anti-CD4 mAb, OKT4; anti-CD2 mAb, OKT11; anti-CD18 mAb, 60.3; anti–ICAM-1, RR1/1.1; anti-CD58, TS2/9; anti-CD11b, OKM1; and anti-CD3, OKT3 were added to the wells at a final concentration of 5 μg/ml. Anti-CD8(OKT8) was used as a control for the antibody to nonpolymorphic class I MHC molecules, W6/32, and an antibody to nonpolymorphic DR molecules, L243. OKT8, W6/32, and L243 were added at a concentration of 1:10 of culture supernatant. The T cells were harvested after 6 days in culture and DNA synthesis was assessed.

monoclonal antibodies specific for these receptors to block significantly (Fig. 22.5). All T-cell interactions that require cell–cell contact are usually dependent on LFA-1–mediated interactions. Several studies have also shown that antibodies to LFA-1 can co-stimulate T-cell activation (1,25,26). It is interesting to note that as with fibronectin-mediated amplification of responses, the characteristic cell–cell adhesion mediated by LFA-1/ICAM-1 that is induced by phorbol esters is not capable of inducing T-cell proliferation.

The role of LFA-1 as a signaling molecule during activation of freshly isolated human peripheral blood T cells was investigated in two different model systems. To analyze the effect of engaging LFA-1 on resting hu-

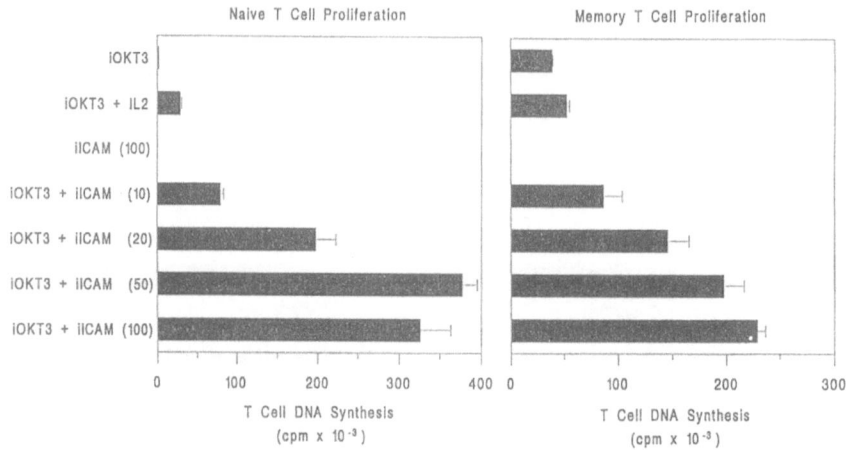

FIGURE 22.6. Effect of immobilized recombinant ICAM-1 on anti-CD3–stimulated T-cell proliferation. OKT3(200 ng/well) was immobilized to microtiter wells, the wells were rinsed, and recombinant ICAM-1 was added to the wells at the concentrations indicated (ng/well). The wells were washed and T cells (1×10^5/well) were added to the wells. Recombinant IL2 (10 U/ml) was also added where indicated. T cells were purified by negative selection panning using a cocktail of antibodies including L243, OKT8, B73.1, OKM1, CD19, and for naive cells UCHL1, and for memory cells 2H4. Naive and memory T-cell DNA synthesis was assessed after a 108-h incubation as described earlier.

man T cells, a recombinant form of one of the natural ligands for LFA-1, ICAM-1 (the generous gift of Dr. Steve Marlin), was co-immobilized in varying concentrations to the bottom of a microtiter well that also contained a suboptimal concentration of the anti-CD3 mAb (OKT3). Recombinant immobilized ICAM-1 was able to augment proliferation of freshly isolated highly purified CD4+ T cells in a concentration-dependent fashion. In the absence of anti-CD3, immobilized ICAM-1 caused no proliferation. This finding was in agreement with studies by Van Seventer et al. that suggested that T-cell responses to immobilized but not soluble OKT3 were augmented by ICAM-1 (27).

To analyze this effect in greater detail, CD4+ T cells were separated into naive and memory subsets in order to determine whether recombinant ICAM-1 could preferentially stimulate a subset of T cells. At the concentration of OKT3 used in these studies, memory CD4+ T cells were able to generate only a minimal response in the absence of co-stimulatory signals and naive CD4+ T cells were completely unable to respond to OKT3 alone. However, both CD4+ T-cell subsets proliferated in response to the combination of immobilized OKT3 and ICAM-1 (Fig. 22.6). It should be noted that immobilized ICAM-1 markedly enhanced the amount of IL2 secreted by each subset, but did not affect the intrinsic

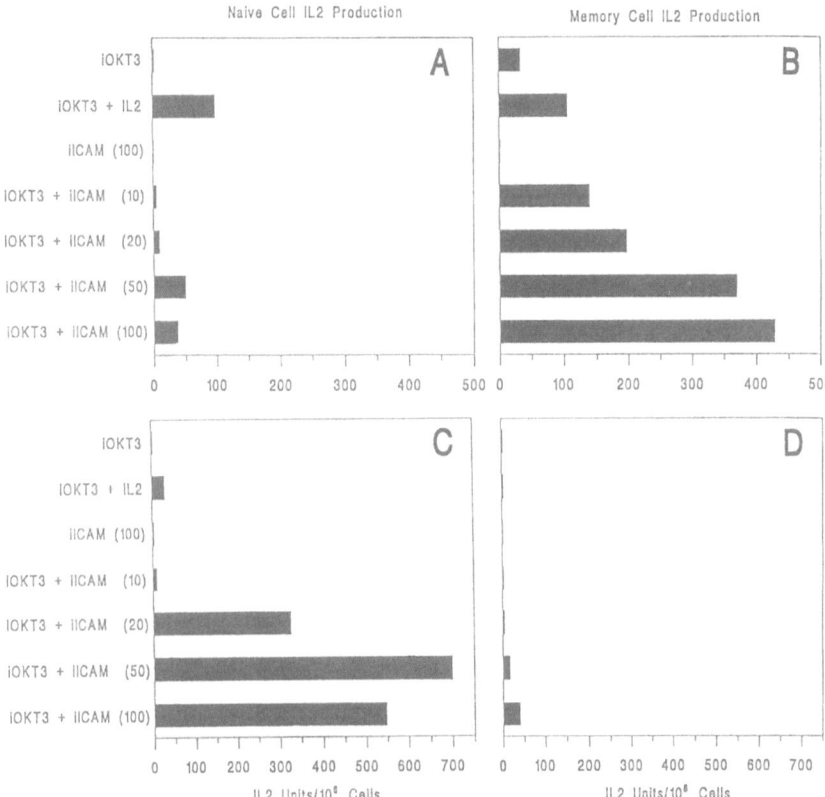

FIGURE 22.7. Augmentation of anti-CD3–stimulated T-cell proliferation does not alter the kinetics of IL2 production by naive and memory cells. Supernatants from the experiment outlined in Figure 22.6 were harvested after 24 h (A and B) or 108 h (C and D) and IL2 content was measured using a CTLL-2 assay and recombinant IL2 (Cetus Corp., Emeryvile, CA) as a standard for quantitation of IL2.

kinetics of IL2 production by naive and memory T cells (Fig. 22.7). In contrast, immobilized ICAM-1 alone had minimal effects on T-cell responses induced by PHA, anti-CD2, anti-CD28 (9.3), or phorbol esters. It should be noted that Van Seventer et al. demonstrated that immobilized ICAM-1 could co-stimulate T-cell responses induced by phorbol ester and a calcium ionophore (27). This result supported the conclusion that under certain circumstances ICAM-1 could provide an activation signal and not just function to increase cellular adherence.

The capacity of LFA-1 to function as a co-stimulatory molecule has been previously explored. Studies by Wacholtz et al. using mAbs of different isotypes had shown that independently crosslinking LFA-1 and CD3 on T-cell clones could induce IL2 production, whereas crosslinking either

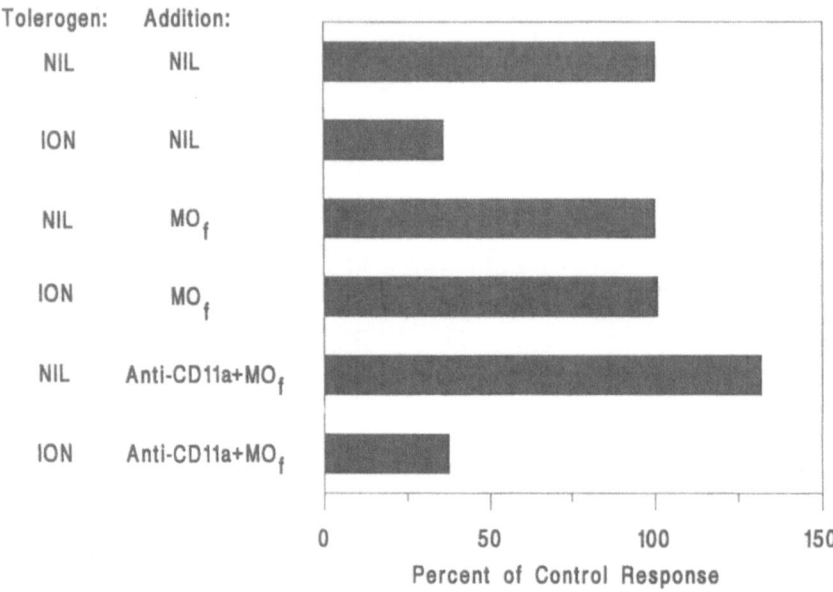

FIGURE 22.8. Effect of anti-CD11a on monocyte reversal of ionomycin-induced tolerance induction. Purified CD4+ T cells (0.5×10^6/ml) were preincubated in the presence or absence of ionomycin ($0.25~\mu$M/ml) to induce nonresponsiveness. Where indicated, paraformaldehyde-fixed monocytes (M_f; 0.25×10^4/well) and/or anti-CD11a (TS1/22; 1:50 supernatant dilution) were added during the preincubation step. Following overnight incubation the cells were washed extensively and stimulated with PHA (0.5 μg/ml) and fresh monocytes (5×10^4/well). The cells were harvested after 3 days in culture and DNA synthesis was assessed.

mAb alone was ineffective. Ligation of LFA-1 with monoclonal antibodies did not induce an increase in intracellular calcium alone, whereas simultaneously but independently ligating CD3 and LFA-1 induced a greater signal than engaging CD3 alone. These results indicate that LFA-1 could function to transduce activation signals to T cells.

A second model system demonstrated that LFA-1 could play a critical role in determining whether a T cell becomes activated or anergic after CD3 ligation. In this system, freshly isolated, highly purified CD4+ T cells were incubated with either an anti-CD3 mAb, 38.1, or the calcium ionophore, ionomycin, overnight in the absence of co-stimulatory signals (28). As a result of this exposure, T cells were rendered hyporesponsive to subsequent stimulation. This anergic state largely reflected their diminished ability to produce IL2 upon subsequent stimulation. The induction of anergy resulted from mobilization of intracellular calcium stores and could be mimicked by the calcium ionophore, ionomycin (28). The optimal concentration of either stimulus for "tolerance induction" was the same that generated maximum T-cell–proliferative responses when accessory cells or phorbol esters were included in the cultures. Of note, incubation for as brief a period as 2 h with ionomycin rendered cells nonresponsive to subsequent stimulation even after 1–3 days of "recovery" in the absence of ionomycin before restimulation with antigen (tetanus toxoid) or mitogen (PHA). Although the T cells expressed near-normal levels of the T-cell antigen receptor–CD3 complex, tolerized T cells generated less than 50% of the proliferation and IL2 production that was observed in control cultures containing optimal numbers of accessory cells (28). However, the addition of supraoptimal numbers of accessory cells could almost completely overcome the effects of tolerance induction.

Additional experiments demonstrated that small numbers of irradiated or paraformaldehyde-fixed monocytes prevented the induction of nonresponsiveness. However, if an anti–LFA-1 mAb (TS1/18) were included in the culture, the monocyte-dependent prevention of anergy induction was blocked and the T cells were once again rendered nonresponsive to subsequent stimulation. The presence of fixed monocytes and/or anti-LFA-1 during the preincubation had no effect on the subsequent responses of control cultures that lacked the tolerogens (Fig. 22.8). These results support the conclusion that LFA-1 is an important signaling molecule on T cells that not only mediates cell–cell contact but also regulates T-cell responses. In the absence of a signal provided by LFA-1, activation stimuli may become tolerogenic.

Conclusions

The integrin family of adhesion molecules plays a critical role in T-cell interactions with other cells and also with matrix proteins. These studies demonstrate that three of the integrin molecules not only mediate T-cell adhesion but also provide signals that co-stimulate T-cell activation. These signals appear to be critical in determining whether antigen recognition leads to activation or anergy.

References

1. Van Noesel, C., F. Miedema, M. Brouwer, M.A. de Rie, L.A. Aarden, and R.A.W. van Lier. 1988. Regulatory properties of LFA-1α and β chains in human T-lymphocyte activation. *Nature* 333:850.
2. Dustin, M.L., and T.A. Springer. 1989. T-cell receptor cross-linking transiently stimulates adhesiveness through LFA-1. *Nature* 341:619.
3. Springer, T.A. 1990. Adhesion receptors of the immune system. *Nature* 346:425.
4. Shimizu, Y., G.A. Van Seventer, K.J. Horgan, and S.Shaw. 1990. Regulated

expression and binding of three VLA (β1) integrin receptors on T cells. *Nature* 345:250.

5. Laffon, A., R. Garcia-Vicuna, A. Humbria, A.A. Postigo, A.L. Corbi, M.O. de Landazuri, and F. Sanchez-Madrid. 1991. Upregulated expression and function of VLA-4 fibronectin receptors on human activated T cells in rheumatoid arthritis. *J. Clin. Invest.* 88:546.

6. Hemler, M.E. 1988. Adhesive protein receptors on hematopoietic cells. *Immunol. Today.* 9:109.

7. Hemler, M.E., C. Huang, and L. Schwartz. 1987. The VLA family: characterization of five distinct cell surface enterodimers each with a common 130,000 molecular weight subunit. *J. Biol. Chem.* 262:3300.

8. Hynes, R.O. 1987. Integrins: A family of cell surface receptors. *Cell* 48:549.

9. Wayner, E.A., A. Garcia-Pardo, M.J. Humphries, J.A. McDonald, and W.G. Carter. 1989. Identification and characterization of the T lymphocyte adhesion receptor for an alternative cell attachment domain (CS-1) in plasma fibronectin. *J. Cell Biol.* 109:1321.

10. Elices, M.J., L. Osborn, Y. Takada, C. Crouse, S. Luhowsky, M.E. Hemler, and R.R. Lobb. 1990. VCAM-1 on activated endothelium interacts with the leukocyte integrin VLA-4 at a site distinct from the VLA-4/fibronectin binding site. *Cell* 60:577.

11. Brown, D.L., D.R. Phillips, C.H. Damsky, and I.F. Charo. 1989. Synthesis and expression of the fibroblast fibronectin receptor in human monocytes. *J. Clin. Invest.* 84:366.

12. Wayner, E.A., and W.G. Carter. 1988. The function of multiple extracellular matrix receptors in mediating cell adhesion to extracellular matrix: preparation of monoclonal antibodies to the fibronectin receptor that specifically inhibit cell adhesion to fibronectin and react with platelet glycoproteins Ic-IIa. *J. Cell Biol.* 107:1881.

13. Conforti, G., A. Zanetti, S. Colella, M. Abbadini, P.C. Marchisio, R. Pytela, F. Giancotti, G. Tarone, L.R. Languino, and E. Dejano. 1989. Interaction of fibronectin with cultured human endothelial cells: characterization of the specific receptor. *Blood* 73:1576.

14. Davis, L.S., N. Oppenheimer-Marks, J.L. Bednarczyk, B.W. McIntyre, and P.E. Lipsky. 1990. Fibronectin promotes proliferation of naive and memory T cells by signaling through both the VLA-4 and VLA-5 integrin molecules. *J. Immunol.* 145:785.

15. Geppert, T.D., and P.E. Lipsky. 1987. Accessory cell independent proliferation of human T4 cells stimulated by immobilized antibodies to CD3. *J. Immunol.* 138:1660.

16. Matsuyama, T., A. Yamada, J. Kay, K. M. Yamada, S.K. Akiyama, S.F. Schlossman, and C. Morimoto. 1989. Activation of CD4 cells by fibronectin and anti-CD3 antibody. A synergistic effect mediated by the VLA-5 fibronectin receptor complex. *J. Exp. Med.* 170:1133.

17. Shimizu, Y., G.A. Van Seventer, K.J. Horgan, and S. Shaw. 1990. Costimulation of proliferative responses of resting CD4+ T cells by the interaction of VLA-4 and VLA-5 with fibronectin or VLA-6 with laminin. *J. Immunol.* 145:59.

18. Clayberger, C., A.M. Krensky, B.W. McIntyre, T.D. Koller, P. Parham, F. Brodsky, D.J. Linn, and E.L. Evans. 1987. Identification and characteriza-

tion of two novel lymphocyte function-associated antigens, L24 and L25. *J. Immunol.* 138:1510.

19. McIntyre, B.W., E.L. Evans, and J.L. Bednarczyk. 1989. Lymphocyte surface antigen L25 is a member of the integrin receptor superfamily. *J. Biol. Chem.* 264:13745.

20. Groux, H., S. Huet, H. Valentin, D. Pham, and A. Bernard. 1989. Suppressor effects and cyclic AMP accumulation by the CD29 molecule ot CD4+ lymphocytes. *Nature* 339:152.

21. Klingemann, H.-G., and S. Dedhar. 1989. Distribution of integrins on human peripheral blood mononuclear cells. *Blood* 74:1348.

22. Nortamo, P., R. Salcedo, T. Timonen, M. Patarroyo, and C.G. Gahmberg. 1991. A monoclonal antibody to the human leukocyte adhesion intercellular adhesion molecule-2. Cellular distribution and molecular characterization of the antigen. *J. Immunol.* 146:2530.

23. Davignon, D., E. Martz, T. Reynolds, K. Kurzinger, and T.A. Springer. 1981. Lymphocyte function-associated antigen one (LFA-1): a surface antigen distinct from Lyt-2/3 that participates in T-lymphocyte-mediated killing. *Proc. Natl. Acad. Sci. USA* 78:4535.

24. Tohma, S., S. Hirohata, and P.E. Lipsky. 1991. The role of CD11a/CD18-CD54 interactions in human T cell-dependent B cell activation. *J. Immunol.* 146:492.

25. Wacholtz, M.C., S.S. Patel, and P.E. Lipsky. 1989. Leukocyte function-associated antigen 1 is an activation molecule for human T cells. *J. Exp. Med.* 170:431.

26. Pircher, H., Groscurth, S. Baumhutter, M. Aguet, R.M. Zinkernagel, and H. Hengartner. 1986. A monoclonal antibody against altered LFA-1 induces proliferation and lymphokine release of cloned T cells. *Eur. J. Immunol.* 16:172.

27. Van Seventer, G.A., Y. Shimizu, K.J. Horgan, and S. Shaw. 1990. The LFA-1 ligand ICAM-1 provides an important costimulatory signal for T cell receptor-mediated activation of resting T cells. *J. Immunol.* 144:4579.

28. Davis, L.S., M.C. Wacholtz, and P.E. Lipsky. 1989. The induction of T cell unresponsiveness by rapidly modulating CD3. *J. Immunol.* 142:1084.

Part 4
In Vivo Models

23
The Role of CD11a/CD18-CD54 Interactions in the Evolution of Animal Models of Tissue Injury

Peter E. Lipsky, Hugo E. Jasin, Ellis Lightfoot, and William J. Mileski

Introduction

Entry of circulating inflammatory cells into tissue plays a central role in the evolution of many forms of tissue injury (1). In many of these conditions, inflammation is a prominent feature, whereas in others tissue injury is the primary manifestation, with inflammation being less apparent. An example of the former is inflammatory arthritis, in which the signs and symptoms of inflammation predominate, although persistent inflammation results in considerable damage to articular structures. On the other hand, in burn injury, tissue injury is readily apparent, whereas an inflammatory component is less prominent. However, in both conditions, it is likely that entry of circulating inflammatory cells into the affected tissue plays a primary role in the progression of tissue injury (2–5).

Inflammatory cells gain access to the tissue by a multistep process that involves adhesive interactions with the endothelial cells of postcapillary venules followed by transendothelial migration (6). A number of receptor–counterreceptor pairs play a role in this process. The members of the Leu-cam family of integrins expressed by circulating inflammatory cells and their counterreceptors on endothelial cells play a particularly important role in the entry of inflammatory cells into sites of tissue injury (5–7). The roles of these receptor–counterreceptor pairs in animal models of inflammatory arthritis and burn injury were, therefore, examined by testing the ability of specific monoclonal antibodies to the adhesion receptors to alter their course and outcome.

Results

Antigen-Induced Arthritis in the Rabbit

Antigen-induced arthritis in rabbits resembles rheumatoid arthritis in its immune pathogenesis, chronicity, and histologic appearance (8,9). In this

TABLE 23.1. Inhibition of the influx of leukocytes into the synovial fluid during acute antigen-induced arthritis by monoclonal antibodies to CD18 but not CD54.

	Treatment		
Exp	mAb	Specificity	Leukocytes in the synovial fluid (WBC $\times 10^{-3}$)
1.	nil	nil	$13,025 \pm 4,895$
	15.7	CD18	$1,500 \pm 341$
2.	nil	nil	$21,588 \pm 4,554$
	RR1/1	CD54	$17,375 \pm 1,705$
	R6.5	CD54	$27,363 \pm 5,475$

Data represent means $\pm$ SEM of the total number of synovial fluid cells in 2 joints of 3 (Exp 1) or 4 (Exp 2) animals. Immunized animals received mAb (anti-CD18, 1 mg/kg; anti-CD54, 2 mg/kg) at 12-h intervals beginning 14 h before administration of intraarticular antigen. Twenty-four hours after receiving intraarticular antigen, the injected joints were lavaged and the total number of leukocytes was determined.

animal model, intraarticular injection of a soluble antigen into animals previously immunized to the same antigen gives rise to an acute arthritis lasting for about 1 wk followed by the establishment of a chronic synovitis which may persist for many months in a majority of the animals. The initial local acute Arthus reaction generated by the introduction of antigen constitutes a significant insult to the articular cartilage. Widespread early chondrocyte death is readily detectable in the superficial layers, and there is a progressive increase in fibrillation and degenerative changes of the collagen fibers (9). The subsequent chronic synovitis is characterized by cellular lining-layer hyperplasia and a prominent cellular infiltrate composed of lymphocytes, macrophages, and plasma cells with a tendency to the formation of lymphoid-follicle-like cellular accumulations (9). This model, therefore, provides a means to examine the interrelationship of acute and chronic inflammation and the role of specific adhesion molecules in mediating each process.

In the initial experiments, a monoclonal antibody directed at CD18 (R15.7) was employed (10). Immunized animals were given five intravenous doses of R15.7 (1 mg/kg) at 12-h intervals commencing 14 h before the intraarticular injection of antigen (11). As shown in Table 23.1, administration of the anti-CD18 mAb markedly decreased the number of leukocytes found in the synovial fluid during the acute phase of arthritis. Differential counts revealed that entry of neutrophils into the synovial fluid was nearly completely prevented. It should be noted that swelling, hyperemia, and an increase in the amount of synovial fluid still developed in the antigen-injected joints, but the infiltration with leukocytes was blocked. In contrast to the effect noted after administration of mAb to CD18, R6.5, A mAb to CD54 (12), had no effect on the infiltration of leukocytes into the synovial fluid of animals during the acute phase of antigen-induced arthritis.

The effect of administration of mAb to CD18 before and during the acute phase on the subsequent development of chronic arthritis was also examined. As a result of the treatment with the anti-CD18 mAb during the acute phase of antigen-induced arthritis, there was marked reduction in the number of leukocytes found in the joint fluid 2 and 4 wk later. At 2 weeks the control animals exhibited a mean of 21.0×10^6 WBCs in synovial fluid vs. 1.7×10^6 WBCs for the anti-CD18–treated animals and at 4 weeks the controls contained 22.1×10^6 synovial fluid WBCs and the anti-CD18–treated 7.4×10^6 WBCs. Histologic evaluation of the tissue also revealed striking differences between the control and experimental groups, with significant decreases in lining-layer hyperplasia, severity of the mononuclear infiltration of the sublining layer, and number and size of lymphoid follicles noted in the animals previously treated with mAb to CD18.

Two characteristics of antigen-induced arthritis were not effected by previous treatment with mAb to CD18. The first was the local production of antibody within the antigen-injected joints. Second, the local deposition of immune complexes, another feature of antigen-induced arthritis (13) was not affected by prior administration of anti-CD18 mAb. Thus, despite the treatment with the anti-CD18 mAb, a local immune response to the injected antigen developed. In the absence of infiltration of leukocytes, and especially neutrophils, during the acute phase of the arthritis and the attendant damage to articular structures, however, the local immune response was unable to produce chronic inflammatory arthritis. These findings imply that migration and local activation of neutrophils during the acute phase of the arthritis play an essential role in the development of chronic arthritis. Of note was the observation that the development of acute arthritis was blocked by mAb to CD18 but not CD54, implying a role of Leu-CAM–mediated neutrophil activation rather than adhesive interactions with ICAM-1 in this process.

Thermal Injury in the Rabbit

Burn injury initially produces an area of irreversible tissue destruction surrounded by a marginal zone with reduced blood flow (14). In the postburn period, ongoing microvascular injury in this zone surrounding the area of initial tissue damage results in extension of the area of tissue necrosis. It has been suggested that infiltration with inflammatory cells, and particularly neutrophils, and their local activation may contribute to microvascular injury following burn injury (2,3). In order to document the role of infiltrating inflammatory cells in the progressive tissue damage characteristic of burn injury and the role of specific adhesion molecules in this process, a model of burn injury in the rabbit was established (15). Under general anesthesia, two sets of three full-thickness 10×30 mm burns separated by two 5×30 mm marginal zones were produced by applying brass probes heated to 100°C for 30 s to the shaved backs of New Zealand white rabbits. Blood flow was measured using a laser doppler blood flow meter equipped with an integrating flow probe (Perimed Inc., Piscataway, NJ). The probe contained seven efferent laser fibers and 14 afferent fibers and measured capillary perfusion in a tissue volume of approximately 1,200 mm^3. Baseline blood flow was measured at designated burn sites, marginal zones, and shaved unburned skin sites and repeated 1, 2, 3, 4, 24, 48, and 72 h after the burn injury. Animals received a single 2-mg/kg dose of monoclonal antibody to CD18 (15.7) or CD54 (R6.5) 30 min before the burn injury or at various times afterward.

All animals developed rapid decreases in blood flow in the burn contact sites to less than 20% of baseline blood flow, which persisted for the subsequent 3 days. By contrast, no changes were noted in blood flow at distant control skin sites. The control animals developed marked decreases

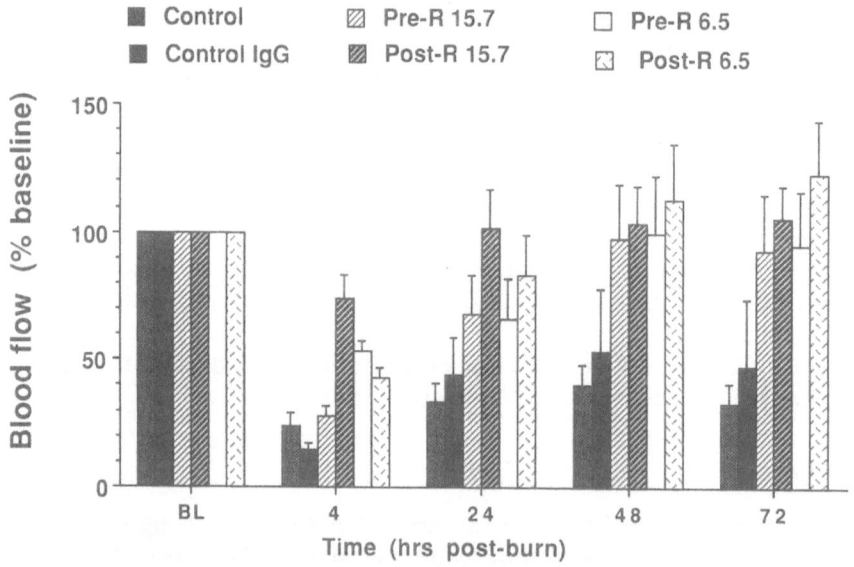

FIGURE 23.1. Monoclonal antibodies to CD18 or CD54 increase blood flow in the marginal zone adjacent to areas of direct burn injury. Relative changes in cutaneous blood flow in the marginal areas as determined by laser doppler analysis are presented as percent of baseline values. The mean (±SEM) data from 12 control animals, 5 animals given R15.7 (1 mg/kg) 30 min before burn injury, 6 animals given R15.7 (1 mg/kg) 30 min after burn injury, 6 animals given R6.5 (2 mg/kg) 30 min before burn injury and 6 animals given R6.5 (2 mg/kg) 30 min after burn injury are shown. BL = baseline.

in blood flow in the marginal sites adjacent to direct burn injury (Fig. 23.1). This was noted within 1–2 h of the burn and persisted for the subsequent 3 days. Although the animals treated with monoclonal antibodies also manifested initial decreases in blood flow in the marginal zone adjacent to the area of direct burn injury, blood flow returned to normal within 4–24 h after the burn and remained normal during the length of the experiment. Similar results were noted after administration of monoclonal antibodies to either CD18 or CD54 and when the monoclonal antibodies were administered before the burn or as late as 3 hours afterward.

Visual evidence of extension of tissue necrosis into the marginal zone to the point that the burn contact sites became confluent was observed in 18 of 44 areas (41%) in the control animals but in only 8 of 76 areas (11%) in the animals treated with monoclonal antibodies. There was no evidence of infection in any of the groups. Histologic evaluation of the marginal zone of control animals revealed marked edema, diffuse infiltration with inflammatory cells, necrosis of the epidermis, loss of dermal appendages, vascular damage, and inflammation in the subdermal muscle. Examination of marginal zones of animals treated with the monoclonal antibodies revealed only edema of the dermis.

In summary, administration of monoclonal antibodies to either CD18 or CD54 either before or soon after burn injury exerts a dramatic protective effect on microvascular injury and progressive tissue necrosis in the marginal zone surrounding the burn contact site. These results strongly imply that infiltration of inflammatory cells into this region mediated by adhesive interactions involving CD18 and CD54 plays a critical role in progressive tissue damage surrounding initial burn injury.

Conclusion

Adhesive interactions involving CD18 and/or CD54 play a critical role in two models of tissue injury in the rabbit. In antigen-induced arthritis, CD18-, but not CD54-, mediated interactions during the acute phase of the process play a critical role in permitting the subsequent local immune response to induce chronic arthritis. In burn injury, adhesive interactions involving both CD18 and CD54 permit infiltration of inflammatory cells leading to microvascular damage and necrosis of tissue adjacent to the original site of burn damage. The results obtained from both of these models suggest that interventions directed at preventing these adhesive interactions could be beneficial in cases of human tissue injury exemplified by these animal models.

Acknowledgments. We are appreciative to Dr. Robert Rothlein for providing the monoclonal antibodies used in these studies.

278   Peter E. Lipsky et al.

References

1. Weiss, S.J. 1989. Tissue destruction by neutrophils. *N. Engl. J. Med.* 320:365–376.
2. Deitch, E.A., Q. Lu, D.Z. Xu, and R.D. Specian. 1990. Effect of local and systemic burn microenvironment on neutrophil activation as assessed by complement receptor expression and morphology. *J. Trauma* 30:259–268.
3. Moore, F.D. Jr., C. Davis, M. Rodrick, J.A. Mannick, and D.T. Fearon. 1986. Neutrophil activation in thermal injury as assessed by increased expression of complement receptors. *N. Engl. J. Med.* 314:948–953.
4. Cush, J., and P.E. Lipsky. 1991. Cellular basis for rheumatoid inflammation. *Clin. Orthopaedics Rel. Res.* 265:9–21.
5. Haynes B.F., L.P. Hale, S.M. Denning, P.T. Le, and K.H. Singer. 1989. The role of leukocyte adhesion molecules in cellular interactions: implications for the pathogenesis of inflammatory synovitis. *Springer Semin. Immunopathol.* 11:163–185.
6. Oppenheimer-Marks, N., L.S. Davis, and P.E. Lipsky. 1990. Human T lymphocyte adhesion to endothelial cells and transendothelial migration. Alteration of receptor use relates to the activation status of both T cell and endothelial cell. *J. Immunol.* 145:140–148.
7. Springer, T.A., M.L. Dustin, T.K. Kishimoto, and S.D. Marlin. 1987. The lymphocyte function-associated LFA-1, CD2, and LFA-3 molecules: cell adhesion receptors of the immune system. *Ann. Rev. Immunol.* 5:223–252.
8. Dumonde D.C., and L.E. Glynn. 1982. The production of arthritis in rabbits by an immunological reaction to fibrin. *Br. J. Exp. Pathol.* 43:373–383.
9. Jasin H.E. 1988. Chronic arthritis in rabbits. Methods in Enzymology. *Edited by G. DiSabato.* 162:379–385.
10. Entman M.L., K. Youker, S.B. Shappell, C. Siegel, R. Rothlein, W.J. Dreyer, F.C. Schmalstieg, and C.W. Smith. 1990. Neutrophil adherence to isolated adult canine myocytes. Evidence for a CD18-dependent mechanism. *J. Clin. Invest.* 85:1497–1506.
11. Jasin H.E., E. Lightfoot, L.S. Davis, R. Rothlein, R.B. Faanes, and P.E. Lipsky. 1992. Amelioration of antigen induced arthritis in rabbits treated with monoclonal antibodies to leukocyte adhesion molecules. *Arthritis Rheum,* 35:541–549.
12. Smith, C.W., R. Rothlein, B.J. Hughes, M.M. Mariscalco, F.C. Schmalsteig, and D.C. Anderson. 1988. Recognition of an endothelial determinant for CD18-dependent human neutrophil adherence and transendothelial migration. *J. Clin. Invest.* 82:1746–1756.
13. Cooke, T.D., E.R. Hurd, M. Ziff, and H.E. Jasin. 1972. The pathogenesis of chronic inflammation in experimental antigen-induced arthritis. II. Preferential localization of antigen-antibody complexes to collagenous tissues. *J. Exp. Med.* 135:323–328.
14. Boykin, J.F., E. Eriksson, and R.N. Pittman. 1980. In vivo microcirculation of a scald burn and the progression of postburn dermal ischemia. *Plastic Reconstruct. Surg.* 66:91–198.
15. Mileski, W., D. Borgstrom, E. Lightfoot, R. Rothlein, R. Faanes, P. Lipsky, and C. Baxter. 1992. Inhibition of leukocyte-endothelial adherence following thermal injury. *J. Surg. Res.* (in press).

24
Mechanisms and Consequences of Leukocyte Adhesion

Robert K. Winn, Sam R. Sharar, William J. Mileski, Charles L. Rice, and John M. Harlan

Introduction

Stimulation of leukocytes in acute inflammation can lead to vascular injury. Neutrophils (PMNs) are the most prominent cells at inflammatory sites during the early stages of inflammation and are thought to be responsible for injury. Vascular injury resulting from activated PMNs might result from one of the two pathways shown schematically in Figure 24.1. The lower pathway shows injury resulting from PMN adherence to endothelial cells, formation of a protected microenvironment, release of injurious inflammatory molecules, and subsequent tissue injury (1). The upper pathway shows PMN–PMN and PMN–endothelial cell adherence leading to vascular occlusion, downstream ischemia, and tissue damage as a result of the ischemia (2). The CD11/CD18 glycoprotein complex on PMNs forms one set of adhesion molecules that can produce both aggregation and adherence following in vitro stimulation (3). Additionally, the CD11/CD18 complex is necessary for PMN emigration into tissue in response to certain stimuli (e.g., PMA, LTB4) in vivo (4). Each of the three elements of the CD11/CD18 glycoprotein complex consists of an α-chain noncovalently linked to a lighter β-chain. There are three distinct α-chains each linked to a β-chain that is identical for each component. The three complexes have been designated CD11a/CD18, CD11b/CD18, and CD11c/CD18 and are commonly called LFA-1; Mac-1 or Mo-1; and p150,95, respectively. CD11a, CD11b, andCD11c are the α-chains and CD18 is the common β-chain. The CD11/CD18 complex is a member of the integrin family of molecules and is sometimes called the β_2 family.

Monoclonal antibodies (MAbs) directed to functional epitopes of the CD11/CD18 glycoprotein complex have been shown to be effective in preventing organ injury in our laboratory as well as other laboratories. We have shown the CD18 MAb 60.3 to be effective in preventing injury following ischemia-reperfusion of rabbit ears (5), following hemorrhagic shock (6–8), and in sepsis (unpublished results). Others have shown tissue protection with CD18 MAbs in bacterial meningitis (9) and ischemia-

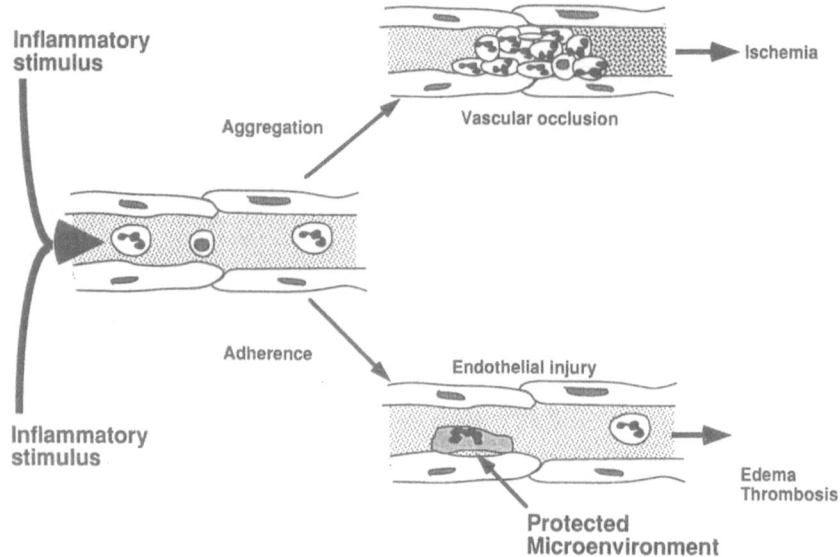

FIGURE 24.1. Schematic representation of possible mechanisms of PMN-mediated organ injury following an inflammatory stimulus. Upper pathway shows PMN–PMN and PMN–endothelial cell adherence resulting in vascular occlusion and ischemia. Lower pathway shows formation of protected microenvironment allowing inflammatory molecules to exceed antiinflammatory molecules.

reperfusion of the cat gut (10). The CD11b MAb 904 was shown to reduce the size of myocardial infarcts (as a % of tissue at risk) in dogs (11). The exact mechanism of protection (blocking aggregation or adherence) has not been established; however, these MAbs are effective in their ability to provide protection.

There are several other molecules that can produce leukocyte adherence in vitro. These include the selectin family of molecules E-selectin, L-selectin, and P-selectin, commonly called endothelial leukocyte adhesion molecule–1 (ELAM-1), leukocyte adhesion molecule–1 (LAM-1, LECCAM-1, MEL-14 antigen, LEU 8), and granular membrane protein–140 (GMP-140, PADGEM, CD62), respectively. The exact function of these molecules has not been completely defined but they appear to mediate leukocyte rolling in vivo (12,13) and in vitro (14); that is, they may initially slow the PMNs before a stronger PMN–endothelial cell interaction occurs via CD11/CD18 that allows leukocyte emigration. The selectins are probably involved in PMN emigration primarily by their influence on rolling although PMN binding to E-selectin has also been implicated in activation of CD11b/CD18 (15). Doershuck et al. (16) have reported a CD18-independent mechanism of PMN emigration into the lung in response to *Streptococcus pneumoniae*. In that study, PMN emigration was CD18-dependent when *S. pneumoniae*–soaked sponges were placed in the abdominal wall but not when the same strain of bacteria was instilled into the lungs. The difference in mechanism of emigration between these two sites could result from differences in pulmonary endothelial cells compared with systemic endothelial cells. Alternatively, the lung is rich in tissue macrophages (Mϕ) and it is possible that these cells, when stimulated, provide a secondary stimulus to PMNs that leads to CD18-independent PMN emigration. Mϕs are known to produce a variety of substances that increase neutrophil or endothelial cell adhesiveness including LTBs, ILs, TNF, PAF, etc. A major component of PMN emigration in response to these molecules has been shown to be CD18-dependent. However, it is possible that one or more of the plethora of mediators produced by Mϕ are capable of inducing CD18-independent emigration. We investigated the role of the Mϕ in CD18-independent PMN emigration resulting from stimulation with *S. pneumoniae*. These experiments were conducted by manipulating the macrophage population in the peritoneal space of rabbits (17) prior to eliciting leukocyte emigration by instillation of bacteria into the peritoneum.

Two additional sets of experiments examining the potential adverse effects of CD18 blockade on host defense also considered the role of Mϕ. These experiments examined the consequences of CD18 blockade in the presence of bacterial infection of skin where there were few resident Mϕ and in the peritoneum where there is a small resident Mϕ population (17,18).

Results (PMN Emigration)

Peritoneal Macrophages (17)

New Zealand White rabbits were studied in these experiments following a variety of stimuli. The majority of the experiments were performed by placing a stimulus in the peritoneal space of rabbits. The stimulus consisted of 10 ml of 10^9 colony-forming units (CFU)/ml of either live *S. pneumoniae* or *E-scherichia coli* instilled into the peritoneum. The *S. pneumoniae* experiments consisted of four groups of rabbits. Group 1 animals had a normal macrophage population in their peritoneum. The macrophage population was increased in group 2 by instilling a solution of 10 g/100 ml of protease peptone dissolved in sterile saline into the peritoneum. This resulted in an increased cell population at 96 h that was greater than 95% macrophages. Experiments in group 2 animals were started 96 h after protease peptone instillation. Group 3 animals had their peritoneum enriched as in group 2 and then the macrophages were removed by lavage with warm saline just prior to starting the experiment. These harvested macrophages were concentrated by centrifugation and then instilled into group 4 animals just prior to the experiment. The *E. coli* experiments consisted of only two groups. The first had normal peritoneal counts and the second were identical to group 2 above.

Peritoneal Emigration (17)

Inflammation was induced in the peritoneal space by instillation of the bacteria (10 ml of 10^9 bacteria/ml) through a small laparotomy. Evaluation of cellular content was determined by peritoneal lavage with 50 ml of normal saline containing 100 U of heparin per ml. Total and differential cell counts were performed on the lavage fluid. Animals within each group described above were assigned to treatment with either saline or CD18 MAb.

The Mϕ population in the peritoneum of normal rabbits (group 1) was relatively small, averaging $16 \pm 3.5 \times 10^3$ Mϕ/ml. Enrichment with protease peptone (group 2) increased this number approximately 20-fold to $316 \pm 70 \times 10^3$ Mϕ/ml. Subsequent reduction by lavage (group 3) reduced the Mϕ population back to approximately normal levels at $10 \pm 1.7 \times 10^3$ Mϕ/ml. Transplantation of Mϕ (group 4) into the peritoneum resulted in $130 \pm 13.9 \times 10^3$ Mϕ/ml or close to 10 times the normal population.

Effect of Bacterial Stimulation on Peritoneal Mϕ

The results of experiments following *E. coli* or *S. pneumoniae* bacterial instillation are shown in Figure 24.2. The PMN population was significantly increased 4 h after instillation in the saline-treated (group 1) animals. MAb 60.3 treatment resulted in a reduction of PMN emigration of approximately 86% in response to *E. coli* and 88% in response to *S. pneumoniae* in these group 1 rabbits. MAb 60.3 was also effective in reducing PMN emigration in the Mϕ-enriched (group 2) rabbits in response

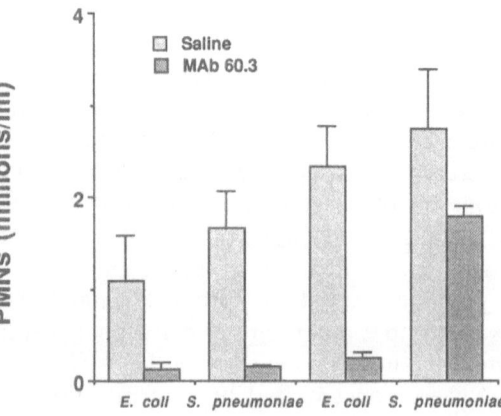

FIGURE 24.2. Peritoneal lavage neutrophils following stimulation with either *E. coli* or *S. pneumoniae* bacteria in normal rabbits (left pair) and in rabbits having their peritoneal macrophage population increased (right pair). Animals were treated with either MAb 60.3 or saline at the time of instillation of bacteria.

to *E. coli* stimulus. This MAb resulted in an 88% reduction when compared with similar group 2 animals treated with saline. The reduction in PMN emigration by MAb 60.3 following *S. pneumoniae* stimulus in the Mφ-enriched peritoneum (group 2) was far less effective. The MAb produced only a 36% inhibition in emigration compared with group 2 animals treated with saline.

Peritoneal Emigration, Mφ-Depleted and Mφ-Transplanted

Live *S. pneumoniae* bacteria were the only stimuli used in these experiments. These bacteria were instilled through a laparotomy as previously described and the number of PMNs in peritoneal lavage fluid was determined. Animals were treated with either saline or MAb 60.3 at the time of instillation of bacteria. The results from Mφ enhancement followed by Mφ depletion (group 3) and Mφ-transplanted (group 4) animals are shown in Figure 24.3 together with the normal (group 1) and Mφ-enriched (group 2) *S. pneumoniae* groups from Figure 24.2. *S. pneumoniae* produced a significant increase in PMNs in the peritoneum of saline-treated animals in both the Mφ-depleted and the Mφ-transplanted groups. There was an 88% reduction in the number of lavage PMNs in group 3 animals receiving MAb 60.3 compared with the saline-treated group. MAb 60.3 resulted in only a 48% reduction in lavage PMNs in group 4 animals compared with equivalent saline-treated animals. The differences were not statistically significant when saline-treated animals were compared with MAb 60.3–treated animals of the Mφ-transplanted groups. In addition, there were no differences between the Mφ-enriched and the Mφ-transplanted groups. However, these groups were significant-

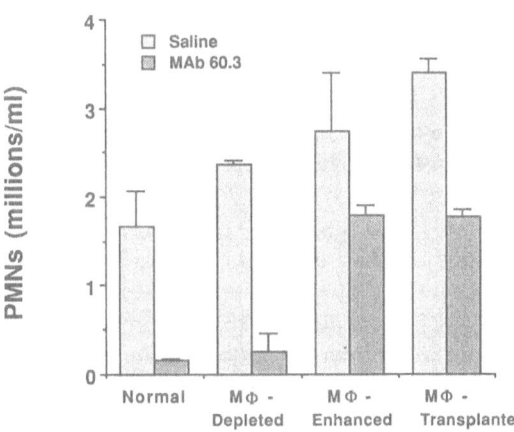

FIGURE 24.3. Peritoneal lavage neutrophils following stimulation with *S. pneumoniae* bacteria. Four groups of animals with different manipulations of their peritoneal macrophage populations are shown. (See text for details.)

ly different from both the Mϕ-enriched-then-depleted group and the normal group.

These data of PMN emigration into the peritoneum show that both *S. pneumoniae* and *E. coli* cause PMN emigration that is largely blocked by the CD18 MAb 60.3 in the absence of Mϕ (group 1 and group 3). However, when Mϕs are present in significant numbers (group 2 and group 4) PMN emigration in response to *S. pneumoniae* can continue even when the MAb 60.3 is given. The emigration toward *E. coli* in the presence of Mϕs (group 2) was blocked by MAb 60.3.

Results (Infections)

Peritonitis (19)

Rabbits were made septic by interrupting the blood supply to their appendix, leaving it in situ, and then performing an appendectomy at 18 h. Treatment with either saline or MAb 60.3 was given at the time of devascularization and again at the time of appendectomy. These animals were followed for 10 days. Infectious complications were monitored and survival curves constructed. Cell counts and bacterial cultures were performed using samples of peritoneal fluid taken at the time of appendectomy. Abscess formation in the peritoneum and wound infections were used to describe infectious complications.

There were ten animals in each of the treatment groups and bacterial cultures from all animals were positive. The half-life of MAb 60.3 is dose-dependent with a dose of 2 mg/kg resulting in a half-life of 12–18 h (20). The plasma concentration resulting from this dose was shown (20) to be effective in preventing PMN emigration into lipopolysaccharide-soaked sponges when evaluated 24 h after implantation. Thus, plasma concentration of MAb 60.3 should exceed saturation levels at the time of appendectomy.

The peritoneum of normal rabbits is essentially free of neutrophils. Devascularization of the appendix resulted in peritoneal fluid containing an average of $22.7 \pm 6.1 \times 10^6$ PMNs/ml. The MAb 60.3–treated animals had a PMN count that averaged $8.7 \pm 3.3 \times 10^6$ PMNs/ml in their peritoneum fluid. There were more cells identified as monocytes in the MAb 60.3–treated group ($20.3 \pm 4.47 \times 10^6$ cells/ml) than in the saline-treated group ($10.5 \pm 5.92 \times 10^6$ cells/ml) at appendectomy. Cell counts were performed on fluid that could be aspirated from the peritoneum, and it was beyond the scope of these studies to investigate the mechanism of emigration. It is clear that MAb 60.3 prevented some, but not all, migration toward the septic focus following appendiceal devascularization.

Mortality was not different between these two groups of animals as de-

FIGURE 24.4. Mortality curves for rabbits after inducing peritoneal sepsis by devascularizing their appendix and leaving it in situ for 19 h. Half of the animals were treated with saline and half with MAb 60.3.

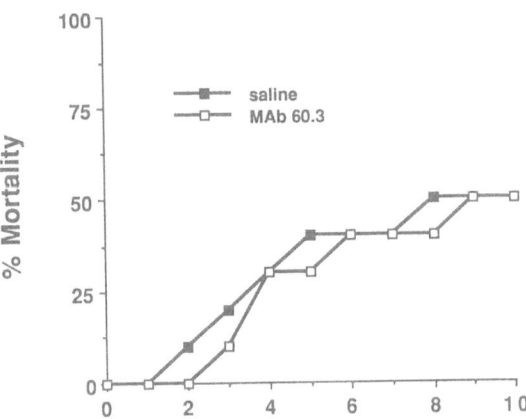

termined by analysis of Kaplan-Meier curves and testing for significance by log-rank test. Mortality curves shown in Figure 24.4 reveal that both groups of animals had a 50% mortality at 10 days and there was no difference in the time of death between the two groups.

Subcutaneous Abscess (18)

Staphylococcus aureus bacteria were suspended in 1 ml of normal saline and injected subcutaneously into four sites on the backs of rabbits. Half of the injections sites received 10^8 colony-forming units (CFU) and the other half received 10^9 CFU. The *S. aureus* was sensitive to the antibiotic cefazolin and rabbits were given 20 mg/kg every 8 h for 72 h. Each rabbit was treated with a single injection of either MAb 60.3 (2 mg/kg) or saline (equal volume) at the time of bacterial injection. These animals were followed for 7 days and then killed. The number and size of the abscesses were then determined.

All saline-treated animals survived the 7 days of the experiment but two of eight MAb 60.3–treated animals died within 48 h of bacterial injection. Postmortem examination of the 10^8 CFU sites revealed a 6.2% incidence of abscess in the saline-treated group and a 66.7% incidence in the MAb 60.3–treated group. At 10^9 CFU there was an incidence of 75% and 100% in the saline- and MAb 60.3–treated groups, respectively. The surface area of abscesses at the 10^8 CFU sites (i.e., the planar area projected by the outline of the abscess) was 0.09 ± 0.15 cm^2 in saline-treated animals and 3.22 ± 4.55 cm^2 in the MAb 60.3–treated group. At 10^9 CFU, abscess area was 2.22 ± 2.48 cm^2 and 41.49 ± 27.47 cm^2 for saline- and MAb 60.3–treated animals, respectively.

Histologic examination of cellular infiltration on day 1 in saline-treated animals revealed a rich subcutaneous accumulation of PMNs and a smaller (approximately 5%) accumulation of mononuclear cells. By way of contrast, the MAb 60.3–treated animals had no PMNs on day 1 and very few mononuclear cells. Thus, MAb 60.3 was effective in preventing emigration of both PMNs and mononuclear cells into skin.

Discussion

The importance of the CD11/CD18 glycoprotein complex in PMN function can be found by examination of the leukocyte-adherence-deficient (LAD) patients (reviewed in 21). These patients are deficient in functional CD11/CD18 complex and suffer from recurrent life-threatening soft-tissue infections that are characterized by the absence of PMNs at the site of infections. PMNs from these patients have essentially identical function to normal PMNs treated with MAbs directed to CD18. It has been assumed that CD11/CD18 adherence is necessary for PMN emigration in response to bacterial stimuli since suppuration does not occur in the soft-tissue infection of the LAD patients. In the experiments described above, PMN emigration following stimulation with live *E. coli* or *S. pneumoniae* resulted primarily from a CD18-dependent pathway in the normal peritoneum. This was altered for *S. pneumoniae* when the number of Mφs was increased by either protease peptone or by transplanting these cells. These manipulations resulted in a significant non-CD18 pathway of neutrophil emigration.

The CD18-independent pathway could have been stimulated by the protease peptone cleavage of a chemotactic product that stimulates a CD18–independent mechanism. However, the transplanted macrophages were washed prior to their being instilled into the peritoneum of recipient animals; thus any molecule that might be carried along with these cells would be significantly reduced in the process. Therefore, it appears that the Mφs produce a molecule that stimulates the CD18-independent pathway.

The CD18-independent pathway may involve a direct stimulation of PMNs or may be secondary to stimulation of endothelial cell adhesion molecules. It seems unlikely that the adhesion molecule involved is a member of the lectin family, as these molecules appear to be responsible for rolling but not emigration. The stimulant may be a monokine produced by Mφs or a breakdown product of the *S. pneumoniae*. This second possibility seems reasonable since *S. pneumoniae*, but not *E. coli*, produced the CD18-dependent process. It is unlikely that the substance stimulating CD18-independent emigration is either tumor necrosis factor (TNF) or interleukin-1 (IL-1) as the Mφs release TNF in response to both *E. coli* and *S. pneumoniae* (unpublished observations) and PMN emigration in response to IL-1 is CD18-dependent (22).

Peritonitis

Adverse effects of transient inhibition of CD18 leukocyte adherence were not seen in experimental peritonitis. There was no difference in mortality or infectious complications in animals treated with MAb 60.3 and those treated with saline. The MAb 60.3–treated animals had an unexpectedly high number of PMNs in their peritoneal space even with MAb in sufficient quantities to block CD18-dependent emigration. In addition, there were significantly more monocytes in the MAb 60.3–treated group compared with saline-treated group. Monocytes are known to have a more complex mechanism of adherence and emigration than PMNs. The β_1 integrin, VLA-4, as well as the β_2 family of integrin receptors, contributes to monocyte cell adherence and emigration. The fact that more monocytes emigrate into the peritoneum in the presence of MAb 60.3 is somewhat surprising, since it was expected that at least a portion of their emigration would result from a CD18 pathway. We have also observed this increased monocyte emigration into the peritoneum in the presence of MAb 60.3 following protease peptone instillation (unpublished results). The increased monocyte emigration in these circumstances is unknown.

Staphylococcus aureus

In saline-treated rabbits subcutaneous injection of S. aureus provoked a brisk infiltration of PMNs as well as some Mϕs. In contrast, the site of infection in the MAb 60.3–treated rabbits did not exhibit any significant monocyte or PMN accumulations 24 h after injection of bacteria. Infiltration of inflammatory cells in saline-treated animals is consistent with previously reported cell migration following injection of bacteria (23). It is clear from these experiments that cellular emigration in response to S. aureus is CD18-dependent in subcutaneous tissue. This site in rabbits is essentially devoid of macrophages under normal conditions (unpublished observations); thus stimulation of CD18- independent emigration such as occurs in Mϕ-rich lung and Mϕ-enriched peritoneum was not possible (16,17). Thus the possible cellular source for the mediator producing CD18-independent emigration was not present. It is apparent that the stimulus for CD18-independent emigration of Mϕs was not present.

Subcutaneous injection of high-dose S. aureus resulted in severe soft-tissue infections that were similar to those seen in the LAD patient. The inoculum of bacteria at 10^8 and 10^9 CFU may not be clinically relevant (18). Nevertheless, inhibition of phagocyte emigration can lead to severe complications with more frequent and larger abscesses in the MAb 60.3–treated animals.

Summary

We have shown that under some conditions PMNs emigrate by a CD18-independent pathway. Doerschuk et al. found emigration in response to *S. pneumoniae* to be CD18-independent in lung but CD18-dependent in subcutaneous tissue (16). They suggested that the difference between emigration could be from differences in the endothelial or resident cell types in the tissue. Mileski et al. documented that a CD18-independent pathway could be induced in the peritoneum by eliciting Mφ, suggesting that Mφ produce a factor that elicits CD18-independent emigration (17).

The results from abdominal sepsis are also consistent with the hypothesis that PMNs can emigrate to the site of infection by a CD18-independent pathway. This pathway may provide sufficient, but much reduced, numbers of PMNs to contain the infectious processes in the peritoneum at least temporarily. The accumulation of Mφ following CD18-independent emigration of monocytes, together with the presence of a small number of Mφs at the start of the infectious process, may provide a signal that initiates CD18-independent emigration of PMNs. On the other hand, subcutaneous tissue has few tissue Mφs and is unable to elicit either monocytes or PMNs in sufficient numbers to protect the host from soft-tissue infections.

The stimulus for CD18-independent emigration of PMNs and the adhesion molecule(s) responsible for emigration are yet to be identified. Their relative importance in performing host defense functions, as well as their role in causing inflammatory cell tissue damage, also remains to be determined.

References

1. Harlan, J.M. 1987. Neutrophil-mediated vascular injury. *Acta Med. Scan. Suppl.* 715:123–129.
2. Barroso-Aranda, J.,G. Schonbein, B. Zweifach, R. Engler. 1988. Granulocytes and no-reflow phenomenon in irreversible hemorrhagic shock. *Circ. Res.* 63:437–447.
3. Schwartz, B.R., H.D. Ochs, P.G. Beatty, J.M. Harlan. 1985. A monoclonal antibody-defined membrane antigen complex is required for neutrophil-neutrophil aggregation. *Blood* 65:1553–1556.
4. Harlan, J.M., B.R. Schwartz, W.J. Wallis, T.H. Pohlman. 1987. The role of neutrophil membrane proteins in neutrophil emigration. In: Movat, H.Z., ed. Leukocyte Emigration and Its Sequelae. Basel: Krager, 94–104.
5. Vedder, N.B., R.K. Winn, C.L. Rice, E. Chi, K.-E. Arfors, J.M. Harlan. 1990. Inhibition of leukocyte adherence by anti-CD18 monoclonal antibody attenuates reperfusion injury in the rabbit ear. *Proc. Natl. Acad. Sci. USA* 81:939–944.
6. Vedder, N.B., B.W. Fouty, R.K. Winn, J.M. Harlan, C.L. Rice. 1989. Role

of neutrophils in generalized reperfusion injury associated with resuscitation from shock. *Surgery* 106:509–516.

7. Vedder, N.B., R.K. Winn, C.L. Rice, E. Chi, K.-E. Arfors, J.M. Harlan. 1988. A monoclonal antibody to the adherence promoting leukocyte glycoprotein CD18 reduces organ injury and improves survival from hemorrhagic shock and resuscitation in rabbits. *J. Clin. Invest.* 81:939–944.

8. Mileski, W.J., R.K. Winn, N.V. Vedder, T.H. Pohlman, J.M. Harlan, C.L. Rice. 1990. Inhibition of CD18-dependent neutrophil adherence reduces organ injury after hemorrhagic shock in primates. *Surgery* 108:205–212.

9. Tuomanen, E.I., K. Saukkonen, S. Sande, C. Cioffe, S.D. Wright. 1989. Reduction of inflammation, tissue damage, and mortality in bacterial meningitis in rabbits treated with monoclonal antibodies against adhesion-promoting receptors of leukocytes. *J. Exp. Med.* 170:959–968.

10. Hernandez, L.A., M.B. Grisham, B. Twohig, K.-E. Arfors, J.M. Harlan, D.N. Granger. 1987. Role of neutrophils in ischemia-reperfusion induced microvascular injury. *Am. J. Physiol.* 253:H699–H703.

11. Simpson, P.J., R.F. Todd. III, J.C. Fantone, J.K. Mickelson, J.D. Griffin, B.R. Lucchesi. 1988. Reduction of experimental canine myocardial reperfusion injury by a monoclonal antibody (Anti-Mo-1, Anti-CD11b) that inhibits leukocyte adhesion. *J. Clin. Invest.* 81:624–629.

12. Ley, K., P. Gaehtgens, C. Fennie, M.S. Singer, L.A. Lasky, S.D. Rosen. 1991. Lectin-like cell adhesion molecule 1 mediates leukocyte rolling in mesenteric venules in vivo. *Blood* 77(12):2553–2555.

13. von Andrian U.H., J.D. Chambers, L.M. McEvoy, R.F. Bargatze, K.E. Arfors, E.C. Butcher. 1991. Two-step model of leukocyte-endothelial cell interaction in inflammation: distinct roles for LECAM-1 and the leukocyte beta 2 integrins in vivo. *Proc. Natl. Acad. Sci. USA* 88(17):7538–7542.

14. Smith, C., K. Takashi, O. Abbass, B. Huges, R. Rothlein, L. McIntire, E. Butcher, D. Anderson. 1991. Chemotactic factors regulate lectin adhesion molecule 1 (LECAM-1)-dependent neutrophil adhesion to cytokine-stimulated endothelial cells in vitro. *J. Clin. Invest.* 87:609–618.

15. Lo, S.K., S. Lee, R.A. Ramos, R. Lobb, M. Rosa, R.G. Chi, S.D. Wright. 1991. Endothelial-leukocyte adhesion molecule 1 stimulates the adhesive activity of leukocyte integrin CR3 (CD11b/CD18, Mac-1, alpha m beta 2) on human neutrophils. *J. Exp. Med.* 173(6):1493–1500.

16. Doerschuk, C.M., R.K. Winn, H.O. Coxson, J.M. Harlan. 1990. CD18-dependent and -independent mechanisms of neutrophil emigration in the pulmonary and systemic microcirculation of rabbits. *J. Immunol.* 144(6):2327–2333.

17. Mileski, W., J. Harlan, C. Rice, R. Winn. 1990. Streptococcus pneumoniae-stimulated macrophages induce neutrophils to emigrate by a CD18-independent mechanism of adherence. *Circ. Shock.* 31(3):259–267.

18. Sharar, S.,R. Winn, C. Murry, J. Harlan, C. Rice. 1991. A CD18 monoclonal antibody increases the incidence and severity of subcutaneous abscess formation after high-dose *Staphylococcus aureus* injection in rabbits. *Surgery* 110:213–220.

19. Mileski, W.J., R.K. Winn, J.M. Harlan, C.L. Rice. 1991. Transient inhibition of neutrophil adherence with the anti-CD18 monoclonal antibody 60.3 does not increase mortality rates in abdominal sepsis. *Surgery* 109:497–501.

20. Price, T., P.G. Beatty, S.R. Corpuz. 1987. In vivo inhibition of neutrophil

function in the rabbit using monoclonal antibody to CD18. *J. Immunol.* 139:4174–4177.

21. Arnaout, M. 1990. Leukocyte adhesion molecule deficiency: its structural basis, pathophysiology and implication for modulating the inflammatory response. *Immunol. Rev.* 114:145–180.

22. Rampart, M., T. Williams. 1988. Evidence that neutrophil accumulation induced by interleukin-1 requires both local biosynthesis and neutrophil CD18 antigen expression in vivo. *Br. J. Pharmacol.* 94:1143–1148.

23. Issekutz, T., A. Issekutz, H. Movat. 1981. The in vivo quantitation and kenetics of monocyte migration into acute inflammatory tissue. 103:47–55.

25
The Role of Cellular Adhesion Molecules in Acute and Chronic Airway Inflammation

Robert H. Gundel, Craig D. Wegner, and L. Gordon Letts

Introduction

The role of pro-inflammatory cells and inflammatory mediators in the pathogenesis of asthma has become a subject of profound investigation. The inflammatory process involves a highly complex series of events occurring at both the tissue and cellular levels in response to a diverse number of stimuli. Of particular interest in recent years has been the up-regulation and enhanced expression of cellular adhesion glycoproteins on vascular endothelium and circulating leukocytes that lead to leukocyte margination, activation, and subsequent migration out of the vessels and into the tissue.

Current views of the pathophysiology of allergic asthma emphasize the importance of an extensive chronic airway inflammatory component and airway hyperresponsiveness in the expression of asthmatic symptoms. This concept has developed, in part, from studies that have demonstrated that the severity of asthma, including diurnal variations in peak flow rates and the extent of therapy required to control asthmatic symptoms, correlates with the degree of airway inflammation and airway hyperresponsiveness (1–5). In addition to the chronic airway inflammation in asthma, it has been recognized for some time that acute asthmatic episodes are associated with the generation and release of potent chemotactic mediators and a transient influx of pro-inflammatory cells into the airways. Thus, cell–cell communications and interaction in the lungs may be of particular importance in the physiological expression of allergic asthma.

In this chapter we discuss the results of studies, using monoclonal antibodies (MAbs) to specific adhesion glycoproteins, designed to determine the role(s) of cellular adhesion molecules in the development of acute and chronic airway inflammation in a primate model of allergic asthma.

Results

Acute Airway Inflammation and Late-Phase Airway Obstruction

Allergen exposure to susceptible asthmatics causes the occurrence of an immediate bronchoconstriction (acute response) that is maximal at 10–20 min and resolves by 1–2 h postexposure. A subgroup of asthmatics also experience a second, more severe late-phase airway obstruction characterized by a slow onset usually beginning 4–5 h after initial allergen exposure and persisting for up to 24 h or more (6,7). The time course and severity of the late-phase response (LPR) is associated with the recruitment of pro-inflammatory cells into the lungs (8,9). These observations suggest an important role for nonresident cells that actively infiltrate and selectively release potent inflammatory mediators that are capable of contributing to the generation of the LPR.

Allergen *(Ascaris suum* extract) inhalation to *Ascaris*-hypersensitive cynomolgus monkeys induces an immediate bronchoconstriction response (acute response) that, in a subgroup of monkeys, is followed 6–8 h later

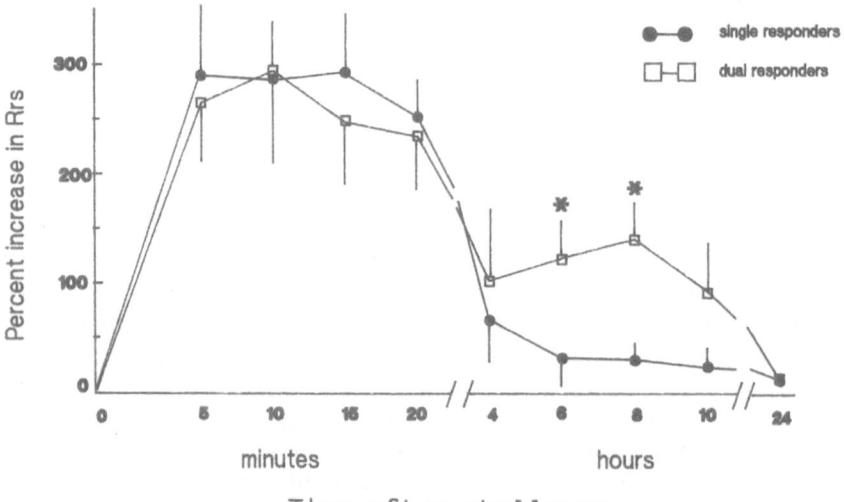

Time after challenge

FIGURE 25.1. Increases in respiratory system resistance (Rrs) during the early and late-phase airway obstruction response following inhaled antigen in single- and dual-responder monkeys. The magnitude of bronchoconstriction during the early response in single- and dual-responder monkeys was similar; however, only the dual-responder monkeys had an increase in Rrs occurring 6–8 h after antigen inhalation. * indicates statistical significance between groups, $p < 0.05$, n = 6 per group.

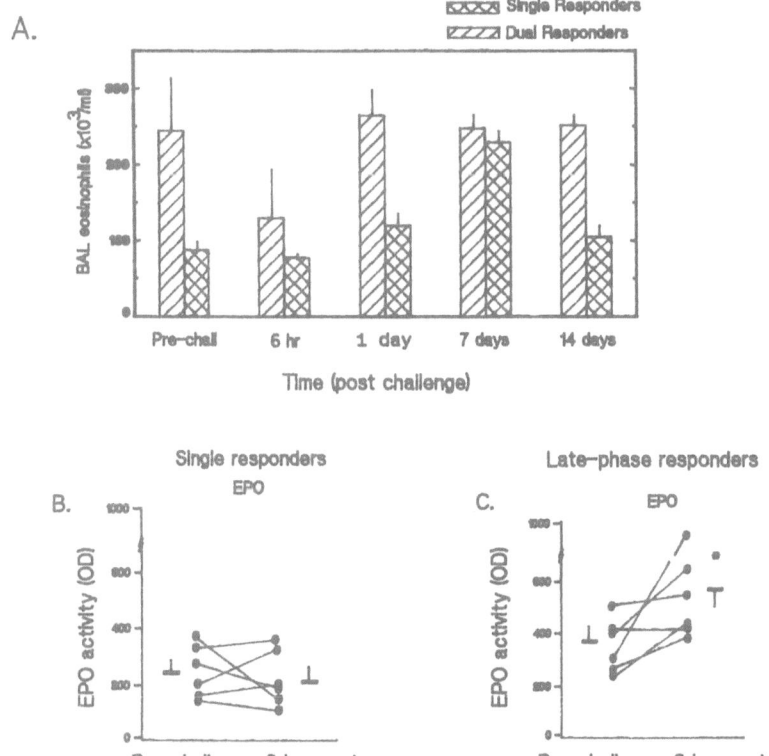

FIGURE 25.2. A. Changes in the number of neutrophils recovered in BAL fluid before and after inhaled antigen challenge in single- and dual-responder monkeys. Both single- and dual-responder monkeys had significant increases in neutrophils 6 h and 1 day after antigen inhalation; however, the magnitude of the neutrophil influx in the dual responders was significantly greater than that occurring in single-responder monkeys. B. Changes in the levels of MPO activity in BAL fluid in single-responder monkeys before and 6 h after antigen inhalation. There was no significant change in MPO activity in single-responder monkeys. C. Changes in the levels of MPO activity in BAL fluid in dual-responder monkeys before and 6 h after antigen inhalation. There was a significant increase in MPO activity 6 h after antigen challenge.

by an LPR (Fig. 25.1). Associated with and significantly correlated to the magnitude of the LPR is an influx of neutrophils into the airways and an increase in myeloperoxidase (MPO) activity in BAL fluid recovered 6 h after antigen inhalation (Fig. 25.2). Immunohistochemical staining for intercellular adhesion molecule–1 (ICAM-1) and endothelial-leukocyte adhesion molecule–1 (ELAM-1) expression and/or up-regulation performed on lung tissue from LPR monkeys (obtained by biopsy) before and 6 h after antigen inhalation indicate that ICAM-1 is constitutively expressed on both airway epithelium and vascular endothelium before antigen challenge and is not consistently upregulated at 6 h postantigen (10). In contrast, little or no staining for ELAM-1 is evident before antigen inhalation; however, 6 h postantigen ELAM-1 staining is clearly increased on vascular endothelium only. In particular, no staining for ELAM-1 was evident on airway epithelium before or after antigen inhalation.

To evaluate the roles of ICAM-1 and ELAM-1 in antigen-induced acute airway inflammation and the LPR, anti–ICAM-1 (R6.5) or anti–ELAM-1 (CL2) MAbs were administered intravenously (2 mg/kg) 1 h before antigen inhalation. Anti–ICAM-1 treatment had no significant effect on the influx of neutrophils into the airways or the associated LPR (Fig. 25.3). In contrast, anti–ELAM-1 treatment significantly reduced both the neutrophil infiltration and the LPR in all animals tested (Fig. 25.4). Thus, antigen-induced acute airway inflammation and associated late-phase airway obstruction appear to have a functional ELAM-1 dependence.

Chronic Airway Inflammation and Airway Hyperresponsiveness

In a series of recent studies we have shown that multiple antigen inhalation challenges induces an intense, prolonged airway inflammation characterized by an up-regulation of adhesion glycoproteins on both pulmonary vascular endothelium and airway epithelium, a striking prolonged airway eosinophilia, and epithelial cell desquamation, all of which are associated with the onset and intensity of airway hyperresponsiveness (11,12). Furthermore, we have shown that ICAM-1 is of particular importance in the induction of antigen-induced airway inflammation and hyperresponsiveness as pretreatment with anti–ICAM-1 MAb (R6.5), either intravenously or by inhalation, prevents the onset of both the airway inflammation and hyperresponsiveness (13,14). Thus, ICAM-1 is important in the signaling process of cells migrating into the lung and in the subsequent induction of airway inflammation and hyperresponsiveness; however, its role in chronic airway inflammation and hyperresponsiveness, as found in persistent asthma, is unclear.

To investigate further the role of ICAM-1 in preexisting chronic airway inflammation, our next series of studies was designed to examine the effects of the anti–ICAM-1 MAb R6.5 treatment in animals with existing persistent airway inflammation and hyperresponsiveness. For comparative purposes, a cross-over study design with dexamethasone was utilized.

Late-phase responder monkeys, as compared to single responders (acute response only), have significantly higher numbers of eosinophils and higher concentrations of eosinophil-derived proteins in bronchoalveolar lavage fluid (BALF) prior to *Ascaris* inhalation (Fig. 25.5). Following

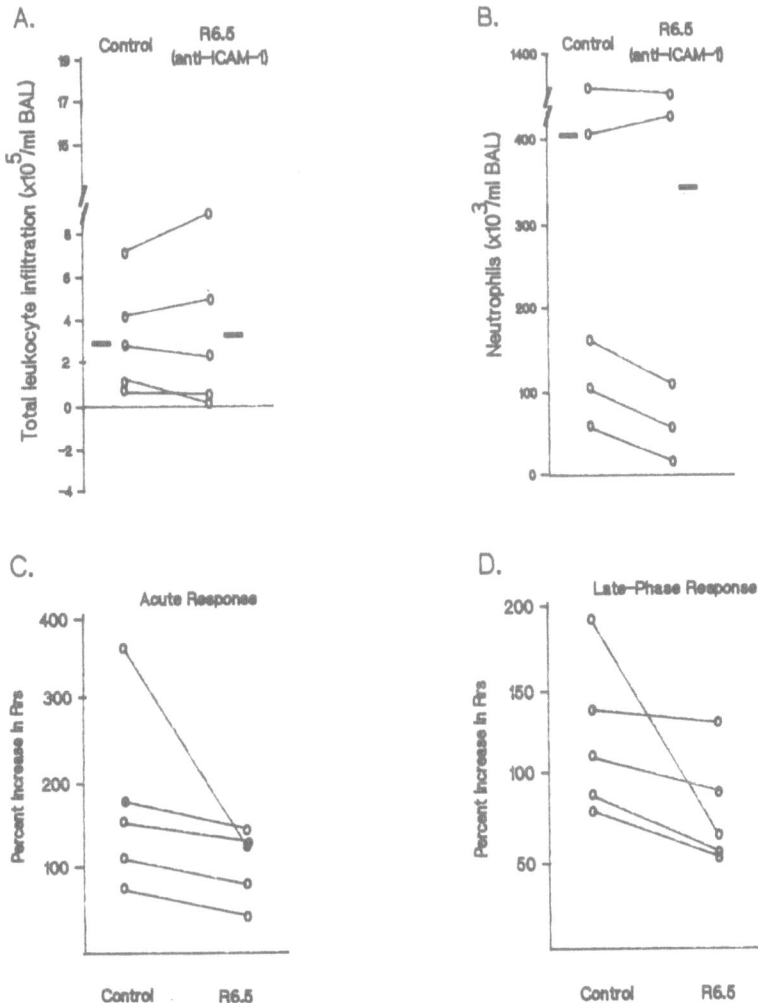

FIGURE 25.3. A. The effects of pretreatment with R6.5 (anti-ICAM-1, 2 mg/kg i.v.) on the total number of infiltrating leukocytes recovered in BAL fluid compared with vehicle control studies. R6.5 treatment had no significant effect on the total number of leukocytes recovered by BAL 6 h after antigen inhalation. B. The effects of R6.5 treatment on the number of neutrophils recovered by BAL 6 h after antigen inhalation compared to control studies. R6.5 had no effect on the number of infiltrating neutrophils during the late-phase response. C. The effects of R6.5 treatment on the acute response occurring immediately after antigen inhalation and (D) the late-phase airway obstruction occurring 6 to 8 hrs after antigen inhalation. R6.5 treatment did not significantly reduce either the acute or late-phase airway response.

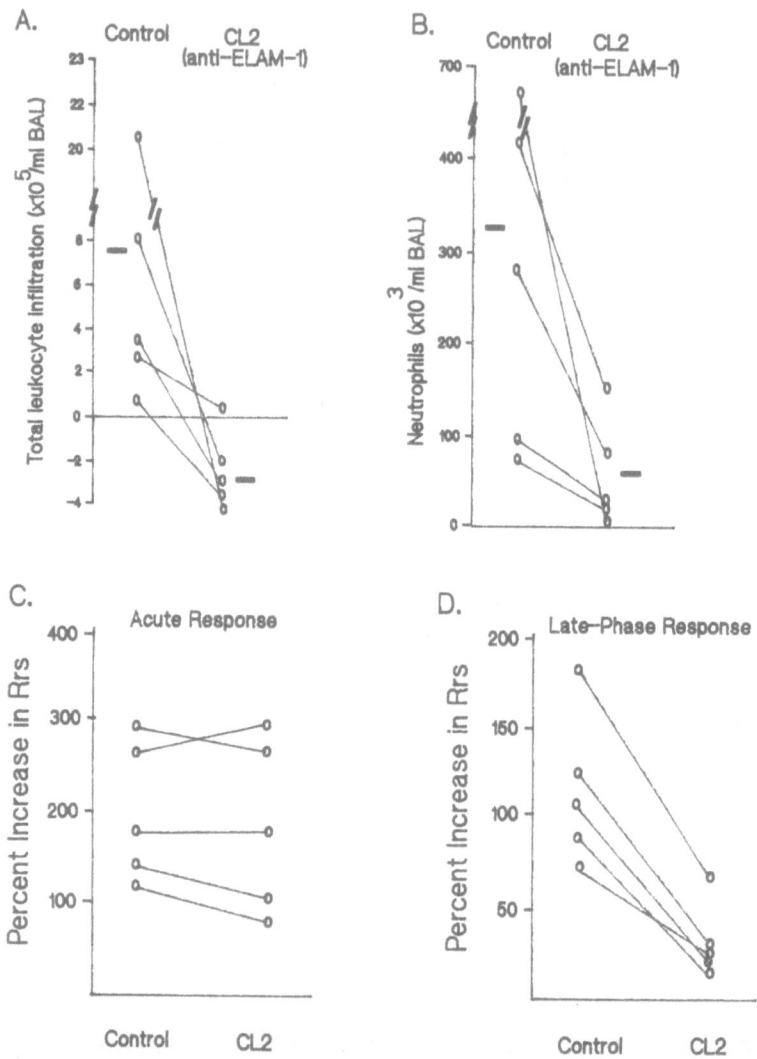

FIGURE 25.4. A. The effects of CL2 (anti–ELAM-1, 2 mg/kg, i.v.) on the total number of leukocytes and (B) the number of neutrophils infiltrating into the airways after antigen inhalation. CL2 significantly reduced both the total number of leukocytes and the number of neutrophils recovered by BAL 6 h after antigen inhalation. C. The effects of CL2 treatment on the acute response and (D) the late-phase response following antigen inhalation. CL2 did not alter the acute bronchoconstriction; however, it significantly reduced the late-phase airway obstruction.

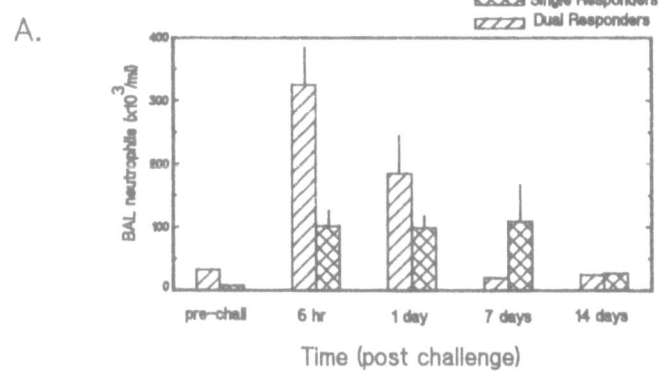

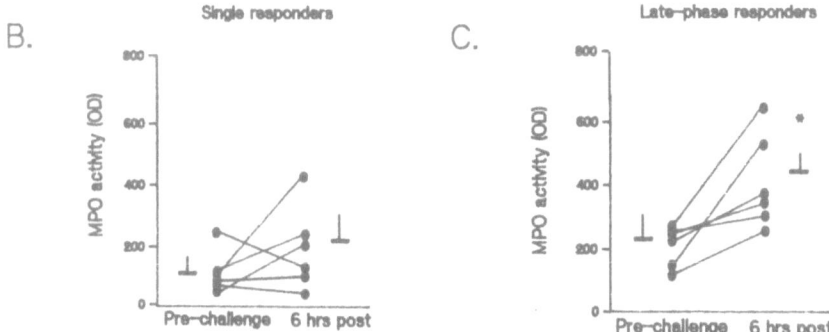

FIGURE 25.5. A. The number of eosinophils recovered by BAL before and after antigen inhalation in single- and dual-responder monkeys. Prior to antigen inhalation, dual-responder monkeys had significantly higher numbers of eosinophils in BAL fluid that remained chronically elevated to day 14 of the study. In contrast, single responders had lower levels of eosinophils before antigen challenge and a significant increase in eosinophils 7 days after challenge. The level of eosinophils returned to baseline levels by day 14 in the single-responder group. B. The level of EPO activity in BAL fluid from single-responder monkeys before and 6 h after antigen inhalation. There was no significant change in EPO activity at the 6-h time point. C. The level of EPO activity in BAL fluid from dual-responder monkeys before and 6 h after antigen inhalation. There was a significant increase in EPO activity occurring 6 h after antigen challenge.

antigen challenge, the number of airway eosinophils decreases acutely concurrent with the occurrence of the LPR, return to baseline (elevated) levels 1 day after challenge, and remain chronically increased. In addition to having chronic airway eosinophilia, LPR monkeys are hyperresponsive to inhaled methacholine when compared with normal animals.

Beginning on day 1, airway cellular composition and airway responsiveness to inhaled methacholine was determined, after which, each animal

A.

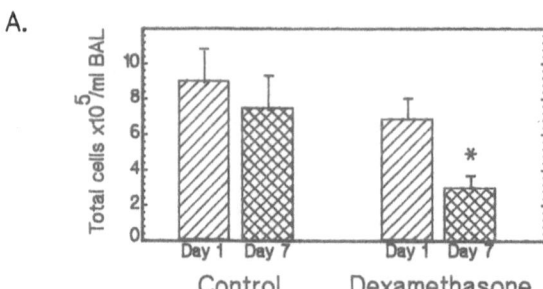

B.

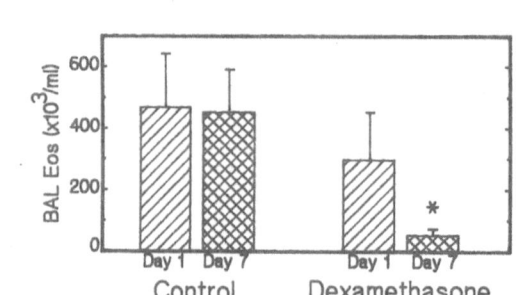

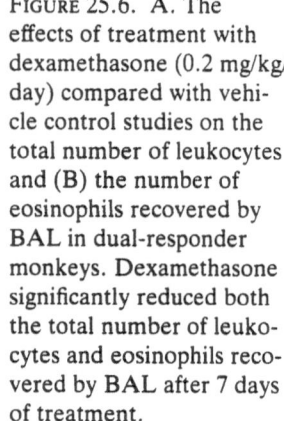

FIGURE 25.6. A. The effects of treatment with dexamethasone (0.2 mg/kg/day) compared with vehicle control studies on the total number of leukocytes and (B) the number of eosinophils recovered by BAL in dual-responder monkeys. Dexamethasone significantly reduced both the total number of leukocytes and eosinophils recovered by BAL after 7 days of treatment.

was dosed with either R6.5 (2 mg/kg/day, i.v.) or dexamethasone (0.2 mg/kg/day, i.m.) for 7 consecutive days. On day 7 of the study, airway cellular composition and airway responsiveness were again assessed for each animal. Treatment with dexamethasone resulted in a significant reduction in airway inflammation as indicated by a decrease in the total number of leukocytes and eosinophils recovered in BAL fluid (Fig. 25.6). The reduction in airway inflammation was associated with a decrease in airway responsiveness to inhaled methacholine (Fig. 25.7). In contrast, treatment with R6.5 had no effect on the existing airway inflammation or airway hyperresponsiveness (Fig 25.8 and 25.9, respectively). Thus, a 7-day treatment with R6.5 did not alter the existing airway inflammation (BAL cell count) or associated airway hyperresponsiveness.

In a separate series of studies, all animals were treated with dexamethasone as decribed above; however, at the end of the 7-day treatment with dexamethasone, each animal received a single inhaled antigen challenge followed by daily injections of R6.5 (2 mg/kg, i.v.) or saline for 7 additional days. Airway cellular composition and airway responsiveness were assessed on days 1, 7, and 14 of this study. Dexamethasone treatment again resulted in a reduction in airway inflammation and hyperresponsiveness (Fig. 25.10 and 25.11). However, cessation of dexamethasone treatment, followed by a single-antigen inhalation challenge, resulted in an increase in the leukocytes and eosinophils recovered by BAL

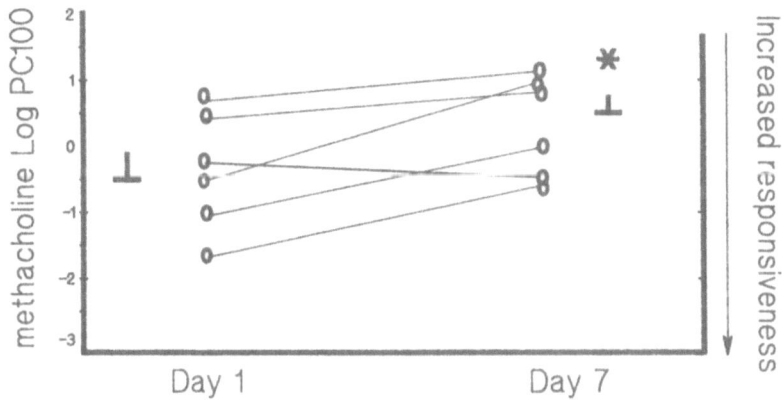

FIGURE 25.7. The effects of dexamethasone treatment on airway responsiveness to inhaled methacholine. Dexamethasone treatment caused a significant increase in methacholine PC100 values indicating that the animals were less sensitive to inhaled methacholine.

FIGURE 25.8. A. The effects of R6.5 treatment (anti–ICAM-1, 2 mg/kg/day) on the total number of leukocytes and (B) the number of eosinophils recovered by BAL. R6.5 treatment did not effect the number of leukocytes or eosinophils recovered by BAL.

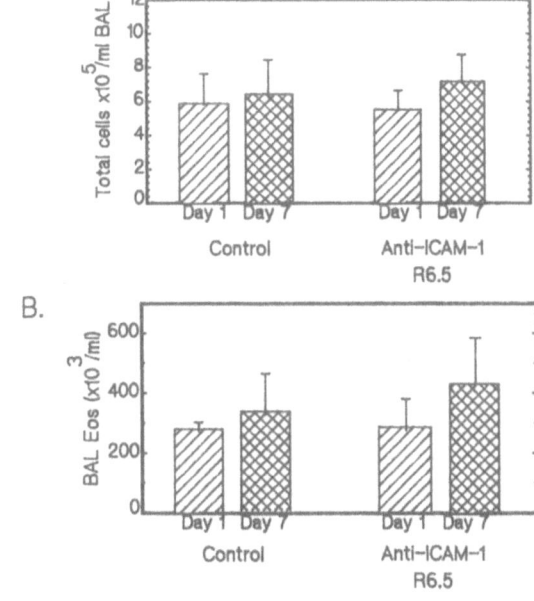

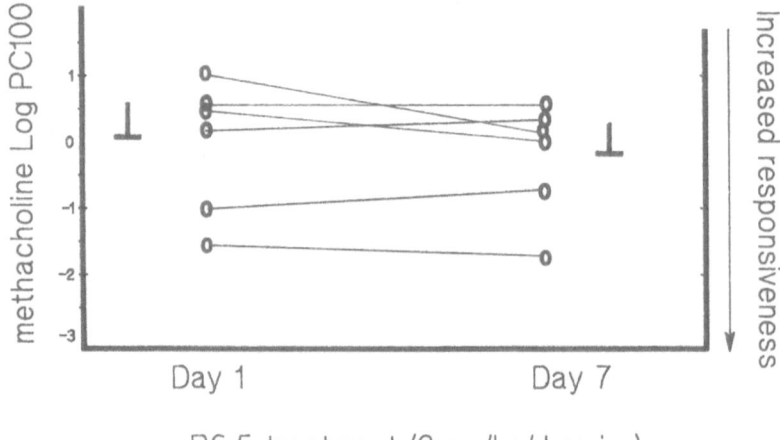

FIGURE 25.9. The effects of R6.5 (anti–ICAM-1) treatment on airway responsiveness to inhaled methacholine. R6.5 treatment did not effect airway responsiveness as indicated by no significant change in methacholine PC100 values.

as well as an increase in airway responsiveness to inhaled methacholine in the saline-treated group at day 14 of the study. In contrast, animals treated with R6.5 during days 7–14 had significantly less leukocytes and eosinophils infiltrating into the airways and no significant increase in airway responsiveness. Thus, ICAM-1 appears to play a pivotal role in the migration of newly recruited eosinophils into the airways.

Discussion

The results of these studies suggest a functional role for both ELAM-1 and ICAM-1 in the onset of antigen-induced acute and chronic airway inflammation, respectively. In allergic primates, the LPR is temporally associated with the development of an acute inflammatory response (a large influx of neutrophils into the airways). This occurs largely via an ELAM-1–dependent mechanism as MAbs against ELAM-1 inhibit both the neutrophil influx and the late-phase airway obstruction response (10). This neutrophil influx has little or no ICAM-1 dependency as MAbs to ICAM-1 had no effect. These results suggest a CD-18–independent mechanism of neutrophil emigration in the development of acute inflammation in response to antigen inhalation and support recent evidence indicating that CD-18–independent mechanisms may, in fact, be involved in neutrophil extravasation and accumulation in certain inflammatory conditions in the lung (15,16). Furthermore, the importance of the selectin family of adhesion molecules is highlighted in in vitro studies of

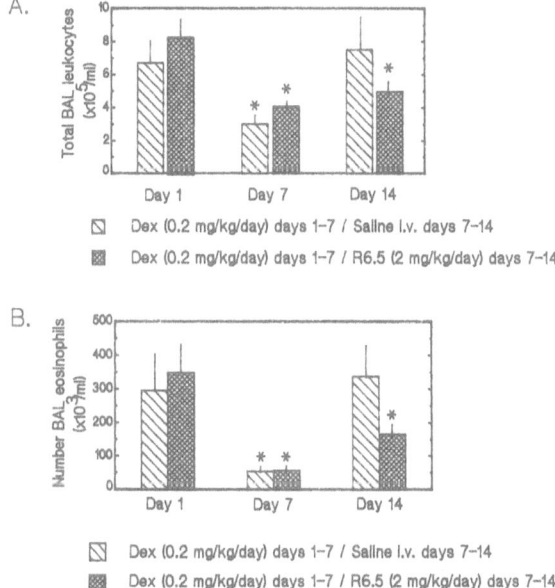

FIGURE 25.10. Animals were treated with dexamethasone for 7 days followed by treatment with R6.5 (crosshatched bars) or vehicle (hatched bars) for an additional 7 days. A. Dexamethasone treatment caused a significant reduction in the total number of leukocytes recovered by BAL at day 7 of the study. During vehicle treatment the number of leukocytes increased back to baseline levels at day 14. Treatment with R6.5 significantly reduced the leukocyte infiltration occurring during day 7 to day 14. B. Dexamethasone treatment significantly reduced the number of eosinophils recovered by BAL at day 7 of the study. Vehicle treatment resulted in a significant increase in the number of eosinophils recovered by BAL at day 14 while treatment with R6.5 significantly inhibited this increase in BAL eosinophils.

CD-18–independent neutrophil–endothelial cell adhesion (17–22). Our results clearly show, in vivo, the involvement of ELAM-1 in acute neutrophil influx into the lung with no functional role for ICAM-1 in this response.

Furthermore, our data strongly suggest an effector cell role for the neutrophil in the late-phase airway response. Evidence to support a role for neutrophils in asthma comes from studies that have demonstrated the presence of high-molecular-weight neutrophil chemotactic activity (HMW-NCA) in the blood of patients after antigen-induced early and late-phase reactions (23,24). Neutrophils have the ability to alter airway function. For instance, neutrophils generate lipid mediators (i.e., prostaglandins, PAF) in response to a variety of stimuli including antigen (25). Neutrophils may also contribute to the LPR by the generation of toxic ox-

A.

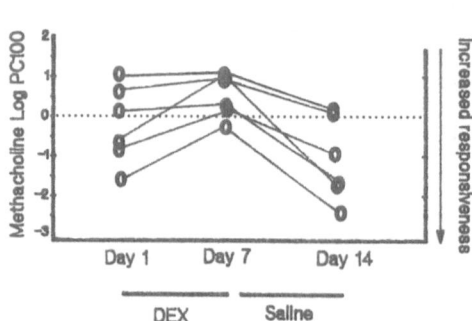

B.

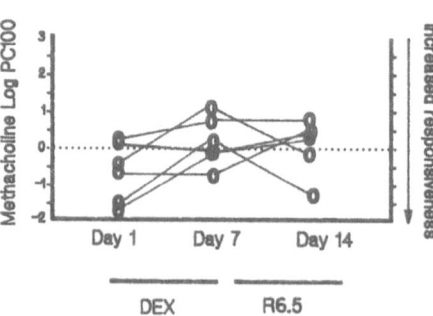

FIGURE 25.11. Animals were treated with dexamethasone for 7 days followed by either vehicle or R6.5 treatment for an additional 7 days. A. Vehicle treatment had no effect on the rebound hyperresponsiveness seen at day 14 of the study. B. In contrast, R6.5 treatment significantly reduced the rebound hyperresponsiveness occurring 7 days after cessation of dexamethasone treatment.

ygen species that can alter airway smooth muscle function and damage airway epithelium leading to impaired lung function (26).

In contrast to the role of ELAM-1 and neutrophils in acute inflammation and transient changes in lung function, ICAM-1 and eosinophils have a major role in chronic allergic airway inflammation and prolonged reductions in airway function. The association between blood and sputum eosinophilia with chronic asthma has been firmly established for many years. More recently, eosinophil influx into the airways and the accumulation of eosinophil-derived proteins have been reported in BAL fluid following antigen inhalation in asthmatic subjects (9). Similarly, Durham and Kay (27) demonstrated an association between peripheral blood eosinophil counts and the degree of nonspecific airway hyperresponsiveness. In addition, several eosinophil granule proteins including the major basic protein (MBP) have been identified on lung tissue taken from patients who have died from asthma (28). MBP is toxic to airway epithelium and is capable of inducing an acute, transient bronchoconstriction and large increase in airway responsiveness (29). Eosinophils can also generate membrane-derived lipid mediators (i.e., LTC_4 and PAF) that have been purported to play a role in the pathogenesis of asthma. It is clear from earlier studies from our laboratory that ICAM-1 plays a pivotal role in the recruitment of eosinophils into the airways and the induction of hyperresponsive airways following multiple antigen inhalation challenges (13) and that ELAM-1 does not play a significant role in this process (30). The results of the present series of studies with dexamethasone support a role for eosinophils in chronic airway hyperresponsiveness. The fact that a 7-day treatment with anti–ICAM-1 (R6.5) did not significantly reduce existing airway inflammation may simply reflect eosinophil survival time in the lung which is most likely not a function of cell adhesion. Experiments in which treatment with murine monoclonal anti–ICAM-1 antibodies are limited to 7 days because of anti-idiotypic responses in the host animals. However, we speculate that a longer treatment time would eventually reduce the airway inflammation as the resident eosinophils die and the influx of newly recruited eosinophils is prevented. Our results from studies in which animals were treated first with dexamethasone followed by anti–ICAM-1 treatment comfirm that newly recruited eosinophils require functional ICAM-1 on both vascular endothelium and airway epithelium to successfully extravasate into the lungs and induce airway hyperresponsiveness.

In summary, current views of asthma emphasize the importance of an extensive chronic inflammatory component (eosinophilic) and airway hyperresponsiveness in the pathophysiology of the disease. In addition to the chronic inflammation and associated hyperresponsiveness, asthmatic subjects also have superimposed, episodic acute inflammation responses early after antigen exposure that are associated with a transient change in lung function. Our studies with a primate model of allergic asthma suggest that different cell types, as well as different adhesion molecules, are playing major roles in acute vs. chronic airway inflammation and airway function. The acute, transient neutrophil influx into the airways and associated late-phase response observed after a single antigen inhalation occur largely via a non-CD18–dependent (ELAM-1) mechanism. The chronic eosinophilic airway inflammation and hyperresponsiveness occurring after multiple antigen inhalations occur via an ICAM-1–dependent process. The potential therapeutic importance of these observations in the pathogenesis of asthma and other inflammatory diseases may lie in our ability to successfully modulate adhesion molecule expression and/or function.

Acknowledgments. The authors would like to acknowledge the excellent technical assistance provided by Ms. C.A. Torcellini, Ms. A.M. LaPlante, and Mr. C.C. Clarke, and Dr. C.W. Smith for providing the anti–ELAM-1 monoclonal antibody.

References

1. Hargreave, F.E., G. Ryan, N.C. Thomson, et al. 1981. Bronchial responsiveness to histamine or methacholine in asthma; measurement and clinical significance. *J. Allergy Clin. Immunol.* 68:347.
2. Boulet, L.P., A. Cartier, N.C. Thomson, R.S. Roberts, J. Dolovich, and F.E. Hargreave. 1983. Asthma and increases in nonallergic bronchial responsiveness from seasonal pollen exposure. *J. Allergy Clin. Immunol.* 71:399.
3. Chan-Yeung, M., S. Lam, and S. Koener. 1982. Clinical features and natural history of occupational asthma due to western red cedar *(Thuja plicata)*. *Am. J. Med.* 72:411.
4. Ryan, G., K. Latimar, J. Dolovich, and F.E. Hargreave. 1982. Bronchial responsiveness to histamine: relationship to diurnal variations of flow rates and improvements after bronchodilation. *Thorax* 37:423.
5. Juniper, E.F., P.A. Frith, and F.E. Hargreave. 1981. Airway responsiveness to histamine and methacholine: relationship to minimum treatment to control symptoms of asthma. *Thorax* 36:575.
6. O'Byrne, P.M., J. Dolovich, and F.E. Hargreave. 1987. Late asthmatic responses. *Am. Rev. Respir. Dis.* 136:740.
7. Larsen, G.L. 1987. The pulmonary late-phase response. *Hosp. Pract.* 22:155.
8. Marsh, W.R., C.G. Irvin, K.R. Murphy, B.L. Behrens, and G.L. Larsen. 1985. Increases in airway reactivity to histamine and inflammatory cells in bronchoalveolar lavage after the late asthmatic response in an animal model. *Am. Rev. Respir. Dis.* 131:875.
9. de Monchy, J.G.R., H.R. Kaufman, P. Venge, G.H. Koeter, H.M. Jansen, H.J., Sluiter, and K. de Vries. 1985. Bronchoalveolar eosinophilia during allergen-induced late asthmatic reactions. *Am. Rev. Respir. Dis.* 131:373.
10. Gundel, R.H., C.D. Wegner, C.A. Torcellini, C.C. Clarke, N. Haynes, R. Rothlein, C.W. Smith, and L.G. Letts. 1991. Endothelial leukocyte adhesion molecule-1 mediates antigen-induced acute airway inflammation and late-phase airway obstruction in monkeys. *J. Clin. Invest.* 88:1407.
11. Gundel, R.H., M.E. Gerritsen, and C.D. Wegner. 1989. Antigen coated sepharose beads induce airway eosinophilia and airway hyper-responsiveness in cynomolgus monkeys. *Am. Rev. Respir. Dis.* 140:629.
12. Gundel, R.H., M.E. Gerritsen, G.J. Gleich, and C.D. Wegner. 1990. Repeated antigen inhalation results in a prolonged airway eosinophilia and airway hyperresponsiveness in primates. *J. Appl. Physiol.* 68:779.
13. Wegner, C.D., R.H. Gundel, P. Reilly, N. Haynes, L.G. Letts, and R. Rothlein. 1990. Intercellular adhesion molecule-1 (ICAM-1) and the pathogenesis of asthma. *Science.* 247:456.
14. Wegner, C.D., R.R. Rothlein, C.C. Clarke, N. Haynes, C.A. Torcellini, A.M. LaPlante, D.R. Averill, L.G. Letts, and R.H. Gundel. 1991. *Am. Rev. Respir. Dis.* 143:A418 (Abstract).
15. Carlos, T.M., and J.M. Harlan. 1990. Membrane proteins involved in phagocyte adherence to endothelium. *Immunol. Rev.* 114:45.
16. Doerschuk, C.M., R.K. Winn, H.O. Coxson, and J.M. Harlan. 1990. CD18-dependent and independent mechanisms of neutrophil emigration in the pulmonary and systemic microcirculation of rabbits. *J. Immunol.* 144:2327.

17. Stoolman, L.M. 1989. Adhesion molecules controlling lymphocyte migration. *Cell.* 56:907.

18. Bevilacqua, M.P., S. Stengelin, Jr., M.A. Gimbrone, and B. Seed. 1989. Endothelial leukocyte adhesion molecule-1: an inducible receptor for neutrophils related to complement regulatory proteins and lectins. *Science.* 243:1160.

19. Geng, J., M.P. Bevilacqua, K.L. Moore, T.M. McIntyre, S.M. Prescott, J.M. Kim, G.A. Bliss, G.A. Zimmerman, and R.P. McEver. 1990. Rapid neutrophil adhesion to activated endothelium mediated by GMP-140. *Nature.* 343:757.

20. Hallman, R., M.A. Jutila, C.W. Smith, D.C. Anderson, T.K. Kishimoto, and E.C. Butcher. 1991. The peripheral lymph node homing receptor, LECAM-1, is involved in CD-18 independent adhesion of human neutrophils to endothelium. *Biochem. Biophys. Res. Commun.* 174:236.

21. Smith, C.W., T.K. Kishimoto, O. Abbassi, B. Hughes, R. Rothlein, L.V. McIntire, E. Butcher, and D.C. Anderson. 1991. Chemotactic factors regulate Lectin Adhesion Molecule 1 (LECAM-1)-dependent neutrophil adhesion to cytokine-stimulated endothelial cells in vitro. *J. Clin. Invest.* 87:609.

22. Luscinskas, R.W., A.F. Brock, M.A. Arnaout, and M.A. Gimbrone, Jr. 1989. Endothelial-leukocyte adhesion molecule-1-dependent and leukocyte (CD11/CD18)-dependent mechanisms contribute to polymorphonuclear leukocyte adhesion to cytokine-activated human vascular endothelium. *J. Immunol.* 142:2257.

23. Atkins, P.C., M. Norman, H. Weiner, and B. Zweiman. 1977. Release of neutrophil chemotactic activity during immediate hypersensitivity reactions in humans. *Am. Intern. Med.* 86:415.

24. Nagy, L., T.H. Lee, and A.B. Kay. 1982. Neutrophil chemotactic activity in antigen-induced late asthmatic reactions. *N. Engl. J. Med.* 306:497.

25. Venge, P. 1985. Eosinophil and neutrophil granulocytes in asthma. *Curr. Clin. Pract.* 25:21.

26. Ward, P.A. G.O. Till, R. Kunkel, and C. Beauchamp. 1983. Evidence for a role of hydroxyl radical in complement and neutrophil dependent tissue injury. *J. Clin. Invest.* 72:789.

27. Durham, S.R., and A.B. Kay. 1985. Eosinophils, bronchial hyperreactivity and late-phase asthmatic reactions. Clin. Allergy 15:411.

28. Filley, W.V., K.E. Holley, G.M. Kephart, and G.J. Gleich. 1988. Identification by immunofluorescence of eosinophil granule major basic protein in lung tissue of patients with bronchial asthma. *Lancet.* ii:11.

29. Gundel, R.H., L.G. Letts, and G.J. Gleich. 1991. Human eosinophil major basic protein induces airway constriction and airway hyperresponsiveness in primates. *J. Clin. Invest.* 87:1470.

30. Wegner, C.D., R.R. Rothlein, C.W. Smith, C.C. Clarke, C.A. Torcellini, N. Haynes, L.G. Letts, and R.H. Gundel. 1991. Contribution of endothelial-leukocyte adhesion molecule-1 (ELAM-1) to airway inflammation and hyperresponsiveness. *Am. Rev. Respir. Dis.* 143:A418 (Abstract).

26
Adhesion Molecules in the Shwartzman Response

LAWRENCE W. ARGENBRIGHT AND RANDALL W. BARTON

Introduction

The classical Shwartzman reaction was elicited in rabbits in intradermal sites that had been "prepared" 24 h previously by the injection of culture filtrate of *Bacillus typhosus* followed by an intravenous "challenge" injection of the same filtrate (1). It has been shown that the active agent in the filtrate that is responsible for the thrombohemorrhagic response that develops within 4 h after challenge is endotoxin (2).

The nature of the preparative and challenge injections has been examined. Endotoxin (2,3), bacteria (2,3), IL-1β (3,4), and TNFα (3) have been used successfully as preparative agents. Although these preparative agents do elicit leukocyte extravasation, other chemotactic agents, such as fMLP, that also elicit leukocyte extravasation, are *not* preparative agents, but they can serve as challenge agents (5). Thus, although the Shwartzman reaction is neutrophil-dependent (6), the effect of the preparative injection involves more than inducing a leukocytic infiltrate.

Neurophil-dependent inflammatory responses have been shown to be mediated by the CD18 family of leukocyte cell-surface glycoproteins (7,8). Stimulation of leukocytes with chemotactic agents such as C5a and fMLP produces rapid up-regulation by plasma membrane expression of preformed molecules contained in cytoplasmic granules (9,10) and increased functional activity of constitutively expressed adhesion molecules (11,12).

More recently, a ligand of the CD18 family of leukocyte adhesion molecules has been indentified, ICAM-1 (8,13). ICAM-1 is inducible on multiple cell types in vitro including endothelial cells by endotoxin, TNFα, IL-1β, IFN-γ or all of these depending on the cell type (14,15). The induced ICAM-1 on endothelial cells mediates both the CD18-dependent adhesion and transendothelial migration of neutrophils (16). The expression of ICAM-1 is also increased at sites of inflammation in vivo (15,17–19). The ability of ICAM-1 to serve as a ligand for the CD18 family on leukocytes is evident by the inhibition of CD18-dependent inflammatory responses by anti–ICAM-1 Mab (19–22).

In the present study we have examined the role of the CD18/ICAM-1 adhesion molecules in the Shwartzman reaction. The results suggest that one effect of a successful preparative agent is the up-regulation of ICAM-1 on vascular endothelium and that the response to the challenge injection is CD18/ICAM-1-dependent.

Results

The intradermal injection of *Salmonella typhosa* endotoxin into rabbit skin produced an acute inflammatory reaction characterized by an accumulation predominantly of neutrophils and some monocytes at 18–20 h after injection. Four to 6 hours after the i.v. challenge injection of zymosan, 10 mg/kg, that systemically activated complement, the endotoxin-injected sites became hemorrhagic whereas saline-injected sites appeared

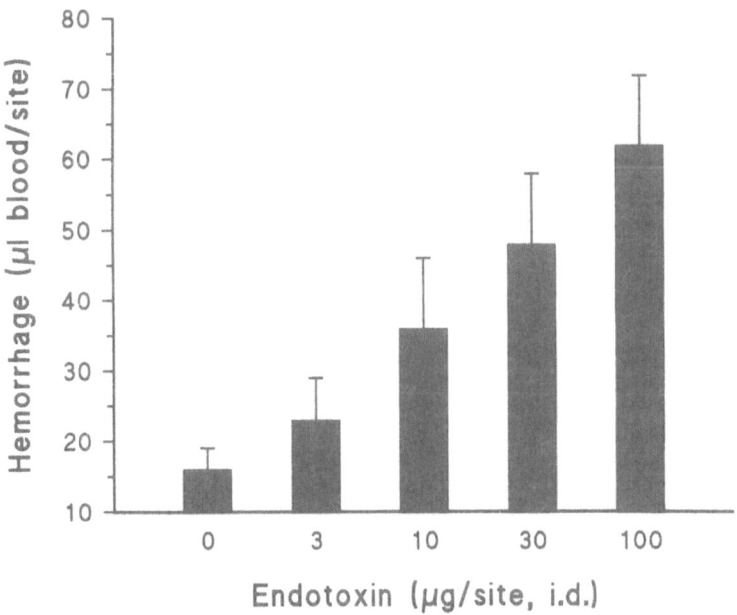

FIGURE 26.1. Quantitation of hemorrhage in Shwartzman reactions in rabbit skin. RBCs were labeled in vivo with 99mTc. Intradermal injection of LPS was performed 20 h before i.v. injection of zymosan. Six hours after zymosan, hemorrhagic reactions had developed. These lesions were removed and the 99mTc activity was measured.

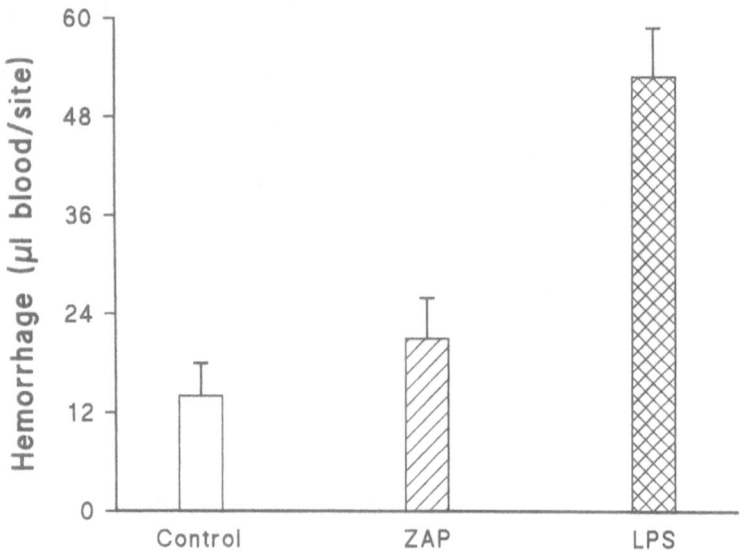

FIGURE 26.2. Comparison of endotoxin and zymosan-activated rabbit plasma (ZAP) to serve as preparatory agents for the Shwartzman response. PBS was injected as a control.

normal. Quantitation of hemorrhage by in vivo labeling of autologous RBCs with 99mTc (200 μCi/kg) revealed a dose-dependent accumulation of RBCs (Fig. 26.1). Histologically, the response to the challenge injection at the previously endotoxin-injected sites was characterized by intravascular leukocyte aggregation, fibrin deposition, and significant hemorrhage. A comparison of the ability of endotoxin or zymosan-activated rabbit plasma to serve as a preparative agent showed that the hemorrhagic response after challenge was only significant in the endotoxin-prepared sites (Fig. 26.2).

The effect of administration of Mabs specific for CD11a, CD11b, CD18, and ICAM-1 (CD54), 2 mg/kg, just prior to the systemic challenge, was evaluated; normal mouse IgG was injected as a control. The results in Figure 26.3 show that greater than 80% inhibition of the hemorrhagic response occurred when either anti-CD18 or anti–ICAM-1 was injected before systemic complement activation. Injection of anti-CD11b (Mac-1) produced marked reduction of hemorrhage whereas injection of anti-CD11a (LFA-1) was not significantly different from control mouse IgG.

Evidence that the challenge injection of zymosan produced neutrophil activation is shown in Figure 26.4. CD18 expression increased approximately threefold after zymosan injection.

Immunohistochemical staining of PBS- and endotoxin-injected sites re-

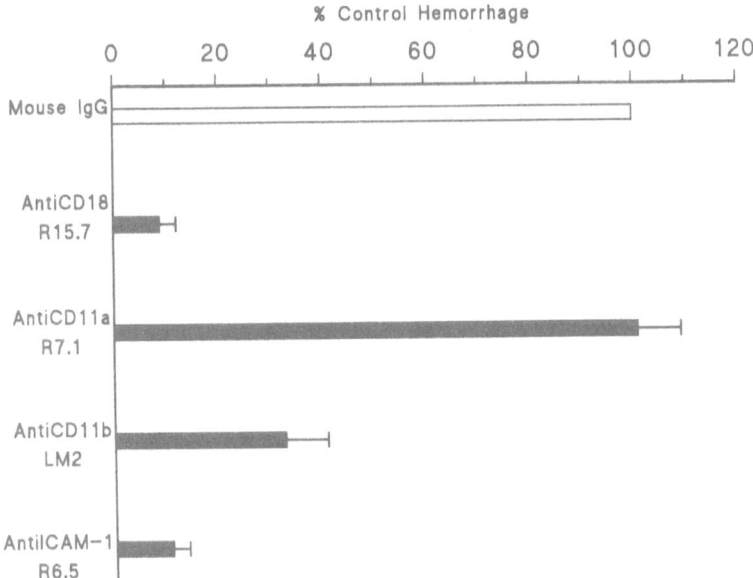

FIGURE 26.3. Effect of anti-adhesion Mabs on the hemorrhage in the Shwartzman response. Mabs reactive with various adhesion molecules were injected i.v., 2 mg/kg, just prior to i.v. zymosan challenge. Six hours after challenge the hemorrhagic response was measured.

vealed that ICAM-1 expression was seen on most vessels in endotoxin-prepared sites whereas little ICAM-1 expression was seen in PBS-injected sites.

Discussion

The results of the present study show that intravenous injection of zymosan to systemically activate complement produced thrombo-hemorrhagic necrosis in "prepared" skin sites that had been injected intradermally with endotoxin 20 h previously. Histologically, this response, characterized by neutrophil aggregation, platelet and fibrin deposition, and massive hemorrhage, is indistinguishable from that of the classic Shwartzman reaction.

The mechanism(s) responsible for the Shwartzman reaction has been unclear. The local preparatory injection is mandatory in order to elicit the response. In skin sites injected with PBS or in normal skin there was no hemorrhage. Thus, systemic complement activation is not sufficent to produce the local response. In addition, although the Shwartzman response has been shown to be neutrophil-dependent (6), intradermal injection of

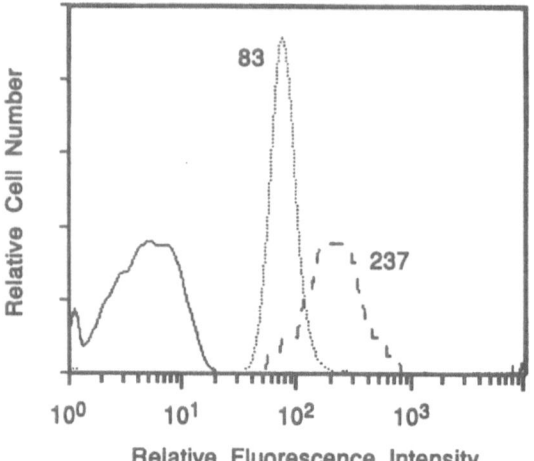

FIGURE 26.4. CD18 expression on rabbit blood neutrophils before and 30 min after i.v. injection of zymosan. (———) control; (········) CD18 before zymosan; (- - -) CD18 after zymosan.

zymosan-activated rabbit plasma, which produced a local neutrophil infiltration, could not be effectively substituted for endotoxin as the preparatory agent.

The inhibition of the Shwartzman response when Mab specific for ICAM-1 was injected just prior to the challenge in concert with the immunhistochemical staining for ICAM-1 in endotoxin- but not PBS-injected sites suggests that ICAM-1 was up-regulated in sites injected with endotoxin as the preparative agent. Bacteria (2,3), IL-1-β (3,4), and TNF-α (3), as well as endotoxin (2,3) have been used as preparative agents for the Shwartzman response. Induction of ICAM-1 expression on endothelial cells in vitro has been accomplished with endotoxin, IL-1, and TNF (14,15). The lack of effect of zymosan-activated rabbit plasma as a preparative agent is consistent with the observation that C5a did not induce ICAM-1 expression on cultured endothelial cells (Dr. Christina Myers, personal communication). Thus, we propose that the only effective preparatory agents in the Shwartzman response are those that induce ICAM-1 expression on the vascular endothelium.

Our data indicate that the Shwartzman response is also CD18-dependent. Mabs specific for CD18 and Mac-1 (CD11b) effectively blocked the hemorrhagic response. The data also suggest that both neutrophil–endothelial cell adhesion and intravascular neutrophil–neutrophil aggregation are required for the response. Other studies have shown that both neutrophil–endothelial cell and neutrophil–neutrophil aggregation are Mac-1–dependent (7,16). Figure 26.5 depicts the proposed mechanism of the development of the ICAM-1-dependent, intravascular-neutrophil–mediated, vasculitic response that characterizes the Shwartzman reaction.

This proposed mechanism may also be responsible for a variety of cli-

"Preparative"   "Challenge"   Hemorrhage
(LPS)   (C5a)

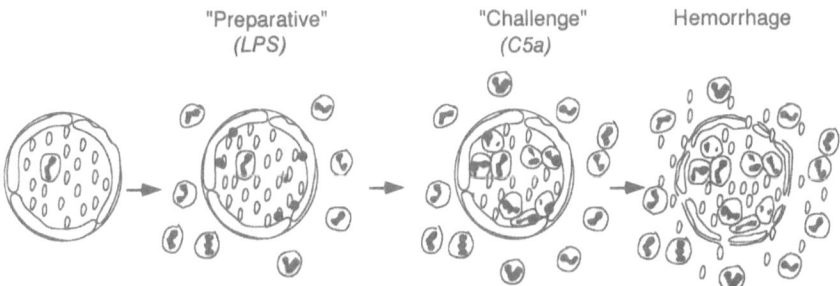

FIGURE 26.5. Proposed mechanism of the development of Shwartzman lesions. The effect of the preparative agent is the induction of ICAM-1 expression (black forms) on the vascular endothelium. The challenge agent then activates the CD18 complex, which binds to vessels now expressing ICAM-1. In addition to CD18/ICAM-1 leukocyte–endothelial adhesion, a leukocyte–leukocyte aggregation also appears to be required.

nical conditions. Indeed, the vascular thrombosis and tissue necrosis of the Shwartzman reaction have been postulated as the pathogenic mechanism of disseminated intravascular coagulation (DIC) (23) and septic shock (24). In addition, the Shwartzman reaction may mimic a variety of vasculitic conditions, including SLE, in which vascular injury may occur in the absence of immune complex deposition (25).

References

1. Shwartzman, G. 1928. A new phenomenon of local skin reactivity to B. typhosus culture filtrate. *Proc. Soc. Exp. Biol. Med.* 25:560.
2. Movat, H.Z., B.J. Jaynes, S. Wasi, K.W. Movat, and M.M. Kopaniak. 1980. Quantitation of the development and progression of the local Shwartzman reaction. In "Bacterial Endotoxins and Host Response." M.K. Agarwal, editor. Elsevier/North-Holland, Amsterdam. pp. 179–201.
3. Movat., H.Z., C.E. Burrows, M.I. Cybulsky, and C.A. Dinarello. 1987. Acute inflammation and a Shwartzman reaction induced by interleukin-1 and tumor necrosis factor. *Am. J. Pathol.* 129:463.
4. Beck, G., G.S. Habicht, J.L. Benach, and F. Miller. 1986. Interleukin-1: a common endogenous mediator of inflammation and the local Shwartzman reaction. *J. Immunol.* 136:3025.
5. Fehr, J., C. Dahinden, and R. Russi. 1984. Formylated chemotactic peptides can mimic the secondary, provoking endotoxin injection in the generalized Shwartzman reaction. *J. Infect. Dis.* 150:160.
6. Stetson, C.A., and R.A. Good. 1951. Shwartzman phenomenon— participation of polymorphonuclear leukocytes. *J. Exp. Med.* 93:49.
7. Anderson, D.C., F.C. Schmalsteig, M.A. Aranout, S. Kohl, M.F. Tosi, N. Dana, G.J. Buffone, B.J. Hughes, B.R. Brinkley, W.D. Dickey, J.S. Abramson, T. Springer, L.A. Boxer, J.M. Hollers, and C.W. Smith. 1984. Abnor-

malities of polymorphonuclear leukocyte function associated with a heritable deficiency of high molecular weight surface glycoproteins (GP138): a common relationship to diminished cell adherence. *J. Clin. Invest.* 74:536.

8. Springer, T.A. 1990. Adhesion receptors of the immune system. *Nature* 346:425.

9. Todd, R.F., M.A. Aranout, R.E. Rosin, C.A. Crowley, W.A. Peters, and B.M. Babior. 1984. Subcellular localization of the large subunit of Mol. a surface glycoprotein associated with neutrophil adhesion. *J. Clin. Invest.* 74:1280.

10. Miller, L.J., D.F. Bainton, N. Borregaard, and T.A. Springer. 1987. Stimulated mobilization of monocyte Mac-1 and p150, 95 adhesion proteins from an intracellular vesicular compartment to the cell surface. *J. Clin. Invest.* 80:535.

11. Buyon, J.P., S.B. Abramson, M.R. Philips, S.G. Slade, G.D. Ross, G. Weissmann, and R.J. Winchester. 1988. Dissociation between increased surface expression of gp165,95 and homotypic neutrophil aggregation. *J. Immunol.* 140:3156.

12. Vedder, N.B., and J.M. Harlan. 1988. Increased surface expression of CD11b/CD18 (Mac-1) is not required for stimulated neutrophil adherence to cultured endothelium. *J. Clin. Invest.* 81:676.

13. Rothlein, R., and T.A. Springer. 1986. The requirement for lymphocyte-function-associated antigen 1 in homotypic leukocyte adhesion stimulated by phorbol ester. *J. Exp. Med.* 163:1132.

14. Pober, J.S., M.A. Gimbrone, Jr., L.A. Lapierre, D.L. Mendrick, W. Fiers, R. Rothlein, and T.A. Springer. 1986. Overlapping patterns of activation of human endothelial cells by interleukin-1, tumor necrosis factor and immune interferon. *J. Immunol.* 137:1893.

15. Dustin, M.L., R. Rothlein, A.K. Bhan, C.A. Dinarello, and T.A. Springer. 1986. Induction by IL-1 and interferon-gamma: Tissue distribution, biochemistry, and function of a natural adhesion molecule (ICAM-1). *J. Immunol.* 137:245.

16. Smith, C.W., S.D. Marlin, R. Rothlein, C. Toman, and D.C. Anderson. 1989. Cooperative interactions of LFA-1 and Mac-1 with intercellular adhesion molecule-1 in facilitating adherence and transendothelial migration of human neutrophils in vitro. *J. Clin. Invest.* 83:2008.

17. Griffiths, C.E.M., and B.J. Nickoloff. 1989. Keratinocyte intercellular adhesion molecule-1 (ICAM-1) expression precedes dermal T lymphocyte infiltration in allergic contact dermatitis (*Rhus* dermatitis). *Am. J. Pathol.* 135:1045.

18. Adams, D.H., S.G. Hubscher, J. Shaw, R. Rothlein, and J.M. Neuberger. 1989. Intercellular adhesion molecule-1 on liver allografts during rejection. *Lancet* 2:1122.

19. Cosimi, A.B., D. Conti, F.L. Delmonico, F.I. Preffer, S.-L., Wee R. Rothlein, R.B. Faanes, and R.B. Colvin. 1990. Therapy with a monoclonal antibody to the adhesion molecule ICAM-1. I. Effects on allograft rejection in non-human primates. *J. Immunol.* 144:4604.

20. Barton, R.W., R. Rothlein, J. Ksiazek, and C. Kennedy. 1989. The effect of anti-intercellular adhesion molecule-1 on phorbol-ester-induced rabbit lung inflammation. *J. Immunol.* 143:1278.

21. Wegner, C.D., R.H. Gundel, P. Reilly, N. Haynes, L.E. Letts, and R. Rothlein. 1990. Intercellular adhesion molecule-1 (ICAM-1) in the pathogenesis of asthma. *Science* 247:456.

22. Argenbright, L.A., L.G. Letts, and R. Rothlein. 1991. Monoclonal anti-bodies to the leukocyte membrane CD18 glycoprotein complex and to inter-cellular adhesion molecule-1 inhibit leukocyte-endothelial cell adhesion in rabbits. *J. Leuk. Biol.* 49:253.
23. Mori, W. 1983. Intravascular coagulation as a clinical manifestation of the Shwartzman reaction. *Biblio. Haematol.* 49:41.
24. Abramson. 1988. Increased levels of plasma anaphylatoxins in systemic lupus erythematosus predict flares of the disease and may elicit vascular injury in lupus cerebritis. *Arthritis Rheum.* 31:632.
25. Fauci, A. S. 1987. The Vasculitic syndromes. In "Principles of Internal Medi-cine." E. Braunwald, K. Isselbacher, R. Petersdorf, J. Wilson, J. Martin, and A. Fauci, editors. McGraw-Hill, New York. pp. 1438–1445.

27
Clinical and Immunological Features Associated With Bovine Leukocyte Adhesion Deficiency

Marcus E. Kehrli Jr., Dale E. Shuster, Mark Ackermann, C. Wayne Smith, Donald C. Anderson, Monique Doré, and Bonnie J. Hughes.

Introduction

Granulocytopathies in dogs (1), humans (2), and cattle (3,4) characterized by persistent progressive neutrophilia in patients affected with severe recurrent bacterial infections and failure to form pus were reported between 1975 and 1987. In all three species, these conditions were determined to be heritable deficiencies of leukocyte surface glycoproteins associated with diminished cell adherence between 1984 and 1990 (5–9). Although published reports of its diagnosis are few, the bovine granulocytopathy syndrome has been diagnosed at veterinary schools throughout the world during the past 8 y. In vitro assessments have identified abnormalities of motile, phagocytic, and oxidative functions of neutrophils which appear to mediate inflammatory deficits in vivo (3,4,7,8). Factors contributing to low frequency of diagnosis of this syndrome may relate to the impracticality of intensive clinical laboratory studies in food-producing animal species.

The human genetic disorder has been extensively described in human patients (termed leukocyte adhesion deficiency or LAD) (6,10–14). Heterogeneous mutations of the gene encoding the common β subunit (CD18) of the β_2 integrins (LFA-1, Mac-1, and p150,95) have been identified as the primary basis for disease in all human cases (13,15–18). Although the pathogenesis of disease in Irish setters has not been completely elucidated, deficient expression of β_2 integrins on leukocyte surfaces suggests a similar underlying defect of the gene encoding canine CD18(9). Until recently the molecular pathogenesis of the bovine granulocytopathy syndrome had remained undetermined. In 1989, a calf with bovine granulocytopathy syndrome was identified as lacking the α subunit of the Mac-1 β_2-integrin in neutrophil plasma membranes (8). The dam and sire of this calf were tested for CD18 expression by flow cytometry and found to express 65–70% of normal levels of CD18 on neutrophils. Eight of 15 paternal half-siblings (both male and female siblings) were also tested and found to express similarly reduced levels of surface CD18, thus indicating a carrier status. Further flow cytometric analysis of members of a three-generation pedigree produced results consistent with an autosomal recessive genetic defect in the gene encoding bovine CD18 (8).

Based upon the absence of the α subunit of the β_2 integrins in isolated neutrophil membranes from the proband and the reduced expression of the β subunit in parents of the proband, we suspected the bovine granulocytopathy syndrome was due to a defect in bovine CD18 expression and proceeded to sequence normal bovine CD18 cDNA (19). In order to to sequence the defective bovine CD18 allele(s), we acquired two Holstein heifers diagnosed with bovine granulocytopathy syndrome. Neutrophils isolated from these heifers were found to express very low levels of CD18 by flow cytometric analysis. Messenger RNA was isolated from leukocytes from one of these heifers and reverse transcribed, and the resulting cDNA for CD18 was sequenced. A point mutation (A → G) resulting in an amino acid substitution (D128G) in CD18 was identified in the first heifer and confirmed in the second heifer. A DNA-PCR test has been developed to identify the genotype of Holstein cattle at this allele (21). We were also successful in reproducing a calf affected with bovine LAD from the mating of the dam and sire of our proband calf. We present here the clinical features and immunological abnormalities observed with bovine leukocyte adhesion deficiency (BLAD).

Results

Clinical Health and History of Three BLAD Patients

Two Holstein heifer calves suspected of having BLAD were purchased at 10.3 (Emilee) and 13.7 (Meg) months of age. Clinically these BLAD patients had a history of poor or stunted growth (50–57% of expected body weight and 92% of expected height), and nonsuppurative peridontal gingivitis with marked recession of the gum line. During a 5-month observation period, Emilee exhibited recurrent loose stools which were associated with a mild fever (+1°C) on one occasion. Upon arrival into our center Meg had a history of periodic mild pneumonitis and bronchitis and had several indurated skin ulcers (1–6 cm in diameter). The skin ulcers were thought to be a result of mechanical removal, by the patient, of warts that were widely dispersed over her head and body. Radiographic examination of the skull of each heifer at 12 and 14 months of age provided further evidence of severe peridontal disease with horizontal and vertical alveolar bone loss of both upper and lower dental arcades. Bone resorption resulted in loss of three premolars in Meg and six premolars in Emilee, as well as severe loosening of the incisors and remaining premolars in each animal. Each animal had gingival regres-

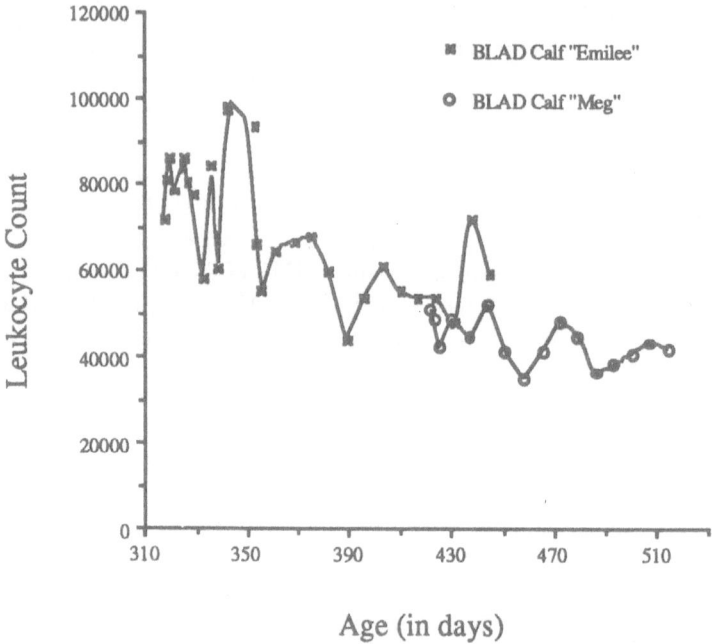

FIGURE 27.1. Total leukocyte count in two BLAD patients housed together in a moderately controlled climate and minimal exposure to bovine pathogens.

sion associated with chronic gingivitis. The peridontal gingivitis in these BLAD patients is identical to that reported in humans with LAD (20). Leukocyte counts in Emilee and Meg since admitted to our center have ranged from 40,000 to 90,000 leukocytes per μ1 (Fig. 27.1).

In order to demonstrate the mode of inheritance of this genetic defect we superovulated and transferred embryos resulting from artificial insemination of a carrier cow (dam of the proband calf) with semen from a known carrier bull (sire of the proband calf). The first calf born as a result of embryo transfer in this experiment was ceasarean-derived and maintained in a gnotobiotic isolator for 6 wk. Leukocytes from the calf (Fred) obtained at birth were found to be essentially free of surface CD18 protein by flow cytometric analysis, and this calf was later identified as homozygous for the D128G allele (21). At 13 days of age gnotobiotic status of the calf was lost when a *Clostridium perfringens* intestinal infection was detected on routine weekly anaerobic culture of rectal swabs. At 23 days of age Fred developed a mild fever (0.8°C greater than the previous 3 days) and leukocytosis (40,800 cells/μ1). Fred was given a single injection of 40 ml of *C. perfringens* antitoxin subcutaneously and 100 mg ceftiofur sodium intramuscularly for 5 days. Fred was also given *Lactobacillus acidophilus* daily per os for 7 days. After *L. acidophilus* was colonized

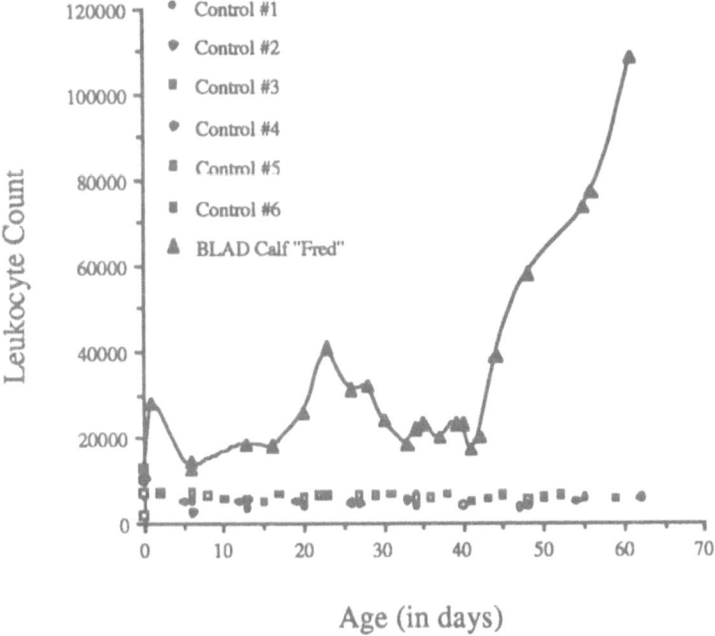

Age (in days)

FIGURE 27.2. Total leukocyte counts in a BLAD patient (Fred) maintained in a gnotobiotic isolator for 6 wk and in an isolation room for 3 additional wk before the calf was euthanatized. Leukocyte counts of this calf compared to those of six age-matched calves maintained in a similar isolator indicate that, in spite of an absence of microflora during the first few days of life, this calf exhibited a leukocytosis (28,000 leukocytes/μl) following ingestion of sterile colostrum and maintained a leukocyte count about two to five times greater than similarly housed calves of normal CD18 expression. Germ-free status was lost on day 13 and the calf was removed from the isolator into a slightly less hygienic environment on day 43.

based upon fecal culture, *Streptococcus fecaelis* was then given per os. Before removal from the gnotobiotic isolator, *L. acidophilus* and *S. fecaelis* were colonized in the gastrointestinal tract as shown by repeated fecal culture. A commercially available mixture of several lactic acid bacteria was also given in an attempt to develop a normal intestinal flora. Fred's thwarted inflammatory responses to the presence of various bacteria (after germ-free status was lost on day 13) and after removal from the isolator into a slightly less hygienic environment are illustrated in Figure 27.2. In all BLAD patients, the leukocytosis was attributable to a mature neutrophilia (75–90% neutrophils).

After a week in a room maintained with a high degree of sanitation, Fred developed a ventral dermatitis and recurrent fevers (as high as 41.2 °C) which were treated with flunixine meglumine and antibiotics. At 56 days of age Fred became reluctant to nurse from a bottle and was fed three times a day by stomach gavage; surprisingly Fred remained hydrated and alert. At 60 days of age, Fred exhibited severe respiratory dyspnea and remained in lateral recumbency for a 12-h period. At this point, Fred was electively euthanatized and a necropsy was performed. Gross pathological observations indicated this calf had a wide spectrum of lesions including a moderate laryngitis, a moderate multifocal bronchopneumonia with neutrophils observed in the airways, chronic ventral dermatitis, and synovitis.

Neutrophil iC3b and Fc Receptor Functions Are Impaired or Absent

Because of recurrent bacterial infections in spite of a striking neutrophilia, defective neutrophil function in bovine granulocytopathy syndrome cattle has been suspected and reported (3,4,7,8). Neutrophils from the BLAD patients reported here had several functional deficits that were consistent with deficient expression of CD11/CD18 glycoprotein (Table 27.1). Defective neutrophil iodination indicates the role of the C3bi receptor in stimulating phagocytosis and the associated respiratory metabolism burst in bovine neutrophils. Further evidence for this role in phagocy-

TABLE 27.1. Selected BLAD patient neutrophil functions expressed as a percentage of four normal adult bovine animals tested on the same days (tested on 17, 13, and 8 different days tested at weekly intervals for each of the three BLAD patients, respectively).

	Iodination[a]	Ingestion[b]	ADNC[c]	AINC[c]
LAD calves (n = 3)	64.8%	20.8%	121%	54.5%

[a]Neutrophil myeloperoxidase-catalyzed iodination was initiated by C3b-coated zymosan particles (1.0 mg/ml), which were phagocytosed by 2.5×10^6 neutrophils in Earle's balanced salt solution with mild agitation at 39°C. Assay was stopped after 20 min by addition of cold 10% (wt/vo) trichloroacetic acid.

[b]Neutrophil Fc-receptor–mediated ingestion of antibody-opsonized ^{125}I-labeled *Staphylococcus aureus*. The *S. aureus* antiserum was heat-inactivated to destroy complement. The standard assay was conducted with 2.5×10^6 neutrophils, bacteria to neutrophil ratio = 60:1, for 10 min at 39°C before stopping ingestion by addition of lysostaphin to digest extracellular *S. aureus*. Neutrophils were washed, and ingested radioactive *S. aureus* was counted in a gamma counter.

[c]Antibody-independent (AINC) and antibody-dependent (ADNC) neutrophil-mediated cytotoxicity toward ^{51}Cr-labeled chicken erythrocytes. Neutrophil-to-erythrocyte ratio of 10:1 (1.25×10^6 neutrophils for 2 h; 39°C) in medium (M199), with a subagglutinating concentration of bovine antichicken erythrocyte serum heat-inactivated at 56°C for 30 min to destroy complement. Specific release of ^{51}Cr into supernatant from labeled chicken erythrocytes was measured in a gamma counter.

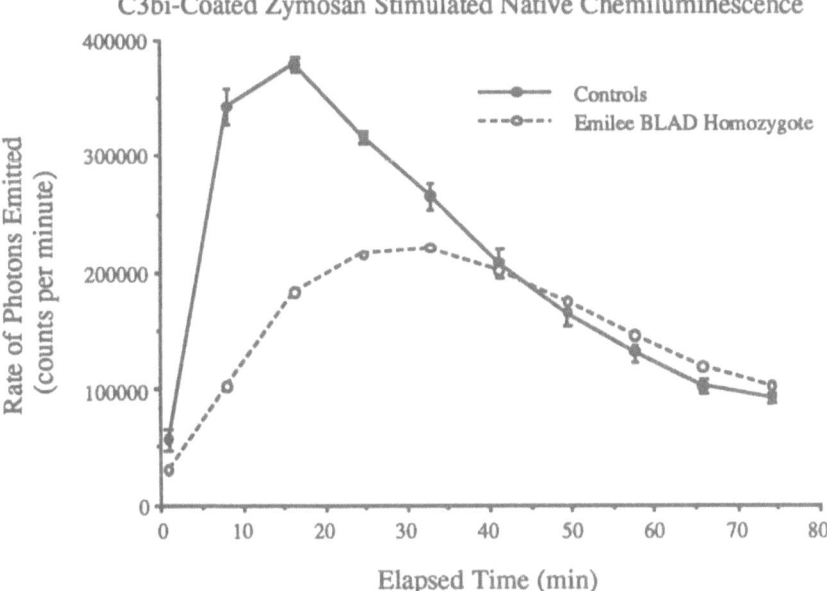

FIGURE 27.3. Phagocytosis-associated neutrophil native chemiluminescence assay kinetics. Neutrophil chemiluminescence determined in a liquid scintillation counter in the out-of-coincidence mode (10^7 neutrophils in 10 ml of Earle's balanced salt solution) was initiated by phagocytosis of C3b-coated zymosan particles (0.2 mg of zymosan/ml). Rate of photon emission from oxidative reactions associated with neutrophil phagocytosis was determined 10 times at ~19 min intervals.

tosis is that C3bi-stimulated native chemiluminescence associated with the phagocytosis-induced burst of respiratory metabolism was impaired and delayed in BLAD patients (Fig. 27.3). Diminished phagocytosis-associated oxidative and secretory functions during ingestion of C3bi opsonized zymosan by neutrophils of BLAD patients are more specifically consistent with a deficiency of the Mac-1 α subunit that contains the complement receptor type 3 (CR-3) epitope reactive with C3bi deposited on test particles such as zymosan (6,10). Endocytosis of IgG-opsonized *S. aureus* by neutrophils of BLAD calves was also profoundly diminished compared with control neutrophils. Impaired Fc-receptor–mediated endocytosis by neutrophils has been previously reported in human beings with LAD, an observation suggesting cooperative interactions of the IgG Fc receptor and CR-3 associated with opsonophagocytosis (22). Several investigators have reported defective antibody-dependent neutrophil cytotoxicity (ADNC) or natural killing (AINC) by neutrophils or mononuclear leukocytes from human patients with LAD (5,11,23–25), whereas studies in selected cases have failed to detect such abnormalities, especially among patients with moderate phenotypes (5,24), Thus, the normal or

TABLE 27.2 Adhesion of isolated bovine neutrophils to keyhole limpet hemocyanin-coated glass.

Bovine (age)	Unstimulated PMN		PAF, 100 ng/ml		PAF + mab R15.7 (10 μg/ml)
	Adh[a]	CD18[b]	Adh	CD18	Adh
Control adult	12	615	77.3	980	16
Neonate control	11.5	607	75.8	925	12
Meg (16 months)	12.5	20	17.3	15	13
Emilee (13 months)	14	14	20.3	17	14
Fred (2 wk)	3	8	6.3	21	6.5

[a] Percent adherent cells. Adhesion to protein-coated glass (26) was studied using glass coverslips coated with keyhole limpet hemocyanin (KLH), a surface shown to mediate Mac-1–dependent adhesion (27), and inserted into an adhesion chamber. Isolated neutrophils were injected into the chamber and allowed to contact the KLH-coated surface for 500 s at room temperature. Chambers were inverted and nonattached cells were allowed to fall off. Results were expressed as percent adherent cells. Platelet activating factor (PAF), used to stimulate neutrophil adherence, was added to cell preparations immediately before the adhesion assay. Anti-CD18 mAb, R15.7 (28), was added 5 min before the stimulus (26).
[b] Indirect immunofluorescent analysis of isolated neutrophils expression of CD18 was performed using murine monoclonal antibody (mAb) and phycoerythrin-conjugated antibody to mouse IgG as previously described (10).

greater-than-normal cytotoxic functions of our BLAD patients' neu-trophils were not inconsistent with deficiency of CD11/CD18.

Neutrophil β_2 Integrin Expression, β_2 Integrin-Dependent Adhesion, and Adhesion-Dependent Functions Are Impaired or Absent.

As shown in Table 27.2, control neutrophil adherence and surface CD18 levels increased following PAF stimulation. These responses were not observed with neutrophil suspensions from the three BLAD patients (Emilee, Meg, and Fred). PAF-enhanced neutrophil adherence was inhibited by treating control neutrophils with anti-CD18 mAb but this mAb was without effect when incubated with BLAD patient neutrophils. These data are consistent with values obtained using this technique with neutrophils from normal human adult donors and patients with CD18 deficiency (6). There appeared to be a small amount of adhesion with the two older BLAD patients, Meg and Emilee, though this may not be significant.

Adherence-dependent hydrogen peroxide production by nertrophils from BLAD patients was undetectable with PAF stimulation and significantly reduced with PMA stimulation compared to control neutrophils.

TABLE 27.3. Adherence-dependent hydrogen peroxide production by bovine neutrophils.

Bovine	PAF	PAF + R15.7	PMA
Control adult	143[a]	13	472
Control neonate	0	0	168
Meg (16 months)	0	0	107
Emilee (13 months)	0	0	110
Fred (2 wk)	0	0	107

Adherence-dependent hydrogen peroxide production (27) was evaluated by allowing isolated neutrophils to contact KLH-coated plastic coincident with chemotactic stimulation, in this case, platelet activating factor (PAF). The reduction in scopoletin fluorescence is directly related to release of hydrogen peroxide into the medium.
[a] pmol $H_2O_2/1.5 \times 10^4$ neutrophils after 2 h.

Adult bovine neutrophils produced hydrogen peroxide in response to PAF (100 μg/ml) and PMA (2 μg/ml), and the PAF response was inhibited by R15.7 mAb (final concentration 5 μg/ml). There was much less hydrogen peroxide production by younger calf neutrophils in response to PMA, and no evidence of response to PAF. It has been reported that CD18-deficient human neutrophils are not reactive in the adherence-dependent assay even though they are capable of producing hydrogen peroxide in response to PMA stimulation (27). Failure of the 5-wk-old calf's cells to respond to PAF is not explained. PMA was effective with adult cells, but less so with young calf cells. This finding is consistent with work showing neonatal bovine neutrophils produce less reactive oxygen during phagocytosis than adults (29,30).

Surface expression of CD18, CD11a, and CD11b was initially evaluated using the following mAbs: R15.7 (anti-CD18), R3.1 (anti-Cd11a), and Leu 15 (anti-CD11b) with neutrophils of Emilee. CD11a and CD11b were detected at 6% and 10% of normal values while CD18 was detected at 15% of normal values (data not shown). However, several subsequent analyses of all three BLAD patients showed CD18 to be expressed at 2.3% of normal levels for resting PMN (range = 1.3–3.3% for resting PMN) to 1.9% of normal levels for PAF-stimulated PMN (range = 1.6–2.2%) (Table 27.2). The initially observed 15% level of CD18 expression compared to controls may have been an artifact of shipping blood samples since we have not seen this high value in over 12 subsequent flow cytometric analyses of leukocytes from the same animal conducted at our location with two separate flow cytometers.

When all three BLAD patients became available for study, additional analysis of surface adherence proteins on neutrophils was conducted (Tables 27.4, 27.5). Surface expression of CD18 and LECAM-1 was evaluated using mAb R15.7 and DREG-56, respectively. Using stained whole blood (Table 27.4) both PMA and PAF elicited some up-regulation

TABLE 27.4. Neutrophil Cd18 and LECAM-1 surface expression using whole blood (phycoerythrin-conjugated antibody used to label the primary antibody).

Bovine	CD18			LECAM-1		
	Resting	PAF	PMA	Resting	PAF	PMA
Control adult	424[a]	533	520	183	126	163
Control neonate (5 wk)	480	547[b]	480[b]	117	102	72
Meg (16 months)	7	11	16	43	35	35
Emilee (13 months)	5	6	7	30	28	25
Fred (2 wk)	2	12	11	87	68	50

[a] Represents mean fluorescence channel values determined after subtraction of values seen with a nonbinding control mAb.
[b] <45% of the cells responded to the stimulus.

of CD18 and some degree of down-regulation of LECAM-1 on normal bovine neutrophils, but these responses were not as marked as reported with human neutrophils. Resting neutrophil LECAM-1 levels in Meg and Emilee were markedly reduced compared to controls, and in Fred, about half of control resting neutrophil levels. Each CD18-deficient calf showed some binding of mAb R15.7 that appeared to be specific, and Fred and Meg showed a slight up-regulation of CD18 following exposure to PAF and PMA (Table 27.4).

Isolation of bovine neutrophils by hypotonic lysis appeared to up-regulate surface expression of CD18 and may have down-regulated LECAM-1 expression on neutrophil surfaces (Table 27.5). Both PMA and PAF stimulated up-regulation of CD18 on isolated control neutrophils to a higher degree than seen in whole blood samples. Unstimulated neutrophils of both Meg and Emilee had low levels of LECAM-1 that did not change with stimulation, while those of Fred had roughly half-normal levels of LECAM-1 that decreased minimally with stimulation. Resting neutrophils of Meg and Emilee showed low surface CD18 which did not increase with stimulation, whereas Fred's neutrophils showed almost undetectable CD18 levels until stimulated with PAF or PMA.

The low levels of LECAM-1 in Meg and Emilee may be a reflection of chronic subclinical infections. The small stimulus-induced down-regulation of LECAM-1 even in normal animals may be due to inadequate activation. This finding requires requires further study since the lack of response in Fred is different than was found in CD18-deficient humans (31) where fMLP caused almost complete loss of LECAM-1. The finding of reduced LECAM-1 on neutrophils of all three BLAD patients is significant in that Fred was maintained in a germ-free state until 13 days of age when *C. perfingens* was isolated from his feces. It is probable that *C. perfringens* was just established in the intestinal tract when Fred was sampled for the data presented in Tables 27.2–27.5 and that no

TABLE 27.5. Neutrophil CD 18 and LECAM-1 surface expression of isolatd neutrophils.

Bovine	CD18			LECAM-1		
	Resting	PAF	PMA	Resting	PAF	PMA
Control adult	610	975	720	139	86	123
Control neonate (5 wk)	600	920	730	170	101	153
Meg (16 months)	15	15	13	27	14	17
Emilee (13 months)	8	12	10	7	10	4
Fred (2 wk)	0	12	10	63	61	67

significant inflammatory response took place until about 3 wk of age (Fig. 27.2) based on only slightly elevated neutrophil counts until 23 days of age.

Conclusions

The granulocytopathy calves observed in veterinary clinics around the world have all been observed to be emaciated or stunted in growth and generally appear unthrifty. Recurrent bacterial infections and persistent, progressive neutrophilia have been noted early in the course of disease. Pyrexia, anorexia, chronic pneumonia, mild enlargement of superficial lymph nodes, ulcerative and granulomatous inflammation of oral mucous membranes, gingivitis, and mild recurrent or chronic diarrhea have all been reported in calves with bovine granulocytopathy syndrome (3,4,7,8).

The clinical, functional, and molecular characteristics of Holstein animals described here indicate that BLAD is similar to that described in humans and Irish setters and this further defines the role of β_2 integrins in the inflammatory host defense. Although our earlier report provided indirect evidence for CD18 deficiency underlying bovine granulocytopathy syndrome in Holstein cattle (8), the present studies on three live animals, including one produced by superovulation and embryo transfer of a carrier cow artificially inseminated with semen from a carrier bull, confirm the molecular basis for this syndrome. The birth of Fred establishes the USDA–National Animal Disease Center as the first facility with a reproducible animal model of human LAD. This research program will benefit studies into gene replacement therapy for human LAD and should provide a wealth of information regarding the basic immunobiology of neutrophil adherence and endothelial cell migration in various infectious disease conditions of cattle.

The value of BLAD as a human LAD model is high since BLAD closely mimics the imunological and clinical features observed in human patients considered to have a moderate to severe phenotype. It appears that there is some CD18 expressed on LAD calves' leukocytes (~2% of normal controls). Furthermore, it is clear that the entire family of β_2 integrins (LFA-1, Mac-1, and gp 150,95) is affected. Surface expression of CD18 on bovine neutrophils from normal (positive staining) and BLAD (no staining) cattle was also illustrated by staining neutophils with the R15.7 mAb followed by a secondary goat antimouse IgG conjugated to 30-nm colloidal gold and then using scanning electron microscopy and backscatter imaging (Ackermann et al., manuscript in preparation). Transmission electron microscopic examination of similarly prepared cells indicated that Mac-1 may be internalized following binding with anti-CD18 mAb.

We hope to produce more BLAD calves since we have additional pregnant surrogates. To date, we have identified five live Holstein cattle (oldest one still alive is 28 months of age) homozygous for the D128G CD18 mutation with a PCR-CNA test (21). We have also confirmed over 15 additional cases of BLAD dating back to 1977 using formalin-fixed tissues from cattle suspected to have bovine granulocytopathy syndrome in Iowa, New York, and Wisconsin. Thus, it is probable that the vast majority (if not all) of BLAD patients represent a homozygous genotype for a single mutant allele.

The dairy industry throughout the world relies on the use of artificial insemination with frozen semen. In the United States, approximately 70% of 10 million dairy cattle are bred through artificial insemination. Testing of approximately 4,000 bulls and cows mated for the purpose of producing future sires used in artificial insemination throughout the world is currently in progress. Results from this testing will provide the necessary information to phase out the use of carriers within 5 y without any significant loss of the existing gene pool necessary for high-quality milk production. It will also be possible to predict with some confidence which animals introduced this defect into the Holstein breed. The extensive (thousands of animals) pedigree information available from the breed registry will contribute greatly to establishing whether BLAD is a recent mutation in Holstein cattle or whether it was present before establishment of the breed.

References

1. Renshaw, H.W., C. Chatburn, G.M. Bruan, R.C. Bartsch and W.C. Davis. 1975. Canine granulocytopathy syndrome: neutrophil dysfunction in a dog with recurrent infections. *J. Am. Vet. Med. Assoc.* 166:443.
2. Crowley, C.A., J.T. Curnutte, R.E. Rosin, J. Andre-Schwartz, J.I. Gallin, M. Klempner, R. Snyderman, F.S. Southwick, T.P. Stossel and B.M. Babior. 1980. An inherited abnormality of neutrophil adhesion: its genetic transmission and its association with a missing protein. *N. Eng. J. Med.* 302:1163.
3. Hagemoser, W.A., J.A. Roth, J. Löfstedt and J.A. Fagerland. 1983. Granulocytopathy in a Holstein heifer. *J. Am. Vet. Med. Assoc.* 183:1093.

4. Nagahata, H., H. Noda, K. Takahashi, T. Kurosawa and M. Sonoda. 1987. Bovine granulocytopathy syndrome: Neutrophil dysfunction in Holstein Friesian calves. *J. Vet. Med. A.* 34:445.

5. Dana, N., I.R.F. Todd, J. Pitt, T.A. Spriger and M.A. Arnaout. 1984. Deficiencies of a surface membrane glycoprotein (Mo1) in man. *J. Clin. Invest.* 73:153.

6. Anderson, D.C., F.C. Schmalstieg, S. Kohl, M.A. Arnaout, B.J. Hughes, M.F. Hollers and C.W. Smith. 1984. Abnormalities of polymorphonuclear leukocyte function associated with a heritable deficiency of high molecular weight surface glycoproteins (GP 138): Common relationship to diminished cell adherence. *J. Clin. Invest.* 74:536.

7. Takahashi, K., K. Miyagawa, S. Abe, T. Kurosawa, M. Sonoda, T. Nakade, H. Nagahata, H. Noda, Y. Chihaya and E. Isogai. 1987. Bovine granulocytopathy syndrome of Holstein-Friesian calves and heifers. *Jpn. J. Vet. Sci.* 49:733.

8. Kehrli, M.E., Jr., F.C. Schmalstieg, D.C. Anderson, M.J. Van Der Maaten, B.J. Hughes, M.R. Ackermann, C.L. Wilhelmsen, G.B. Brown, M.G. Stevens and C.A. Whetstone. 1990. Molecular definition of the bovine granulocytopathy syndrome: Identification of deficiency of the Mac-1 (CD11b/CD18) glycoprotein. *Am. J. Vet. Res.* 51:1826.

9. Giger, U., L.A. Boxer, P.J. Simpson, B.R. Lucchesi and R.F.T. III. 1987. Deficiency of leukocyte surface glycoproteins Mo1, LFA-1, and Leu M5 in a dog with recurrent bacterial infections: an animal model. *Blood.* 69:1622.

10. Anderson, D.C., F.C. Schmalstieg, M.J. Finegold, B.J. Hughes, R. Rothlein, L.J. Miller, S. Kohl, M.F. Tosi, R.L. Jacobs, T.C. Waldrop, A.S. Goldman, W.T. Shearer and T.A. Springer. 1985. The severe and moderate phenotypes of heritable Mac-1, LFA-1 deficiency: Their quantitative definition and relation to leukocyte dysfunction and clinical features. *J. Infect. Dis.* 152:668.

11. Anderson, D.C. and T.A. Springer. 1987. Leukocyte adhesion deficiency: An inherited defect in the Mac-1, LFA-1 and p150,95 glycoproteins. *Ann. Rev. Med.* 38:175.

12. Springer, T.A., W.S. Thompson, L.J. Miller, F.C. Schmalstieg and D.C. Anderson. 1984. Inherited deficiency of the Mac-1, LFA-1, p150,95 glycoprotein family and its molecular basis. *J. Exp. Med.* 160:1901.

13. Kishimoto, T.K., N. Hollander, T.M. Roberts, D.C. Anderson and T.A. Springer. 1987. Heterogenous mutations in the β subunit common to the LFA-1, Mac-1, and p150,95 glycoproteins cause leukocyte adhesion deficiency. *Cell.* 50:193.

14. Todd, R.F., III and D.R. Freyer. 1988. The CD11/CD18 leukocyte glycoprotein deficiency. *Hematol. Oncol. Clin. N. Am.* 2:13.

15. Marlin, S.D., C.C. Morton, D.C. Anderson and T.A. Springer. 1986. LFA-1 immunodeficiency disease: Definition of the genetic defect and chromosomal mapping of alpha and beta subunits by complementation in hybrid cells. *J. Exp. Med.* 164:855.

16. Dana, N., L.K. Clayton, D.G. Tennen, M.W. Pierce, P.J. Lachmann, S.A. Law and M.A. Arnaout. 1987. Leukocytes from four patients with complete or partial Leu-CAM deficiency contain the common beta-subunit precursor and beta-subunit messenger RNA. *J. Clin. Invest.* 79:1010.

17. Kishimoto, T.K., K. O'Connor and T.A. Springer. 1989. Leukocyte adhesion deficiency: Aberrant splicing of a conserved integrin sequence causes a moderate deficiency phenotype. *J. Biol. Chem.* 264:3588.
18. Hibbs, M.L., A.J. Wardlaw, S.A. Stacker, D.C. Anderson, A. Lee, T.M. Roberts and T.A. Springer. 1990. Transfection of cells from patients with leukocyte adhesion deficiency with an integrin β subunit (CD18) restore lymphocyte function-associated antigen-1 expression and function. *J. Clin. Invest.* 85:674.
19. Shuster, D.E., Brad T. Bosworth and M.E. Kehrli Jr. 1992. Sequence of bovine CD18 cDNA: comparison with human and murine sequences. *Gene* 114:267.
20. Waldrop, T.C., D.C. Anderson, W.W. Hallmon, F.C. Schmalstieg and R. L. Jacobs. 1987. Peridontal manifestations of the heritable Mac-1, LFA-1, deficiency syndrome. Clinical, histopathologic and molecular characteristics. *J. Periodontol.* 58:400.
21. Shuster, D.E., M.E. Kehrli Jr., R.O. Gilbert and M.R. Ackermann. 1992. A prevalent point mutation responsible for leukocyte adhesion deficiency in Holstein dairy cattle. *P.N.A.S.* 89:(In press).
22. Arnaout, M.A., J. Pitt, H.J. Cohen, J. Melamed, F.S. Rosen and H.R. Colten. 1982. Deficiency of a granulocyte-membrane glycoprotein (gp150) in a boy with recurrent bacterial infections. *N. Eng. J. Med.* 306:693.
23. Kohl, S., T.A. Springer and F.C. Schmalstieg. 1984. Defective natural killer cytotoxicity and polymorphonuclear leukocyte antibody dependent cellular cytotoxicity in patients with LFA-1/OKM-1 deficiency. *J. Immunol.* 133:2972.
24. Kohl, S., L.S. Loo, F.C. Schmalstieg and D.C. Anderson. 1986. The genetic deficiency of leukocyte surface glycoprotein Mac-1, LFA-1 p150,95 in humans is associated with defective antibody-dependent cellular cytotoxicity in vitro and defective protection against herpes simplex virus in vivo. *J. Immunol.* 137:1688.
25. Krensky, A.M., S.J. Mentzer, C. Clayberger, D.C. Anderson, F.C. Schmalsteig, S.J. Burakoff and T.A. Springer. 1985. Heritable lymphocyte function-associated antigen-1 deficiency: Abnormalities of cytotoxicity and proliferation associated with abnormal expression of LFA-1. *J. Immunol.* 135:3102.
26. Smith, C.W., J.C. Hollers, R.A. Patrick and C. Hassett. 1979. Motility and adhesiveness in human neutrophils: effects of chemotactic factors. *J. Clin. Invest.* 63:221.
27. Shappell, S.B., C. Toman, D.C. Anderson, A.A. Taylor, M.L. Entman and C.W. Smith. 1990. Mac-1 (CD11b/CD18) mediates adherence-dependent hydrogen peroxide production by human and canine neutrophils. *J. Immunol.* 144:2702.
28. Entman, M.L., K. Youker, S.B. Shappell, R. Rothlein, W.J. Dreyer, F.C. Schmalsteig and C.W. Smith. 1990. Neutrophil adherence to isolated adult canine myocytes: Evidence for a CD18-dependent mechanism. *J. Clin. Invest.* 85:1497.
29. Dore, M., D.O. Slauson and N.R. Neilsen. 1990. Membrane NADPH oxidase activity and cell size in bovine neonatal and adult neutrophils. *Pediatr. Res.* 28:327.
30. Dore, M., D.O. Slauson and N.R. Neilsen. 1991. Decreased respiratory burst

activity in neonatal bovine neutrophils stimulated by protein kinase C agonists. *Am. J. Vet. Res.* 52:375.
31. Smith, C.W., T.K. Kishimoto, O. Abbassi, B.J. Hughes, R. Rothlein, L.V. McIntire, E. Butcher and D.C. Anderson. 1991. Chemotactic factors regulate lectin adhesion molecule 1 (LECAM-1) dependent neutrophil adhesion to cytokine-stimulated endothelial cells in vitro. *J. Clin. Invest.* 87:609.

28
The Role of Leukocyte Adhesion Molecules During Ischemia and Reperfusion

THOMAS F. LINDSAY, JAMES HILL, C.R. VALERI, D. SHEPRO, AND
HERBERT B. HECHTMAN

Introduction

Neutrophils have been implicated in tissue injury following ischemia (1). Initial evidence of their involvement was based largely on experiments using neutrophil depletion to modify ischemic injury (2). Recent evidence in heart, liver, intestine, and skeletal muscle has suggested that neutrophil interaction in this setting involves the leukocyte integrin molecules (3–6).

Cytokines, components of ischemic injury, have been shown in vitro to stimulate endothelium to increase adhesion of unstimulated neutrophils in a mechanism dependent upon neutrophil CD11a/CD18 (LFA-1) and CD11b/CD18 (MAC-1) and endothelial intercellular adhesion molecule (ICAM) (7,8). Cytokine-stimulated endothelium will up-regulate ICAM and endothelial leukocyte adhesion molecule (ELAM) expression by a protein-synthesis–dependent mechanism over different time courses (8). The demonstration of tumor necrosis factor (TNF) release following liver and intestinal ischemia suggests that cytokine stimulation of the endothelium occurs in vivo following some ischemic injuries (9,10).

Lower torso ischemia and reperfusion lead to local muscle injury, the sequestration of PMNs within the lung vasculature, and increased lung permeability (11). The remote lung injury has been associated with an early thromboxane-stimulated neutrophil activation, with increased metabolic oxidative activity (12). The present studies focus on the role of the leukocyte integrins in hindlimb tourniquet ischemia and intestinal ischemia particularly with regard to the activity of cytokines in inducing lung neutrophil sequestration in these settings. We sought to determine if the lung endothlium was cytokine stimulated and whether this was the mechanism for leukosequestration and lung injury. We examined the role of these adhesion molecules in mediating injury to the lungs themselves following atelectasis.

Results

Following a 3-h period of bilateral hindlimb tourniquet ischemia in the rabbit, at 10 min of reperfusion a significant neutropenia to $3,560 \pm 490/$ mm^3 developed that resolved at 1 h of reperfusion. The administration of R15.7 (a mAb against CD18) 10 min prior to reperfusion prevented this neutropenia (13).

In these same experiments the lungs progressively sequestered neutrophils during 4 h of reperfusion as assessed by quantative histology. This increase was prevented in animals treated with R15.7. After 4 h of reperfusion lung permeability rose as quantified by bronchoalveolar lavage (BAL) protein content and increased lung wet-to-dry-weight ratio. The increased lung permeability was significantly reduced by treatment with R15.7.

A separate series of experiments in the rat was conducted to assess the role of cytokines in the lung injury that follows hindlimb ischemia (11). Increased serum levels of TNF were noted, using an enzyme-linked immunosorbent assay, in three out of six animals tested following 4 h of bilateral hindlimb ischemia and 4 h of reperfusion. During the 4-h period of reperfusion there was a progressive increase in lung neutrophil sequestration (250% of sham) (Fig. 28.1). This was significantly reduced although

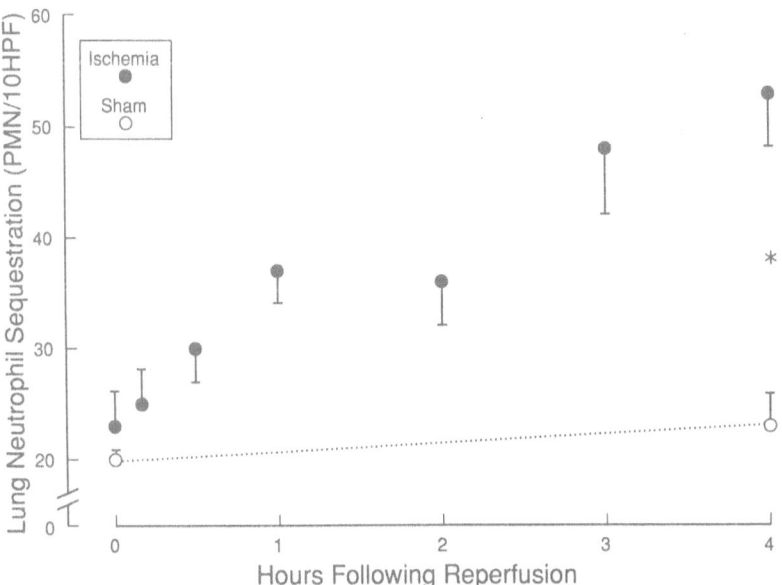

Figure 28.1. Following hindlimb ischemia there was a progressive increase in lung neutrophil sequestration. * indicates $p < 0.05$ relative to sham. [Reprinted with permission from *J. Appl. Physiol.* (11).]

not to normal (183% of sham) using a polyclonal TNF antiserum given 30 min prior to reperfusion. In addition to the reduction in neutrophil sequestration the increased lung permeability was modified as shown by a reduction in BAL protein and lung wet-to-dry-weight ratio. It is concluded that lower torso ischemia and reperfusion lead to the generation of TNF and that it contributes to the lung injury via direct neutrophil stimulation or via the activation of pulmonary endothelial cells or both.

A more impressive TNF release occurred within the first hour of reperfusion after 1 h of small bowel ischemia in the rat (14). In this setting the prominent lung neutrophil sequestration and permeability were completely prevented by treatment with TNF antiserum.

In vitro experiments have described the up-regulation of the CD11/CD18 complex on the neutrophil following contact with various chemotactic stimuli, such as leukotriene B_4 (LTB_4), FMLP, and complement fragment C5a. However, evidence suggests that increased surface expression of CD11/CD18 is not required for increased neutrophil adhesion (15). Welbourn et al. examined this issue following hindlimb ischemia in the rabbit (16). The postischemic rise in plasma LTB_4 levels (10^{-9}–10^{-10} M) was not reduced following administration of R15.7. Further, these LTB_4 levels were found to be insufficient to cause an up-regulation of CD18. Additional in vivo experiments were conducted following 3 h of ischemia and 10 min of reperfusion, at which time a significant leukopenia developed in the rabbits which could be prevented with R15.7. When control and ischemic plasma harvested at 10 min of reperfusion were incubated with neutrophils from a normal rabbit, no increase in the surface expression of CD18 was noted by flow cytometry. Also no increase in cell-surface expression of CD18 was noted in vivo in rabbits given R15.7 prior to hindlimb ischemia when measured before ischemia, or at 10 min pre- or postreperfusion. Reperfusion plasma resulted in marked neutrophil diapedesis in rabbit dermabrasion chambers in a dose-dependent manner, while ischemic plasma (collected prior to reperfusion) did not result in diapedesis (Fig. 28.2). This diapedesis was prevented by treatment of rabbits with intravenous R15.7 prior to application of the reperfusion plasma to the dermabrasion chambers. Finally, diapedesis induced by the ischemic/reperfused plasma was unaffected by the inhibition of protein synthesis. Thus, plasma harvested following bilateral hindlimb ischemia and reperfusion was able to produce significant neutrophil diapedesis although it did not increase the expression of the CD18 complex on the surface of neutrophils. This series of in vivo experiments confirms that the activity of the CD11/CD18 integrin appears to be due to a translational or conformational change and not due to up-regulation. This conformational change of the CD11/CD18 complex is sufficient to induce neutrophil adhesion and diapedesis as well as the leukopenia observed at 10 min of reperfusion.

The role of Mac-1 has been investigated in another series of ischemia

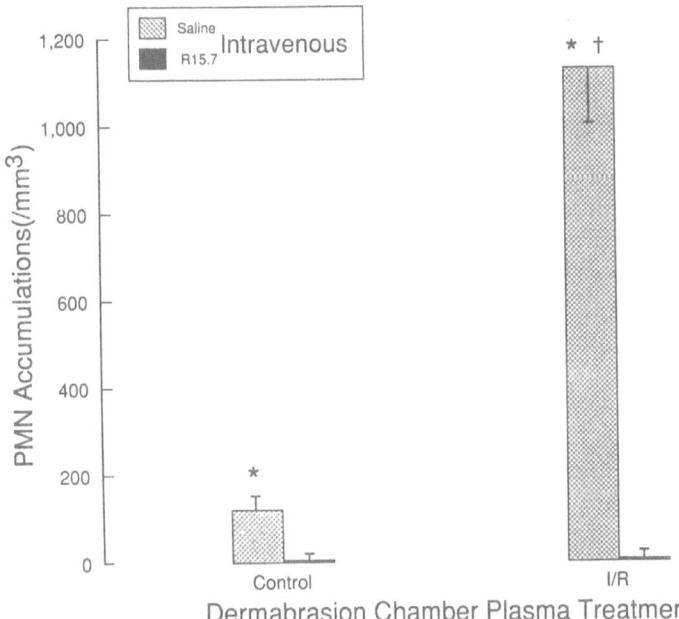

FIGURE 28.2. Introduction of ischemia/reperfusion plasma into dermabrasion chambers resulted in diapedesis 3 h later. Treatment with R15.7 prevented diapedesis for both ischemia/reperfusion and control plasma. * indicates $p < 0.01$ relative to mAb R15.7 in each group. † indicates $p < 0.01$ relative to control plasma. [Reprinted with permission from *J. Immunol.* (16).]

and reperfusion experiments. In rats, 1 h of small bowel ischemia followed by 3 h of reperfusion lead to significant lung injury (14). In animals treated with AB 17 (an mAb against the CD11b subunit) the increased lung permeability was reduced to sham levels (20). However no reduction in lung neutrophil sequestration was observed when compared to animals treated with a control antibody. In a further series of observations, neutralization of cytokine action with either antiserum to TNF or with the interleukin-1 receptor antagonist protein resulted in a reduction of lung myeloperoxidase levels and lung permeability to those of sham controls.

Atelectasis was induced by ventilation of the lungs with 100% oxygen followed by occlusion of the right middle lobe bronchus for 1 h (unpublished observations). Twenty minutes after reexpansion leukopenia was noted, $2{,}870 \pm 210$ WBC/mm^3 compared to control animal values of $6{,}500$ WBC/mm^3. Three hours following reexpansion neutrophils were sequestered in both the atelectatic and nonatelectatic lung segments. The lung that contained the previously atelectatic region had increased permeability as demonstrated by a rise in BAL protein content and lung wet-to-dry-weight ratio. Administration of nitrogen mustard to render the rabbits neutropenic abolished the lung leukosequestration and limited both the BAL protein accumulation and the increased wet-to-dry-weight ratio. In animals given mAb R15.7 the initial leukopenia was prevented and the later lung neutrophil sequestration and permeability were markedly reduced.

Discussion

Administration of an anti–CD18 mAb following hindlimb ischemia prevented the neutropenia, lung neutrophil sequestration, and permeability. Other experimental work has demonstrated a rise in neutrophil oxidative metabolism following this ischemic injury (12). These two experiments would suggest that during reperfusion there is activation of the CD11/CD18 integrins on the neutrophils and that these adhesion molecules then may interact with pulmonary endothelial cell ICAM resulting in lung leukosequestration and increased permeability. Supporting findings were noted following treatment of animals whose hindlimbs had been made ischemic, and they were administered TNF antiserum, which limited the lung neutrophil sequestration and increased permeability. Although TNF antiserum incompletely reduced the lung neutrophil sequestration while R15.7 completely abolished the PMN sequestration, we conclude that an important TNF-dependent component is operative. One may speculate that the TNF antiserum was not completely effective in neutralizing TNF because of its synthesis prior to tourniquet release, via a nonoccluded medullary artery (12). This was not the case with intestinal ischemia, where vascular occlusion was complete, as was the effect of the TNF antiserum.

It is likely that the neutrophil–endothelial cell adhesive interaction is based upon CD11a/CD18 binding to ICAM (7). As the neutrophil responds to cytokine stimulation, an increased role for CD11b/CD18 would be expected although this integrin is not a determinant of adhesion in the intestinal ischemia model (20). The reduction in lung permeability noted with R15.7 might be expected to be due largely to inhibition of CD11a/CD18-mediated PMN adhesion and secondly to the inhibition of CD11b/CD18 adhesion-dependent hydrogen peroxide production (17). These same mechanisms of injury are assumed to also apply to the atelectatic lung.

Following small intestinal ischemia, lung permeability and leukosequestration appear totally dependent upon the action of TNF and IL-1. While it was not possible to separate the effect of the cytokines on the pulmonary endothelium and the neutrophils, nevertheless one may suspect that both cell types are activated. The inability of the MAC-1 mAb to significantly reduce lung leukosequestration would suggest that adhesion occurs secondary to the interaction of LFA-1 (CD11a/CD18) with ICAM and perhaps a neutrophil glycoprotein expressing sialyl-Lewis X interacting with ELAM. The fact that an mAb against MAC-1 prevents permeability following neutrophil adhesion indicates that a second step following adhesion is required to induce increased lung permeability. The second step may be MAC-1–dependent hydrogen peroxide production and granule release (15).

The inability of AB 17 to reduce lung PMN sequestration indicates a dissociation between adhesion and microvascular permeability (20). These finding are not unique. In a study of lung injury induced by endotoxin and FMLP, increased lung ^{125}I albumen flux was reduced by a platelet-activating factor antagonist despite a failure to reduce lung neutrophil sequestration (18). In another study of hepatic injury, mAbs against CD11b and CD18 prevented the injury but neutrophil accumulation was reduced by only 50% (19). Finally, in vitro experiments have demonstrated that mAbs against CD11b inhibited the respiratory burst of platelet-activating-factor–stimulated neutrophils without reducing their adhesion to endothelial cells (17).

In summary, following hindlimb or intestinal ischemia and lung atelectasis an increase in lung neutrophil sequestration and permeability was observed, events mediated by CD11/CD18. Following intestinal but not hindlimb ischemia, cytokines could be shown to be the major intermediate responsible for lung PMN sequestration and injury. Further, MAC-1 was responsible for the increased lung permeability, but not the neutrophil adhesion.

References

1. Lucchesi, B.R., and K.M. Mullane. 1986. Leukocytes and ischemia-induced myocardial injury. *Ann. Rev. Pharmacol. Toxicol.* 26:201.
2. Klausner, J.M., H. Anner, I.S. Paterson, L. Kobzik, C.R. Valeri, D.Shepro, and H.B. Hechtman. 1988. Lower torso ischemia-induced lung injury is leukocyte dependent. *Ann. Surg.* 208:761.
3. Entman, M.L., L. Michael, R.D. Rossen, W.J. Dreyer, D.C. Anderson, A.A. Taylor, and C.W. Smith. 1991. Inflammation in the course of early myocardial ischemia. *FASEB J.* 5:2529.
4. Jaeschke, H., A. Farhood, and C.W. Smith. 1990. Neutrophils contribute to ischemia/reperfusion injury in rat liver in vivo. *FASEB J.* 4:3355.
5. Hernandez, J.M., M.B. Grisham, B.Twohig, K.E. Arfors, J.M. Harlan, and D.N. Granger. 1987. Role of neutrophils in ischemia-reperfusion-induced microvascular injury. *Am. J. Physiol.* 253:H699.
6. Korthius, R.J., M.B. Grisham, and D.N. Granger. 1988. Leukocyte depletion attenuates vascular injury in postischemic skeletal muscle. *Am. J. Physiol.* 254:H823.
7. Smith, C.W., S.D. Marlin, R. Rothlein, C. Toman, and D.C. Anderson. 1989. Cooperative interactions of LFA-1 and Mac-1 with intercellular adhesion molecule-1 in facilitating adherence and transendothelial migration of human neutrophils in vitro. *J. Clin. Invest.* 83:2008.

8. Luscinskas, F.W., M.I. Cybulsky, J.M. Kiely, C.S. Peckins, V.M. Davis, and M.A. Gimbrone. 1991. Cytokine-activated human endothelial monolayers support enhanced neutrophil transmigration via a mechanism involving both endothelial leukocyte adhesion molecule-1 and intercellular adhesion molecule-1. *J. Immunol.* 146:1617.

9. Colleti, L.M., D.G. Remick, G.D. Burtch, S.L. Kunkel, R.M. Streiter and D.A. Campbell. 1991. Role of tumor necrosis factor alpha in the pathophysiologic alterations after hepatic ischemia/reperfusion injury in the rat. *J. Clin. Invest.* 85:1936–1943.

10. Caty, M.G., K.S. Guice, T.K. Oldham, D.G. Remick, and S.L. Kunkel. 1990. Evidence for tumor necrosis factor induce pulmonary microvascular Injury after intestinal ischemia-reperfusion. *Ann. Surg.* 212:694

11. Welbourn, R., G. Goldman, M. O'Riordain, T.F. Lindsay, I.S. Paterson, L. Kobzik, C.R. Valeri, D. Shepro, and H.B. Hechtman. 1991. Role of tumor necrosis factor as mediator of lung injury following lower torso ischemia. *J. Appl. Physiol.* 70:2645.

12. Paterson, I.S., J.M. Klausner, G. Goldman, L. Kobzik, R. Welbourn, C.R. Valeri, D. Shepro, and H.B. Hechtman. 1989. Thromboxane mediates the ischemia induced neutrophil oxidative burst. *Surgery* 106:224.

13. Welbourn, R., G. Goldman, J. Hill, T.F. Lindsay, C.R. Valeri, D. Shepro, and H.B. Hechtman. 1991. Lung injury following hindlimb ischemia is mediated by neutrophil CD18 adherence receptors. *FASEB J.* 5:6502 (abst.).

14. Hill, J., T.F. Lindsay, L. Kobzik, C.R. Valeri, D. Shepro, and H.B. Hechtman. 1991. Tumor necrosis factor alpha mediates intestinal ischemia/reperfusion induced liver and lung injury. *FASEB J.* 5:7259 (abst.).

15. Schileiffenbaum, B., R. Moser, M. Patarroyo, and J. Fehr. 1989. The cell surface glycoprotein Mac-1 (CD 11b/CD18) mediates neutrophil adhesion and modulates degranulation independently of its quantitative cell surface expression. *J. Immunol.* 142:3537.

16. Welbourn, R., G. Goldman, L. Kobzik, I.S. Paterson, C.R. Valeri, D. Shepro, and H.B. Hechtman. 1990. Neutrophil adherence receptors (CD 18) in ischemia; dissociation between quantitative cell surface expression and diapedesis mediated by leukotriene B_4. *J. Immunol.* 145:1906–1911.

17. Shappell S.B., C. Toman, D.C. Anderson, A.A. Taylor, M.L. and C. Wayne Smith. 1990. Mac-1 (CD 11b/CD 18) mediates adherence-dependent hydrogen peroxide production by human and canine neutrophils. *J. Immunol.* 144:2702.

18. Anderson, B.O., R.S. Poggetti, P.F. Shanley, D.D. Bensard, J.M. Pitman, D.W. Nelson, G.J.R. Whitman, A. Banerjee, and A.H. Harken. 1991. Primed neutrophils injure rat lung through a platelet-activating factor-dependent mechanism. *J. Surg. Res.* 50:510.

19. Jaeschke H., A. Farhood, and C.W. Smith. (1991). Neutrophil-induced liver cell injury in endotoxin shock is a CD11b/CD18-Dependent Mechanism *AMS Physiol.* 261 G1051–G1056

20. Hill, J., Lindsay, T., Rusche, J., Valeri, C.R., Shepro, D., Hechtman, H.B. A Mac-1 antibody reduces liver and lung injury but not neutrophil sequestration following intestinal ischemia-reperfusion. *Surgery.* 1992 (in press).

Part 5
Clinical

29
Abnormalities of LECAM-1– and Mac-1–Dependent Neutrophil–Endothelial Cell Adhesion in the Developing Host

Donald C. Anderson, Omid Abbassi, James D. Fortenberry, Joyce Koenig, and C. Wayne Smith

Introduction

Incompletely defined inflammatory deficits contribute to infectious morbidity and mortality in human neonates and neonatal animals (1). Previous studies in neonatal animal models have demonstrated diminished or delayed emigration of blood neutrophils into extravascular inflammatory sites (2–4) which, in turn, may reflect deficits of neutrophil chemotactic functions as demonstrated by numerous in vitro investigations (5,6). To understand the molecular and functional basis for these (presumably) developmental deficits, recent studies have focused on the molecular events contributing to neutrophil–endothelial interactions in vitro and in vivo. As described in this report, significant abnormalities of CD18-dependent as well as CD18-independent adherence mechanisms appear to underlie diminished neutrophil localization and/or function (7–12). These abnormalities may contribute to impaired host defense capabilities or, alternatively, may serve a physiologic role against inflammatory injury of neonatal tissues in the postpartum setting. These possibilities will be addressed in future investigations which should exploit the neonate as a model of inflammation, to allow a further definition of its molecular basis.

Results

Studies of CD18-Dependent Mechanisms of Neutrophil–Endothelial Cell Adherence In Vitro

Comparative Studies of Transendothelial Migration by Neonatal or Healthy Adult Neutrophils

A visual neutrophil adherence assay employing human umbilical vein endothelial cell (HUVEC) monolayers under static (1 g) conditions was used to demonstrate significant deficits of transendothelial migration by

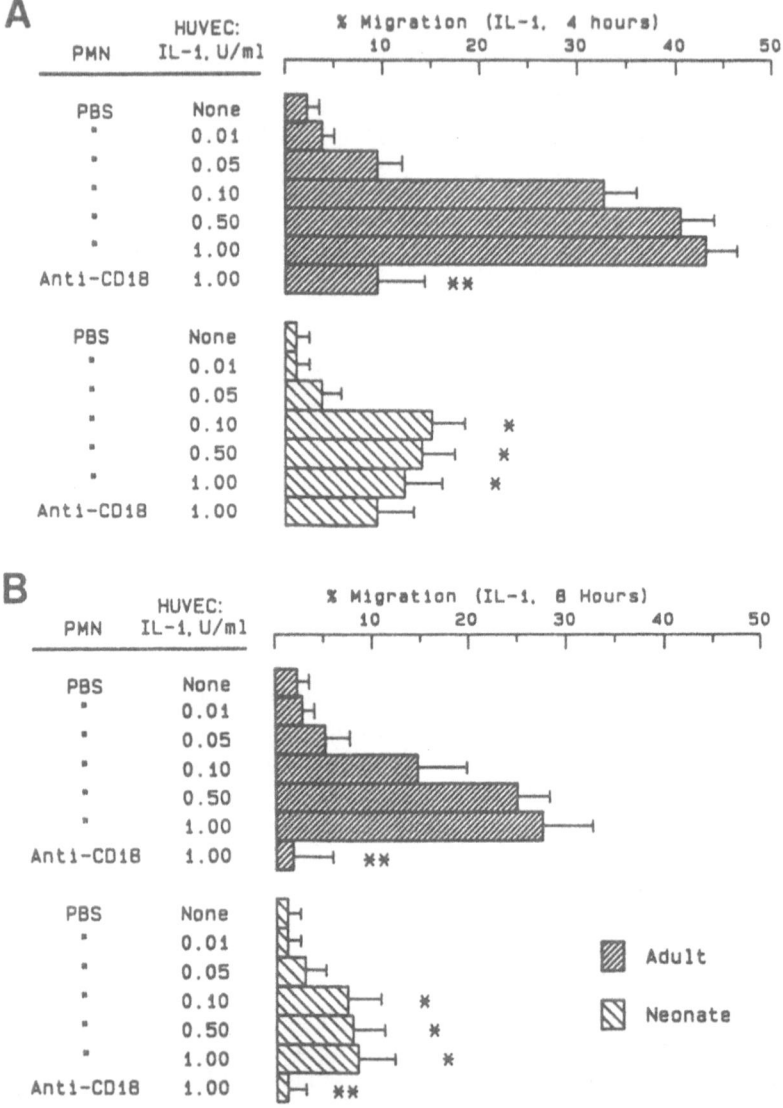

FIGURE 29.1. Transendothelial migration by neonatal or adult neutrophils. Confluent endothelial cell monolayers were preincubated with varying concentrations of IL-1 for 4 h (A) or 8 h (B) at 37°C and washed twice with PBS. Neutrophils incubated in either PBS or PBS with 5 μg/ml of anti-CD18 MAb (TS1/18) were injected into adhesion chambers and incubated for an additional 1,000 s at 21°C. The percentage of cells migrating through the monolayers was visually determined. Values shown represent the mean ± SD of 6–23 separate determinations among neonatal or healthy adult test groups and are expressed as the percentage of neutrophils originally contacting the monolayer; (*) p < 0.001 compared with mean values for healthy adult cells studies under the same experimental conditions; (**) p < 0.001 compared with mean values for healthy adult cells studies under the same experimental conditions in the absence of anti-CD18 MAb.

neonatal neutrophils (13) (Fig. 29.1). As shown, adult but not neonatal neutrophils demonstrate a dose-dependent increase in migration following incubation of HUVEC with IL-1 (or LPS) for 4 h. That abnormal CD18-dependent mechanisms underlie diminished migratory properties of neonatal neutrophils was shown by the inability of anti-CD18 monoclonal antibodies (MAb) to significantly inhibit their migration, as was shown for adult neutrophils. Moreover, in related studies using HUVEC substrates incubated with IL-1 for 8 or 24 h (conditions which select for CD18-dependent migration), neonatal neutrophils demonstrated profoundly diminished migration values.

Since previous studies have defined cooperative interactions of Mac-1 (CD11b/CD18) and LFA-1 (CD11a/CD18) in facilitating neutrophil adherence to and migration through HUVEC monolayers (14), a series of

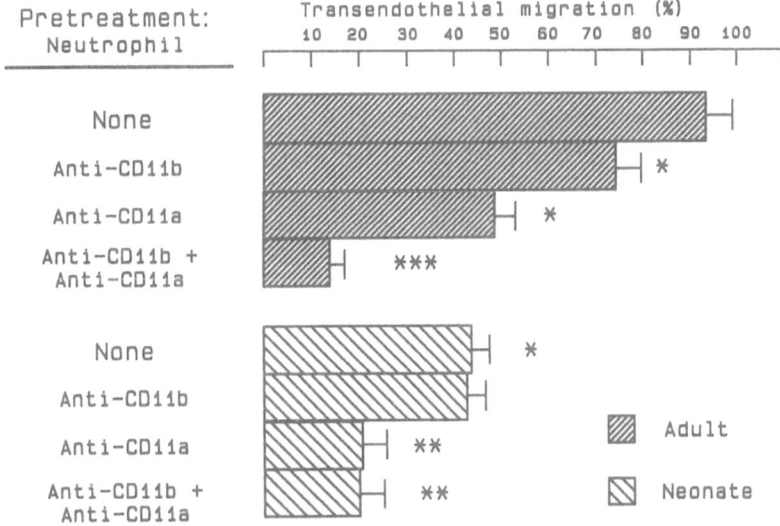

FIGURE 29.2. Effects of anti-Mac-1 (CD11b) or LFA-1 (CD11a) MAbs on neutrophils migration. Endothelial cell monolayers were preincubated with IL-1 for 4 at 37°C and washed twice with PBS. Neonatal or healthy adult neutrophils were incubated in either PBS or PBS containing 5 μg/ml anti-CD11b MAB (904) and/ or anti-CD11a MAb (R3.1). Cell suspensions were injected into adhesion chambers and allowed to contact monolayers for 1,000 s at 37°C. Migration was visually evaluated as described (13). Values shown represent the mean ± SD of determinations for 12 neonatal (coarsely hatched bars) and six healthy adult (finely hatched bars) neutrophil suspensions and are expressed as a percentage of cells originally contacting the monolayer; (*) $p < 0.01$ compared with healthy adult neutrophils, no antibody pretratment; (**) $p < 0.001$ compared with neonatal neutrophils, no antibody pretreatment; (***) $p < 0.001$ compared with adult neutrophils + anti-CD11a MAb alone.

studies was carried out to evaluate the relative importance of each of these integrins with respect to diminished migration by neonatal cells. Initially, subunit-specific MAbs were incorporated in comparative migration assays as shown in Figure 29.2. For adult neutrophils, anti-CD11a and anti-CD11b MAbs were inhibitory when used alone, and a significantly greater inhibitory effect was observed when these MAbs were used in combination. In contrast, migration by neonatal cells was significantly inhibited by anti-CD 11a but not anti-CD11b MAbs. Moreover, the degree of inhibition evident when neonatal cells were preincubated with a combination of these MAbs was not significantly different than that observed with anti-CD11a MAbs alone. These findings suggest abnormalities of Mac-1 but not LFA-1 underlie diminished migration by neonatal cells.

Influence of Chemotactic Stimuli on Neutrophil Adhesion

Augmentation of neutrophil adhesion to unstimulated HUVEC by chemotactic factors is known to involve quantitative and/or functional activation of Mac-1 on the neutrophil surface (14,15). Moreover, our previous

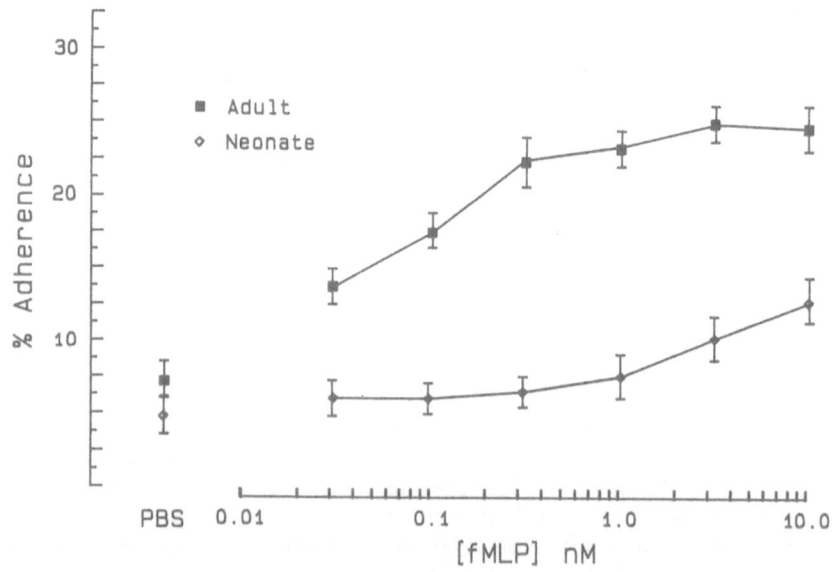

FIGURE 29.3. Effects of chemotactic stimulation on the adhesion of neonatal or adult neutrophils to unstimulated endothelial cells. Neutrophils were exposed to the concentration of fMLP indicated for 300 s at 37°C before contacting endothelial cell monolayers in adhesion chambers. Results shown represent the mean ± SD values of determinations on six neonates and six adult cell suspensions tested and are expressed as the percent of neutrophils adherent to the monolayers.

studies have shown that chemotactic factors fail to normally elicit an up-
regulation of surface Mac-1– or Mac-1–dependent adhesion when incubated with neonatal neutrophils (11,16). Thus, a series of protocols was
carried out to evaluate the contribution of Mac-1 to endothelial cell adhesion by neonatal cells. In dose-response studies using F-Met-Leu-Phe
(fMLP) and unstimulated HUVEC substrates, neonatal neutrophils demonstrated profoundly diminished responses as compared with adult cells
over a broad concentration range of this chemotactic factor (Fig. 29.3). In
separate protocols, adhesion of fMLP-stimulated neonatal or control
neutrophils was evaluated on HUVEC preincubated with IL-1 for 8 h, a
condition which selects for CD18-dependent endothelial ligands (13). As
shown in Figure 29.4, fMLP failed to enhance adherence of neonatal cells
under these conditions. Moreover, the increment of control cell adhesion
elicited by fMLP was completely abrogated by anti-CD11b MAbs, a

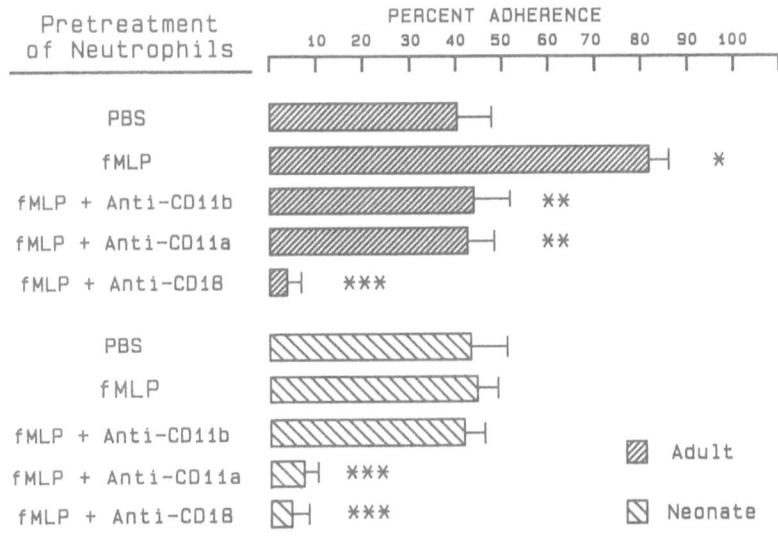

FIGURE 29.4. Comparative effects of anti-CD11a, -CD11b, and -CD18 MAbs on
adherence of neutrophils to IL-1–stimulated endothelial cells. Endothelial cell
monolayers were preincubated with IL-1 for 8 h at 37°C and washed before assay.
Neonatal or adult neutrophils were incubated with 904 (anti-CD11b) (10 μg/ml),
R3.1 (anti-CD11a) (10 μg/ml), or R15.7 (anti-CD18) (10 μg/ml) MAb in the presence of fMLP (10 nmol/l) for 5 min before contacting the monolayer. Adherence
was evaluated as described (13). Results shown represent mean ± SD determinations for studies of five neonatal (finely hatched bars) and four healthy adult
(coarsely hatched bars) cell suspension; (*) p ≤ 0.01 compared with unstimulated
neonatal or adult suspension; (**) p ≤ 0.01 compared with stimulated adult neutrophils; (***) p ≤ 0.01 compared with stimulated adult or neonatal suspensions
containing anti-CD11b.

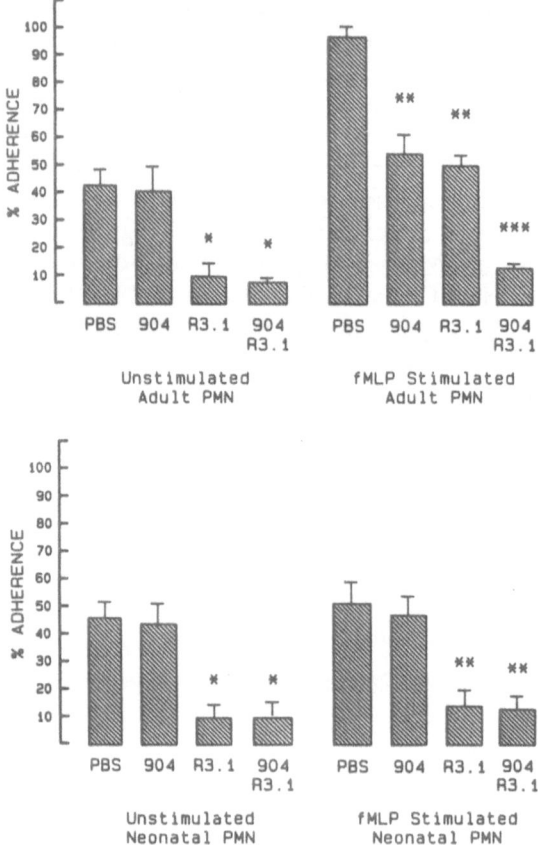

FIGURE 29.5. Adhesion of neonatal or adult neutrophils to purified ICAM-1 in artificial planar membranes. Utilizing techniques previously described (13) the adherence of unstimulated or fMLP-stimulated (10 nM, 5 min, 37°C) neutrophils was assessed in the presence of absence of MAb directed against CD11b (904), CD11a (R3.1) or a combination of these MAb. Values shown represent the mean ± 1 SD of 8–12 determinations on neonatal or adult cell suspensions tested; (*) p ≤ .001 compared with unstimulated adult or neonatal cells, no antibody treatment; (**) p ≤ .005 compared with stmulated adult or neonatal cells, no antibody treatment; (***) p < .001 compared to adult cells treated with anti CD116 (904) or anti CD11a Mr6.

finding not observed with neonatal cells in the presence of this chemotactic factor. Anti-LFA-1 (CD11a) MAb significantly inhibited neutrophil adhesion among both test groups, but in contrast to findings with adult neutrophils, this MAb was as inhibitory as anti-CD18 MAb when tested with neonatal cells. These findings indicate that chemotactic factors fail to elicit Mac-1–dependent adhesion of neonatal to endothelial cells, a process that is required for normal transendothelial migration.

Adhesion of Neutrophils to Purified ICAM-1 Substrates

ICAM-1 represents the major counter-receptor on endothelial cells for CD11/CD18 glycoproteins (14,17). Thus, studies summarized in Figure 29.5 were carried out to evaluate the functional interactions of chemotactically stimulated Mac-1 on neonatal neutrophils with purified ICAM-1 in planar membrane substrates. Adherence of unstimulated neonatal neutrophils to ICAM-1 was comparable to that of adult control cells. That

LFA-1 is the principal determinant in this process was shown by the inhibitory effect of anti–LFA-1 MAb (but not anti–Mac-1 MAb) when incubated with neutrophils among both test groups. Chemotactic stimulation significantly enhanced adhesion of adult control neutrophils but not neonatal cells. Moreover, the increments of adhesion exhibited by control cells were almost totally inhibitable by anti-CD11b MAb, which had no effect on fMLP-stimulated neonatal neutrophils suspensions. Further, a combination of anti-CD11a and anti-CD11b MAb was significantly more inhibitory then when either of these MAbs were reacted alone with control neutrophils, whereas this combination of MAbs was no more inhibitory than anti–LFA-1 MAb alone in studies of neonatal cells. These studies indicate that fMLP fails to normally elicit Mac-1–dependent adhesion of neonatal neutrophils to purified ICAM-1 or ICAM-1 expressed on cytokine-treated HUVEC. These findings imply that deficits of these adhesive interactions underlie diminished transendothelial migration in vitro and possibly in vivo.

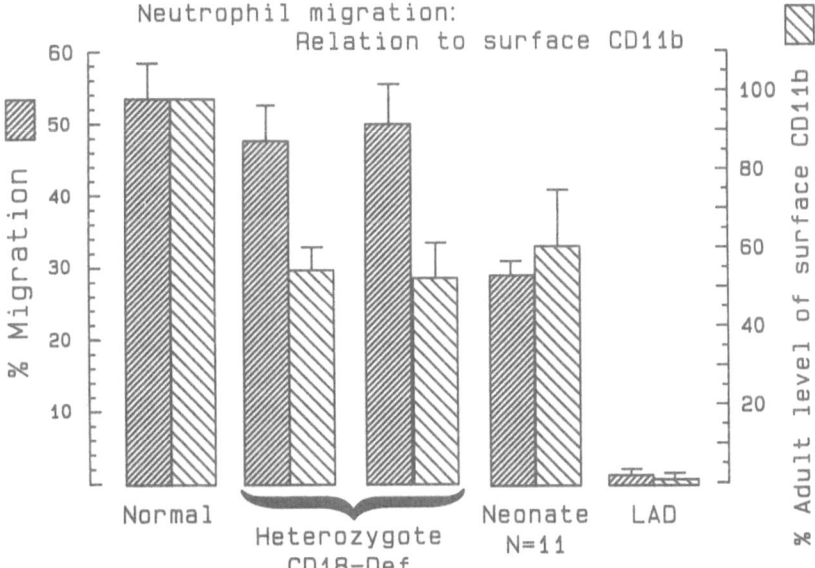

FIGURE 29.6. Relation of neutrophil expression of CD11b to migratory function. Neutrophil suspensions of healthy term neonates, CD11/CD18-deficient patients, or heterozygotes and adult control donors were studied with respect to: (1) transendothelial migration through IL-1–stimulated HUVEC as described (13), and (2) expression of CD11b (904 MAb binding) following fMLP exposure (10 nM, 15 min, 37°C). Values shown represent the mean ± SD of determinations among 4 CD11/CD16-deficient patients, 6 CD11/CD18-deficient heterozygotes, 11 neonates, and 2 healthy adult donars; (*) $p < 0.01$ compared with adult or heterozygote migration values.

Relation of Mac-1 Expression to Transendothelial Migration

Although diminished levels of surface Mac-1 on fMLP-, C5a-, or PMA-elicited neonatal neutrophils have been previously reported (8,16,18), comparative studies shown in Figure 29.6 indicate that functional activation in addition to quantitative deficits of Mac-1 on neonatal cells contribute to their diminished migratory function. In contrast to the impaired capacity of CD11/CD18 genetically deficient cells to transmigrate, neutrophils of individuals heterozygous for this trait demonstrate normal levels of migration even though their cells express only intermediate levels of Mac-1 when stimulated with fMLP (i.e., $54 \pm 7\%$ of normal, $n = 8$). Neonatal cells in this study expressed somewhat higher levels of surface Mac-1 when studied under these conditions and yet showed significantly diminished migration as compared to normal or heterozygote cells.

Studies of CD18-Independent Mechanisms of Neutrophil–Endothelial Cell Adhesion In Vitro

Comparative Studies of Neutrophil Adherence and Migration Under Conditions of Flow

The adhesion of neutrophils isolated from cord blood or term vaginally delivered neonates to IL-1–stimulated endothelial cells was studied using a parallel plate flow chamber technique previously described (19,20). Adherence values were determined at a wall shear stress of 1.85 dynes/cm^2, conditions under which CD18-independent mechanisms predominate. As shown in Figure 29.7, the level of adhesion by neonatal cells when studied under these conditions was significantly diminished as compared to values for adult suspensions studied on the same date. Using the same experimental procedures, the transendothelial migration by neonatal or cord blood neutrophils through IL-1–stimulated HUVEC was also shown to be significantly diminished (Fig. 29.7). This was consistently observed among 20 neonatal suspensions tested in which mean migration values were expressed as phase-bright cells/mm^2 or as a percentage of values determined for adult samples studied on the same test dates. When computed as a percentage of the total adherent cell population undergoing migration, neonatal neutrophil migration was also significantly diminished compared to controls. These findings indicate that the overall diminished capacity of cord blood neutrophils to migrate through endothelial monolayers reflects, in part, a decreased proportion of cells capable of normal adhesion to HUVEC under conditions of flow. They further indicate that the migratory capacity of an adherent subpopulation of neonatal neutrophils is also reduced.

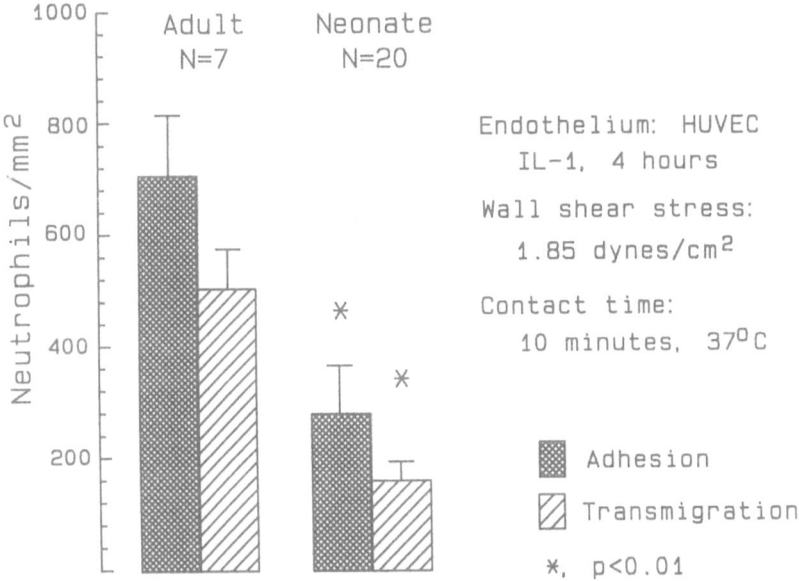

FIGURE 29.7. Neutrophil adhesion to or transendothelial migration through IL-1–stimulated (5 u/ml, 4 h) HUVEC monolayers under conditions of flow at a wall shear stress of 1.85 dynes/cm^2 for 10 min. Values shown represent the mean $\pm$ SE determinations for 20 neonatal or 7 adult suspensions studied and are expressed as the number of adherent (phase-bright) cells/mm^2 or the number of cells undergoing migration through the endothelial cell monolayer (phase-dark cells) 10 min after injection of neutrophils into parallel plate chambers. For these study populations, the mean $\pm$ SD percent of total adherent cells undergoing migration (i.e., phase-dark/phase-bright + phase-dark cells) was 41 ± 15 (5 min) or 60 ± 17 (10 min) for neonatal suspensions and 57 ± 12 (5 min) or 74 ± 8 (10 min) for adult control suspensions ($p < 0.003$ at both time intervals).

LECAM-1–Dependent Neutrophil Adhesion Under Conditions of Flow

The findings of diminished adhesion of neonatal neutrophils to endothelial monolayers in the presence of shear forces where neutrophil LECAM-1 has been shown to function in adhesion (19) suggested a possible abnormality in this surface glycoprotein. Therefore, comparative studies (Fig. 29.8) were performed to evaluate the relative inhibitory effects on neutrophil adhesion of the anti–LECAM-1 MAb DREG 56 (21). Pretreatment of adult control unstimulated neutrophils with this MAb significantly inhibited their adherence to IL-1–stimulated HUVEC when assessed at a wall shear stress of 1.85 dynes/cm^2, findings indicating a major contribution of LECAM-1 to endothelial cell adhesion under these conditions. Overall levels of adhesion for untreated cord blood cells were

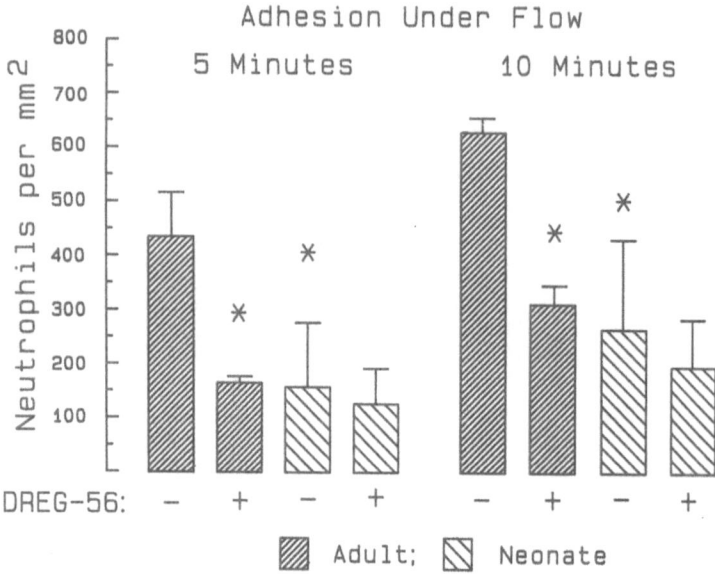

FIGURE 29.8. Effects of MAb, DREG-56, on adhesion of cord blood and adult neutrophils to endothelial cell monolayers studied at a wall shear stress of 1.85 dynes/cm². Isolated neutrophils were preincubated in the presence or absence of MAb DREG-56 (10 μg/ml) and immediately perfused over the endothelial cell monolayer in the parallel plate flow chamber as described (19). The numbers of neutrophils adhering were determined after 5 or 10 min of flow. Values shown are the mean ± SD of determinations performed on six neonatal (coarsely hatched bars) and paired adult control (finely hatched bars) samples; *p ≤ 0.005 compared to untreated adult control.

comparable to those of most adult control suspensions treated with DREG-56 MAb. Moreover, only minimal (and statistically insignificant) inhibition of adhesion of cord blood cells by DREG-56 MAb was evident in most studies among a large population of neonates (Fig. 29.8). Thus, the reduced ability of cord blood neutrophils to adhere under flow conditions appears to reflect the diminished level of LECAM-1–dependent adhesion.

Immunofluorescence Analysis of LECAM-1 on the Surface of Cord Blood, Neonatal, or Healthy Adult Neutrophils

The above findings of diminished LECAM-1–dependent neutrophil adhesion of cord blood neutrophils prompted quantitative evaluations of their surface LECAM-1, the results of which are shown in Figure 29.9. Isolated neutrophils were incubated in the presence or absence of the chemotactic peptide fMLP and then reacted with MAb DREG-56. As shown, MAb

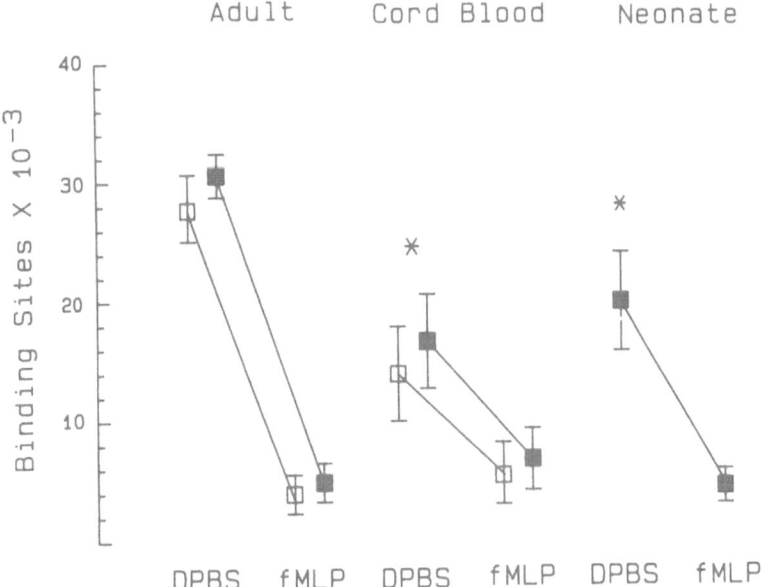

FIGURE 29.9. Immunofluorescence analysis MAb DREG-56 binding to neonatal and adult neutrophils. Isolated neutrophils (open squares) or whole blood samples (solid squares) were incubated in the presence or absence of the chemotactic factor fMLP (10 nM, 15 min, 37°C) and then reacted with the MAb DREG-56 and FITC-labeled antimouse IG. Surface levels of DREG-56 are expressed as antibody binding sites per cell. Values shown represent mean ± SE determinations performed on 12 cord blood, 15 neonatal (ages 12–48 h), and 7 adult neutrophil and/or whole blood preparations. *p < 0.01 compared to unstimulated adult control.

binding of untreated cord or neonatal cells was significantly reduced as compared to adult control samples processed simultaneously. To confirm that this reduction was not induced by cell separation procedures alone, whole blood samples were assessed in a similar manner (data also shown in Fig. 29.9). Among cord blood samples obtained from 36 vaginally delivered neonates evaluated on 10 separate test dates, DREG-56 MAb binding was markedly diminished compared to experimental controls or overall mean values for adult control suspensions. Similar low levels of binding were observed for infants delivered by caesarean section without general anesthesia and for neonates at 12–48 h of age.

As previously reported, "down-regulation" or "shedding" of neutrophil LECAM-1 by fMLP stimulation was apparent among all control neutrophil suspensions (21). In some cases, MAb binding by unstimulated cord blood or neonatal cells was comparable to the low levels of binding by fMLP-stimulated adult cells, and fMLP elicited minimal or no "down-

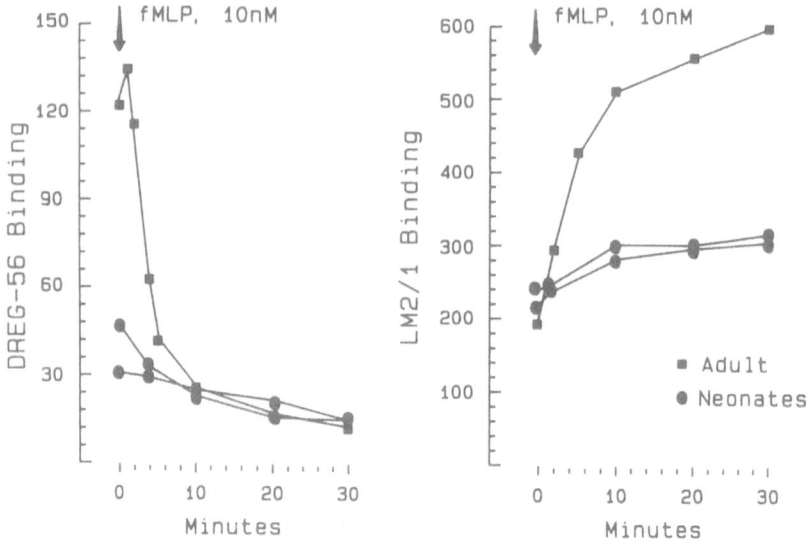

FIGURE 29.10. Expression of LECAM-1 and Mac-1 (CD11b/CD18) on cord blood or healthy adult neutrophils; kinetics of regulation by fMLP. Isolated neutrophils were preincubated with fMLP (10 nM, 37°C) for the various time intervals shown and then reacted at 4°C with MAb DREG-56 (anti–LECAM-1) or anti-CD11b MAb LM2/1 before staining with FITC-labeled antimouse Ig. Surface Ag expression for each cell suspension and time inteval is shown as the mean fluorescence channel of 5,000 cells collected from each suspension. Representative tracings for two neonatal suspensions are compared to that of a healthy adult suspension studied on the same test date.

regulation" of DREG-56 binding on neutrophils of neonatal suspensions. Essentially no diminution of MAb binding after fMLP exposure was evident in 2 of 15 whole blood samples or 4 of 21 neutrophil suspensions among the neonatal populations tested. Representative comparisons of the kinetics of neutrophil surface LECAM-1 regulation by fMLP are shown in Figure 29.10. Expression of LECAM-1 on stimulated adult neutrophils was transiently elevated (at 30 s) after which it rapidly down-regulated to levels 10–20% of baseline values within 10–30 min of incubation. In contrast, the low levels of DREG-56 binding to cord blood neutrophils were only minimally influenced by fMLP stimulation over the 30-min incubation interval. The regulation of surface Mac-1 (CD11b) was simultaneously assessed in the same protocols. As shown (Fig. 29.10), the kinetics of CD11b up-regulation elicited by fMLP on adult neutrophils was inversely related to the regulation of LECAM-1. Somewhat different relationships were observed in studies of neonatal cells. The degree of fMLP-elicited up-regulation of CD11b expression was diminished as compared to controls as previously reported (8,11). However, unlike the

FIGURE 29.11. Comparisons of neutrophil adhesion under conditions of flow with MAb DREG-56 binding to cord blood or adult control neutrophils. Isolated neutrophils were preincubated in the presence or absence of fMLP (10 nM, 15 min room temperature) and aliquots of these suspensions were concurrently assessed for adhesion to IL-1–elicited HUVEC monolayers under conditions of shear stress (1.85 dynes/cm^2) and surface MAb DREG-56 binding (mean fluorescence channel). (A.) Values shown represent the mean determinations from paired experiments (n = 4) for neonatal (coarsely hatched bars) or adult control (finely hatched bars) suspensions. *p < 0.001 compared to unstimulated adult control. (B.) Linear regression of all data (including neonatal, adult, with and without stimulation) obtained where both adhesion under flow and DREG-56 binding were determined on the same cell preparations (r^2 = 0.316, p < 0.01).

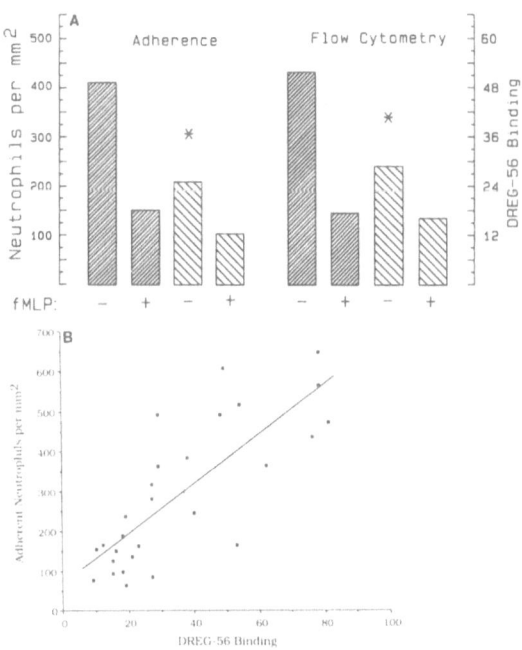

often-profound abnormalities of LECAM-1 surface expression observed on unstimulated neonatal neutrophils, baseline levels of CD11b on these cells were only slightly elevated or comparable to those of control cells. Among 36 cord blood suspensions assessed for both surface CD11b and LECAM-1 expression, levels of surface CD11b were within a normal range (mean ± 2 SD of adult control) in 32 of 36 cases. In contrast, 33 of 36 of these suspensions demonstrated diminished levels of LECAM-1 in many cases below the entire range of adult control determinations.

Relationship of Neutrophil Surface LECAM-1 to Adhesion Under
Conditions of Shear Stress

Direct comparisons of the levels of MAb DREG-56 binding and adhesion
by unstimulated or fMLP-stimulated neutrophils are shown in Figure
29.11. For both adult control and cord blood neutrophils, levels of adhesion under conditions of shear stress were directly related to levels of
surface LECAM-1. fMLP pretreatment diminished adhesion of control
cell suspensions from 515 ± 68 cells/mm^2 to 152 ± 15 cells/mm^2. Similar
relationships were evident in studies of cord blood neutrophils. However,
the mean level of adhesion of untreated cord blood cells was only approximately 39% of adult neutrophils and the corresponding binding of
DREG-56 MAb was 40% of adult levels ($p < .001$ for each compared to
adult control).

Conclusion

Results of these in vitro investigations indicate that abnormalities of two
distinct molecular mechanisms contribute to impaired adhesive interactions of neonatal neutrophils with vascular endothelium (1) The present
studies and previous reports indicate quantitative and functional abnormalities of Mac-1 (CD11b/CD18) including diminished Mac-1 expression
on fMLP-elicited neonatal neutrophils (8,10,11,16) and diminished Mac-
1–dependent adhesion to protein-coated substrates (5,16), endothelial
cells (8), and purified ICAM-1 (8). Transendothelial migration by neonatal neutrophils in vitro is diminished but not absent, probably because
expression of LFA-1 on neonatal cells is normal and this molecule contributes to both adhesion and migration especially under static conditions
(8,16). (2) The present studies and those reported elsewhere also document significant abnormalities of LECAM-1 including diminished surface
levels of LECAM-1 on unstimulated neonatal neutrophils and markedly
diminished levels of neutrophil adhesion to cytokine-elicited endothelial
cells, especially when assessed under conditions of shear stress in which
CD18 integrins appear to contribute little to adhesion. Furthermore,
MAbs recognizing LECAM-1 fail to significantly inhibit adhesion of
neonatal neutrophils under these conditions in contrast to their potent
antiadhesive effects when incubated with adult neutrophils.

Observations in these investigations and other studies of neonatal models have helped to define emerging concepts of the molecular mechanisms
involved in neutrophil–endothelial interactions in vitro and the mechanisms by which neutrophils emigrate at inflammatory sites in vivo (5–
12,22). Based on in vitro investigations and intravital animal studies, it is
reasonable to propose a multistep, dynamic, and integrated functional
model by which neutrophils interact with vascular endothelium, localize,

and emigrate at inflammatory sites (19,21,23–26). Initial adhesion of un-stimulated (circulating) neutrophils to vascular endothelium is dependent on the up-regulation of adhesive determinants on the endothelium (e.g., ELAM-1) that react with LECAM-1. This interaction appears to facilitate vascular rolling in vivo (24) as well as adhesion under conditions of phy-siologic shear stress in vitro (19,26). Subsequent steps involve the streng-thening of the adhesion and transendothelial migration resulting from the engagement of CD18-dependent mechanisms. This appears to require the reciprocal regulation of LECAM-1 and Mac-1 by chemotactic factors, presumably generated intravascularly (e.g., C5a) or elaborated locally by endothelial or other cell types (e.g., GM-CSF, IL-8, TNFα). Initial LECAM-1–mediated endothelial attachment presumably down-regulates, thus allowing for a subsequent migration of neutrophils into extravascular sites. The latter response appears to require the induction of ICAM-1 on endothelial cells by cytokines or LPS (as reviewed in 25) and involves the cooperative interactions of both LFA-1 and Mac-1 with ICAM-1 and possibly other endothelial ligands (14). More specifically, engagement of Mac-1 with ICAM-1 appears to require chemotactic activation of neu-trophils, suggesting that quantitative or biochemical alterations of Mac-1, or alterations of its density distribution on the cell surface, serve to "up-regulate" and facilitate its recognition of ICAM-1 (8,14,24,27–29).

The mechanisms underlying deficits of Mac-1 up-regulation or dimi-nished levels of LECAM-1 on neonatal neutrophils are uncertain but de-serve further study. Jones et al. previously showed that diminished Mac-1 expression on chemotactically stimulated neonatal cell suspension involves an impaired intracellular translocation of Mac-1 contained in gelatinase-rich secondary granules (11). The present studies clearly indi-cate that functional activation of Mac-1 (required for ligation with ICAM-1) is also deficient in neonatal cells. Smith et al. have reported diminished chemotactic induction of Mac-1 on human neonatal eosinophils (12), and diminished up-regulation of Mac-1 by chemotactic agonists on neutrophils of neonatal rabbits has also been recently reported by Fortenberry et al. (9).

It is reasonable to propose two explanations for diminished LECAM-1 on neonatal neutrophils. (1) An acquired abnormality may exist due to shedding of LECAM-1 induced by stimulating factors present in cord blood or neonatal plasma such as GM-CSF. This factor has been reported to be elevated in cord blood (30), and rGM-CSF stimulates normal neu-trophils to down-regulate LECAM-1 in vitro (31). Other factors known to induce down-regulation of this molecule include TNFα, LTB$_4$, C5a, and PAF (19,21,31); these factors have not yet been studied in plasma of neonatal subjects. (2) A second possible explanation for reduced neu-trophil LECAM-1 is that production on inflammatory cells may be de-velopmentally delayed at least during the first 48h of life (22). Such a possi-bility would be selective for myeloid cells since levels of LECAM-1 are actually elevated on blood lymphocytes of neonates (22). A recent report by Smith et al. indicates that human neonatal eosinophils (as well as neutrophils) demonstrate significantly diminished surface LECAM-1 (32). Confirmatory studies in neonatal rabbits also indicate significantly diminished LECAM-1 in 1 and 7-day-old animals (9). Moreover, levels of this determinant on neutrophils of neonatal (<1 month) holstein calves are also profoundly diminished (M. Kerhli et al., unpublished observation). LECAM-1 is expressed on myeloid cell precursors in bone marrow, cells that would not be expected to recirculate through peripheral lymph nodes or accumulate at inflammatory sites (31). Thus, the finding of high levels of progenitor cells in blood of normal neonates may be linked to reduced levels of LECAM-1 on these cells. This possibility is of particular interest since progenitor cell attachment to bone marrow stromal cells may involve C-type lectins evolutionarily related to LECAM-1 (33).

The physiologic or potential pathologic implications of the present studies are unclear and deserve further study. However, preliminary studies by Fortenberry et al. (10) in a model of thioglycolate-induced peritonitis in neonatal rabbits indicate: (1) significantly diminished neutrophil accumulation 4h after thioglycolate installation (as compared to responses in adult rabbits) and (2) minimal net inhibitory effects on neutrophil emigration by either anti-CD18 MAb (R15.7) or anti LECAM-1 MAb (DREG-200) when administered systemically to neonatal animals as compared to potent inhibitory effects observed in adult animals. These in vivo findings are consistent with the observed low levels of LECAM-1 on unstimulated neutrophils and the diminished levels of Mac-1 on fMLP-stimulated neutrophils isolated from neonatal rabbits. Additive inhibitory effects of anti-CD18 and LECAM-1 MAbs on peritoneal exudation were observed in adult but not neonatal rabbits. These findings provide further evidence for cooperative interactions of LECAM-1 and CD11b/CD18 in neutrophil localization in vivo, at least in selected tissues. This interpretation is consistent with that of Jutila et al. (34) and von Andrian et al. (24). Studies by each of these investigators demonstrated independent and additive inhibitory effects by systemically administered anti–LECAM-1 and anti-CD11b (or CD18) MAbs in murine or lapine models of inflammation, respectively. Collectively, these observations and the results of our present studies support the possibility that abnormalities of LECAM-1– and CD11/CD18-dependent neutrophil adhesion contribute to an attenuated inflammatory response in human or animal neonates. These alterations may contribute to impaired inflammatory host defenses or alternatively may represent a physiologic mechanism protective against injury of neonatal tissues in the postpartum setting.

Acknowledgements. The work was supported by NIH grant AI19031. The expert secretarial assistance of Juli Hammond and Michelle Swarthout is acknowledged.

References

1. Wilson, C.B. 1986. Immunologic basis for enhanced susceptibility of the neonate to infection. *J. Pediatr.* 108:1–12.
2. Schuit, K.E., and L. Homisch. 1984. Inefficient in vivo neutrophil migration in neonatal rats. *J. Leuk. Biol.* 35:583–586.
3. Martin, T.R., C.E. Rubens, and C.B. Wilson. 1988. Lung antibacterial defense mechanisms in infant and adult rats: Implications for the pathogenesis of Group B streptococcal infections in the neonatal lung. *J. Infect. Dis.* 157:91–100.
4. Cheung, A.T.W., G. Kurland, M.E. Miller, E.W. Ford, S.A. Avin, and E.M. Walsh. 1986. Host defense deficiency in newborn nonhuman primate lungs. *J. Med. Primatol.* 15:37–47.
5. Anderson, D.C., B.J. Hughes, and C.W. Smith. 1981. Abnormal mobility of neonatal polymorphonuclear leukocytes. Relationship to impaired redistribution of surface adhesion sites by chemotactic factor or colchicine. *J. Clin. Invest.* 68:863–874.
6. Anderson, D.C. 1989. Neonatal neutrophil dysfunction. *Am. J. Ped. Hem. Onc.* 11:224–226.
7. Anderson, D.C., O. Abbassi, T.K. Kishimoto, J.M. Koenig, L.V. McIntire, and C.W. Smith. 1991. Diminished lectin-, epidermal growth factor-, complement binding domain-cell adhesion molecule-1 on neonatal neutrophils underlies their impaired CD18-independent adhesion to endothelial cells in vitro. *J. Immunol.* 146:3372–3379.
8. Anderson, D.C., R. Rothlein, S.D. Marlin, S.S. Krater, and C.W. Smith. 1990. Impaired transendothelial migration by neonatal neutrophils: Abnormalities of Mac-1 (CD11b/CD18)-dependent adherence reactions. *Blood* 78:2613–2621.
9. Fortenberry, J.D., M.M. Mariscalco, J.R. Marolda, C.W. Smith, and D.C. Anderson. 1991. Diminished surface levels of LECAM-1 on neonatal lapine neutrophils. *Pediat. Res. (abst)* 29(4):274A.
10. Fortenberry, J.D., M.M. Mariscalco, J.R. Marolda, C.W. Smith, and D.C. Anderson. 1991. Diminished CD18-dependent neutrophil emigration in neonatal lapine peritonitis. *Pediat. Res. (abst)* 29(4):275A.
11. Jones, D.H., F.C. Schmalstieg, K. Dempsey, S.S. Krater, D.D. Nannen, C.W. Smith, and D.C. Anderson. 1990. Subcellular distribution and mobilization of Mac-1 (CD11b/CD18) in neonatal neutrophils. *Blood* 75:488–498.
12. Smith, J.B., D.E. Campbell, S.D. Douglas, B.Z. Garty, A. Ludomirsky, R.A. Polin, and M.C. Harris. 1988. Fetal neutrophils have distinct subpopulations which differ in complement and FC receptor expression. *Ped. Res.* 23:1599. Abstract.
13. Smith, C.W., R. Rothlein, B.J. Hughes, M.M. Mariscalco, F.C. Schmalstieg, and D.C. Anderson. 1988. Recognition of an endothelial determinant for CD18-dependent human neutrophil adherence and transendothelial migration. *J. Clin. Invest.* 82:1746–1756.
14. Smith, C.W., S.D. Marlin, R. Rothlein, C. Toman, and D.C. Anderson. 1989. Cooperative interactions of LFA-1 and Mac-1 with intercellular adhesion molecule-1 in facilitating adherence and transendothelial migration of human neutrophils in vitro. *J. Clin. Invest.* 83:2008–2017.
15. Jones, D.H., F.C. Schmalstieg, H.K. Hawkins, B.L. Burr, H.E. Rudloff,

S.S. Krater, C.W. Smith, and D.C. Anderson. 1989. Characterization of a new mobilizable Mac-1 (CD11b/CD18) pool that co-localizes with gelatinase in human neutrophils. In *Leukocyte Adhesion Molecules: Structure, Function, and Regulation*. T.A. Springer, D.C. Anderson, A.S. Rosenthal, and R. Rothlein, eds. Springer-Verlag, New York, P.106.

16. Anderson, D.C., K.L.B. Freeman, B. Heerdt, B.J. Hughes, R.M. Jack, and C.W. Smith. 1987. Abnormal stimulated adherence of neonatal granulocytes: Impaired induction of surface Mac-1 by chemotactic factors or secretagogues. *Blood* 70:740–750.

17. Staunton, D.E., S.D. Marlin, C. Stratowa, M.L. Dustin, and T.A. Springer. 1988. Primary structure of intercellular adhesion molecule 1 (ICAM-1) demonstrates interaction between members of the immunoglobulin and integrin supergene families. *Cell* 52:925–933.

18. Bruce, M.C., J.E. Bailey, K. Medvik, and M. Berger. 1987. Impaired surface membrane espression of C3bi, but not C3b receptors in neonatal neutrophils. *Pediatr. Res.* 21:306–311.

19. Smith, C.W., T.K. Kishimoto, O. Abbassi, B.J. Hughes, R. Rothlein, L.V. McIntire, E. Butcher, and D.C. Anderson. 1991. Chemotactic factors regulate lectin adhesion molecule 1 (LECAM-1)-dependent neutrophil adhesion to cytokine-stimulated endothelial cells in vitro. *J. Clin. Invest.* 87:609–618.

20. Lawrence, M.B., L.V. McIntire, and S.G. Eskin. 1987. Effect of flow on polymorphonuclear leukocyte/endothelial cell adhesion. *Blood* 70:1284–1290.

21. Kishimoto, T.K., M.A. Jutila, E.L. Berg, and E.C. Butcher. 1989. Neutrophil Mac-1 and MEL-14 adhesion proteins inversely regulated by chemotactic factors. *Science* 245:1238–1241.

22. Koenig, J.M., D.C. Anderson, and C.W. Smith. 1991. Surface levels of LECAM-1 on neonatal neutrophils are further diminished at 24 hours of life. *Pediat. Res. (abst)* 29(4):276A.

23. Kishimoto, T.K., R.A. Warnock, M.A. Jutila, E.C. Butcher, C.L. Lane, D.C. Anderson, and C.W. Smith. 1991. Antibodies against human neutrophil LECAM-1 (LAM- 1/Leu-8/DREG-56 antigen) and endothelial cell ELAM-1 inhibit a common CD18-independent adhesion pathway in vitro. *Blood* 78:805–811.

24. von Andrian, U.H., J.D. Chambers, L.M. McEvoy, R.F. Bargatze, K.-E. Arfors, and E.C. Butcher. 1991. Two step model of leukocyte-endothelial cell interaction in inflammation: Distinct roles for LECAM-1 and the leukocyte beta-2 integrins in vivo. *Proc. Natl. Acad. Sci. USA* 88:7538–7542.

25. Smith, C.W., and D.C. Anderson. 1991. PMN adhesion and extravasion as a paradigm for tumor dissemination. In *Cancer and Metastasis Review*. P. Frost, B. Kerbel, and R. Greig, eds. Kluwer Academic Publishers. 10:61–78

26. Abbassi, O., C.L. Lane, S.S. Krater, T.K. Kishimoto, D.C. Anderson, L.V. McIntire, and C.W. Smith. 1991. Canine neutrophil margination mediated by lectin adhesion molecule-1 (LECAM-1) in vitro. *J. Immunol.* 147:2107–2115

27. Diamond, M.S., D.E. Staunton, A.R. deFougerolles, S.A. Stacker, J. Garcia-Aguilar, M.L. Hibbs, and T.A. Springer. 1990. ICAM-1 (CD54): A counter-receptor for Mac-1 (CD11b/CD18). *J. Cell Biol.* 111:3129–3139.

28. Buyon, J.P., S.G. Slade, J. Reibman, S.B. Abramson, M.R. Philips, G. Weissman, and R. Winchester. 1990. Constitutive and induced phosphorylation of the alpha and beta-chains of the CD11/CD18 leukocyte integrin family. *J. Immunol.* 144:191–197.

29. Detmers, P.A., S.K. Lo, E. Olsen-Egbert, A. Walz, M. Baggiolini, and Z.A. Cohn. 1990. Neutrophil-activating protein 1/interleukin 8 stimulates the binding activity of the leukocyte adhesion receptor CD11b/Cd18 on human neutrophils. *J. Exp. Med.* 171:1155–1162.
30. Laver, J., E. Duncan, M. Abboud, C. Gasparetto, I. Sahdev, D. Warren, J. Bussel, P. Auld, R.J. O'Reilly, and M.A.S. Moore. 1990. High levels of granulocyte and granulocyte-macrophage colony-stimulating factors in cord blood of normal full-term neonates. *J. Pediatr.* 116:627–632.
31. Griffin, J.D., O. Spertini, T.J. Ernst, M.P. Belvin, H.B. Levine, Y. Kanakura, and T.F. Tedder. 1990. Granulocyte-macrophage colony-stimulating factor and other cytokines regulate surface expression of the leukocyte adhesion molecule-1 on human neutrophils, monocytes, and their precursors. *J. Immunol.* 145:576–584.
32. Smith, J.B., R.D. Kunjummen, T.K. Kishimoto, and D.C. Anderson. 1991. Neonatal eosinophils and neutrophils have similar abnormalities of expression of leukocyte-endothelial cell adhesion molecule-1 (LECAM-1). *Ped. Res.* 29.278. Abstract.
33. Tavassoli, M., and C.L. Hardy. 1990. Molecular basis of homing of intravenously transplanted stem cells to the marrow. *Blood* 76:1059–1070.
34. Jutila, M.A., L. Rott, E.L. Berg, and E.C. Butcher. 1989. Function and regulation of the neutrophil MEL-14 antigen in vivo: Comparison with LFA-1 and MAC-1. *J. Immunol.* 143:3318–3324.

30
Expression of ICAM-1 in Human Liver

JAMES NEUBERGER, DAVID ADAMS, AND STEFAN HUBSCHER

Introduction

Immune mechanisms are involved in the pathogenesis and progression of liver damage in a number of different disorders. Perhaps the best example of immune mechanisms causing liver damage occurs in allograft rejection. Liver transplantation has now become an accepted form of treatment for end-stage liver disease. As with other organ grafts, rejection remains a significant problem. Overimmunosuppression results in increased susceptibility to infection and underimmunosuppression leads to progressive graft damage due to immune-mediated rejection. Of the different forms of liver allograft rejection, acute rejection is the commonest, occurring in up to 70% of patients (1,2). Over 80% of instances respond to a short course of high-dose corticosteroids. The characteristic histological features of acute allograft rejection include bile duct damage; portal tract inflammation with infiltration by lymphocytes, eosinophils, neutrophils, and monocytes; and venous endothelialitis. The other form of rejection is chronic/ductopenic rejection. This occurs in 10% of allograft recipients. Clinically, the onset is usually apparent within the first 6 months after transplantation. In the early stages there is an intense lymphocyte infiltration around the portal tract, but as the disease progresses, the infiltration becomes less. The hallmark of the condition is the progressive loss of bile ducts from the portal tracts and a vasculopathy with foamy cells in the intima of the arteries. This form of rejection is almost invariably progressive and retransplantation remains the only therapeutic option.

The bile ducts are also the major site of damage in two other chronic liver diseases which are presumed to have an immune-mediated basis. Primary biliary cirrhosis (PBC) is a progressive disease which typically occurs in middle-aged women and is characterized by a lymphocytic infiltration around the middle-sized intrahepatic bile ducts leading to progressive bile duct damage and loss (3,4). This ultimately leads to cirrhosis, chronic liver failure, and death. There are a number of features suggesting that immune mechanisms are important including increased levels of immunoglobulins, antimitochondrial auto-antibodies, abnormal T-cell responses, and infiltration of the liver by lymphocytes and granulomas. In primary sclerosing cholangitis (PSC) the entire biliary tree is involved in a diffuse inflammatory process. Involvement of immune mechanisms is suggested by the presence of increased immunoglobulins, antineutrophil antibodies, association with the IILA phenotype DR4, and lymphocytic infiltration of the biliary tree (5,6).

Because of the importance of immune mechanisms in these three liver conditions (allograft rejection, PBC, PSC), we have looked for evidence of ICAM expression on the liver in these conditions and in controls (7,8).

Patients and Methods

Liver Transplant Patients

Fifty-two biopsy samples from 50 allograft recipients were studied. Patients were maintained on standard immunosuppression with prednisolone, azathioprine, and cyclosporin as described in detail elsewhere (9).

Fifteen samples were available from patients with acute allograft rejection before receiving treatment with high-dose cortico-steroids. Five samples were available 4 wk after institution of high-dose steroid treatment (either oral prednisolone 200 mg daily or methylprednisolone 1 g intravenously per day for 3 days) which in all cases resulted in clinical, biochemical, and histological resolution of rejection. Seven samples were available from patients with chronic/ductopenic rejection.

Nonrejection Complications

Twenty-one samples were taken from patients with other causes of graft dysfunction (pure cholestasis/biliary obstruction, 11), ischemia-preservation injury (9), and bacterial cholangitis (1). In none of these patients was there any histological or clinical evidence of rejection.

Stable Patients

Four samples were taken from patients well at least 1 yr after transplantation and in all cases the liver chemistry and histology were normal.

Other Liver Diseases

Chronic Liver Disease

Thirteen samples were obtained from patients with primary biliary cirrhosis; in all cases the diagnosis was made on the basis of clinical, serological, and histological findings. Seven samples were available from patients with primary sclerosing cholangitis; in all cases the diagnosis was made on clinical, biochemical, histological, and radiological findings showing typical features.

As a comparative group, samples were taken from patients with other causes of cirrhosis (cryptogenic cirrhosis, 5), autoimmune chronic active hepatitis (3), alpha-1-antitrypsin deficiency (3), Wilson's disease (1), hemochromatosis (1), and Budd-Chiari syndrome (1).

Acute Liver Disease

Seven samples were available from patients with acute/subacute liver failure due to presumed non-A, non-B viral infection.

Normal Liver

There were 3 samples from patients with noncirrhotic liver disease (hepatocellular carcinoma) and 7 samples were available from patients with "normal" liver, taken from donor liver at the time of transplantation.

None of the controls was taking immunosuppressive drugs except the three patients with autoimmune chronic active hepatitis, who were receiving prednisolone.

Immunohistochemistry

Liver material was frozen in liquid nitrogen immediately after biopsy and stored at −70°C. After thawing, sections were fixed in acetone and stained for ICAM-1 using a three-step immunoalkaline-phosphatase method. Sections were incubated with a mouse monoclonal antibody recognizing ICAM-1 (RR0001—a gift from R. Rothlein), diluted in 1:5,000 buffered saline (PBS) at pH 7.6, at room temperature for 30 min, washed and then incubated with a polyclonal rabbit antimouse immunoglobulin (Dakopatts, High Wycombe, Bucks), diluted 1 in 25, for 30 min. After three further washes, the antibody was detected using three-step immunoalkaline phosphatase technique (Dakopatts). As a negative control, mouse ascites not containing any immunoglobulin to human antigens was used.

Assessment of Histology

Sections were examined by two observers (D.A. and S.H.). Structures were analyzed without knowledge of the patient's clinical status. The intensity of staining was graded on a semiquantative scale on a 0–3 scale (3 being the most intense). This method has been validated elsewhere using scanning confocal laser microscopy (8).

For the liver transplant sections, the bile ducts, endothelial cells, and hepatocytes were examined and scored. For those sections from patients with other chronic liver diseases, the following structures were assessed. Bile ducts were examined according to size: proliferating bile ductules (the smallest bile duct branches with a generally inconspicuous lumen usually confined to the marginal zones in the portal areas), interlobular ducts (duct with diameters between 20 and 100 μm), and medium-sized septal ducts (those with duct diameters greater than 100 μm).

Hepatocytes were assessed according to the intensity of staining, their distribution, and the approximate proportion of cells staining positive. Venous and arteriolar endothelial cell staining, sinusoidal lining cells, and infiltrating cells were also assessed.

Results

The results are shown in Tables 30.1 and 30.2.

TABLE 30.1. Scoring of intensity of ICAM-1 staining in liver transplant recipients (7)

Disease	N	Median score		
		Hepatocyte	Endothelium	Bile ducts
Normal controls	13	0	1.5	0
Transplant stable	4	0.5	1.5	0
Acute rejection	16	2.0	2.5	2.0
Resolving rejection	5	0.5	2.0	0.5
Chronic rejection	7	3.0	2.25	3.0
Other complications	20	2.75	1.0	0.6

TABLE 30.2. Mean scoring of intensity of ICAM-1 staining in other liver diseases (18)

	Hepatocyte	Endothelium	Bile ducts		
			Ductules	Lobular	Medium
Normal liver	0.6	1.0	N/A	0.3	0.9
(n)	(11)	(11)	(0)	(11)	(11)
PBC	2.3	0.96	2.9	2.9	0.05
(n)	(14)	(14)	(14)	(12)	(11)
PSC	1.8	1.0	2.6	2.7	0
(n)	(6)	(6)	(6)	(3)	(4)
Cirrhosis	1.5	0.5	0.4	0.1	0.04
(n)	(14)	(14)	(11)	(14)	(14)
Acute hepatitis	2.3	0.58	0.75	0.75	0.67
(n)	(6)	(7)	(7)	(7)	(6)

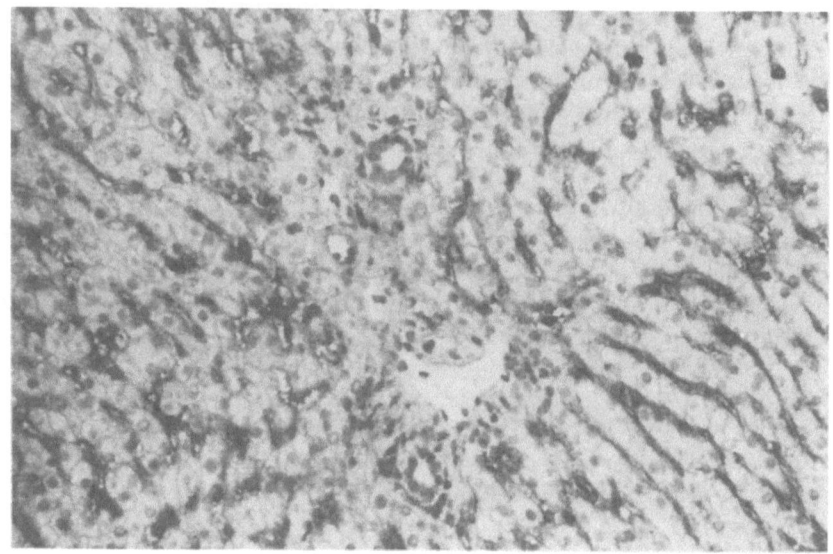

FIGURE 30.1. Normal human liver stained for ICAM-1 showing minimal staining of biliary epithelial cells, with mild staining of endothelial cells, portal infiltrate, and hepatocytes. Sinusoidal cells are well stained. (Magnification ×150)

Normal Livers

Biliary epithelial cells showed minimal staining. There was mild staining of the endothelial cells, portal infiltrate and hepatocyte membranes. Sinusoidal lining cells stained positive (Fig. 30.1).

Liver Transplant Biopsies

Biopsies from patients with acute rejection showed marked staining of the bile duct endothelium and perivenular hepatocytes (Fig. 30.2.). The staining scores were significantly greater than those in the nonrejecting groups. In those with treated rejection, bile duct and hepatocyte staining was significantly less than in those with acute rejection and indeed did not differ significantly from the control subjects or those with stable graft function. The intensity of endothelial staining was similar to that seen in rejection.

In those with chronic rejection there was very intense staining of the residual bile ducts and hepatocytes. This was stronger than that seen in acute rejection. With the one exception of the patient with bacterial cholangitis, those with other complications following transplantation showed staining intensities of bile ducts, hepatocytes, and endothelium similar to that of control patients.

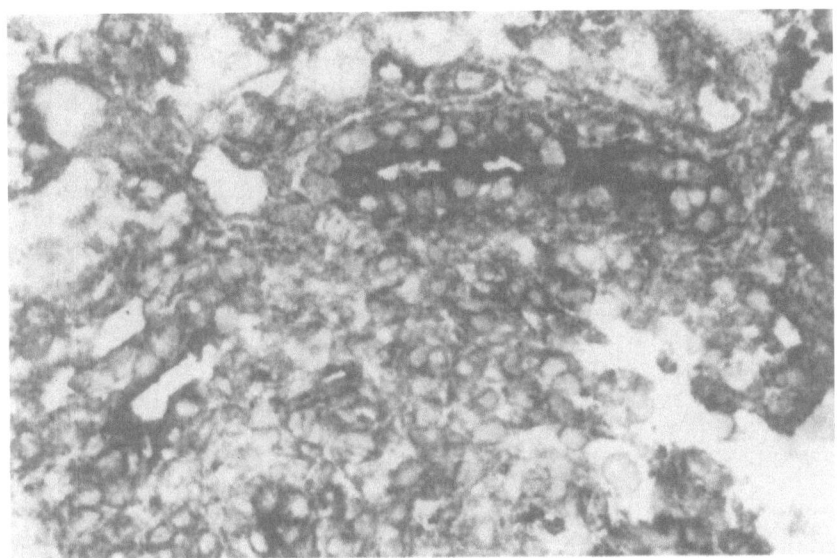

FIGURE 30.2. Liver biopsy from a patient with acute rejection stained for ICAM-1, showing strong staining of bile ducts, endothelium, and perivenular hepatocytes. (Magnification ×350)

Chronic Liver Diseases

Primary Biliary Cirrhosis

In primary biliary cirrhosis there was intense staining of the interlobular and proliferating bile ductules which was significantly greater than that seen in patients with other liver diseases. The medium-sized bile ducts were negative in PBC even in the presence of active inflammation. The hepatocytes showed a marked increase in ICAM expression which was predominantly periseptal in distribution (Fig. 30.3).

Primary Sclerosing Cholangitis

A pattern of ICAM staining similar to PBC was seen in those with PSC with increased staining of the septal bile ducts. Although hepatocyte staining was slightly less than that observed in PBC, staining was predominantly periseptal in distribution.

Other Liver Conditions

There was little staining of bile duct cells of any sized ducts in any of the other cirrhotics or in those with acute hepatitis. However, there was periseptal staining of hepatocytes especially in the periseptal areas adja-

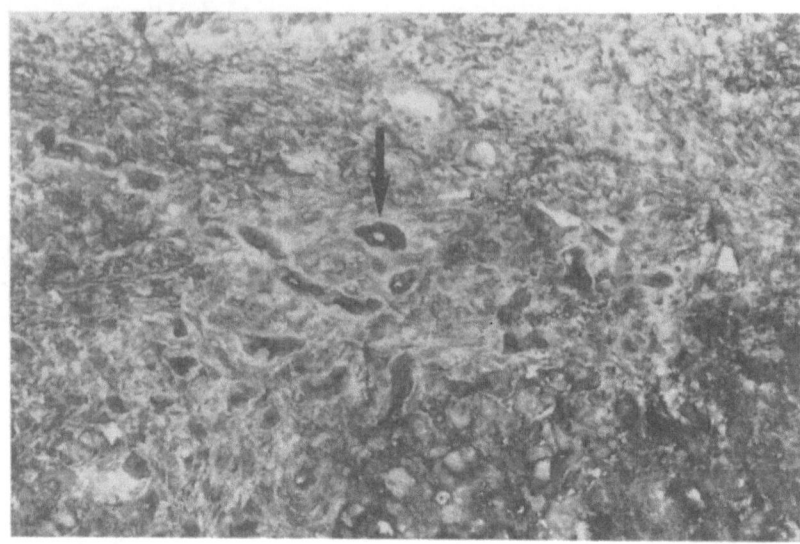

FIGURE 30.3. Section of PBC liver stained for ICAM-1, showing positive staining of proliferating bile ductules and the surviving lobular bile duct (arrowed). (Magnification ×150)

cent to areas of fibrogenesis in those with nonbiliary cirrhosis (Fig. 30.4). In patients with subacute failure there was increased intensity of hepatocyte staining when compared with controls, which was similar to that seen in biliary cirrhosis; the pattern of staining was diffuse and was particularly marked around those areas of hepatocyte necrosis. In two of these, these were areas of fibrosis with again periseptal staining of the hepatocytes.

Discussion

These results confirm the early work by Dustin et al. (10) to the effect that there is moderate expression of ICAM on sinusoidal lining cells but only weak staining on portals and hepatic endothelial cells. This study shows strongly increased expression of ICAM on those biliary epithelial cells which are the main site of damage in both acute and chronic allograft rejection and the biliary cirrhotic diseases PBC and PSC. Again, in graft rejection, PBC, and PSC there was increased expression of ICAM in those structures which show increased HLA expression (11,12). However, in allograft rejection, while ICAM expression resolves rapidly after institution of high-does corticosteroids, the bile duct expression of HLA DR is maintained at a high level for longer (11,12).

In primary biliary cirrhosis, there was strong ICAM expression on the

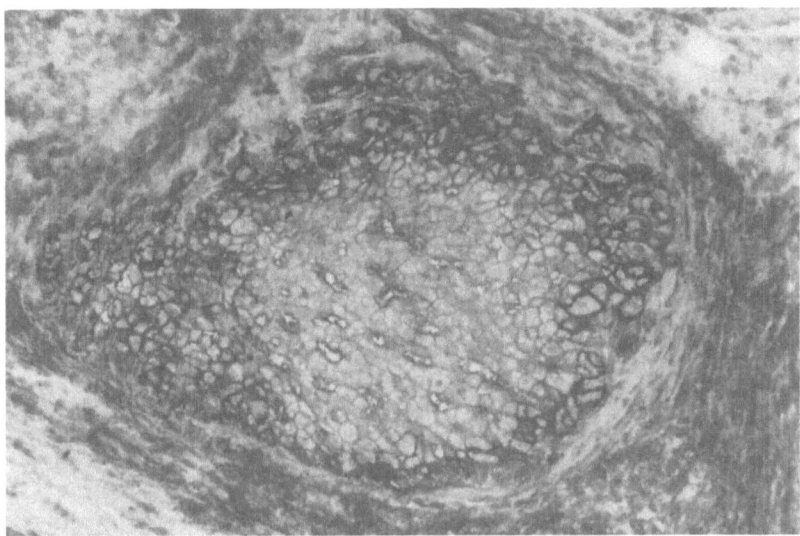

FIGURE 30.4. Section of a cirrhotic liver stained for ICAM-1 showing intense membrane staining of the periseptal hepatocytes. (Magnification ×150)

small bile ducts while the medium-sized ones were negative. This observation may be of importance in understanding the pathogenesis of the disease since these small ducts are the target of the immune-mediated damage whereas the medium-sized ducts were spared (13,14). MHC expression in PBC is enhanced on both the medium and small bile ducts (15–17). Thus, in both allograft rejection and PBC, ICAM expression seems more closely related to structural damage than MHC antigen expression. In PSC most of the medium-sized bile ducts have been destroyed by the disease process; thus, it was not possible to comment on the differential bile duct expression of ICAM or HLA products.

The mechanism of ICAM induction remains to be elucidated. Activated lymphocytes are present at the site of damage in all three conditions. Activated T-cells release pro-inflammatory cytokines such as interferon-γ, interleukin 1, and TNF which increase ICAM expression on a variety of cell lines studied in vitro (10,18). We have shown that incubation of isolated biliary epithelial cells with pro-inflammatory cytokines results in increased ICAM expression which is not abolished by co-culture with corticosteroids (19). Additional factors are likely to be involved since in several of the PBC biopsies examined, ICAM-1 expression on the medium-sized bile ducts was not seen even in the presence of local inflammation while the interlobular bile ducts showed positive ICAM staining. MHC expression may also be enhanced by similar stimuli.

Proliferating bile ductules in both PBC and PSC showed strong ICAM expression. In contrast, proliferating bile ductules were only weakly or negatively stained in those with cirrhosis due to other causes and those with fulminant hepatitis. ICAM expression, therefore, is not a feature of bile duct proliferation per se.

It is unlikely that the increased ICAM expression on bile ducts is a consequence of liver damage because in those with cirrhosis due to causes other than PBC and PSC, there was no increase in ICAM expression. It is unlikely too that the increase in ICAM expression is a result of cholestasis since those patients who were cholestatic following liver transplantation from causes other than rejection did not have increased ICAM expression. In vitro studies (Ayres, unpublished observations) have shown that bile salts do not induce ICAM expression in vitro on biliary epithelial cells. The pathogenesis of both PBC and PSC is uncertain. One hypothesis is that PBC could result as a consequence of infection (20); the putative agent could induce local release of cytokines and expression of ICAM.

Some of the sections of liver from patients with cirrhosis regardless of etiology showed a striking pattern of membranous hepatocyte staining which was concentrated in the periseptal regions adjacent to areas of fibrosis. This was observed sometimes in the absence of an inflammatory infiltrate. It is possible that the distribution of hepatocyte staining may be related to the fibrogenesis observed in cirrhosis. Fibroblasts can be stimulated in vitro to express ICAM-1 and activated fibroblasts do secrete a number of cytokines which may be responsible for induction of ICAM-1 expression. If ICAM-1 expression is important for the progress of fibrogenesis, this does have implications for the treatment and prevention of cirrhosis.

These observations suggest that induction of ICAM-1 expression on target structures is of importance in the process of allograft rejection, but other adhesion molecules are likely to be involved (21). Nevertheless, a central role of ICAM-1 in rejection is suggested by both animal and human studies when therapeutic administration of anti–ICAM-1 has been shown to be of therapeutic benefit (22).

Acknowledgments. We are grateful to Dr. R. Rothlein for the generous supply of antibody to ICAM-1, to Mrs. J. Shaw for technical assistance, and to Miss M. Calcutt for secretarial help.

References

1. Adams D.H. and J. Neuberger. 1990. Patterns of graft rejection following liver transplantation. *J. Hepatol.* 10:11–119.
2. Demetris A.J., Jaffe R., and T.E. Starzl. 1987. A review of adult and paediatric post-transplant liver pathology. *Pathol. Ann.* 2:347–386.
3. Gershwin M.E. and MacKay I.R. 1991. Primary biliary cirrhosis: Paradigm or

paradox for auto-immunity. *Gastroenterology* 100:822–833.
4. Kaplan M.M. 1987. Primary biliary cirrhosis. *New Eng. J. Med.* 316:521–527.
5. Weisner R.H., Grambsch P.M., Dickinson E.R., and J. Ludwig et al. 1989. Primary sclerosing cholangitis. *Hepatology* 10:430– 436.
6. Chapman R.W.G., Arborgh B.A.M., and J.M. Rhodes. 1980. Summerfield J.A., Dick R., Scheuer B., Sherlock D. *Primary sclerosing cholangitis. Gut.* 21:870–877.
7. Adams D.H., Hubscher S.G., Shaw J., Rothlein R., and J. Neuberger. 1989. Intercellular adhesion molecule-1 on liver allografts during rejection. *Lancet.* 2:1122–1125.
8. Adams D.H., Hubscher S.G., Shaw J., Johnson G.D., Rothelin R., and J. Neuberger. 1991. Increased expression of intercellular adhesion molecule-1 (ICAM-1) on bile ducts in primary biliary cirrhosis and primary sclerosing cholangitis. *Hepatology* 14:426–431.
9. Kirby R.M., McMaster P., and D. Clements, et al. 1987. Orthotopic liver transplantation: post-operative complications and their management. *Br. J. Surg.* 74:3–11.
10. Dustin M.L., Rothlein R., Bhan A.C., Dinarello C., and M.I. Springer. 1986. Induction by Il-1 and interferon, tissue distribution, biochemistry and function of a natural adherence molecule (ICAM-1). *J. Immunol.* 137:245–254.
11. Steinhoff G., Wongeit K., and R. Pichlmayr. 1988. Analysis of sequential changes in major histocompatibility complex expression in human liver grafts after transplantation. *Transplantation* 45:394–401.
12. Gouw A.S.H., Houthoff H.S., Huitema S., Beeklen J.M., Gips C.H., and S. Poppema. 1987. Expression of MHC antigens and replacement of donor cells by recipient ones in human liver allograft. *Transplantation* 43:291–296.
13. Weisner R.H., La Russo N., and J. Ludwig, et al. 1985. Comparison of the clincopathological features of primary sclerosing and primary biliary cirrhosis. *Gastroenterology* 88:108–114.
14. Rubin E., Schaffner F., and H. Popper. 1965. Primary biliary cirrhosis: chronic non-suppurative destruction cholangitis. *Am. J. Pathol.* 46:387–407.
15. Nagafuchi Y. and P.J. Scheuer. 1986. Hepatic beta-2-microglobulin distribution in primary biliary cirrhosis. *J. Hepatol.* 2:73–80.
16. Ballardini G., Mirakian R., and F.B. Bianchi, et al. 1984. Aberrant expression of HLA DR molecules on bile duct epithelium in primary biliary cirrhosis. *Lancet.* 2:1009–1013.
17. Van Den Ord J.J., Sciot R., and V.J. Desmet. 1986. Expression of MHC products by normal and abnormal bile duct epithelium. *J. Hepatol.* 3:310–317.
18. Rothlein R., Czajkowski M., O'Neil M.M., Marlin S., Mainolfi E., and V.J. Merluzzi. 1988. Induction of intercellular adhesion molecule-1 on primary and continuous cell lines by pro-inflammatory cytokines. *J. Immunol.* 141:1665–1669.
19. Ayres R.C.S., Neuberger J., Shaw J., and D. Adams. 1991. In vitro Induction of ICAM-1 and MHC antigens on human biliary epithelial cells by proinflammatory cytokines. *Hepatology* 14:97A
20. Hopf U., Moller B., Stemerowicz R., Lobeck H., Rodloff A., Freudenberg

M., Galanos L., and D. Huitn. 1989. Relation between E coli R (rough) forms in gut, lipid A in liver and primary biliary cirrhosis. *Lancet* 2:1419–1422.

21. Steinhoff G., Behrend M., and R. Pichlmayr. 1991. Sinusoidal endothelia lack vascular adhesion molecules (VCAM- 1, ELAM-1, GMP-140) expressed on portal tract endothelia. *J. Hepatol.* 13:S73.

22. Flavin T., Ivens K., and R. Rothlein. 1991. Monoclonal antibodies against intercellular adhesion molecule prolong cardiac survival in cynomologous monkeys. *Transplant Proc.* 23:533–534.

31
Detection and Characterization of Circulating ICAM-1

Elizabeth A. Mainolfi, Steven D. Marlin, and
Robert Rothlein

Introduction

Over the past several years, the expression of intercellular adhesion molecule–1 (ICAM-1) on cells from normal and diseased tissue has been extensively studied. Almost as soon as ICAM-1 was identified, it was found to be a cell-surface adhesion molecule that was inducible in vitro with inflammatory cytokines such as IL-1, TNFα, and IFNγ on a multitude of cell types. Furthermore, ICAM-1's expression on cells from normal and diseased tissue has been and continues to be a focus of much research from many laboratories. To date, it has been reported that there is a low constitutive expression of ICAM-1 on venule endothelial cells; however, its expression is markedly increased at inflammatory sites (1). Furthermore, there is an increased ICAM-1 expression on multiple cell types including keratinocytes in inflammatory skin lesions (2,3), transplanted liver bile duct and perivenular hepatocytes during rejection (4), and in cases of primary biliary cirrhosis (5), endothelium in the brain surrounding MS plaques and EAE lesions in man and rodent, respectively (6–12), transplanted kidney glomeruli and tubules during rejection (13), as well as lung epithelial cells following antigen provocation (14). Increased ICAM-1 is also found on melanoma cells following metastasis (15,16) and recently ICAM-1 expression has been shown to be expressed on placental macrophages and decidua during pregnancy.

In terms of structure, ICAM-1 belongs to the immunoglobulin gene superfamily. It has five extracellular immunoglobulin like domains, a single transmembrane region, and a short cytoplasmic tail (17,18). In addition to mediating CD18-dependent leukocyte adhesion, ICAM-1 has been identified as the obligate receptor for major group rhinoviruses (19–21). In fact, a genetically engineered form of ICAM-1 lacking the cytoplasmic tail and transmembrane region has been shown to inhibit rhinovirus infection in vitro (21).

ICAM-1 has been shown to be locally up-regulated during inflammatory processes, and since several other inducible molecules on leukocytes

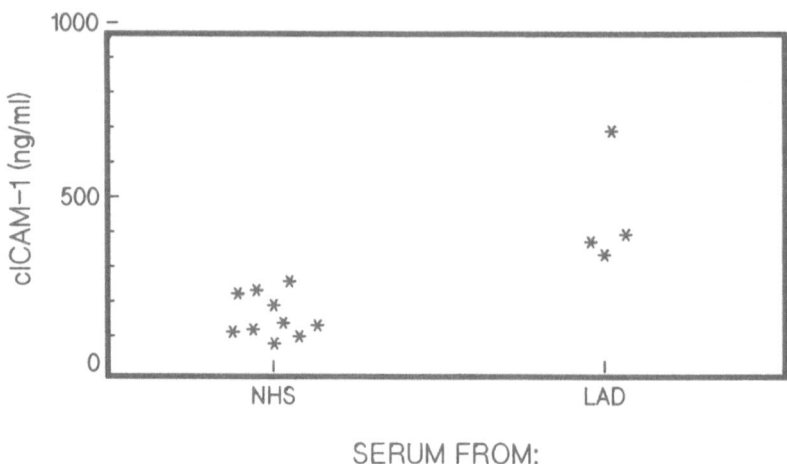

FIGURE 31.1. Serum levels of cICAM-1 detected by ELISA with genetically engineered soluble ICAM-1 used as standard material. Sera were obtained from 10 normal individuals (NHS) and four individuals with leukocyte adhesion deficiency (LAD).

have been found to exist in a soluble form in sera [the IL-2 receptor (22) and the low-affinity IgE receptor (CD23) (23)], we investigated whether a circulating form of ICAM-1 existed in serum or other bodily fluids and if so, whether there were increased levels of cICAM-1 expressed during inflammation. The following is a summary of the results of these studies.

We developed an ICAM-1–specific "sandwich" ELISA using anti–ICAM-1 MAb CL203 directed against domain 4 of ICAM-1 as the trapping reagent and biotinylated anti–ICAM-1 MAb R6.5 directed against domain 2 of ICAM-1 (24) as the detecting MAb. Using this assay, we found sera from normal healthy donors to have an average of 156 ng/ml of a circulating form of ICAM-1 (Fig. 31.1) based on a standard curve of known concentrations of recombinant sICAM-1 lacking both the transmembrane region and the cytoplasmic tail (21). The presence of cICAM-1 in normal serum prompted an examination of sera from patients with leukocyte adhesion deficiency (LAD): These patients fail to express the LFA-1 heterodimer due a defect in the gene encoding the CD18 subunit (25). We hypothesized that LAD patients might have elevated levels of cICAM-1, either due to the absence of this receptor for ICAM-1, or due to the accompanying disease process. The four LAD patient sera tested had an average of 439 ng/ml of cICAM-1, and one serum had greater than 700 ng/ml (Fig. 31.1).

The cellular source of cICAM-1 was assessed by screening supernatants of various cell lines or primary cell cultures incubated with or without agents known to induce ICAM-1 expression on cell surfaces (24). Results of these assays revealed that cICAM-1 was not detected in supernates from human umbilical vein endothelial cells (HUVEC) stimulated with LPS for 24 h; from the adenocarcinoma cell line A549 or the fibrolast cell line IMR-90 with or without stimulation with IL-1 from 1 to 5 days; and in supernates from the myelomonocytic cell line U937 that was driven toward monocytic differentiation and ICAM-1 expression by addition of the phorbol ester PMA. Furthermore, the non–ICAM-1–expressing T lymphoma SKW3 also failed to express cICAM-1.

In contrast, detectable levels of cICAM-1 were found in culture supernates from the EBV-transformed B-cell line JY as well as in supernates from peripheral blood mononuclear cells after 2–3 days in culture.

Structurally, cICAM-1 had intact epitopes on domain 1, domain 2, domain 4, and domain 5 as determined by antibody reactivity with MAbs RR1, R6.5, Cl203, and CA7, respectively, suggesting that cICAM-1 contains substantial structural similarities with authentic ICAM-1, and contains all or part of domains D1, D2, D4, and D5 that comprise the ectodomain. This conclusion is supported by molecular weight determination of cICAM-1 using size exclusion chromotography. cICAM-1 from normal and LAD sera eluted in the same fraction as recombinant sICAM-1 (24).

cICAM-1 was also found to retain true functional activity in that it was able to mediate LFA-1–dependent leukocyte adhesion when it was immobilized on a plate. These data suggest that cICAM-1 retains all, or most, of the structural features necessary for binding to LFA-1 and is in fact potentially active as an LFA-1 ligand (24).

In addition to the LAD patients, we have evaluated sera samples from patients with other diseases associated with an inflammatory response. cICAM-1 sera levels from 52 patients receiving cardiac allografts were followed over a 3-wk posttransplant period (26). In this case, while the cICAM-1 concentrations did not correlate with the histological diagnosis of rejection, there was a correlation between cICAM-1 levels and patient survival. Survivors had lower cICAM-1 levels than nonsurvivors (266+/−137 vs. 434+/−220). Patients with levels greater than 375 ng/ml had a reduced 3-yr survival (49% vs. 83%, $p < 0.01$). Furthermore, patients with both elevated cICAM-1 and sIL-2R levels had the poorest survival rate (25% vs. 82%, $p < 0.001$). It was concluded that elevated levels of cICAM-1 in the first 3 wk posttransplant were prognostic for increased mortality but not increased rejection as determined by endomyocardial biopsy.

Increased expression of ICAM-1 on bile ducts, venous endothelium, and infiltrating lymphocytes has been immunohistologically demonstrated in the rejection of human liver transplants (4). In patients receiving liver allografts it was determined that cICAM-1 sera levels were elevated in patients with acute rejection (744 ng/ml). The levels were also increased in those with graft dysfunction due to infection (760 ng/ml) or other complications leading to graft dysfunction (630 ng/ml). Stable transplant recipients had sera levels of 250 ng/ml while normal healthy individuals had levels of 156 ng/ml.

In the bile of stable transplant recipients, no cICAM-1 was detected. In patients with complications, there was a mode of 0 ng/ml; however, 6 of the 14 patients did have cICAM-1 detected in the bile. In acute rejection, the cICAM-1 concentration was significantly higher with a median of 505 ng/ml. Furthermore, the increased bile:sera ratio of cICAM-1 during acute rejection suggests that cICAM-1 is not accumulating in the bile by passive diffusion from the circulation but is being produced locally in the liver (27,28).

Several liver transplant patients were followed posttransplant and serial sera and bile samples were evaluated. Bile and sera cICAM-1 levels increase 48 h prior to an acute rejection episode and there was a subsequent decrease of cICAM-1 after successful treatment with high-dose corticosteriod. The stable engraftments did not have elevated biliary cICAM-1. The cICAM-1 profile in these patients correlated with sIL2R levels in both sera and bile. Because of this correlation and also the in vitro cell culture studies mentioned earlier it appears that the cICAM-1 in the bile is produced locally in the liver by infiltrating lymphocytes during acute rejection.

Increased expression of ICAM-1 has been reported on melanoma cells (15). Correlation of increased expression and increased metastasis has been reported (16,29). Twenty-two normal human sera samples and 56 patients with malignant melanoma were assayed for cICAM-1 (30). The control normal human sera had a mean of 155 ng/ml. Stage I melanoma patients had a significantly elevated cICAM-1 mean concentration of 406 ng/ml. Stage II patient levels were lower but still significantly elevated at 233 ng/ml, and stage III patients had elevated levels of 302 ng/ml. An increased cICAM-1 concentration was inversely correlated with disease-free survival as compared to low cICAM-1 patients ($25.8 +/- 5.4$ vs. $38.8 +/- 2.2$ months). Furthermore, stage III patients with increased cICAM-1 concentrations showed a significant decrease in survival time ($33.6 +/- 6.3$ vs. $64.3 +/- 13.1$ months). There was no correlation of cICAM-1 serum levels with clinical symptoms of malignancy such as primary tumor characterstics or degree of metastisis, nor with patient physical characteristics such as age or gender.

While we continue to look for cICAM-1 levels in various clinical syndromes, it is becoming increasingly clear that cICAM-1 levels are elevated with increased inflammation and that quantitation of cICAM-1 levels may have prognostic and/or diagnostic value.

References

1. Dustin, M.L., R. Rothlein, A.K. Bhan, C.A. Dinarello, and T.A. Springer. 1986. Induction by IL-1 and interferon, tissue distribution, biochemistry, and

function of a natural adherence molecule (ICAM-1). *J. Immunol.* 137:245.

2. Vejlsgaard. G.L., E. Ralfkiaer, C. Avnstorp, M. Czajkowski, S.D. Marlin, and R. Rothlein. 1989. Kinetics and characterization of intercellular adhesion molecule-1 (ICAM-1) expression on keratinocytes in various inflammatory skin lesions and malignant cutaneous lymphomas. *J. Amer. Acad. Dermtol.* 20:782.

3. Griffiths, C.E.M. and B.J. Nickoloff. 1989. Keratinocyte intercellular adhesion molecule-1 (ICAM-1) expression precedes dermal T lymphocytic infiltration in allergic contact dermatitis (Rhus dermatitis). *Am. J. Pathol.* 135:1045.

4. Adams, D.H., S.G. Hubscher, J. Shaw, R. Rothlein, and J.M. Neuberger. 1989. Intercellular adhesion molecule 1 on liver allografts during rejection. *Lancet* 2:1122.

5. Adams, D.H., S.G. Hubscher, J. Shaw, G.D. Johnson, C. Babbs, R. Rothlein, and J.M. Neuberger. 1991. Increased expression of intercellular adhesion molecule 1 on bile ducts in primary biliary cirrhosis and primary sclerosing cholangitis. *Hepatology* 14:426.

6. Sobel, R.A., M.E. Mitchell, and G. Fondren. 1990. Intercellular adhesion molecule-1 (ICAM-1) in cellular immune reactions in the human central nervous system. *Am. J. Pathol.* 136:1309.

7. Frohman, E.M., T.C. Frohman, M.L. Dustin, B. Vayuvegula, B. Choi, A. Gupta, S. van den Noort, and S. Gupta. 1989. Induction of ICAM-1 expression on human fetal astrocytes by interferon-gamma, tumor necrosis factor-alpha and interleukin-1: relevance to intracerebral antigen presentation. *J. Neuroimmunol.* 23:117.

8. Raine, C.S., S.C. Lee, L.C. Scheinberg, A.M. Duijvestijn, and A.H. Cross. 1990. Adhesion molecules on endothelial cells in the central nervous system: An emerging area in the neuroimmunology of multiple sclerosis. *Clin. Immunol. Immunopathol.* 57:173.

9. Cannella, B., A.H. Cross, and C.S. Raine. 1990. Upregulation and coexpression of adhesion molecules correlate with relapsing autoimmune demyelination in the central nervous system. *J. Exp. Med.* 172:1521.

10. Cannella, B., A.H. Cross, and C.S. Raine. 1991. Adhesion-related molecules in the central nervous system: Upregulation correlates with inflammatory cell influx during relapsing experimental autoimmune encephalomyelitis. *Lab. Invest.* 65:23.

11. Raine, C.S. 1991. Multiple sclerosis: A pivotal role for the T cell in lesion development. *Neuropathol. Appl. Neurobiol.* 17:265.

12. Wilcox, C.E., A.M.V. Ward, A. Evans, D. Baker, R. Rothlein, and J.L. Turk. 1990. Endothelial cell expression of the intercellular adhesion molecule-1 (ICAM-1) in the central nervous system of guinea pigs during acute and chronic relapsing experimental allergic encephalomyelitis. *J. Neuroimmunol.* 30:43.

13. Cosimi, A.B., D. Conti, F.L. Delmonico, F.I. Preffer, S.-L. Wee, R. Rothlein, R. Faanes, and R.B. Colvin. 1990. In vivo effects of monoclonal antibody to ICAM-1 (CD54) in nonhuman primates with renal allografts. *J. Immunol.* 144:4604.

14. Wegner, C.D., R.H. Gundel, P. Reilly, N. Haynes, L.G. Letts, and R. Rothlein. 1990. Intercellular adhesion molecule-1 (ICAM-1) in the pathogenesis of asthma. *Science* 247:456.

15. Matsui, M., M. Temponi, and S. Ferrone. 1987. Characterization of a

monoclonal antibody-defined human melanoma-associated antigen susceptible to induction by immune interferon. *J. Immunol.* 139:2088.

16. Natali, P., M.R. Nicotra, R. Cavaliere, A. Bigotti, G. Romano, M. Temponi, and S. Ferrone. 1990. Differential expression of intercellular adhesion
molecule 1 in primary and metastatic melanoma lesions. *Cancer Res.* 50:1271.

17. Staunton, D.E., S.D. Marlin, C. Stratowa, M.L. Dustin, and T.A. Springer.
1988. Primary structure of intercellular adhesion molecule 1 (ICAM-1) demonstrates interaction between members of the immunoglobulin and integrin
supergene families. *Cell* 52:925.

18. Simmons, D., M.W. Makgoba, and B. Seed. 1988. ICAM, an adhesion
ligand of LFA-1, is homologous to the neural cell adhesion molecule NCAM.
Nature 331:624.

19. Greve, J.M., G. Davis, A.M. Meyer, C.P. Forte, S.C. Yost, C.W. Marlor,
M.E. Kamarck, and A. McClelland. 1989. The major human rhinovirus
receptor is ICAM-1. *Cell* 56:839.

20. Staunton, D.E., V.J. Merluzzi, R. Rothlein, R. Barton, S.D. Marlin, and
T.A. Springer. 1989. A cell adhesion molecule, ICAM-1, is the major surface
receptor for rhinoviruses. *Cell* 56:849.

21. Marlin, S.D., D.E. Staunton, T.A. Springer, C. Stratowa, W. Sommergru-
ber, and V.J. Merluzzi. 1990. A soluble form of intercellular adhesion
molecule-1 inhibits rhinovirus infection. *Nature* 344:70.

22. Rubin, L.A., C.C. Kurman, M.E. Fritz, W.E. Biddison, B. Boutin, R. Yar-
choan, and D.L. Nelson. 1985. Soluble interleukin receptors are released
from activated human lymphoid cells in vitro. *J. Immunol.* 135:3172.

23. Gordon, J., J.A. Cairns, M.J. Millsum, G.R. Guy, and S. Gillis. 1988. Interleukin and soluble CD23 as progression factors for human B-lymphocytes.
Eur. J. Immunol. 18:1561.

24. Rothlein, R., E.A. Mainolfi, M. Czajkowski, and S.D. Marlin. 1991. A form
of circulating ICAM-1 in human serum. *J. Immunol.* 147:3788.

25. Kishimoto, T.K. and T.A. Springer. 1989. Human leukocyte adhesion deficiency: Molecular basis for a defective immune response to infections of the
skin. In Current Problems in Dermatology. 18:106.

26. Ballantyne, C., E.A. Mainolfi, J.B. Young, N.T. Windsor, B. Cocanougher,
E.C. Lawrence, D.C. Anderson, and R. Rothlein. 1991. Prognostic value of
increased levels of circulating intercellular adhesion adhesion molecule-1 after
heart transplant. *Clin. Res.* 39#2:285a.

27. Adams, D.H., E.A. Mainolfi, J. Neuberger, E. Elias, and R. Rothlein. 1991.
Secretion of circulating ICAM-1 into bile during liver allograft rejection. *Gut*
32:A576.

28. Adams, D.H., P. Burra, E.A. Mainolfi, J. Nueberger, E. Elias, and R. Rothlein. 1991. Circulating ICAM-1 in chronic liver disease: evidence for lymphocyte activation in PBC and PSC. *J. Hepatol.* 13:s3.

29. Johnson, J.P., B.G. Stade, B. Holzmann, W. Schwable, and G. Riethmuller.
1989. De novo expression of intercellular-adhesion molecule 1 in melanoma
correlates with increased risk of metastasis. *Proc. Natl. Acad. Sci. U. S. A.*
86:641.

30. Harning, R., E.A. Mainolfi, J.C. Bystryn, M. Henn, V.J. Merluzzi, and R.
Rothlein. 1991. Serum levels of circulating intercellular adhesion molecule 1
in human malignant melanoma. *Cancer Res.* 51:5003.

32
Phase I Clinical Trial of Anti-ICAM-1 (CD54) Monoclonal Antibody Immunosuppression in Renal Allograft Recipients

A. Benedict Cosimi, Robert Rothlein, Hugh Auchincloss Jr.,
Francis L. Delmonico, Nina Tolkoff-Rubin, Linda Scharschmidt,
Stephen H. Norris, Craig Haug, and Robert B. Colvin

Introduction

Rejection and the complications of immunosuppressive therapy continue to be major causes of allograft and recipient loss following all types of clinical transplantation. Although recent advances in the management of immunosuppressed allograft recipients have improved early graft survival rates, most centers continue to report a 1-yr allograft or patient loss of approximately 20% and a subsequent relentless attrition rate of 3–5% each year (1–3). Much of this morbidity and mortality is due to the relatively nonspecific suppression of immune responses induced by currently available therapeutic protocols.

Our efforts to develop more selective approaches to immunosuppression have concentrated primarily upon the use of antibodies directed against the particular host cells presumed to be involved in rejection. Initially, the polyclonal agent ATG (4,5) and more recently the monoclonal antibody OKT3 were successfully incorporated into treatment regimens (6,7). T cells were selected as the target for these approaches since they are responsible for most early graft rejection (8). More recent studies of the T-cell antigen recognition process have further clarified the mechanisms involved in the rejection reaction. In addition to the well-established role of the T-cell antigen receptor complex (TCR/CD3), which binds to MHC antigens, it is now recognized that several antigen-nonspecific accessory molecules promote attachment of T cells to their targets (or antigen-presenting cells) and transduce essential regulatory signals to the T cell. LFA-1 and ICAM-1 (CD54) molecules form one such critical adhesive receptor–ligand pair.

ICAM-1, a surface glyoprotein in the Ig superfamily, is induced on endothelial cells, fibroblasts, macrophages, epithelial cells, and activated lymphocytes by various inflammatory mediators. Interaction of LFA-1 and ICAM-1 is required for optimal T-cell function, as judged by the inhibitory effects on leukocyte adhesion and function induced by antibodies to either of these components in vitro [reviewed in (8)]. These observations suggest that ICAM-1 may be important in both the stimulatory and the effector stages of T-cell–mediated allograft responses. To evaluate the possibility that allograft rejection might be blocked by interfering with LFA-1/ICAM-1–mediated leukocyte adhesion, the anti–ICAM-1 murine mAb, BIRR1 (R6.5), which is cross-reactive with monkeys, was first studied in cynomolgus recipients of renal allografts (9).

In 16 cynomolgus renal allograft recipients, a 12-day course of BIRR1 was administered during the peritransplant period as the sole immunosuppressive agent (0.01–2 mg/kg/day). Survival in 14 recipients with technically successful grafts was significantly prolonged (24.2 ± 2.4 vs. 9.2 ± 0.6 days for controls; $p < 0.001$). We had observed that ICAM-1 was expressed on vascular endothelium in normal kidneys and other organs in monkeys in a pattern similar to that in humans. During cellular rejection in controls, ICAM-1 expression increased on endothelial cells, infiltrating mononuclear leukocytes, and tubular cells. Biopsies during BIRR1 administration showed decreased T-cell infiltration compared with controls and decreased arterial endothelial inflammation. No BIRR1-induced changes were detected in circulating T cells, aside from variable coating with mouse IgG. These studies demonstrated for the first time that ICAM-1 is an important molecule in the pathogenesis of rejection and that allograft survival can be prolonged by an approach which does not result in the depletion of lymphocyte populations that is characteristic of most conventional immunosuppressive agents. An interesting observation made in these cynomolgus studies was that there appeared to be a decrease in the severity of acute tubular necrosis (ATN) that is normally manifested in this model by a transient rise in serum creatinine during the first several days following transplantation. Decreased ATN could be explained by BIRR1 inhibition of neutrophil adhesion to the vascular endothelium, which has been shown to be an important factor in the injury associated with ischemia and shock (10).

Based upon these encouraging in vitro and in vivo observations, we have undertaken a phase I clinical study to evaluate the toxicity, dosage requirements, and potential efficacy of BIRR1 in cadaver donor renal allograft recipients. Because of the suggestive evidence that BIRR1 might be useful in limiting postischemia reperfusion injury as well as rejection, our initial trial selected recipients at high risk for delayed graft function. Such patients include those receiving a second or greater allograft; those who are highly sensitized (PRA: >60%); and those receiving an allograft that was retrieved from an unstable donor or that had been preserved ex vivo for greater than 36 h. Long-term (1-yr) allograft survival is inferior in recipients with delayed graft function, generally being reported as only 55–65% in contrast to the 80–85% survival observed if early function is achieved.

TABLE 32.1. Dosage modifications during phase I BIRR1 clinical trial.

Recipient	Dosage[a]	Therapeutic BIRR1 serum levels during treatment	DGF	Reject
1	(1)	No	+	+
2	(1)	No	+	+
3	(1)	No	+	+
4	(2)	Yes	−	−
5	(2)	No	+	+
6	(2)	No	+	+
7	(2)	No	+	+
8	(3)	Yes	−	−
9	(3)	Yes	+	−
10	(3)	Yes	−	−
11	(3)	Yes	−	+
12	(3)	No	−	+
13	(3)	Yes	+	+
14	(4)	Yes	+	−
15	(4)	Yes	−	−
16	(4)	Yes	−	−
17	(5)	Yes	−	+
18	(5)	Yes	−	−

[a] *Dosage*: (1) 20-mg loading dose followed by 10 mg/day × 13.
      (2) 40-mg loading dose followed by 20 mg/day × 13.
      (3) 80-mg loading dose followed by 60 mg/day followed by 40 mg/day × 4.
      (4) 160-mg loading dose followed by 80 mg/day × 5.
      (5) 160-mg loading dose followed by 40 mg/day × 5.

Results

Clinical Observations

Eighteen patients fulfilling the inclusion criteria were enrolled. The initial protocol was designed to treat the patients with low-dosage BIRR1 (20-mg loading dose followed by a 2-wk course at 10 mg/day). This dosage by weight is comparable to the lowest dose found effective in the nonhuman primate studies. Significant BIRR1 serum levels were not measurable in the first three patients studied and all three suffered rejection episodes requiring OKT3 rescue (Table 32.1). The next four patients were, therefore, treated with twice the dosage (40-mg loading, 20 mg/day × 13). Serum BIRR1 levels were again quite variable, often being unmeasurable during the 14-day treatment period. These observations indicate that there are considerably more CD54 combining sites available in dialyzed, uremic patients than in the normal, nonhuman primates we had previously treated. Therefore, we modified the protocol further by administering the entire 300-mg BIRR1 dosage over a 6-day interval (80-mg loading, followed by 60 mg/day × 1 and 40 mg/day × 4). Six patients received this dosage without incident. Because initial 24-h BIRR1 trough levels were still not felt to be adequate, however, the dosing regimen was adjusted further in the final five patients. The optimal regimen seeking to maintain biologically efficacious (as defined by in vitro inhibition of cell aggregation) BIRR1 serum levels of >10 μg/ml was found to require a 160-mg loading dose followed by either 40 or 80 mg/day over the subsequent 5 days.

One patient (recipient 9) who had received an allograft with unusually severe ischemic damage had a prolonged period of ATN. Because of the persistent ATN, OKT3 rather than CyA was administered for another 2 wk following the anti-CD54 course. Although excellent allograft function returned, this recipient expired 3 months later after developing widespread lymphoma. In addition, one recipient died at 3 months following complications of a myocardial infarction; one recipient developed recurrent membranous glomerulonephritis 10 months after transplantation; and one recipient, who received the lowest dosage of BIRR1 and subsequently required OKT3, died of *Aspergillus* pneumonitis at 3 months after transplantation. Thus with a follow-up of 14–27 months, 14 of 18 allografts (78%) continue with good to excellent function. This is encouraging in a recipient group preselected for the relatively poor prognosis of the allograft. (Three of the contralateral kidneys from the same donors were lost to primary nonfunction and one had been discarded because of concern regarding the preservation history. Currently, overall allograft survival of the contralateral kidneys transplanted into conventionally treated patients is 56%).

Immunopathology

In the monkey studies, the most striking site of deposition of the BIRR1 was on the surface of the graft endothelium (9). This led to the hypothesis that interference with vascular rejection was relevant to the beneficial effect. In the human recipients who received low doses in the initial phase, the intraoperative graft biopsies obtained 1 h after allograft reperfusion showed little or no detectable mouse IgG along the endothelium. These findings further supported the need for higher doses, which resulted in more dramatic endothelial deposition in small vessels and glomeruli, similar to those previously observed in the monkey. In general, there was no evident inflammation associated with the accumulation of mAb along endothelial surfaces, which increased in serial samples over the first 7–10 days.

The initial human allograft biopsies showed features typical of ischemic injury, as expected in this selected group of kidneys with a prolonged preservation time. The tubular lesions persisted for up to a week. Transient glomerular inflammation appeared in biopsies obtained at days 7–8 (leukocytes, endothelial swelling) in three of the initial seven patients. This had not been observed in the monkey renal biopsies. We tentatively attributed this to an immune response to the deposited mIgG or to an alteration in the graft rejection process (acute allograft glomerulopathy). This observation contributed to the decision to modify to a higher dose administered for a shorter time in order to limit the possibility of an immune-complex type of reaction. No glomerular lesions have been detected in the 11 patients treated with the most recent dosages.

Immunologic Monitoring

Peripheral blood T-cell subsets were monitored in these patients as they were in the monkey recipients (9). Again, CD54 antibody-coated peripheral blood monocytes and lymphocytes were detected but the cells did not clear from the circulation. No appreciable changes in the blood lymphocyte populations (reactive with mAbs to CD3, -4, -8, -16, -20) were observed.

The sera of all recipients were serially analyzed by ELISA (11) to detect human antimouse antibodies (HAMA). Sixteen of the 18 patients developed HAMA to the BIRR1 despite concomitant administration of conventional immunosuppressive agents. As in our experience with other mAb protocols, the antimurine antibodies were first detected between 3 and 14 days following completion of the BIRR1 therapy course and usually were no longer detectable after 3–6 months. Of particular importance for clinical protocols, our studies confirmed the idiotypic specificity of the HAMA response, suggesting it would be possible to successfully treat patients with sequentially administered mAbs directed to different epitopes. Indeed, this hypothesis has been validated as we have successfully treated four of these patients with OKT 3 after they had developed HAMA to BIRR1, even though both mAbs have the Ig2a isotype.

Toxicity

All dosing modifications have been well tolerated. Three patients complained of chills and nausea associated with the initial infusion. Since this was typically given together with other medications in the immediate preoperative period, no definite association with the BIRR1 administration was established. Clear-cut "first-dose" reactions, such as those reported with some mAb infusions (12), were not observed in these 18 patients. No evidence of hemolysis, thrombocytopenia, serum sickness, increased susceptibility to infection, or impaired wound healing was identified. As noted earlier, one patient developed widespread lymphoma and one patient developed *Aspergillus* pneumonia, both after being treated with consecutive courses of BIRR1 and OKT3 mAb therapy.

Conclusions

This initial clinical trial demonstrates that inhibition of leukocyte adhesion by BIRR1 therapy is well tolerated and may be efficacious in controlling allograft rejection and possibly the severity of reperfusion injury. This was not a controlled study, in that the contralateral kidneys were not randomly assigned and were generally transplanted at other institutions. Nevertheless, one of those kidneys was discarded by the recipient hospital because of the prolonged preservation history, and three suffered primary nonfunction (PNF). This contrasts with 100% utilization of the donor kidneys and no instances of PNF in the 18 BIRR1-treated renal allograft recipients.

As indicated in Table 32.1, our initial observations suggest a minimal serum BIRR1 level must be maintained to achieve efficacy. Of the 11 patients who had consistent biologically efficacious levels (defined as >10 μg/ml), 3 have had an acute rejection episode; 3 had DGF [requirement for more than one dialysis in the first posttransplant week (13)]; and 8 had early graft function. In contrast, 6 of the 7 patients who did not have biologically efficacious serum levels of BIRR1 experienced DGF; only 1 had early graft function. All of the 7 patients without satisfactory levels of BIRR1 have experienced at least one acute rejection episode.

Antibodies to cell-surface components can affect cell function in several ways, including cell lysis, blockage or modulation of receptors, and signal transduction. We hypothesize that blockage of lymphoid or endothelial function by BIRR1, through direct interference or transduction of a negative signal, is the most plausible explanation for the results observed. We found no evidence for cytotoxicity of BIRR1 to the cells that bound it in vivo (endothelial cells, lymphocytes, monocytes, dendritic cells). BIRR1 could block sensitization by binding to dendritic cells or endothelial cells in the graft, both of which are known to express CD54. An alternative hypothesis, also supported by our preclinical studies, is that BIRR1 acts by blocking the effector phase, particularly the attack on endothelial cells. Finally, the mAb therapy could interfere with cell recruitment and emigration. Our studies, to date, do not allow us to reach a final conclusion regarding which of these mechanisms is primarily operative. The observations do, however, further confirm the importance of accessory molecules in T-cell function and, thus, the timeliness of more extensive studies evaluating the immunosuppressive efficacy of antibodies such as BIRR1 directed against these structures. Accordingly, a prospectively randomized trial is now planned.

References

1. Kriett JM, Tarazi RY, Kaye MP. The Registry of the International Society for Heart Transplantation. In: Terasaki PI (ed). Clinical Transplants. Los

Angeles: UCLA Tissue Typing Laboratory. 1990;21–28.

2. Cecha JM, Terasaki PI. The UNOS Scientific Renal Transplant Registry. In: Terasaki PI (ed). Clinical Transplants. Los Angeles: UCLA Tissue Typing Laboratory. 1990;1–10.

3. Belle SH, Detre KM, Beringer KC, Murphy JB, Vaughn WK. Liver Transplantation in the United States: 1988–1989. In: Terasaki PI (ed). Clinical Transplants. Los Angeles: UCLA Tissue Typing Laboratory. 1990;11–19.

4. Cosimi AB, Wortis HH, Delmonico FL, Russell PS. Randomized clinical trial of antithymocyte globulin in cadaver renal allograft recipients: importance of T cell monitoring. *Surgery* 80:155–163, 1976.

5. Shield CF, Cosimi AB, Tolkoff-Rubin N, Rubin RH, Herrin J, Russell PS. Use of antithymocyte globulin for reversal of acute allograft rejection. *Transplantation* 28:461–464, 1979.

6. Cosimi RB, Burton RC, Rubin RH, Goldstein G, Kung PC, Hansen WP, Delmonico FL, Russell PS. Use of monoclonal antibodies to T-cell subsets for immunologic monitoring and treatment in recipients of renal allografts. *N. Engl. J. Med.* 305:308–314, 1981.

7. Cosimi AB, Burton RC, Colvin RB, Goldstein G, Delmonico FL, LaQualia MP, Tolkoff-Rubin NE, Rubin RH, Herrin JT, Russell PS. Treatment of acute renal allograft rejection with OKT3 monoclonal antibody. *Transplantation* 32:535–539, 1981.

8. Colvin RB. Cellular and molecular mechanisms of allograft rejection. *Ann. Rev. Med.* 41:361–376, 1990.

9. Cosimi AB, Conti D, Delmonico FL, Preffer FI, Wee, SL, Rothlein R. Faanes R, Colvin RB. In vivo effects of monoclonal antibody to ICAM-1 (CD54) in nonhuman primates with renal allografts. *J. Immunol.* 144:4604–4612, 1990.

10. Weiss SJ. Tissue destruction by neutrophils. *N. Engl. J. Med.* 320:365–375, 1989.

11. Delmonico FL, Fuller TC, Russell PS, Colvin RB, Cosimi AB. Variation in patient response associated with different preparations of murine monoclonal antibody therapy. *Transplantation* 47:92–95, 1989.

12. Chatenoud L, Ferran C, Legendre C, Thouard I, Merite S, Reuter A, Gevaert Y, Kreis H, Franchimont P, Bach JF. In vivo cell activation following OKT3 administration. *Transplantation* 49:697–702, 1990.

13. Halloran PF, Aprile MA, Farewell V, Ludwin D, Smith EK, Tsai SY, Bear RA, Cole EH, Fenton SS, Cattran DC. Early function as the principal correlate of graft survival. *Transplantation* 46:223–228, 1988.

33
Adhesion Molecules in Human Renal Allograft Rejection: Immunohistochemical Analysis of ICAM-1, ICAM-2, ICAM-3, VCAM-1, AND ELAM-1

VOLKER NICKELEIT, MOLLY MILLER, A. BENEDICT COSIMI, AND ROBERT B. COLVIN

Adhesion molecules, such as the intercellular adhesion molecules (ICAM), vascular cell adhesion molecule (VCAM), and endothelial–leukocyte adhesion molecule (ELAM), are believed to play important roles in immunologic responses in vivo (1–3). One of the clearest and clinically most important examples of immunologic injury is graft rejection, mediated by T cells that recognize foreign alloantigens (4).

A key mechanism of graft rejection is the attachment of T cells to the endothelium with consequent injury to the endothelium, termed endothelialitis or endovasculitis (5). The molecules that mediate this injury are not defined, although the constitutive and induced adhesion molecules are logical candidates. ICAM-1 (CD54) was the first ligand discovered to interact with the integrin receptor, LFA-1 (CD11a/CD18), primarily expressed by cells of bone marrow origin, including circulating T cells (1). Further studies led to the discovery of additional LFA-1 ligands. ICAM-2 is constitutively expressed on endothelial cells, in contrast to ICAM-1, which is induced by interferon-γ and other cytokines (6). ICAM-3 is expressed primarily on leukocytes (7). VCAM-1 is a distinct, cytokine-inducible, endothelial adhesion molecule that binds to a different integrin receptor VLA-4 (CD49d/CD29) expressed on leukocytes (2). VLA-4 also binds a non-RGD site in fibronectin (CS1). A third member of the adhesion molecule family is now termed ELAM-1 (3). On endothelium this molecule is induced rapidly but transiently by cytokines in vitro. ELAM-1 binds to sialylated Lewis x carbohydrate and promotes neutrophil adhesion to endothelium (8).

We have reported that a monoclonal antibody (MAb) to ICAM-1 (BIRR1, also known as R6.5) prolonged the survival of renal allografts in nonhuman primates, and noted that the endovasculitis and interstitial infiltrate was reduced (9). While this study demonstrated that of the various adhesion molecules, ICAM-1 is a key player, we thought that a comprehensive study of the expression of other candidate molecules in acute and

TABLE 33.1. Monoclonal antibodies to adhesion molecules.

ICAM-1	BIRR1 (R6.5)	Rothlein
ICAM-2	CBR-IC2/2	Springer
ICAM-3	CBR-IC3/2	Springer
LFA-1	R3.1	Rothlein
VCAM-1	4B9	Lobb
VLA-4	HP1/2	Lobb
ELAM-1	CL26C/10B7	Rothlein

chronic graft rejection would be potentially useful in planning future therapeutic strategies.

Methods

Frozen samples from biopsies of human renal allografts from the files of the Department of Pathology at the Massachusetts General Hospital were stained with MAbs to the adhesion molecules listed in Table 33.1, using previously published avidin biotin complex (ABC) immunoperoxidase techniques (10). Biopsies were divided into three diagnostic groups based on diagnoses made on formalin-fixed paraffin-embedded tissue. Those (n = 13) biopsies classified as acute cellular rejection showed dense, patchy, interstitial mononuclear leukocytic infiltrates; edema; and infiltration of tubules with mononuclear cells; over half also had endovasculitis. Those (n = 9) with chronic rejection showed interstitial fibrosis with focal tubular atrophy, a patchy mononuclear cell infiltrate, and focal arterial intima hyperplasia. A group of controls (n = 6) was taken from unremarkable renal parenchyma from nephrectomies with renal cell carcinoma or pretransplant biopsies of donor kidneys. Staining intensity was scored on coded sections over a range of 0 (for no staining) to 3+ (for marked staining).

Results

Arterial and Capillary Staining

In controls the endothelium of arteries/arterioles showed widespread but not intense staining (1+) for ICAM-1 and ICAM-2 (Table 33.2). The peritubular and glomerular endothelium was more markedly stained (2+). VCAM-1 was detected focally in peritubular capillaries over a range of intensities from 0 to 2+; arteries and arterioles did not stain. There was no staining for ICAM-3, LFA-1, VLA-4, or ELAM-1.

In acute and chronic rejection, ICAM-1 expression became more pro-

TABLE 33.2. Vascular endothelium.

	Controls		Acute rejection		Chronic rejection	
	A	C	A	C	A	C
ICAM-1	1+	2+	2–3+	2+	2+	2+
ICAM-2	1+	2+	1+	1+	1+	1+
ICAM-3	0	0	0	0	0	0
VCAM-1	0	0–2+	1–2+	0–2+	1–2+	1+
ELAM-1	0	0	0	0	0	0
VLA-4	0	0	0	0	0	0
ELAM-1	0	0	0	0	0	0

Abbreviations: A, arterial/arteriolar endothelium; C, peritubular capillary endothelium.

nounced in arterial endothelium (2–3+) but stayed unchanged in the peritubular and glomerular capillaries (2+) (Fig. 33.1). ICAM-2 showed no striking change; during rejection, however, capillary endothelial cells displayed somewhat less staining (1+) than the controls. Arterial endothelium in acute and chronic rejection showed definite, but focal, staining for VCAM-1 (1–2+) (Fig.33.2). In peritubular capillaries VCAM-1 was present more uniformly in chronic rejection (1+). In acute rejection it remained focally positive (0–2+). Glomeruli did not stain (except for the epithelium of Bowman's capsule). No ICAM-3, LFA-1, VLA-4, or ELAM-1 was detected in the peritubular capillary or arteriolar endothelium, although the mononuclear cells in arterioles sometimes stained with LFA-1 and/or VLA-4. ELAM-1 did stain positive control tissue (appendicitis).

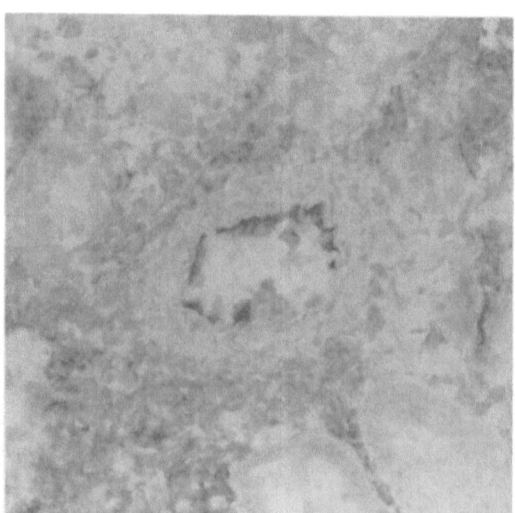

FIGURE 33.1. Acute cellular rejection, immunoperoxidase stain using BIRR1 (anti–ICAM-1). A small artery has strongly positive endothelium with adherent mononuclear cells. The interstitial infiltrate is diffusely positive, obscuring any peritubular capillary staining. Tubules show little or no positive reaction. ×313.

FIGURE 33.2. Acute cellular rejection, immunoperoxidase stain using 4B9 (anti–VCAM-1). A small artery has positive endothelium with adherent mononuclear cells. The interstitial infiltrate is negative. Tubular epithelium is focally positive, in some areas having a basal distribution. ×500.

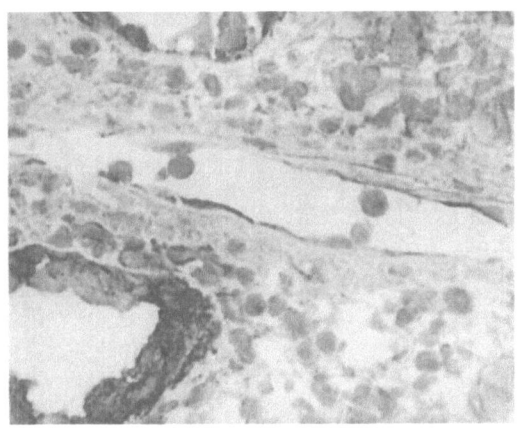

TABLE 33.3. Tubular epithelium.

	Controls	Acute rejection	Chronic rejection
ICAM-1	0	2+	2+
ICAM-2	0	0	0
ICAM-3	0	0	0
VCAM-1	1+	2+	2+
LFA-1	0	0	0
VLA-4	0	0	0
ELAM-1	0	0	0

Tubular Epithelium

Renal tubules in control tissue showed no staining for the examined adhesion molecules except for focal VCAM-1 (Table 33.3). In acute cellular rejection and chronic rejection, tubules were extensive and intensely stained for both ICAM-1 and VCAM-1. The cellular polarization of VCAM and ICAM-1 was strikingly different in the tubular cells. VCAM-1 was primarily basal, and ICAM-1 was primarily apical, in the brush border. ICAM-2, ICAM-3, LFA-1, VLA-4, and ELAM-1 did not stain tubules.

Inflammatory Infiltrate

The inflammatory infiltrate in all kidneys with rejection stained with ICAM-1, ICAM-3, LFA-1, and VLA-4 (Table 33.4) (Fig. 33.3). The differences in pattern were more in the staining distribution than in intensity. LFA-1 and VLA-4 gave a much more homogeneous distribution in

TABLE 33.4. Inflammatory infiltrate.

	Acute rejection	Chronic rejection
ICAM-1	2+	2+
ICAM-2	0	0
ICAM-3	2+	2+
VCAM-1	0	0
LFA-1	2+	2+
VLA-4	2+	2+
ELAM-1	0	0

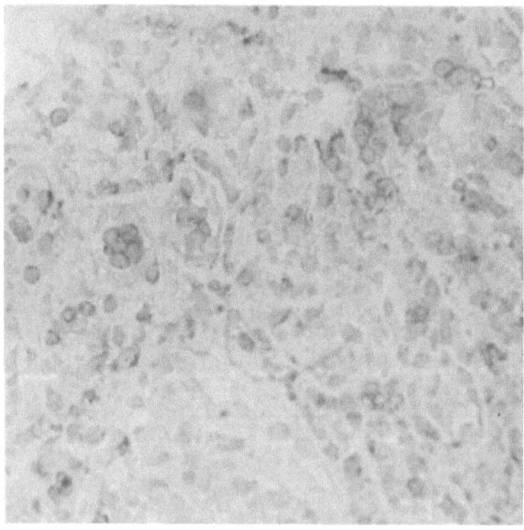

FIGURE 33.3. Acute cellular rejection, immunoperoxidase stain using CBR-IC3/2 (anti–ICAM-3). The interstitial infiltrate is a mixture of discretely positive cells and negative cells. No vascular or tubular staining is detectable. ×313.

cells in the inflammatory infiltrate than ICAM-3 and ICAM-1, which typically presented as patchy staining of the monouclear infiltrates. The infiltrate had no appreciable staining for ICAM-2 and VCAM-1, or ELAM-1.

Discussion

We found that ICAM-1 was upregulated during all phases of rejection and ICAM-2 stayed unchanged. Bishop and Hall (11) reported that endothelial staining for ICAM-1 in large vessels and capillaries stayed unchanged with rejection whereas tubular staining was increased. In our series, ICAM-1 staining of arterial/arteriolar endothelium in rejection was more pronounced than endothelial staining in control tissue. Several previous reports indicated that ICAM-1 was up-regulated on endothelial cells (arteries or veins) during cellular rejection in different allografts (kidneys, liver, heart) (9,12–14). The unchanged staining pattern of ICAM-2 is concordant with the in vitro data of de Fougerolles, who did not find substantial up-regulation of ICAM-2 with inflammatory mediators (6).

A new member of the intercellular adhesion molecule family (ICAM-3) shows an entirely different distribution in grafts. Neither vascular endothelium nor tubular epithelium was stained; however, the lymphocytic infiltrate showed focal positive cells. The physiological importance of ICAM-3 is unclear, but its presence on naive, nonactivated cells has suggested a role in the initial phase of sensitizization (7).

Our findings of VCAM-1 are in concordance with previous reports. Briscoe (13) found marked staining of venular endothelium in cardiac allografts showing rejection (arteries and arterioles were not described) and Rice demonstrated an upregulation of VCAM-1 in venules during inflammatory reactions (15,16). Increased expression of VCAM-1 was also noted in endothelial cells in inflamed rheumatoid synovium (17).

In our series, the VCAM-1 staining of endothelial cells in control tissue was not very uniform. Staining ranged from none to focal moderate capillary endothelial staining. Data published in the literature on VCAM-1 staining of endothelium of normal/control tissue are not uniform either. Rice (16) found only infrequent weak capillary endothelial staining. Briscoe (13) found faint staining of capillary and venule endothelium in controls during their studies of cardiac allografts. Furthermore, cases of cardiac allografts with little CD3+ infiltrates, classified as rejection, showed a range from no staining to moderate staining of endothelium. The different results of endothelial staining for VCAM-1 might be due to the fact that the VCAM-1 expression is caught in different phases of induction due to various specific or nonspecific underlying conditions stimulating the release of cytokines. The results published would thus represent a spectrum of different phases of activation.

We hypothesize that the pathological changes seen in the vascular component of renal allograft rejection involve principally ICAM-1 (and to a lesser degree ICAM-2) with their receptor LFA-1. Both are present constitutively in the human microvasculature and thus may play a role in the initial afferent sensitization phase as well as the effector phase. Transendothelial migration of T cells is more dependent on ICAM-1 than VCAM-1 in models using umbilical vein endothelium (18).

VCAM-1 and its receptor VLA-4 are expressed and potentially also important in the pathogenesis of acute or chronic endothelial injury in grafts. In addition, the staining pattern of VCAM-1 in the basal portion of tubular epithelial cells during rejection suggests that VCAM-1 is more accessible to infiltrating T cells than ICAM-1, which is detected largely in the brush border. Class II histocompatibility antigens have a similar basolateral distribution (19). Thus VCAM-1 might play a more important role in tubular changes during rejection than ICAM-1. ELAM-1 has no detectable presence in graft rejection, although a role in the very early stage cannot be excluded. Van Seventer showed that both ICAM-1 and VCAM-1, but not ELAM- 1, served as a co-stimulatory signal in activation of resting T cells (20). The present study supports the notion that therapies directed at ICAM-1, ICAM-2, LFA-1, VCAM-1, and VLA-4 molecules may lead to new methods of controlling the vascular injury of graft rejection.

Acknowledgments. The authors are grateful to Robert Rothlein (Boerhinger Ingelheim), Roy Lobb (Biogen), and Timothy Springer (Harvard Medical School) for kindly supplying antibodies for this study and to Charles Alpers (Seattle) for discussion of unpublished data. Supported in part by USPHS NIH grant P01-HL-18646.

References

1. Springer TA. 1990. Adhesion receptors of the immune system. *Nature* 346:425.
2. Elices MJ, Osborn L, Takada Y, Crouse C, Luhowsky S, Hemler ME, Lobb RR. 1990. VCAM-1 on activated endothelium interacts with the leukocyte integrin VLA-4 at a site distinct from the VLA-4 fibronectin binding site. *Cell* 60:577.
3. Bevilacqua MP, Stengelin S, Gimbrone MA, Jr, Seed B. 1989. Endothelial-leukocyte adhesion molecule 1; an inducible receptor for neutrophils related to complement regulatory proteins and lectins. *Science* 243:1160.
4. Colvin RB. 1990. Cellular and molecular mechanisms of allograft rejection. *Ann. Rev. Med* 41:361.
5. Colvin RB. 1991. The pathogenesis of vascular rejection. *Transplant. Proc* 23:2052.
6. de Fougerolles AR, Stacker SA, Schwarting R, Springer TA. 1991. Characterization of ICAM-2 and evidence for a third counter-receptor for LFA-1. *J. Exp. Med* 174:253.
7. de Fougerolles AR, Springer TA. 1992. Intercellular adhesion molecule 3, a third adhesion counter receptor for lymphocyte function associated molecule 1 on resting lymphocytes, *J. Exp. Med.* 175:185.
8. Picker LJ, Warnock RA, Burns AR, Doerschuk CM, Berg EL, Butcher EC. 1991. The neutrophil selectin LECAM-1 presents carbohydrate ligands to the vascular selectins ELAM-1 and GMP 140. *Cell* 66:921.
9. Cosimi AB, Conti D, Delmonico FL, Preffer FI, Wee S-L, Rothlein R, Fannes R, Colvin RB. 1990. In vivo effects of monoclonal antibody to ICAM-1 (CD54) in non-human primates with renal allografts. *J. Immunol.* 144:4604.
10. Tuazon TV, Schneeberger EE, Bhan AK, McCluskey RT, Cosimi AB, Schooley RT, Rubin RH, Colvin RB. 1987. Mononuclear cells in acute allograft glomerulopathy. *Am. J. Pathol.* 129:119–132.
11. Bishop GA, Hall BM. 1989. Expression of leukocyte and lymphocyte adhesion molecules in the human kidney. *Kidney International* 36:1078.

12. Adams DH, Shaw J, Hubscher SG, Rothlein R. 1989. Intercellular adhesion molecule 1 on liver allografts during rejection. *Lancet* 2:1122.
13. Briscoe DM, Schoen FJ, Rice GE, Bevilacqua MP, Ganz P, Pober JS. 1991. Induced expression of endothelial-leukocyte adhesion molecules in human cardiac allografts. *Transplantation* 51:537.
14. Kanagawa K, Ishikura H, Takahashi C, Tamatani T, Miyasaka M, Togashi M, Koyanagi T, Yoshiki T. 1991. Identification of ICAM-1 positive cells in the nongrafted and transplanted rat kidney an immunohistochemical and ultrastructural study, *Transplantation* 52:1057.
15. Rice GE, Munro JM, Bevilacqua MP. 1990. Inducible Cell Adhesion 110 (INCAM-110) is an endothelial receptor for lymphocytes. *J. Exp. Med.* 171:1369.
16. Rice GE, Munro JM, Corless C, Bevilacqua MP. 1991. Vascular and non-vascular expression of INCAM-110. *Am. J. Pathol.* 138:385.
17. van Dinther Janssen AC, Horst E, Koopman G, Newmann W, Scheper RJ, Meijer CJ, Pals ST. 1991. The VLA 4/VCAM-1 pathway is involved in lymphocyte adhesion to endothelium in rheumatoid synovium. *J. Immunol.* 147:4207.
18. Oppenheimer-Marks N, Davis LS, Bogue DT, Ramberg J, Lipsky PE. 1991. Differential utilization of ICAM-1 and VCAM-1 during the adhesion and transendothelial migration of human T lymphocytes, *J. Immunol.* 147:2913.
19. Benson EM, Colvin RB, Russell PS. 1985. Induction of IA antigens in murine renal transplants. *J. Immunol.* 134:7.
20. van Seventer GA, Newman W, Shimizu Y, Nutman TB, Tanaka Y, Horgan KJ, Gopal TV, Ennis E, O'Sullivan D, Grey H, et al. 1991. Analysis of T cell stimulation by superantigen plus major histocompatibility complex class II molecules or by CD3 monoclonal antibody: costimulation by purified adhesion ligands VCAM-1, ICAM-1, but not ELAM-1, *J. Exp. Med.* 174:901.

Index